Foundations of Engineering Mechanics

Series Editors

Vladimir I. Babitsky, School of Mechanical, Electrical and Manufacturing Engineering, Loughborough University, Loughborough, Leicestershire, UK

Jens Wittenburg, Karlsruhe, Germany

The series "Foundations of Engineering Mechanics" includes scientific monographs and graduate-level textbooks on relevant and modern topics of application-oriented mechanics. In particular, the aim of the series is to present selected works of Russian and Eastern European scientists, so far not published in Western countries, drawing from the large pool of experience from major technological research projects. By contributing to the long tradition of enrichment of Western science and teaching by Eastern sources, the volumes of the series address to scientists, institutional and industrial researchers, lecturers and graduate students.

More information about this series at https://link.springer.com/bookseries/3582

Nikolai Nikolaevich Polyakhov ·
Petr Evgenievich Tovstik ·
Mikhail Petrovich Yushkov ·
Sergey Andreevich Zegzhda
Authors

Petr Evgenievich Tovstik
Editor

Rational and Applied Mechanics

Volume 2. Special Problems and Applications

 Springer

Authors
Nikolai Nikolaevich Polyakhov
St. Petersburg, Russia

Petr Evgenievich Tovstik
Faculty of Mathematics and Mechanics
Saint Petersburg State University
St. Petersburg, Russia

Mikhail Petrovich Yushkov 🔾
Faculty of Mathematics and Mechanics
Saint Petersburg State University
St. Petersburg, Russia

Sergey Andreevich Zegzhda
Saint Petersburg State University
St. Petersburg, Russia

Editor
Petr Evgenievich Tovstik
Faculty of Mathematics and Mechanics
St. Petersburg State University
St. Petersburg, Russia

Translated by
A. R. Alimov 🔾
Division of Mathematics
Moscow State University
Moscow, Russia

A. K. Belyaev
St. Petersburg State University
St. Petersburg, Russia

D. A. Lisachenko
St. Petersburg State University
St. Petersburg, Russia

G. A. Sinilshchikova
Saint Petersburg State University
St. Petersburg, Russia

E. L. Belkind
Saint Petersburg State University
St. Petersburg, Russia

A. S. Kovachev
St. Petersburg State University
St. Petersburg, Russia

V. A. Shelkovina
St. Petersburg State University
St. Petersburg, Russia

ISSN 1612-1384 ISSN 1860-6237 (electronic)
Foundations of Engineering Mechanics
ISBN 978-3-030-64120-7 ISBN 978-3-030-64118-4 (eBook)
https://doi.org/10.1007/978-3-030-64118-4

The Review on the Textbook of the Editor of the Series "Foundations of Engineering Mechanics" of Springer Publishing

This fundamental and thoroughly prepared course in rational and applied mechanics reflects the high level of teaching this discipline by professors of St. Petersburg State University. The main feature of the course proposed is a wide discussion along with the classical problems of mechanics of new areas of study borne by developing applications. This provides rich material for the formation of numerous special courses as well. Although the main content of the course is designed for post-graduate students, many sections can be studied at the undergraduate student level. Lecturers will also find rich materials here.

In terms of the amount of material presented, the course has no analogues in English literature. Being published the books will find an important place in the libraries of the major world universities. As an alternative title, you can offer to the authors: Rational and Applied Mechanics. Theory and Applications. Such title will be better perceived in an English-speaking environment. The content of the manuscript corresponds well to the topics of the series "Foundations of Engineering Mechanics". I recommend accepting the manuscript for publication.

V. I. Babitsky

The authors express their sincere gratitude to Prof. V. I. Babitsky for his review of the Russian translation of the textbook.

Introduction to the Second Volume

As was stated in the Introduction to the first volume of the book, the material of the general course "Rational mechanics" and a number of special courses delivered by the authors in the faculty of mathematics and mechanics at Leningrad–Saint Petersburg State University over many years, is divided into three sections.

The first volume contained the sections "Kinematics" and "Dynamics. General aspects of rational mechanics. Fundamentals of analytical mechanics". They incorporated chapters dedicated to the particle kinematics, kinematics of the rigid solid, composite motion, particle dynamics, system dynamics, constrained motion, small oscillations of systems, dynamics of the rigid solid, variational principles in mechanics, statics, integration of equations in mechanics, elements of the special relativity theory.

The second volume of the book, offered to the readers, contains the third section "Dynamics. Some applied problems of rational mechanics". The material of the chapters of this section corresponds to the main contents of a number of special courses delivered at the chair of theoretical and applied mechanics at Saint Petersburg University. Each volume has its own continuous numeration of chapters and can be considered as a self-sufficient book. Basic literature in the second volume is given in the footnotes in each chapter. Each chapter title is followed by the names of the authors.

In **Chapter** 1, we define the Lyapunov stability for a disturbing motion and give Lyapunov's theorems on the stability, asymptotic stability, and instability. Next, we formulate Lagrange, Lyapunov, and Chetaev's theorems on the stability of positions of equilibrium and stationary motions of conservative systems. The effect of dissipative and gyroscopic forces on the stability of an equilibrium position of a conservative system (Thompson and Tait's theorems) is discussed. The stability of the equilibrium position is examined from the linear approximation. Routh–Hurwitz's and Mikhailov's criteria for negativity of the real parts of the roots of a polynomial are given. The stability of periodic motions of nonautonomous systems is studied from the linear approximation. The stability of the zero solution of the Mathieu equation is considered, to which one may reduce oscillations of a pendulum with a vibrating suspension point.

In **Chapter** 2, a special emphasis is given to approximate methods of solution of nonlinear equations (the small-parameter method, asymptotic methods). A relation between the Bubnov–Galerkin method and the Gauss principle is established. A detailed exposition is given to P. L. Kapitsa's method of direct separation of motions, which at present is not adequately presented in textbooks. Theoretical results are clarified by solving a number of new examples. The last subsections are dedicated to strange attractors and the two-scale expansion method.

In **Chapter** 3, a loaded platform is studied by two different techniques—application of classical theorems of rational mechanics and using a special form of motion equations introduced in Chapter 8 of Vol. I. Kinematics of the platform of a testing bench, differential equations of the loaded platform, solving direct and inverse dynamical problems, standard motions of the platform, introducing a feedback, providing stable oscillation of a loaded Stewart platform, workspaces of the platform in the 6D space of generalized coordinates are considered. The third possible method—the Lagrange equations of the second kind—is used for stabilization of the equilibrium state of the platform of a testing bench.

Chapter 4 discusses the simplest models of rotor systems with a finite number of degrees of freedom. Various types of rotor oscillations due to their imbalance, the unequal elastic characteristics of the shaft or bearings, as well as the influence of internal friction and structural damping, are studied. Issues of balancing of rotors equipped with passive automatic ball balancers are investigated.

In **Chapter** 5, we state problems of control theory and give a brief survey of some methods of their solution. Problems in the control theory can be subdivided into two large classes. The first class is related to the choice of a control which is optimal in a sense, while the second one is related to the problem of the confinement of motion on a selected path or near it. The solution to some problems of control theory is given. The concepts of controllability, stabilizability, and observability are introduced.

Chapter 6 is divided into two parts. The first part formulates the generalized Chebyshev problem and presents the two theories of motion for its solution. For the creation of these theories, the apparatus of nonholonomic mechanics with high-order constraints is developed. In the first theory, a compatible system of differential equations is constructed for the determination of unknown generalized coordinates and the Lagrange multipliers. The second theory is based on using the generalized Gauss principle. Applications of the theories are illustrated by solving the problem of the satellite motion with a constant magnitude acceleration.

In the second part of the chapter, it is offered to use the second theory of motion for nonholonomic systems with high-order constraints for solving one of the most important problems of the control theory—the problem of choice of an optimal control force transferring a mechanical system during the given time from one phase state to another.

It is shown that when solving the problem formulated with the help of the Pontryagin maximum principle, a high-order constraint is realized continuously in minimization of the function of a control force squared. That is why it is convenient to apply for solving the same problem the generalized Gauss principle that is appropriate to the theory of motion for high-order nonholonomic systems. This makes it possible to construct a control force in the form of a polynomial in time. The use of the theory offered is illustrated by solving a model problem of oscillation suppression of a trolley with pendulums. The extended boundary problem, in which the values of accelerations are also given at the beginning and at the end of motion, is formulated and solved. As a result, one can find a control force without the jumps, which are appropriate to the solution obtained by means of the Pontryagin maximum principle. At the end of the chapter damping, the oscillation of a flexible "arm" of a manipulator is considered.

Chapter 7 gives a brief account of the methods of determination of probabilistic characteristics of motion of mechanical systems subject to random forces. In the introductory subsections, we give the required definitions of random variables and processes. The auxiliary material presented here is not a treatise on the probability theory. In the definition of the probabilistic characteristics, we shall be mostly concerned with the correlation level when the expectations and correlation function are determined under the condition that these characteristics are given for the exterior forces. For the stationary processes, the Fourier transform is used and the spectral densities are defined. For statistically linear systems, it proves possible to find an exact solution. However, nonlinear systems can be treated only by approximate methods—these are the methods of statistical linearization and statistical modeling. The method of the solution of the Fokker–Planck–Kolmogorov equation, mentioned at the end of this chapter, gives an exact solution; however, the area of its practical application is very narrow.

Chapter 8 concerns the classical impact theory, although it starts with the Hertz theory of the impact of elastic spheres. We do it in order to clearly outline the scope of the problems where the basic preconditions of the classical impact theory are valid. We discuss in detail the concept of the restitution coefficient introduced by Newton. We give a new deduction of the algebraic system of the Lagrange equations of the first and second kind, corresponding to the classical impact theory of mechanical systems with ideal constraints. It is essential that this system of equations takes a particularly simple form in a number of tasks due to the use of quasivelocities. As an example, we consider the impact on a straight chain of rods and on a chain situated on a circular arc. In these problems, the Lagrange equations written in terms of quasivelocities have the same form that the finite-difference equations. This allowed us to build an analytical solution of the two given problems. We also consider other important examples of application of the considered methods of classical impact theory.

In **Chapter** 9 the classical Euler results on nonlinear statical deformation of a longitudinally compressed rod, the results of the work by M. A. Lavrent'ev and A. Yu. Ishlinskiy on the buckling in the dynamical longitudinal compression, and modern results relating in general to the study of interaction of longitudinal and lateral vibration are presented. The chapter introduces the readers to fundamental investigative methods—the D'Alembert and Fourier methods, the methods studying the parametric resonances, the asymptotic two-scale expansion method.

In **Chapter** 10, basic coordinate systems used in flight dynamics are introduced, the motion equations of an aircraft in the body-axes system are investigated, the forces acting on an aircraft are discussed, the questions concerned with the motion of systems of variable mass are considered, inter alia, the formula for calculation of the thrust of a jet engine is given, the methods of nonholonomic mechanics for aircraft targeting are applied. It is noted that flight dynamics studies the motion of airplanes and launch vehicles in the Earth's atmosphere. Methods of flight dynamics can be applied to the study of motion of submarines and surface vessels.

Just like in volume I, while relating the material a double numeration of formulae is being used in each chapter, with the first digit indicating the number section. When it is necessary to refer to a formula from another chapter, we give the verbal identification of the number of the chapter and the volume where the required formula is located.

The monograph is successively divided into Parts, Chapters, Sections, and subsections.

Contents of volume II is translated into English by A. R. Alimov (chapters 1, 2, 3, 5, 6, 10), A. K. Belyaev (chapter 9), A. S. Kovachev (chapter 4), D. A. Lisachenko (chapter 8), V. A. Shelkovina (chapter 7). The general editing of the English text (excluding chapter 9) is fulfilled by A. R. Alimov, E. L. Belkind, and G. A. Sinilshchikova.

The authors will be grateful to the readers who will take the trouble of sending their notes concerning this book.

Petr Evgenievich Tovstik

Mikhail Petrovich Yushkov

yushkovmp@mail.ru

Contents

Part III
Dynamics. Some Applied Problems of Rational Mechanics

The last Part of the monograph deals with some branches of higher mathematics that have a direct relationship to rational mechanics (Chaps. 1, 2, 5, and 7). The material of these chapters and of the first volume help to represent a number of supplementary chapters of analytical mechanics (Chaps. 3, 4, 8, 10) in a new way, these chapters being of great practical importance. In turn, this makes it possible both to investigate and extend the solutions of some important mechanical problems (Chap. 9) and to offer new methods for calculation of practically important problems as well (for example, Chap. 6).

The material of this Section of the textbook can be used for creating special courses.

Chapter 1
Stability of Motion

P. E. Tovstik and M. P. Yushkov ⓘ

In this chapter, we define the Lyapunov stability for a disturbed motion and give Lyapunov's theorems on the stability, asymptotic stability and instability. Next, we formulate Lagrange, Lyapunov and Chetaev's theorems on the stability of positions of equilibrium and stationary motions of conservative systems. The effect of dissipative and gyroscopic forces on the stability of an equilibrium position of a conservative system (Thompson and Tait's theorems) is discussed. The stability of the equilibrium position is examined from the linear approximation. Routh–Hurwitz's and Mikhailov's criteria for negativity of the real parts of the roots of a polynomial are given. The stability of periodic motions of nonautonomous systems is studied from the linear approximation. The stability of the zero solution of the Mathieu equation is considered, to which one may reduce oscillations of a pendulum with vibrating suspension point.

1 Differential Equations of Disturbed Motion

We shall assume that the system of Lagrange equations of the second kind with $y_\sigma = q^\sigma$, $y_{s+\sigma} = \dot{q}^\sigma$, $\sigma = \overline{1, s}$, $n = 2s$, can be written as

$$\frac{dy_k}{dt} = Y_k(y_1, y_2, \ldots, y_n, t), \quad k = \overline{1, n}. \tag{1.1}$$

We recall that the method of changing from Lagrange equations of the second kind to system (1.1) with $n = 3$ was treated in Sect. 6 of Chap. 4 of Vol. I.

Under the action of forces with potential, the motion equations reduce to the system of canonical Eqs. (6.10) of Chap. 4 of Vol. I, which in new notation can be also transformed to equations of form (1.1).

© Springer Nature Switzerland AG 2021
N. N. Polyakhov et al., *Rational and Applied Mechanics*,
Foundations of Engineering Mechanics,
https://doi.org/10.1007/978-3-030-64118-4_1

We shall consider the phase coordinates $y_1, y_2, \ldots, y_n$ as a *representative point M* in the n-dimensional Euclidean space.[1] Now the system of Eqs. (1.1) can be put in the form

$$\frac{dy}{dt} = Y(y, t), \qquad y = (y_1, y_2, \ldots, y_n)^T, \quad Y = (Y_1, Y_2, \ldots, Y_n)^T. \qquad (1.2)$$

The system of differential equations (1.2) is solved under the initial conditions

$$y(0) = y_0. \qquad (1.3)$$

The solution $y(t)$ corresponding to these initial conditions is called an *undisturbed motion*. Any motion corresponding to other solutions of Eq. (1.2) will be called a *disturbed motion*. We let $\widetilde{y}(t)$ denote the disturbed motion; it is also a solution of Eq. (1.2), but with the initial data $\widetilde{y}(0) = \widetilde{y}_0$, which are different from (1.3).

In the study of Eq. (1.2) one may always put

$$\widetilde{y}(t) = y(t) + x(t), \quad \widetilde{y}(0) = y_0 + x_0,$$

where $x(t)$ is the *perturbation*, and x_0 is the *initial perturbation*. Besides,

$$\frac{dx}{dt} = X(x, t), \quad x(0) = x_0,$$
$$X(x, t) \equiv Y(y(t) + x(t), t) - Y(y(t), t). \qquad (1.4)$$

It is clear that $X(0, t) \equiv 0$, and hence Eq. (1.4) has the zero solution: $x(t) \equiv 0$. By this the study of properties of an undisturbed solution of Eq. (1.2) is reduced to that of the zero solution of Eq. (1.4). The motion $x(t) \equiv 0$ can be interpreted as a state of rest of system (1.4).

If the vector function X does not depend explicitly on time t, then Eq. (1.4) is called *autonomous*.

Definition 1 If, for any arbitrarily small positive number ε, one may find a positive number δ such that $\|x(t)\| < \varepsilon$ if $\|x_0\| < \delta$ for any time t, then an undisturbed motion is called *Lyapunov-stable*. Otherwise, it is called *Lyapunov-unstable*.

Definition 2 A stable motion is called *asymptotically stable* if there exists such a number $\gamma > 0$ such that $x(t) \to 0$ as $t \to \infty$ if $\|x_0\| < \gamma$.

Some further remarks are worth making in connection with these definitions.

It is appropriate to assume that the phase coordinates x_k, $k = \overline{1, n}$ are dimensionless, for otherwise one will have to deal with adding quantities of different dimensions.

---

[1] It should not be confused with the Hertz representative point introduced in Chap. 5 of Vol. I.

By $\|z\|$ we denote a norm of a vector z (in the definitions one may use any norm). The most frequently used are the following norms:

$$\|z\|_1 = \sqrt{\sum_{k=1}^{n} z_k^2} \quad \text{and} \quad \|z\|_2 = \max_{k=\overline{1,n}} |z_k|,$$

the first one is the Euclidean length of a vector z, and the second one is the maximal departure of phase coordinates from zero.

It is assumed that there exists $\varepsilon > 0$ such that, for $\|x_0\| < \varepsilon$, Eq. (1.4) has a bounded solution on an infinite time interval. Otherwise, stability is not defined. For example, the Cauchy problem for the first-order equation $\dot{x} = x^2$, $x(0) = \varepsilon$ has a bounded solution $x(t) = 1/(\varepsilon^{-1} - t)$ only for $t < \varepsilon^{-1}$.

In this chapter we shall not be concerned with building a disturbed motion. We shall only speak about stability or instability of an undisturbed motion.

To clarify the **geometrical meaning** of the above concepts, we consider the motion of a representative point, taking for definiteness the norm $\|z\|_1$. In the course of time, the representative point draws in the n-dimensional space a curve, which is called the *integral curve* or the *phase trajectory*. We consider two the spheres $\|x\| = \delta$ and $\|x\| = \varepsilon$, which we will call, respectively, the ε-sphere and the δ-sphere. Now a motion is stable if any integral curve starting at $t_0 = 0$ at a point M_0 inside the δ-sphere (Fig. 1) remains in the ε-sphere. If a motion is unstable, then at least one trajectory of the representative point M will eventually cross the ε-sphere even if M_0 is arbitrarily close to the origin. If the point M tends to the origin as $t \to \infty$, then the system is asymptotically stable.

As an example, we consider small oscillations described by the Cauchy problem

$$\ddot{x} + 2h\dot{x} + \omega^2 x = 0, \quad x(0) = a, \quad \dot{x}(0) = v.$$

Fig. 1 δ-sphere and ε-sphere

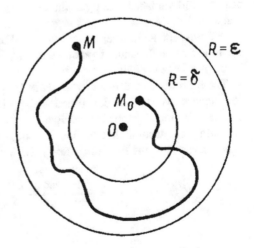

For $h = 0$ the problem has the solution $x(t) = a \cos \omega t + (v/\omega) \sin \omega t$. Clearly, taking sufficiently small initial values of a and v, we get the functions $x(t)$ and $\dot{x}(t)$ have arbitrarily small values for any t. Hence, for $h = 0$ the zero solution is stable. In the same manner, it is easily checked that for $h > 0$ we have $x(t) = Ce^{-ht} \cos(kt + \alpha)$, which gives the asymptotic stability of the zero solution.

The study of the stability of solutions of Eq. (1.2) is closely related with that of a motion in accordance with a given program. Assume one needs to ensure the motion according to the law $y = y(t)$. To this aim it is necessary that the initial conditions would fit accurately the values in (1.3). However, this is unfeasible in practice, and so the real motion is disturbed. An important problem is to determine the effect of initial perturbations on the departure of the real motion from the program motion.

2 The Direct Lyapunov Method

A. M. Lyapunov put forward two methods for solving motion stability problems. The first one is the *method of characteristic exponents of solutions* of equations of disturbed motion, the second one is capable of assessing the motion stability from the behaviour of some auxiliary functions.

Let us consider the second method, which is called the *Lyapunov's Direct Method*. We require several definitions.

A *Lyapunov function* is a function $V(x, t)$, $\|x\| < R$, $0 \leqslant t < \infty$, continuously differentiable in both arguments and vanishing at the origin ($V(0, t) = 0$).

A Lyapunov function $V(x, t)$ is called *positive (negative)* if $V(x, t) \geqslant 0$ ($V(x, t) \leqslant 0$) for all x, t.

In accordance with these definitions, the function $V(x, t) \equiv 0$ is both positive and negative.

A time-independent Lyapunov function $V(x)$ is called *positive definite* if it is positive and vanishes exactly at the origin (more precisely, there exists $\varepsilon > 0$ such that $x = 0$ if $V(x) = 0$ for $\|x\| < \varepsilon$).

A time-dependent Lyapunov function $V(x, t)$ is called *positive definite* if there exists a time-independent positive definite Lyapunov function $W(x)$ such that $V(x, t) \geqslant W(x)$ for all t.

The definition of a *negative definite* Lyapunov function is similar. Below, for definiteness, we shall be concerned with positive Lyapunov functions, the negative ones are obtained by the sign change.

Taking into account that the function $x(t)$ satisfies Eq. (1.4), we calculate the total time derivative of the function $V(x(t), t)$. We have

$$V'(x, t) = \frac{dV}{dt} = \frac{\partial V}{\partial x} \cdot \frac{dx}{dt} + \frac{\partial V}{\partial t} = \sum_{k=1}^{n} \frac{\partial V}{\partial x_k} X_k(x, t) + \frac{\partial V}{\partial t}.$$

The function $V'(x, t)$ is called *the derivative of a Lyapunov function $V(x, t)$ calculated in accordance with the equation of disturbed motion*. Since $X(0, t) \equiv 0$, the function $V'(x, t)$ itself is a Lyapunov function.

Theorem 1 (Lyapunov's stability theorem) *If there exists a positive definite Lyapunov function $V(x, t)$, whose derivative $V'(x, t)$ calculated in accordance with the equation of disturbed motion is negative (or is identically zero), then the undisturbed motion $x = 0$ is stable.*

Proof By the definition of Lyapunov stability, we take an arbitrary $\varepsilon > 0$ and try to find $\delta > 0$. Let

$$a = \min_{|x| = \varepsilon} W(x) > 0, \quad V(x, t) \geqslant W(x) > 0, \quad |x| = \varepsilon.$$

Here, the minimum of the continuous function $W(x)$ is searched on the closed set $|x| = \varepsilon$, which for the norm $\|x\|_1$ is the sphere of radius $R = \varepsilon$ (see Fig. 1). By Weierstrass theorem, this minimum exists and is positive, because the function $W(x)$ is positive definite. The function $V(x, 0)$ is continuous and $V(0, 0) = 0$. Hence, there exists $\delta > 0$ such that $V(x, 0) < a$ for all $|x| < \delta$. This is the required value of δ. Indeed, we take an arbitrary point M_0 inside the sphere of radius $R = \delta$ and consider a trajectory emerging from this point (see Fig. 1). Along this trajectory we integrate in time the equation

$$\frac{dV}{dt} = V'(x(t), t), \quad V(t) - V(0) = \int_0^t V'(x(t), t) dt. \tag{2.1}$$

The function V' is negative, and hence $V(t) \leqslant V(0)$. We have $a > V(0) \geqslant V(t) \geqslant W(t)$. The trajectory may not intersect the sphere of radius $R = \varepsilon$, because on this sphere $W \geqslant a$, which contradicts the inequality $W < a$.

Theorem 2 (Lyapunov's asymptotic stability theorem) *If there exists a positive definite Lyapunov function $V(x, t)$ with infinitely small upper bound, whose derivative $V'(x, t)$ in accordance with the equation of disturbed motion is negative definite, then the undisturbed motion $x = 0$ is asymptotically stable.*

Theorem 3 (Lyapunov's instability theorem) *If there exists a Lyapunov function $V(x, t)$ with infinitely small upper bound, whose derivative $V'(x, t)$ in accordance with the equation of disturbed motion is positive definite, and if the function $V(x, 0)$ assumes only positive values for all sufficiently small $|x|$, then the undisturbed motion $x = 0$ is unstable.*

Theorems 2 and 3 will be given without proof. We give some explanations.

The term 'function with infinitely small upper bound' was introduced by Lyapunov. It is equivalent to the uniform continuity of a function $V(x, t)$ in x as $0 \leqslant t < \infty$. Namely, for any $\varepsilon > 0$ there exists $\delta > 0$ such that $|V(x, t)| < \varepsilon$ for

$|x| < \delta$ and for all t. For example, the function $V(x, t) = x^2/(1 + t)$ has not infinitely small upper bound.

For the asymptotic stability it is required that $V(x, t) \to 0$ as $t \to \infty$, which follows from formula (2.1) in the case of a negative definite function V'. If a function $V(x, t)$ has infinitely small upper bound, then $V(x, t) \to 0$ implies that $x \to 0$.

By Theorem 3 one may take an initial point M_0 at which $V(x, 0) > 0$ is arbitrarily close to 0. According to (2.1), if V' is positive definite, then the function $V(t)$ is increasing and the trajectory will leave the ε-sphere, which indicates the instability.

Example 1 Assume that the equations of disturbed motion reads as

$$\dot{x}_1 = x_2 + \mu x_1^3, \quad \dot{x}_2 = -x_1 + \mu x_2^3. \tag{2.2}$$

The function $V = (1/2)(x_1^2 + x_2^2)$ is positive definite. Its total time derivative in accordance with the equations of disturbed motion (2.2) is as follows:

$$V' = \frac{dV}{dt} = x_1\dot{x}_1 + x_2\dot{x}_2 = \mu \left(x_1^4 + x_2^4\right).$$

In this example, the derivative V' have the same sign as the parameter μ. Hence, for $\mu = 0$ by Lyapunov's Theorem 1 the motion is stable. For $\mu < 0$ Theorem 2 shows that the system is asymptotically stable, and for $\mu > 0$ the motion is unstable by Theorem 3.

Unfortunately, there are no general methods of finding the function V. We give only some recommendations for searching thereof. In mechanical problems it is convenient to take as a Lyapunov function an expression similar to that in form of the total mechanical energy of the system. Sometimes one manages to construct the required function by adding the right-hand sides of the motion equations multiplied by coordinates. In some cases it is expedient to employ various quadratic forms with the use of some tests for positivity or negativity of quadratic forms (like, for example, Sylvester's law of inertia).

If, for the equations of disturbed motion, the integrals

$$F_k(x) = C_k, \quad k = \overline{1, m}$$

are known, then N. G. Chetaev proposed to search a Lyapunov function as the combination

$$V(x) = \sum_{k=1}^{m} \lambda_k \left[F_k(x) - F_k(0)\right] + \sum_{k=1}^{m} \varkappa_k \left[F_k^2(x) - F_k^2(0)\right], \tag{2.3}$$

which is called *bundle of integrals*. If in this case one manages to find $\lambda_k, \varkappa_k, k = \overline{1, m}$, so that the function V is positive definite, then the motion is stable by Lyapunov's theorem. For convenience, one of the coefficients may be taken to be 1. Sometimes it proves possible to construct a function V with $\varkappa_k = 0, k = \overline{1, m}$.

In the general case, a function $V(x)$ may be sought in the form

$$V(x) = f(F_1(x) - F_1(0), F_2(x) - F_2(0), \ldots, F_m(x) - F_m(0)),$$

where f is any function.

Example 2 As an illustration of the use of the bundle of integrals (2.3) we consider the stability of permanent rotations of a rigid body around a fixed point (the centre of inertia). The equations of rotational motions read as

$$A\frac{d\omega_x}{dt} + (C - B)\omega_y\omega_z = 0, \quad B\frac{d\omega_y}{dt} + (A - C)\omega_z\omega_x = 0,$$

$$C\frac{d\omega_z}{dt} + (B - A)\omega_x\omega_y = 0, \tag{2.4}$$

where A, B, C are the principal moments of inertia (we assume that $A > B > C$), ω_x, ω_y, ω_z are the projections of the angular velocity on the principal axes of inertia.

System (2.4) has two integrals (the energy and angular momentum integrals)

$$A\omega_x^2 + B\omega_y^2 + C\omega_z^2 = C_1, \quad A^2\omega_x^2 + B^2\omega_y^2 + C^2\omega_z^2 = C_2.$$

Let us prove the stability of permanent rotation about the x-axis with the largest moment of inertia $\omega_x = \omega_x^0$, $\omega_y = \omega_z = 0$. We introduce the perturbations x_k and define $\omega_x = \omega_x^0 + x_1$, $\omega_y = x_2$, $\omega_z = x_3$. Then the above motion integrals will assume the form

$$F_1 = A(\omega_x^0 + x_1)^2 + Bx_2^2 + Cx_3^2, \quad F_2 = A^2(\omega_x^0 + x_1)^2 + B^2x_2^2 + C^2x_3^2.$$

It is easily checked that the bundle of integrals

$$V(x_1, x_2, x_3) = AF_1 - F_2 + (F_1 - A(\omega_x^0)^2)^2$$

is a positive definite function.

In the same fashion one may prove the stability of motion about the z-axis with the smallest moment of inertia. A rotation about the y-axis with mean moment of inertia is unstable, which will be shown below by using the linear approximation. The method employing the bundle of integrals is incapable of proving the instability of motion.

3 Stability of Equilibrium and Stationary Motions of Conservative Systems

Stability of equilibrium position of the system. We recall that earlier (see Sect. 5 of Chap. 5 of Vol. I) we formulated *Lagrange's theorem* to the effect that the equilibrium position is stable if the potential energy of a conservative holonomic stationary system has an isolated minimum in this position. This result can be easily proved by using Lyapunov's Theorem 1.

Indeed, Lagrange equations of the second kind, when written in the normal form, or canonical Hamilton equations can be looked upon as equations of disturbed motion if one takes that in an equilibrium position the generalized coordinates and the generalized velocities are all zero. Since a conservative system is subject to scleronomic constraints, there exists the energy integral

$$T + \Pi = \text{const.}$$

Taking

$$V = T + \Pi \tag{3.1}$$

as a Lyapunov function, it is easily shown that V is a positive definite function of generalized coordinates and generalized velocities. Indeed, by definition the kinetic energy T is a positive definite quadratic form. The potential energy Π can be found up to a additive term, and hence one may assume that it vanishes at the origin. But then one may assert that $\Pi(q^1, q^2, \ldots, q^s) \geqslant 0$, because it is assumed that at the origin the potential energy has an isolated minimum. Hence, the above function (3.1) is positive definite, its total time derivative in accordance with the equations of distributed motion is identically zero, because $T + \Pi = \text{const.}$ Thus, by Lyapunov's first theorem, the equilibrium position is stable.

It is worth noting that conditions for the instability of equilibrium positions are much more difficult to find. We give some **tests for instability**. To do so we expand the potential energy near an equilibrium position as a series in generalized coordinates by gathering terms of different orders of smallness

$$\Pi = \Pi_2 + \Pi_3 + \Pi_4 + \ldots, \qquad \Pi_2 = \sum_{k,l=1}^{n} c_{kl} q_k q_l,$$

$$\Pi_3 = \sum_{k,l,m=1}^{n} c_{klm} q_k q_l q_m, \ldots. \tag{3.2}$$

The following result holds.

An equilibrium position is unstable if the potential energy in the equilibrium position

− *has no minimum value; this follows from the expression* for Π_2 (A. M. Lyapunov);

Fig. 2 Beam equilibrium on
a log

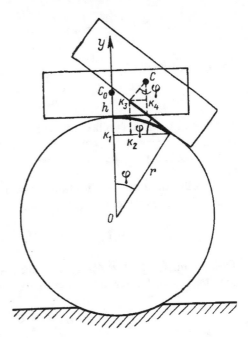

– *has a strict minimum; this follows from the consideration of the lowest terms*
 ($m \geqslant 2$) *in expansion* (3.2) (A. M. Lyapunov);
– *is a homogeneous function of the generalized coordinates and the equilibrium*
 position has no minimum value (N. G. Chetaev).

Example 3 On a fixed-fixed log there is a beam of length $2h$, its centre of mass C_0
lies on the extension of the log vertical radius (Fig. 2). The beam rolls over the log
without slip. Find conditions for the stability of an equilibrium position.

Let φ be the rotation angle of the beam. The length of the corresponding arc is
$r\varphi$, where r is the log radius. Let us calculate the potential energy of the beam due
to gravity. According to Fig. 2,

$$\Pi = mg(y_c - h - r) = mg(OK_1 + K_2K_3 + K_4C - h - r) =$$
$$= mg(r\cos\varphi + r\varphi\sin\varphi + h\cos\varphi - h - r).$$

A necessary and sufficient condition for an equilibrium is that the generalized
forces acting on the system be zero. In the case under consideration,

$$\frac{\partial\Pi}{\partial\varphi} = mg(r\varphi\cos\varphi - h\sin\varphi) = 0,$$

whence $\varphi = 0$ is the equilibrium position. To test the stability of this position, we
find the second derivative:

$$\frac{\partial^2 \Pi}{\partial \varphi^2} = mg(r \cos \varphi - r \varphi \sin \varphi - h \cos \varphi).$$

For $\varphi = 0$ we have

$$\frac{\partial^2 \Pi}{\partial \varphi^2} = mg(r - h).$$

It follows that in an equilibrium position for $r > h$ the potential energy has a minimum, and for $r < h$, it assumes a maximum value. In the first case, the system is stable, and in the second case, is unstable.

Stability of stationary motions. Assume that $q^1, q^2, \ldots, q^r$ are cyclic coordinates. The corresponding Lagrange function reads as

$$L = L\left(t, q^{r+1}, q^{r+2}, \ldots, q^s, \dot{q}^1, \dot{q}^2, \ldots, \dot{q}^s\right).$$

The coordinates $q^{r+1}, q^{r+2}, \ldots, q^s$ are called *position coordinates*. Given r cyclic coordinates, the Lagrange equations have r integrals:

$$\partial L / \partial \dot{q}^i = C_i, \quad i = \overline{1, r}. \tag{3.3}$$

If the arbitrary constants C_i, $i = \overline{1, r}$, are found from the initial conditions, then r generalized velocities can be determined from Eqs. (3.3):

$$\dot{q}^i = \widehat{\dot{q}}^i \left(t, q^{r+1}, q^{r+2}, \ldots, q^s, \dot{q}^{r+1}, \dot{q}^{r+2}, \ldots, \dot{q}^s, C_1, C_2, \ldots, C_r\right),$$
$$i = \overline{1, r}.$$

This is possible, because

$$\det \left[\frac{\partial^2 L}{\partial \dot{q}^i \partial \dot{q}^j} \right] \neq 0, \quad i, j = \overline{1, r}.$$

Substituting these functions into the expression for kinetic energy, we exclude the dependence on the cyclic coordinates and velocities:

$$\widehat{T} = \widehat{T}\left(t, q^{r+1}, q^{r+2}, \ldots, q^s, \dot{q}^{r+1}, \dot{q}^{r+2}, \ldots, \dot{q}^s, C_1, C_2, \ldots, C_r\right).$$

According to Routh, from the existence of integrals (3.3) it follows that the system of Lagrange equations of second kind (I.6.4) can be reduced to the equations

$$\frac{d}{dt} \frac{\partial R}{\partial \dot{q}^p} - \frac{\partial R}{\partial q^p} = -\frac{\partial \Pi}{\partial q^p}, \quad R = \widehat{T} - \sum_{i=1}^{r} C_i \dot{q}^i,$$
$$p = \overline{r+1, s}, \tag{3.4}$$

where R is the *Routh function*. Equations (3.4) contain only position coordinates and velocities; they are called *Routh equations*.

In the function R one may single out the following terms:

$$R = R_2 + R_1 + R_0, \quad R_2 = \frac{1}{2} \sum_{j,k=r+1}^{s} a_{jk}\dot{q}^i\dot{q}^k,$$

$$R_1 = \sum_{j=r+1}^{s} a_j\dot{q}^j.$$

Here, a_{jk}, a_j, R_0 are functions of position coordinates and integration constants $C_1, C_2, \ldots, C_r$.

We have $\partial R_0 / \partial \dot{q}^p = 0$ and

$$\frac{d}{dt}\frac{\partial R_1}{\partial \dot{q}^p} - \frac{\partial R_1}{\partial q^p} = \frac{da_p}{dt} - \sum_{j=r+1}^{s}\frac{\partial a_j}{\partial q^p}\dot{q}^j =$$

$$= \sum_{j=r+1}^{s}\frac{\partial a_p}{\partial q^j}\dot{q}^j - \sum_{j=r+1}^{s}\frac{\partial a_j}{\partial q^p}\dot{q}^j = \sum_{j=r+1}^{s}g_{pj}\dot{q}^j,$$

where

$$g_{pj} = \frac{\partial a_p}{\partial q^j} - \frac{\partial a_j}{\partial q^p},$$

and hence Eqs. (3.4) can be written as

$$\frac{d}{dt}\frac{\partial R_2}{\partial \dot{q}^p} - \frac{\partial R_2}{\partial q^p} = -\frac{\partial W}{\partial q^p} - \sum_{j=r+1}^{s}g_{pj}\dot{q}^j,$$

$$p = r+1, s.$$

(3.5)

Here,

$$W = \Pi - R_0.$$

(3.6)

With Eqs. (3.5) one may associate some *reduced system*, in which the functions R_2 and W play the role of the kinetic and potential energy. As was already pointed out in Sect. 5 of Chap. 7 of Vol. I, the terms of the sum $\sum_{j=r+1}^{s}(-g_{pj}\dot{q}^j)$, are gyroscopic forces, because they are typical for systems containing gyroscopes. These forces have a skew symmetric matrix, because $g_{pj} = -g_{jp}$. An important property of gyroscopic forces is that they work is zero for the actual displacement of the system. As a result, there exists the energy integral

$$R_2 + W = R_2 + \Pi - R_0 = \text{const.}$$

A motion of a system will be called *stationary* if all the position coordinates and cyclic velocities are stationary; that is, if the following conditions are satisfied:

$$q^p(t) = q_0^p, \quad p = \overline{r+1, s}, \quad \dot{q}^i = \dot{q}_0^i, \quad i = \overline{1, r}.$$

From system (3.5) it follows that to a stationary motion of the initial system there corresponds the state of rest of the reduced one. Besides, the generalized forces becomes zero; that is,

$$\left(\frac{\partial W}{\partial q^p}\right)_{q_0^p} = 0, \quad p = \overline{r+1, s},$$

or

$$\left(\frac{\partial \Pi}{\partial q^p}\right)_{q_0^p} = \left(\frac{\partial R_0}{\partial q^p}\right)_{q_0^p}, \quad p = \overline{r+1, s}. \tag{3.7}$$

Thus, the study of a stationary motion is reduced to that of stability of a state of rest of the reduced system. In this case, *Routh's theorem* can be looked as a stability criterion: *if in a stationary motion the function $W = \Pi - R_0$ has a minimum and if perturbations do not disturb the cyclic integrals (3.3), then this motion is stable with respect to the position coordinates and velocities.*

Example 4 Let us examine the stability of motion of a conical pendulum (see Sect. 2 of Chap. 6 of Vol. I). Expressions for the kinetic and potential energies appear as:

$$T = \frac{ml^2}{2}\left(\dot{\theta}^2 + \dot{\psi}^2 \sin^2 \theta\right), \quad \Pi = -mgl \cos \theta.$$

Here, ψ is the cyclic coordinate and θ is the position coordinate. From the cyclic integral

$$\frac{\partial T}{\partial \dot{\psi}} = ml^2 \dot{\psi} \sin^2 \theta = C \tag{3.8}$$

we find that $\dot{\psi} = C / (ml^2 \sin^2 \theta)$. Substituting this into the expression for kinetic energy, this gives

$$\widehat{T} = \frac{ml^2}{2}\dot{\theta}^2 + \frac{C^2}{2ml^2 \sin^2 \theta}.$$

In this case, the Routh function reads as

$$R = \widehat{T} - C\widehat{\dot{\psi}} = \frac{ml^2}{2}\dot{\theta}^2 - \frac{C^2}{2ml^2 \sin^2 \theta},$$

which gives

$$R_2 = \frac{ml^2}{2}\dot{\theta}^2, \quad R_1 = 0, \quad R_0 = -\frac{C^2}{2ml^2 \sin^2 \theta}.$$

Let us write function (3.6):

$$W = \Pi - R_0 = -mgl \cos \theta + \frac{C^2}{2ml^2 \sin^2 \theta}.$$

As before, for a stationary motion, θ and $\dot{\psi}$ will be denoted, respectively, as α and ω. From condition (3.7) we have

$$mgl \sin \alpha = \frac{C^2 \cos \alpha}{ml^2 \sin^3 \alpha}.$$

Using the cyclic integral (3.8) for the stationary motion to exclude the constant C, we have $\omega^2 l \cos \alpha = g$.

For $\theta = 0$ the function W has a minimum, because

$$\left. \frac{dW}{d\theta} \right|_{\theta=\alpha} = 0, \quad \left. \frac{d^2W}{d\theta^2} \right|_{\theta=\alpha} = mgl \cos \alpha + \frac{C^2(3\cos^2 \alpha + \sin^2 \alpha)}{ml^2 \sin^4 \alpha} > 0.$$

Here, it was taken into account that the angle α varies in the range $0 < \alpha < \pi/2$.

Thus, from Routh's theorem it follows that the motion of a conical pendulum is stable.

4 Thompson and Tait's Theorems

In this subsection we discuss the effect of dissipative and gyroscopic forces on the stability of an equilibrium of a conservative system. We shall be concerned with a linear system with n degrees of freedom, which can be written in the vector form as

$$A \cdot \ddot{q} + B \cdot \dot{q} + \Gamma \cdot \dot{q} + C \cdot q = 0, \quad q = (q_1, q_2, \ldots, q_n)^T. \quad (4.1)$$

Here, q is the vector of generalized coordinates, A, B, Γ, C are constant square matrices, the matrices A, B, C are symmetric, the matrix Γ is skew symmetric, the matrix A is positive definite, and the matrix B is positive or is zero. The terms in (4.1) describe in succession the inertia forces, dissipative forces, gyroscopic forces, and potential forces.

If $B = \Gamma = 0$ and if the matrix C is positive definite, then by Lagrange's theorem the equilibrium $q = 0$ is stable. Below, we discuss the effect of the matrices B and Γ on the stability of an equilibrium. The results are given in the form of four theorems due to Thompson and Tait.

Theorem 1 *If the equilibrium position of a conservative system is stable (that is, the matrix **C** is positive definite), then the addition of dissipative and gyroscopic forces does not violate the stability.*

Theorem 2 *If under the hypotheses of Theorem 1 the dissipative forces with complete dissipation (that is, the matrix **B** is positive definite), then the equilibrium position is asymptotically stable.*

Theorem 3 *If the degree of instability of the equilibrium position of a conservative system is even, then the addition of gyroscopic forces may result in stability loss of the equilibrium position. If the degree of instability is odd, then the addition of gyroscopic forces may not result in stability.*

Theorem 4 *If the degree of instability is even and if the equilibrium position becomes stable (see Theorem 3) as the result of addition of gyroscopic forces, then the addition of dissipative forces destroys the stability.*

All Thompson and Tait's theorems are stated without proofs, we shall just give some discussions.

Let us consider the Lyapunov function

$$V(q, \dot{q}) = \frac{1}{2}\dot{q}^T \cdot A \cdot \dot{q} + \frac{1}{2}q^T \cdot C \cdot q,$$

which equals the total energy of the system. Under the hypotheses of Theorems 1 and 2 this function is positive definite. Its derivative, which is calculated by virtue of system (4.1), is as follows:

$$V' = -\dot{q}^T \cdot B \cdot \dot{q}. \tag{4.2}$$

The gyroscopic forces do not work, and hence the matrix Γ does not enter into expression (4.2).

Under the hypotheses of Theorem 1 the function V' is negative, and hence the conclusion of Theorem 1 immediately follows from Lyapunov's Theorem 1 on stability.

The function V' is not negative definite function of $2n$ arguments $q, \dot{q}$, hence Lyapunov's Theorem 2 on asymptotic stability does not apply in this setting. From formula (4.2) it follows that the total energy of the system decreases in time. Hence, the conclusion of Theorem 2 is as expected.

For a proof of Theorem 2 we shall search a solution of system (4.1) in the form $q(t) = ue^{\mu t}$. For the amplitude vector u, we have the algebraic system of equations

$$D(\mu) \cdot u = 0, \qquad D(\mu) = \mu^2 A + \mu B + \mu \Gamma + C. \tag{4.3}$$

Let μ be one of the $2n$ roots of the characteristic equation

$$\det(D(\mu)) = 0, \tag{4.4}$$

and let $\boldsymbol{u} = \boldsymbol{u}_1 + i\boldsymbol{u}_2$ be the corresponding amplitude vector, in which the real and imaginary parts are separated. Multiplying Eq. (4.3) by the vector $(\boldsymbol{u}_1 - i\boldsymbol{u}_2)^T$ and transforming μ, this gives the quadratic equation

$$a\mu^2 + (b + i\gamma)\mu + c = 0, \tag{4.5}$$

where

$$a = \boldsymbol{u}_1^T \cdot \boldsymbol{A} \cdot \boldsymbol{u}_1 + \boldsymbol{u}_2^T \cdot \boldsymbol{A} \cdot \boldsymbol{u}_2, \quad b = \boldsymbol{u}_1^T \cdot \boldsymbol{B} \cdot \boldsymbol{u}_1 + \boldsymbol{u}_2^T \cdot \boldsymbol{B} \cdot \boldsymbol{u}_2,$$
$$c = \boldsymbol{u}_1^T \cdot \boldsymbol{C} \cdot \boldsymbol{u}_1 + \boldsymbol{u}_2^T \cdot \boldsymbol{C} \cdot \boldsymbol{u}_2, \quad \gamma = 2\boldsymbol{u}_1^T \cdot \boldsymbol{\Gamma} \cdot \boldsymbol{u}_2.$$

From the properties of the matrices $\boldsymbol{A}$, $\boldsymbol{B}$, $\boldsymbol{C}$ we have a, b, $c > 0$. Now one may show that, for any real γ, both roots of Eq. (4.5) have negative real parts.

In Theorems 3 and 4 (as distinct from Theorems 1 and 2) it is not assumed that the matrix $\boldsymbol{C}$ is positive definite. By the symmetry of the matrix $\boldsymbol{C}$, all roots λ_k of the characteristic equation $\det(\lambda\boldsymbol{A} + \boldsymbol{C}) = 0$ are real. The number m of positive roots of this equation is known as the *degree of instability* of the system. The case of zero roots is not considered.

Using the linear nonsingular transformation $\boldsymbol{q} = \boldsymbol{H} \cdot \boldsymbol{p}$ we reduce the conservative part of system (4.1) to the principal coordinates

$$\ddot{p}_k - \lambda_k p_k = 0, \quad k = \overline{1, n}. \tag{4.6}$$

Equations (4.6) with $\lambda_k > 0$ correspond to unstable motions. Assume that the degree of instability m is even. We split the unstable motions into pairs and show that by adding gyroscopic forces the unstable motions of a pair may become stable. Let us consider one of the pairs:

$$\ddot{p}_1 + \gamma\dot{p}_2 - \lambda_1 p_1 = 0, \quad \ddot{p}_2 - \gamma\dot{p}_1 - \lambda_2 p_2 = 0, \quad \lambda_1, \lambda_2 < 0. \tag{4.7}$$

All roots of the characteristic equation $\mu^4 + (\gamma^2 - \lambda_1 - \lambda_2)\mu^2 + \lambda_1\lambda_2 = 0$ of system (4.7) are purely imaginary with $|\gamma| > \sqrt{\lambda_1} + \sqrt{\lambda_2}$.

If the degree of instability m is odd, then one of the positive roots λ_k has no counterpart. As a result, the motion of system (4.1) will be unstable with any gyroscopic forces. Let us consider the characteristic Eq. (4.4) $f(\mu) = \det(\boldsymbol{D}(\mu)) = 0$ with $\boldsymbol{B} = 0$. The function $f(\mu)$ is an even function of μ, because, from one side, the transpose of the matrix $\boldsymbol{D}(\mu)$ does not change its determinant, and on the other hand, the gyroscopic terms with factor μ change the sign. We set $\mu^2 = \lambda$ and rewrite Eq. (4.4) in the form $f_1(\lambda) = f(\mu) = 0$. This equation has a positive root $\lambda > 0$, because the sign of $f_1(0)$ agrees with the negative sign of the determinant $\det(\boldsymbol{C})$ (for odd m), and for $\lambda \to \infty$ the sign of $f_1(\lambda)$ agrees with the positive sign of the determinant $\det(\boldsymbol{A})$.

In discussing Theorem 4 we shall consider only system (4.7), which includes the dissipative forces

$$\ddot{p}_1 + b_1\dot{p}_1 - \gamma\dot{p}_2 - \lambda_1 p_1 = 0, \quad \ddot{p}_2 + b_2\dot{p}_2 + \gamma\dot{p}_1 - \lambda_2 p_2 = 0, \quad \lambda_1, \lambda_2 < 0.$$
$$(4.8)$$

The characteristic equation of system (4.8)

$$\mu^4 + a_1\mu^3 + a_2\mu^2 + a_3\mu + a_4 = 0, \quad a_3 = -b_1\lambda_2 - b_2\lambda_1$$

has negative coefficient a_3, which speaks about instability of the zero solution of system (4.8).

Real systems always contain resistance forces, and hence the stability achieved by the action of gyroscopic forces is called *temporal Lyapunov's stability*, as distinct from the usual Lyapunov stability, which is called *secular stability* in gyroscopic systems.

Example 5 Small oscillations near the vertical position of a symmetric spinning top are described by the approximate system of equations

$$A\ddot{\alpha} + C\omega\dot{\beta} - Pz_c\alpha = 0, \quad A\ddot{\beta} - C\omega\dot{\alpha} - Pz_c\beta = 0, \quad (4.9)$$

where A is the moment of inertia of the top with respect to the horizontal axis passing through the support point, C is the moment of inertia with respect to the top axis, α, β are the angles of deflection of the top axis with respect to the vertical line in the xz, yz planes, ω is the angular velocity of the top, P is the top weight, and $OC = z_c$ is the distance of the centre of mass from the support point (see Fig. 3).

The degree of instability of system (4.9) with $\omega = 0$ is 2, which corresponds to the possibility of fall in the xz, yz planes. For

$$\omega > \frac{2\sqrt{APz_c}}{C}$$

Fig. 3 Vertical position stability of a symmetric rotating top

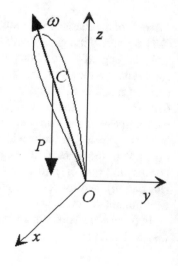

the vertical position of the top becomes stable, but this stability is deteriorated by the action of the friction forces.

5 Study of the Stability from the Linear Approximation

In many cases the stability of the zero solution of an equation under perturbation can be assessed from the consideration of the linear approximation. Let us consider the autonomous equation

$$\frac{dx}{dt} = X(x), \quad X(0) = 0. \tag{5.1}$$

Assuming that the vector function $X(x)$ is holomorphic in x, we expand it into a series and consider its linear approximation

$$\frac{dx}{dt} = A \cdot x + \theta(x), \quad A = \frac{dX}{dx}\Big|_{x=0} = (a_{km}),$$

$$a_{km} = \frac{\partial X_k}{\partial x_m}\Big|_{x=0}, \quad k, m = \overline{1, n},$$

where the vector function $\theta(x)$ contains only the terms of the second and higher orders of smallness as $x \to 0$. However, the holomorphicity assumption of the function $X(x)$ is very restrictive, but the theorems stated in this subsection remain valid also with $\theta(x) = o(\|x\|)$.

In parallel with system (5.1) or (5.2), we shall consider its linear approximation

$$\frac{dy}{dt} = A \cdot y. \tag{5.2}$$

Seeking the solution of system (5.2) in the form $y = ue^{\lambda t}$, we have the system of linear homogeneous algebraic equations for the unknowns u

$$(A - \lambda E) \cdot u = 0, \tag{5.3}$$

where E is the unit matrix of order n.

System (5.3) has a nontrivial solution if the determinant composed of its coefficients is zero,

$$\Delta(\lambda) = \det(A - \lambda E) = \begin{vmatrix} a_{11} - \lambda & \dots & a_{1n} \\ \dots & \dots & \dots \\ a_{n1} & \dots & a_{nn} - \lambda \end{vmatrix} = 0. \tag{5.4}$$

Next, we write the characteristic Eq. (5.4) as the equality to zero of the polynomial of degree n:

$$\Delta(\lambda) = a_0\lambda^n + a_1\lambda^{n-1} + \cdots + a_{n-1}\lambda + a_n = 0. \tag{5.5}$$

Let $\lambda_1, \ldots, \lambda_n$ be the roots of the characteristic Eq. (5.5). As follows from the three theorems stated below, the stability of the zero solution of both the linear system (5.2) and the nonlinear system (5.1) depends substantially on these roots.

Theorem 1 (on asymptotic stability) *If all roots of the characteristic equation have negative real parts, then the zero solution of both the linear system (5.2) and of the nonlinear system (5.1) is asymptotically stable.*

Theorem 2 (on instability) *If at least one root of the characteristic equation has positive real part, then the zero solution of both the system (5.2) and of the nonlinear system (5.1) is unstable.*

Theorem 3 (the critical case) *Assume that among the roots of the characteristic equation there are not roots with positive real part and there are zero and/or purely imaginary ones. Then the zero solution of the first approximation Eqs. (5.2) is stable if to the roots with zero real part there correspond simple elementary divisors of determinant (5.4), and is unstable if to the multiple roots with zero real part there correspond multiple elementary divisors. In the critical case, the asymptotic stability, stability or instability zero solution of the nonlinear equation (5.1) depends on the nonlinear terms $\theta(x)$.*

For a proof of Theorems 1–3 in the part of the linear system (5.2) it suffices to recall that its general solution is composed of the partial solutions of the form $u_k e^{\lambda_k t}$ in the case of simple elementary divisors of matrix (5.4) and of the partial solutions of the form

$$\left(u_k^{(0)} + u_k^{(1)}t + \cdots + u_k^{(m-1)}t^{m-1}\right)e^{\lambda_k t}$$

in the case of elementary divisors of multiplicity m.

For a proof of Theorems 1 and 2 for the nonlinear system (5.1), we shall be concerned only with the particular case when all roots of Eq. (5.5) are simple, real, and among them there are no zero root (even though Theorems 1 and 2 also hold without these restrictions). Let us consider a nonsingular linear transform of the unknowns $x = H \cdot z$, which brings the matrix A to diagonal form. Now the system of Eqs. (5.2) reads as

$$\dot{z}_k - \lambda_k z_k + \widehat{\theta}_k(z) = 0, \quad k = \overline{1,n}, \tag{5.6}$$

where, as before, $\widehat{\theta}_k(z)$ are nonlinear functions satisfying the condition $\widehat{\theta}_k(z) = o(|z|)$ as $z \to 0$.

To prove Theorem 1 we consider the following positive definite Lyapunov function

$$V(z) = \frac{1}{2}\sum_{k=1}^{n} z_k^2.$$

Its derivative by virtue of Eq. (5.6) is

$$V'(z) = \sum_{k=1}^{n} \lambda_k z_k^2 + P(z), \quad P(z) = -\sum_{k=1}^{n} z_k \widehat{\theta}_k(z) = o(|z|^2).$$

For all $\lambda_k < 0$ the function $V'(z)$ is negative definite, because the correction term $P(z)$ is of high order of smallness as $z \to 0$ and has no effect on the sign. Hence, system (5.6) satisfies the hypotheses of Lyapunov's asymptotic stability theorem. The zero solution of system (5.1) is also asymptotically stable, because the transformation $x = H \cdot z$ is nonsingular.

To prove Theorem 2 we consider the Lyapunov function

$$V(z) = \frac{1}{2} \sum_{k=1}^{n} \lambda_k z_k^2.$$

Its derivative by virtue of Eq. (5.6) is as follows:

$$V'(z) = \sum_{k=1}^{n} \lambda_k^2 z_k^2 + Q(z), \quad Q(z) = -\sum_{k=1}^{n} \lambda_k z_k \widehat{\theta}_k(z) = o(|z|^2).$$

By the hypotheses of Theorem 2 the function $V'(z)$ is positive definite, the function $V(z)$ can assume positive values for arbitrarily small values of $|z|$. Hence, the hypotheses of Lyapunov's instability theorem are satisfied.

In connection with Theorem 3 for the nonlinear system (5.1) let us go back to the discussion of Example 1, which was concerned with system of Eq. (2.2):

$$\dot{x}_1 = x_2 + \mu x_1^3, \quad \dot{x}_2 = -x_1 + \mu x_2^3.$$

The characteristic equation of linear approximation

$$\Delta(\lambda) = \begin{vmatrix} \lambda & 1 \\ -1 & \lambda \end{vmatrix} = \lambda^2 + 1 = 0$$

has purely imaginary roots $\lambda_{1,2} = \pm 1$. In Example 1 it was shown that depending on the parameter μ the zero solution is asymptotically stable ($\mu < 0$), is stable ($\mu = 0$) or is unstable ($\mu > 0$).

A. M. Lyapunov examined the critical cases in the case of one zero root of the characteristic Eq. (5.4), in the case of two nonzero roots, and in the case of a pair of purely imaginary roots. The research of critical cases under various assumptions is still continuing at the present time.

In the remaining part of the subsection we shall formulate some tests that all roots of Eq. (5.5) have negative real parts.

Let us first formulate **the necessary condition**.

A necessary condition that all the roots of Eq. (5.5) *have negative real parts* for $a_0 > 0$ *is that* $a_k > 0, k = \overline{1, n}$.

Assume that all roots of Eq. (5.5) have negative realparts. We factor the left-hand side of (5.5) as follows:

$$\Delta(\lambda) = a_0 \prod_{k=1}^{n} (\lambda - \lambda_k). \tag{5.7}$$

For a real root we have $\lambda - \lambda_k = \lambda + \alpha_k$, $\alpha_k > 0$. For a pair of complex-conjugate roots $\lambda_{1,2} = -\alpha \pm \beta$, we have $(\lambda - \lambda_1)(\lambda - \lambda_2) = \lambda^2 + 2\lambda\alpha + \alpha^2 + \beta^2$. Hence, and product (5.7) consists of factors with positive coefficients.

The Routh–Hurwitz criterion. Consider an infinite matrix composed of the coefficients of Eq. (5.5)

$$\begin{pmatrix} a_1 & a_3 & a_5 & a_7 & \dots \\ a_0 & a_2 & a_4 & a_6 & \dots \\ 0 & a_1 & a_3 & a_5 & \dots \\ 0 & a_0 & a_2 & a_4 & \dots \\ 0 & 0 & a_1 & a_3 & \dots \\ \cdot & \cdot & \cdot & \cdot & \cdot \end{pmatrix}. \tag{5.8}$$

A necessary and sufficient condition that all roots of Eq. (5.5) *have negative real parts is that* (for $a_0 > 0$) *the following n inequalities be satisfied:*

$$\Delta_1 = a_1 > 0, \quad \Delta_2 = \begin{vmatrix} a_1 & a_3 \\ a_0 & a_2 \end{vmatrix} > 0,$$

$$\Delta_3 = \begin{vmatrix} a_1 & a_3 & a_5 \\ a_0 & a_2 & a_4 \\ 0 & a_1 & a_3 \end{vmatrix} > 0, \dots, \Delta_n > 0. \tag{5.9}$$

The determinants in (5.9) are composed from the upper left part of matrix (5.8). The entries a_k in (5.9) are replaced by zeros for $k > n$. The Routh–Hurwitz criterion is given without proof.[2]

Polynomials (5.5) whose all roots have negative real parts are called *Hurwitz polynomials*.

Consider the following particular cases.

For $n = 1$, the Routh–Hurwitz criterion gives the condition $a_1 > 0$.

For $n = 2$, we have the conditions $a_1 > 0$ and $a_2 > 0$.

For $n = 3$, the positiveness of the coefficients of Eq. (5.5) is already insufficient. In addition to the positiveness of the coefficients, the following condition must be additionally satisfied: $a_1 a_2 > a_0 a_3$. This condition was first obtained by I. A. Vyshnegradskii in the study of the operational stability of Watt's governor. After changing variable, the third-order equation $a_0 \lambda^3 + a_1 \lambda^2 + a_2 \lambda + a_3 = 0$ becomes

---

[2] For a proof, see, for example, Sect. XV. 12 of the book: *F. R. Gantmacher*. The theory of matrices. Vols. 1, 2. New York: Chelsea Publishing Co. 1959.

Fig. 4 Evaluation of the asymptotic stability region using the Routh–Hurwitz criterion

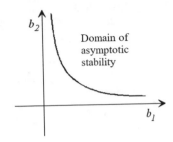

Domain of asymptotic stability

Fig. 5 Mikhailov diagram

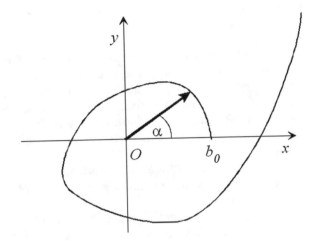

$z^3 + b_1 z^2 + b_2 z + 1 = 0$, and so, together with the conditions $b_1 > 0$, $b_2 > 0$, the additional condition reads as $b_1 b_2 > 1$. On the plane b_1, b_2, the region of asymptotical stability lies above the parabola $b_1 b_2 = 1$ (see Fig. 4).

Mikhailov's criterion. In stating this criterion we write Eq. (5.5) in the form

$$f(\lambda) = b_0 + b_1\lambda + b_2\lambda^2 + \cdots + b_n\lambda^n = 0, \quad b_0 > 0. \qquad (5.10)$$

We set $\lambda = i\omega$. Then

$$f(i\omega) = f_1(\omega) + i f_2(\omega),$$
$$f_1(\omega) = b_0 - \beta_2\omega^2 + b_4\omega^4 - \ldots, \quad f_2(\omega) = b_1\omega - \beta_3\omega^3 + b_5\omega^5 - \ldots.$$

On the complex plane $z = x + iy$, we construct the parametric curve

$$x = f_1(\omega), \quad y = f_2(\omega), \quad 0 \leqslant \omega < \infty. \qquad (5.11)$$

It is assumed that Eq. (5.10) has no purely imaginary roots. Then curve (5.11) does not pass through the origin, and one may introduce the radius vector connecting a current point on the curve with the origin (see Fig. 5).

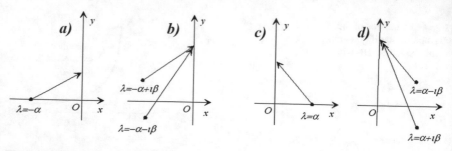

Fig. 6 On the number of roots of the characteristic equation with negative and positive real parts

We shall be concerned with the angle of rotation of this vector when the parameter ω varies from 0 to infinity. The total variation $\Delta\alpha$ of this angle is proportional to $\pi/2$ and equals $\Delta\alpha = k\pi/2$, where k is an integer number, because the angle $\alpha = 0$ with $\omega = 0$ and α tends to 0 or to $\pm\pi/2$ as $\omega \to \infty$ depending on the parity of n. The last assertion follows from the formula $\tan\alpha = f_2(\omega)/f_1(\omega)$, and depending on the parity of n, the power of the numerator or the denominator is larger. In the example in Fig. 5 $k = 5$.

Now we may state **Mikhailov's criterion**:

A necessary and sufficient condition that all roots of Eq. (5.10) have negative real parts is that $\Delta\alpha = n\pi/2$.

The following side result appears in the construction of the Mikhailov diagram:

If $k = n - 2m$, then Eq. (5.10) has $n - m$ roots with negative real part and m roots with positive real part.

For a *proof*, we factor polynomial (5.10) as

$$f(\lambda) = a_0(\lambda - \lambda_1)(\lambda - \lambda_2)\ldots(\lambda - \lambda_n)$$

and use the fact that the argument of the product of several complex numbers is equal to the sum of the arguments of the factors.

Let us find the contribution of single cofactors into the total increment of the angle $\Delta\alpha$ with variation of the parameter ω from 0 to $+\infty$. The terminus of the vector advances along the y-axis from 0 to $+\infty$. From Fig. 6a it is clear that the contribution of the factor with negative root $\lambda = -\alpha$ is $\pi/2$, while the contribution of two factors corresponding to a pair of purely imaginary roots $\lambda = \alpha \pm i\beta$ with negative real part is π (see Fig. 6b). The contributions of the roots with positive real part are, respectively, $-\pi/2$ (Fig. 6c) and $-\pi$ (Fig. 6d). The proof is completed by adding the contributions.

6 Stability of Periodic Solutions from Consideration of the Linear Approximation

Consider the equation with periodic right-hand side

$$\frac{dy}{dt} = Y(y, t), \quad Y(0) = 0, \quad Y(x, t + T) \equiv Y(x, t). \tag{6.1}$$

Assume that Eq. (6.1) has the periodic solution $y(t) = \varphi(t)$, $\varphi(t + T) = \varphi(t)$. We set $y(t) = \varphi(t) + x(t)$. Then, for the perturbation $x(t)$, we have an equation with the linear approximation

$$\frac{dx}{dt} = A(t) \cdot x + \theta(x, t), \quad A(t) = \left. \frac{dY}{dy} \right|_{y=\varphi(t)} = (a_{km}(t)),$$

$$a_{km}(t) = \left. \frac{\partial Y_k}{\partial y_m} \right|_{y=\varphi(t)}, \quad k, m = \overline{1, n}, \tag{6.2}$$

where the vector function $\theta(x, t) = o(\|x\|)$. As distinct from Eq. (5.2), here the matrix $A(t)$ and the vector function $\theta(x, t)$ are periodic in t with period T.

Qualitatively, the results on the stability of the zero solution of Eq. (6.2) repeat verbatim those for the autonomous Eq. (5.2). The fact is that there exists a nonsingular linear transformation $z = H(t) \cdot x$, under which Eq. (6.2) assumes the form

$$\frac{dz}{dt} = A_*(t) \cdot z + \theta_*(z, t), \quad \theta_*(x, t) = o(\|x\|)$$

with the constant matrix A_*. By this, the problem under consideration is reduced in the linear approximation to the autonomous one.

Let us now get back to Eq. (6.2) and consider the equation of linear approximation

$$\frac{dx}{dt} = A(t) \cdot x, \quad A(t + T) = A(t). \tag{6.3}$$

We shall seek a solution of this equation satisfying the relation

$$x(t + T) = \rho x(t).$$

By the periodicity, $x(t + kT) = \rho^k x(t)$, and hence the parameter ρ, which is called the *characteristic exponent*, serves to measure the solution growth rate in time. It is clear that $|x(t)| \to 0$ for $|\rho| < 1$ and $|x(t)| \to \infty$ for $|\rho| > 1$.

To calculate the characteristic exponents we write the fundamental matrix of solutions $X(t)$ of Eq. (6.3)

$$\frac{dX}{dt} = A(t) \cdot X, \quad X(0) = E. \tag{6.4}$$

Any solution of Eq. (6.3) reads as

$$x(t) = X(t) \cdot C, \quad C = (C_1, C_2, \ldots, C_n)^T, \tag{6.5}$$

where C is the vector of arbitrary constants. Substituting (6.5) into (6.3) with $t = 0$, we get the equation $(X(T) - \rho E) \cdot C = 0$, whence follows the equation for the characteristic exponents

$$\det(X(T) - \rho E) = 0. \tag{6.6}$$

The matrix $X(T)$ is called matricant. Equation (6.6) has n roots $\rho_1, \rho_2, \ldots, \rho_n$.

Theorems 1 and 2 of the previous subsection hold almost verbatim also for the periodic Eqs. (6.2) and (6.3):

Theorem 1 (on asymptotic stability) *If all the characteristic exponents are smaller than one in absolute value, then the zero solution of systems (6.2) and (6.3) is asymptotically stable.*

Theorem 2 (on instability) *If at least one characteristic exponent is greater than one in absolute value, then the zero solution of systems (6.2) and (6.3) is unstable.*

Theorem 3 (the critical case) *If all the characteristic exponents are not greater than one in absolute value and if the characteristic exponents with $|\rho| = 1$ are simple roots of Eq. (6.6), then the zero solution of the linear equation (6.3) is stable. As before, the stability of the zero solution of the nonlinear equation depends on the function $\theta(x, t)$.*

Unfortunately, in the general setting, the matricant $X(T)$ can be constructed only via numerical integration of the Cauchy problem (6.4) on the interval $0 \leqslant t \leqslant T$.

If the periodic component of the matrix $A(t)$ is small, that is,

$$A(t) = A_0 + \mu A_1(t), \quad \mu \ll 1,$$

then in order to construct the matricant and examine the stability, one may involve methods of expanding into series in powers of the small parameter; this will be illustrated below.

7 Oscillations of a Pendulum with a Vibrating Suspension Point. The Mathieu Equation

The equation of oscillations of the mathematical pendulum in the vertical plane reads as

$$\ddot{\varphi} + \frac{g}{l} \sin \varphi = 0,$$

where $\varphi(t)$ is the angle of deflection of the pendulum with respect to the vertical line, l is the pendulum length, g is the acceleration of gravity. In the case when

Fig. 7 Pendulum with an
oscillating suspension point

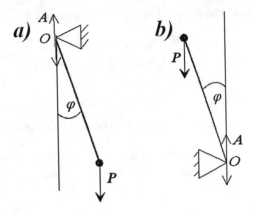

the suspension point vibrates in the vertical direction, one needs to assume $g = g_0 + A\omega^2 \sin \omega t$, where A and ω are, respectively, the amplitude and frequency of oscillations (Fig. 7a).

In the dimensionless notation the equation of oscillations in linear approximation can be written in the standard form of the *Mathieu equation* as

$$x'' + (q + \mu \cos 2\tau)x = 0, \quad (\cdot)' = \frac{d(\cdot)}{d\tau}, \tag{7.1}$$

where

$$\tau = \frac{\omega t}{2}, \quad q = \frac{4g_0}{l\omega^2}, \quad \mu = \frac{4A}{l}.$$

For $\mu \ll 1$ the linearly independent solutions of Eq. (7.1) are sought in the form

$$\begin{aligned}
&x_1(\tau, \mu) = x_{10}(\tau) + \mu x_{11}(\tau) + \mu^2 x_{12}(\tau) + \cdot, \\
&x_1(0, \mu) = 1, \quad x_1'(0, \mu) = 0, \\
&x_2(\tau, \mu) = x_{20}(\tau) + \mu x_{21}(\tau) + \mu^2 x_{22}(\tau) + \cdot, \\
&x_2(0, \mu) = 0, \quad x_2'(0, \mu) = 1.
\end{aligned} \tag{7.2}$$

In the zero approximation, we have $x_{10} = \cos(q_1\tau)$, $x_{20} = \sin(q_1\tau)/q_1$, $q_1 = \sqrt{q}$, the successive approximations are found from the recurrent Cauchy problems

$$x_{km}'' + q x_{km} + x_{k,m-1} \cos 2\tau = 0, \quad x_{km}(0) = 0, \quad x_{km}'(0) = 0,$$
$$k = 1, 2, \quad m = 1, 2, \ldots,$$

whose solution is expressed in terms of the Duhamel integral

$$x_{km}(\tau) = -\frac{1}{q_1} \int_0^\tau \sin(q_1(\tau - \tau_1)) x_{k,m-1}(\tau_1) \cos 2\tau_1 d\tau_1.$$

The function in Eq. (7.1) has period π, and hence Eq. (6.6) assumes the form

$$\begin{vmatrix} x_1(\pi, \mu) - \rho & x_1'(\pi, \mu) \\ x_2(\pi, \mu) & x_2'(\pi, \mu) - \rho \end{vmatrix} = 0 \quad \text{or} \quad \rho^2 - 2a\rho + \Delta = 0, \qquad (7.3)$$

where

$$2a = x_1(\pi, \mu) + x_2'(\pi, \mu), \quad \Delta = x_1(\pi, \mu)x_2'(\pi, \mu) - x_1'(\pi, \mu)x_2(\pi, \mu).$$

It is easily checked that $\Delta' = 0$, and hence $\Delta = 1$ in view of the initial conditions (7.2), and so Eq. (7.3) assumes the form

$$\rho^2 - 2a\rho + 1 = 0. \qquad (7.4)$$

For $|a| > 1$, both roots ρ_1 and ρ_2 of Eq. (7.4) are real, one of them is greater than one in absolute value, and hence, the zero solution of Eq. (7.1) is unstable. If $|a| < 1$, the roots ρ_1 and ρ_2 are both complex, and besides, $|\rho_{1,2}| = 1$. Hence, the zero solution is stable. The case $|a| = 1$ defines the boundary of the instability region on the plane of parameters (q, μ).

Using formulas (7.2), this gives

$$a = \cos(q_1\pi) + \mu^2 \frac{\pi \sin(q_1\pi)}{16q_1(q_1^2 - 1)} + O(\mu^3),$$

whence it follows that for $\mu \ll 1$ an instability is possible if q_1 is close to one of the integers $1, 2, \ldots$. Let $q_1 = 1 + \varepsilon$, then

$$a = -1 + \frac{\pi}{2}\left(\varepsilon^2 - \frac{\mu^2}{16}\right) + O(\mu^3). \qquad (7.5)$$

This gives us the principal instability region $|\varepsilon| < \mu/4 + O\mu^2$ or $|q - 1| < \mu/8 + O\mu^2$. To construct the remaining instability regions, one needs to retain larger number of terms in expansion (7.5). We give the boundaries of the first three instability regions (with terms of order μ^3), which are taken from handbooks on special functions:

$$1 - \frac{\mu}{8} - \frac{\mu^2}{32} + \frac{\mu^3}{512} < q^{(1)} < 1 + \frac{\mu}{8} - \frac{\mu^2}{32} - \frac{\mu^3}{512},$$

$$4 - \frac{\mu^2}{48} < q^{(2)} < 4 + \frac{5\mu^2}{48},$$

$$9 + \frac{\mu^2}{64} - \frac{\mu^3}{512} < q^{(3)} < 9 + \frac{\mu^2}{64} + \frac{\mu^3}{512}.$$

The instability regions follow the axis $\mu = 0$ at the points $q^{(n)} = n^2$ (Fig. 8). For the principal instability region, the tangent lines to its boundaries at $\mu = 0$ form

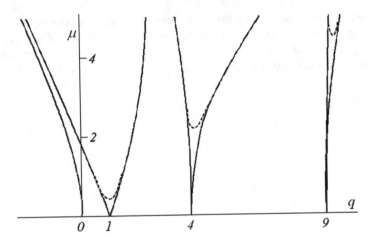

Fig. 8 Ince–Strutt diagram

a finite angle (this is the *wide* instability region). The remaining instability regions are *narrow*, because the tangent lines to their boundaries agree for $\mu = 0$.

Above we have examined oscillations of a pendulum without consideration of friction forces. The dashed lines in Fig. 8 indicate the instability regions with friction forces proportional to the velocity. Now these regions do not follow the axis $\mu = 0$, and an instability in more narrow regions may excite for larger values of μ (than in broad regions).

Let us now consider the stability of the vertical position of an inverted pendulum (Fig. 7b), when the point of support is in the bottom. This position is unstable is there are no oscillations of the point of support (as the pencil standing on its point is unstable); but according to Kapitsa[3] it may be stable in the case of oscillations.

Small oscillations of an inverted pendulum are described by the same Eq. (7.1), but with $q < 0$. Setting $q = -q_2^2$ and calculating the parameter a by the above method, we find

$$a = \cosh(q_2\pi) -$$
$$- \mu^2 \frac{q_2\pi(4 - 5q + q^2)\cosh(q_2\pi) + (2(2 + q^2) + \pi(4 - 5q + q^2))\sinh(q_2\pi)}{32q_2(1 - q)^2(4 - q)},$$

whence we get the boundary q^* of the stability region $q > -q^*$, where

$$q^* = \mu^2 \frac{2 + \pi^{-1}}{16} + O(\mu^3).$$

This region also contains negative q and is quite narrow (see Fig. 8).

---

[3] See the paper: *P. L. Kapitsa.* Dynamical stability of a pendulum with vibrating suspension point Journal of Experimental and Theoretical Physics. **21** (5). 1951. Pp. 588–598 [in Russian].

Figure 8 shows a fragment of the *Ince–Strutt diagram*. Considering this diagram in the large, we note that for $\mu \ll 1$ the regions of unstable values of q are relatively small. Instability regions increase with increasing μ, and on the other hand, for $\mu \gg 1$ the majority of q fall in instability regions.

Chapter 2
Nonlinear Oscillations

P. E. Tovstik and M. P. Yushkov (ORCID)

In this chapter, a special emphasis is given to approximate methods of solution of nonlinear equations (the small-parameter method, asymptotic methods). A relation between the Bubnov–Galerkin method with the Gauss' principled is established. A detailed exposition is given to P. L. Kapitsa's method of direct separation of motions, which at present is not adequately presented in textbooks. Theoretical results are clarified by solving a number of new examples. The last subsection is dedicated to strange attractors.

1 Basic Properties of Nonlinear Systems

In the study of oscillations of a point (see Sect. 7 of Chap. 4 of Vol. I) we have considered a solution of a linear differential equation, which was linear because the elastic force acting on a point is subject to Hooke's law and the resistance force is proportional to the point velocity. However, for sufficiently large deformations, Hooke's law does not hold, and in the general case the elastic force F is some nonlinear function of x. This force is frequently given in the form

$$F = ax + bx^3, \quad a > 0.$$

For $b > 0$, the elastic force has rigid characteristic, and for $b < 0$, it has soft characteristic. With such restoring force, the equation of motion of a point may be written as the *Duffing equation*

$$\ddot{x} + \omega^2 x + \mu x^3 = 0, \quad \omega^2 = \frac{a}{m}, \quad \mu = \frac{b}{m}. \tag{1.1}$$

© Springer Nature Switzerland AG 2021
N. N. Polyakhov et al., *Rational and Applied Mechanics*,
Foundations of Engineering Mechanics,
https://doi.org/10.1007/978-3-030-64118-4_2

Fig. 1 Variation of the force of dry friction by Coulomb's law

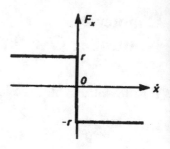

The nonlinearity of an elastic force may be manifested even when Hooke's law is satisfied. To support this assertion we consider a body which, during oscillations, is subject to force from various number of forces. In this case, the dependence of the elastic force on the displacement looks like a broken line.

The resistance force may also be nonlinear with respect to the velocity. For example, for sufficiently large velocities, the aerodynamic resistance force varies according to a nonlinear law (see Chap. 10). The dependence on the velocity is also fairly involved for the resistance force, which reflects the dissipation of energy inside a deformable system (usually, this quantity is estimated in terms of the hysteresis loop, which characterizes the departure from Hooke's law under cyclic deformations of an elastic body). The dependence of the force of dry friction on the velocity of displacement is also nonlinear. The simplest law of variation of the velocity due to C. A. de Coulomb is depicted in Fig. 1.

This being so, nonlinear equations reflecting the various character of forces acting on a point can arise in substantially different problems of mechanics. We also note that even weakly nonlinear systems (that is, for systems in which the nonlinear terms of the motion equation are proportional to the small parameter μ) have a number of specific properties that are not amenable to the linear theory of oscillations. Thus, in the analysis of oscillations in linear systems, all possible types of motion are characterized with the help of particular processes, which is impossible in nonlinear system, because the principle of superposition of particular solutions does not hold for a nonlinear equation. In a nonlinear system, only may not independently study the behaviour of single oscillation modes, because being substituted in the differential equation the sums of two different modes have an effect on each other. Similarly, if one expands the perturbing force into a Fourier series, then for a nonlinear equation its action on the system will not be equal to the sum of actions of all terms of this series.

In linear systems, the characteristic frequency of oscillations is independent of the initial conditions; that is, the oscillations are tautochronic. This property fails to hold in the nonlinear setting (see Sect. 2 of Chap. 6 of Vol. I).

In a linear system, the forced oscillations excited by a perturbing harmonic force have the same frequency as this force. However, even in weakly nonlinear systems there may appear oscillations whose frequencies are either multiples of the frequency of the exciting force or fractional with respect to it.

Fig. 2 The energy balance diagram

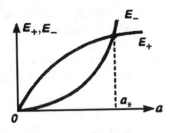

In linear systems with resistance, steady oscillations may appear only when subject to an external periodic force. However, in nonlinear systems, similar oscillations may also appear when the external force is not periodic in time. Their appearance may be caused by the presence in the system of radiators or absorbers of energy, whose performance is controlled by the oscillating system itself. Periodic motions appearing in a similar autonomous system are called *self-excited oscillations*. A classical example of a self-excited system is a clock's pendulum. For small deviations of a pendulum, which are insufficient for triggering the clock's mechanism, the pendulum executes linear damped oscillations. If, however, the pendulum is given a deviation for which the clock's mechanism is engaged, then it will execute self-excited oscillations, which can be explained only in the framework of the nonlinear theory. These oscillations are stable and tautochronic.

As an example of a self-excited system we consider a mass on a spring, which lays on an infinite conveyer belt moving with constant velocity. The force exerted from the band to the mass is made up of the swinging force and the force that retards the system. The origination of self-excited oscillations in this case can be explained as follows.

Assume that the energy balance diagram is as shown in Fig. 2, which depicts the dependences of the energy E_+ entering the system and the energy E_- consumed by it on the oscillation amplitude of the mass a. It is seen from the diagram that for $a < a_*$ the energy influx exceeds its consumption, and hence the amplitude increases. For $a > a_*$, the oscillation amplitude decreases, because the energy consumption exceeds its influx. If now $a = a_*$, then the system executes stable self-excited oscillations with this amplitude, the influx and consumption of the energy compensating.

Nonlinear systems frequently encounter *parametric oscillations* due to periodic variation of some of its parameters. Under certain relations on the characteristics of a system, the solution may be unstable, as a result of which small perturbations will be responsible for an increase in its oscillations. Such a phenomenon is known as the *parametric resonance*. Finding the conditions for the parametric resonances to occur usually requires the study of a linearized differential equation, whose coefficients are periodic in time. So, the study of nonlinear oscillations frequently calls for the necessity in the simultaneous study of linear systems with time-dependent coefficients. Note that parametric oscillations may also appear in linear systems.

It is worth noting that some unexpected phenomena were revealed in the numerical integration of differential equations of nonlinear systems with nonsmall values of the nonlinearity parameter. It turned out that depending on the initial conditions in a nonlinear mechanical system there may appear several stable solutions. Besides, in the phase space there may appear sets to which solutions of the system unboundedly approach (are attracted) with increasing time. These sets were called *strange attractors* (from the English word *attraction*).

2 Particular Cases of Nonlinear Oscillations

Analytic solution of some nonlinear equations. The study of the behaviour of nonlinear systems usually calls for the use of involved mathematical machinery. However, one succeeds in finding an exact analytic solution only for a small class of nonlinear equations.

In frequently happens that a nonlinear problem can be subdivided into a sequence of linear problems, of which each one has a solution. Under this approach, the data characteristic of the completion of motion on one step are taken as the initial conditions of the next step. This method for construction of solutions is sometimes called the *step-by-step method* (the method of piecewise linear approximation).

Consider a mass m on a rough plane support moving horizontally according to the law ($k = \text{const}$, $J_0 = \text{const}$)

$$\eta = \eta_1 + \eta_2 = \frac{J_0}{k} t - \frac{J_0}{k^2} \sin kt . \tag{2.1}$$

The relative displacement x is described by the nonlinear differential equation

$$m\ddot{x} = -m\ddot{\eta} - F \, \text{sign} \, \dot{x} , \tag{2.2}$$

where F is the force of dry friction. At each step of motion, on which the sign of velocity $\dot{x}$ is unchanged, this equation is linear. Assume that $|m\ddot{\eta}| > F$ at the instants when $\dot{x} = 0$. Hence, for the zero initial conditions, the solution of Eq. (2.2) can be written in the form

$$x(t) = f(t, t_0) , \quad t_0 < t < t_1 ,$$

$$x(t) = \sum_{\nu=1}^{n-1} f(t_{\nu+1}, t_\nu) + f(t, t_n) , \quad t_n < t < t_{n+1} , \tag{2.3}$$

$$n = 1, 2, \ldots ,$$

$$f(t, t_n) = J_0 k^{-2} [\sin kt - \sin kt_n - k(t - t_n) \cos kt_n + 0.5 \, (-1)^n \mu k^2 (t - t_n)^2] ,$$

$$n = 0, 1, 2, \ldots , \quad \mu = F/(mJ_0) .$$

Here, t_n, $n = 1, 2, \ldots$, are instants of time at which $\dot{x} = 0$. They are found from the recursive relation

$$\cos kt_{n+1} = \cos kt_n - (-1)^n \mu k (t_{n+1} - t_n),$$

the motion starts at time t_0 satisfying the equation $\sin kt_0 = \mu$.

The solution (2.3) was constructed under the assumption that

$$|m\ddot{\eta}(t_n)| > F; \quad \text{that is} \quad |\sin kt_n| > \mu, \quad n = 1, 2, \ldots.$$

One may show that this condition is satisfied for $\mu \ll 1$.

When a material point moves in a straight line in a constant field of force, when the motion equation reads as

$$m\ddot{x} + f(x) = 0, \tag{2.4}$$

some information about the motion may be obtained form the law of conservation of mechanical energy

$$m\dot{x}^2/2 + \Pi(x) = h, \tag{2.5}$$

where the potential energy $\Pi(x)$ is related with the force $f(x)$ by the relation

$$\Pi(x) = \int_{x^*}^{x} f(x)\, dx. \tag{2.6}$$

Here, x^* characterizes the position at which the potential energy is zero.

From (2.5) it follows that the time t, during which the point moves from x_0 to x is given by the integral

$$t = \pm \int_{x_0}^{x} \frac{\sqrt{m}\, dx}{\sqrt{2}\sqrt{h - \Pi(x)}}. \tag{2.7}$$

Here, the integral with 'minus' ('plus') sign corresponds to the case when $x > x_0$ (respectively, $x < x_0$).

We apply formula (2.7) to the Duffing equation (1.1) for the soft characteristic ($\mu < 0$). In the example under study, we have by (1.1), (2.4) and (2.6)

$$m = 1, \quad f(x) = \omega^2 x + \mu x^3, \quad x^* = 0, \quad \Pi(x) = \frac{\omega^2 x^2}{2} + \frac{\mu x^4}{4}.$$

Assume that at the initial time $x = a$ and $\dot{x} = 0$. Hence, from the law of conservation of mechanical energy (2.5), we get

$$h = \frac{\omega^2 a^2}{2} + \frac{\mu a^4}{4} .$$

(2.8)

If a is such that $f(a) > 0$, that is,

$$-\mu a^2 < \omega^2 ,$$

(2.9)

then for $t > 0$ the point moves towards the origin, and, hence, $x < a = x_0$ in the integral (2.7). As a result, we have

$$t = -\int_a^x dx \bigg/ \left(\sqrt{2} \sqrt{\frac{\omega^2 a^2}{2} + \frac{\mu a^4}{4} - \frac{\omega^2 x^2}{2} - \frac{\mu x^4}{4}} \right) =$$

$$= \int_x^a dx \bigg/ \left(\omega \sqrt{(a^2 - x^2) \left[1 + \frac{\mu}{2\omega^2} (a^2 + x^2) \right]} \right) .$$

Changing to the new variable α_1 by the formula

$$x = a \sin \alpha_1 ,$$

(2.10)

and putting

$$k^2 = -\frac{\mu a^2}{2\omega^2} \left(1 + \frac{\mu a^2}{2\omega^2} \right)^{-1} ,$$

we get

$$\tau \equiv \omega \sqrt{1 + \frac{\mu a^2}{2\omega^2}}\, t = \int_{\alpha_1}^{\pi/2} \frac{d\alpha}{\sqrt{1 - k^2 \sin^2 \alpha}} =$$

$$= \int_0^{\pi/2} \frac{d\alpha}{\sqrt{1 - k^2 \sin^2 \alpha}} - \int_0^{\alpha_1} \frac{d\alpha}{\sqrt{1 - k^2 \sin^2 a}} = K(k) - F(\alpha_1, k) ,$$

where we have used the customary notation for complete and incomplete elliptic integrals of the first kind. Setting $\alpha_1 = 0$, we determine the time t_* of motion to the origin

$$t_* = K(k) \bigg/ \sqrt{\omega^2 + \frac{\mu a^2}{2}} .$$

If the same value of h, as given by formula (2.8), corresponds to the initial conditions $x(0) = 0$, $\dot{x}(0) = \sqrt{2h}$, then the time of motion from the origin to the position $x = a \sin \alpha_1$ is given by the integral

$$\tau \equiv \omega \sqrt{1 + \frac{\mu a^2}{2\omega^2}}\, t = \int_0^{\alpha_1} d\alpha / \sqrt{1 - k^2 \sin^2 \alpha} = F(\alpha_1, k) .$$

Reversing the integral, we obtain

$$\alpha_1 = \psi(\tau, k),$$

and further since

$$\sin \alpha_1 = \sin \psi(\tau, k) = \operatorname{sn}(\tau, k),$$

it follows in view of (2.10) that the law of motion of a point can be written as

$$x = a \operatorname{sn}(\tau, k) = a \operatorname{sn}\left(\sqrt{\omega^2 + \frac{\mu a^2}{2}}\, t,\, k\right).$$

The last function is τ-periodic with period $4K(k)$, and hence, the motion in time has the period

$$T = 4K(k) \Big/ \left(\omega\sqrt{1 + \frac{\mu a^2}{2\omega^2}}\right) = 4t_* . \tag{2.11}$$

This formula is obtained in view of (2.9). From this formula one may find the dependence of the period of oscillations on the initial conditions. For the corresponding linear problem one should put $\mu = 0$. Then the modulus k of the elliptic integral also vanishes. Hence, since $K(0) = \pi/2$ if follows from (2.11) that

$$T = 2\pi/\omega.$$

If a varies from zero to the limiting possible value $a_* = \omega/\sqrt{-\mu}$, then $\sqrt{1 + \mu a^2/(2\omega^2)}$ ranges from 1 to $\sqrt{1/2}$, and the modulus k varies from 0 to 1. Hence, the period T increases from $2\pi/\omega$ to infinity, because $K(1) = \infty$.

Analysis of motion with the use of the phase plane. For a qualitative study of the motion of nonlinear systems it is convenient to use the phase plane. This is particularly helpful in consideration of a conservative autonomous system. In this case, instead of a given Eq. (2.4), one introduces the system of two first-order equations

$$\frac{dx}{dt} = y, \quad \frac{dy}{dt} = -f(x) \tag{2.12}$$

(in what follows, we put for convenience $m = 1$).

The plane of the variables x, y is called the *phase plane*, and the point with coordinates (x, y) is called *representative*. The curve along which moves the representative point is the *phase trajectory* or the *integral trajectory* (see Sect. 1 of Chap. 1). Its equation is easily obtained from system (2.12):

$$\frac{dy}{dx} = -\frac{f(x)}{y}. \tag{2.13}$$

The singular points of equation are found from the conditions

$$f(x) = 0, \quad y = 0; \tag{2.14}$$

that is, they all lie on the x-axis. Since $y = dx/dt$, it follows that for $y = 0$ the velocity of the material point is zero. In the singular points (2.14) we have, moreover, $f(x) = 0$, and hence the force vanishes. Hence, the singular points of the differential equation (2.13) correspond to the equilibrium cases of the material system.

The velocity (the *phase velocity*) of the representative point is as follows:

$$v = \sqrt{\left(\frac{dx}{dt}\right)^2 + \left(\frac{dy}{dt}\right)^2} = \sqrt{y^2 + f^2(x)}.$$

As a result, the velocity of the representative point never vanishes, except at the singular points.

In order to find the phase trajectories we integrate the equation with separable variables (2.13). We have

$$y^2/2 + \Pi(x) = h, \tag{2.15}$$

where h is an arbitrary constant. This expression has a simple physical significance: it is the integral of energy (2.5). We rewrite the equation of phase trajectories (2.15) as

$$\pm y = \sqrt{2}\sqrt{h - \Pi(x)}. \tag{2.16}$$

It is seen that all phase trajectories are symmetric about the x-axis. According to expression (2.13), the tangents to the integral curves at the points of their intersection with the x-axis are parallel to the y-axis, with the exception of the singular points (2.14). At the points (x, y), where x satisfies the equation $f(x) = 0$ and $y \neq 0$ is given by (2.16), the tangents to the integral curves are parallel to the x-axis.

The motion along the phase trajectories in the upper half-plane is from left to right (since the positiveness of y means an increase of the function x), and in the lower half-plane the motion is from right to left, as indicated by arrows on the integral curves.

Consider the phase trajectories of the Duffing equation (1.1) for the rigid characteristic ($\mu > 0$). On the *energy balance plane* (Fig. 3) we consider the curve

$$\Pi(x) = \frac{\omega^2 x^2}{2} + \frac{\mu x^4}{4}; \tag{2.17}$$

Fig. 3 Phase trajectories of
the Duffing equation for the
soft characteristic

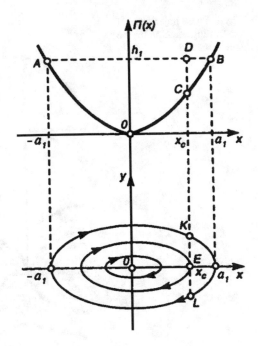

this is a parabola of the fourth order. We assume that the energy is constant $h = h_1$, which corresponds to certain initial conditions. According to the energy conservation law (2.15), we have $\Pi(x) = h_1$ for $y = 0$. Let a_1 and $-a_1$ be the real roots of the equation $\Pi(x) - h_1 = 0$. To the points $A = (-a_1, h_1)$, $B = (a_1, h_1)$ on curve (2.17) on the phase plane there correspond the points $(-a_1, 0)$, $(a_1, 0)$. The integral curve passing through these points can be constructed as follows: we set the abscissa x_C, and then, on the energy balance plane, we measure the length of the interval $CD = h_1 - \Pi(x_C)$, and from this length using the intervals $EK = EL = \sqrt{2}\sqrt{h_1 - \Pi(x_C)}$ we find on the phase plane the symmetric points K and L of the integral curve.

So, to the arc AOB on the phase plane there corresponds a closed curve on the energy balance plane. Varying h one may build the family of integral curves. For $h = 0$ the integral curve degenerates into the point $(0, 0)$. Near this point the phase trajectories look like ellipses, because for small x the entropy integral can be approximately written in the form

$$\frac{y^2}{2} + \frac{\omega^2 x^2}{2} = h.$$

Hence, the motion near the singular point, which corresponds to the minimum of the potential energy, is characterized by the presence of phase trajectories, which are closed nested ellipse-like curves. Such a singular point is called a *centre*. The motion of the representative point along such trajectories is periodic, which corresponds to a periodic motion of a material point. In this case, the singular point corresponds

Fig. 4 Phase trajectories of
the Duffing equation for the
soft characteristic

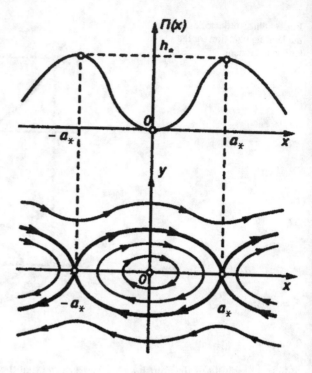

to a stable equilibrium position. The period T of motion with $h = h_1$ can be found
from formula (2.7). Calculating T as the doubled travel time along the upper portion
of the curve from the point with the abscissa $-a_1$ to the point with the abscissa a_1,
we obtain

$$T = \int_{-a_1}^{a_1} \sqrt{2}\, dx \big/ \sqrt{h_1 - \Pi(x)}.$$

From this formula it is seen that the oscillations are nonisochronous. As a result,
this motion cannot be considered Lyapunov stable, because given two representative
points, which start their motion from two nearby points and move along nearby
trajectories, eventually move apart at some finite distance due to the difference of
the periods. Here, the proximity of trajectories means that in this case the system is
orbitally stable.

We now illustrate, using the phase plane, the motion described by the Duffing
equation (1.1) for the soft characteristic ($\mu < 0$). The potential energy of this system
for $x = 0$ has a minimum, and for $x = \pm a_* = \pm \omega/\sqrt{-\mu}$ it has maxima (Fig. 4). The
origin on the phase plane, corresponding to the minimum of the potential energy, is
again a centre-type singular point surrounded by closed phase trajectories.

Let us examine the character of phase trajectories near the singular points $(a_*, 0)$
and $(-a_*, 0)$ corresponding to the maxima of the potential energy. The equations
of phase trajectories can be found from the equation of conservation of the total

Fig. 5 Inflexion of the curve
of the potential energy

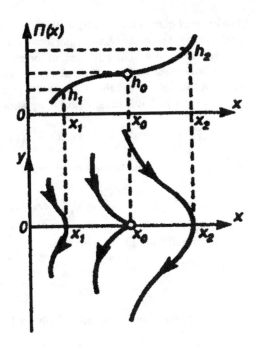

mechanical energy

$$\frac{y^2}{2} + \frac{\omega^2 x^2}{2} + \frac{\mu x^4}{4} = h .$$

If one puts $x = a_* + \xi$, then for sufficiently small ξ, this equation can be approximately written in the form

$$\frac{y^2}{2} - \omega^2 \xi^2 = h - h_* , \qquad h_* = \frac{\omega^2 a_*^2}{2} + \frac{\mu a_*^4}{4} = -\frac{\mu a_*^4}{4} .$$

It follows that near the singular point under study the integral curves have the following form: for $h = h_*$ these are two lines passing through the singular point, and for $h > h_*$ and $h < h_*$ these are hyperbolas intersecting, respectively, the vertical and horizontal axes. Such a particular point is called a *saddle*. From Fig. 4 it is seen that this equilibrium position is unstable. An integral curve with self-intersection passing through singular points is called a *separatrix*.[1] In the figure they are shown in bold lines. Finding separatrices is of particular interest, because they separate the integral curves into the regions corresponding to the phase trajectories of different types.

Let us now consider the singular point of the differential equation (2.13) corresponding to the inflexion of the curve of the potential energy with x_0; here we shall assume that at this point the tangent to this curve is horizontal (Fig. 5). In this case, for $h = h_0$ the phase trajectory at the point $(x_0, 0)$ has a first cusp of the first kind.

---

[1] From the French word *séparer*, to separate.

The integral curves for $h = h_1 < h_0$ or $h = h_2 > h_0$ may be easily depicted on the phase plane. From the phase trajectories it is seen that the motion of the system is Lyapunov unstable. The equilibrium for $x = x_0$ also proves unstable.

We now consider the motion of an autonomous system, when in parallel with the conservative force $f(x)$ there is a nonconservative force $\varphi(x, \dot{x})$, the latter depends on the velocity. In this case, with the equation of motion

$$\ddot{x} + \varphi(x, \dot{x}) + f(x) = 0, \quad m = 1$$

one may associate the first-order system

$$\frac{dx}{dt} = y, \quad \frac{dy}{dt} = -\varphi(x, y) - f(x)$$

and the equations of phase trajectories

$$\frac{dy}{dx} = -\frac{\varphi(x, y) + f(x)}{y}. \tag{2.18}$$

As above, to the singular points, which are determined from the equations

$$y = 0, \quad \varphi(x, 0) + f(x) = 0, \tag{2.19}$$

there correspond equilibrium positions of the mechanical system.

Nonlinear nonconservative dissipative autonomous systems are grouped into two categories: *dissipative* and *self-excited* systems. In dissipative systems no periodic motion is possible. In these systems one observes scattering (dissipation) of energy, which results in damped oscillations. A motion of self-excited systems is also accompanied with energy loss, but these losses are automatically compensated by the influx of energy from some external source and are controlled by the system itself. Some self-excited systems were considered in Sect. 1. In the motion of dissipative systems, the singular points (2.19) are usually stable or unstable foci or nodes. In the stable case, such points are approached by phase trajectories, and in the unstable case, these curves diverge from singular points.

As an example of a motion of a dissipative system, we investigate the oscillations of a conventional linear oscillator with unit mass under Coulomb friction **F**. The motion of the mass is described by the following nonlinear equation

$$\ddot{x} + \omega^2 x - F_x = 0, \quad \omega^2 = \text{const} .$$

The projection of the force of dry friction to the x-axis is like a curve in Fig. 1. For its analytical representation it is convenient to employ the function sign $\dot{x}$. However, for $\dot{x} = 0$ this function is zero. Under the same condition, the projection F_x may assume any value from $-r$ to r. Hence, the motion of a material point with dry friction is characterized by the presence of a *stagnation zone*, in which the point is in equilibrium if the elastic force acting on it is majorized by the friction force. The stagnation zone on the phase plane is the interval $[-r/\omega^2, \ r/\omega^2]$. This interval consists of singular points of Eq. (2.18).

So, to the motion of a material point there corresponds the differential equation

$$\ddot{x} + \omega^2 x + r \operatorname{sign} \dot{x} = 0 . \tag{2.20}$$

One could obtain the solution of this equation by the step-by-step method. However, for this purpose Lienard's method also proves useful.

Lienard's method. The phase trajectories for any differential equation $dy/dx = f(x, \ y)$ can be approximately constructed by the method of isoclines known from a course of differential equations. Lienard's method is capable of building phase trajectories corresponding to the equation of motion of the form

$$\ddot{x} + \varphi(\dot{x}) + x = 0 . \tag{2.21}$$

Consider the phase plane. The phase trajectories will be found from the equation

$$\frac{dy}{dx} = -\frac{\varphi(y) + x}{y} .$$

Next, consider the curve on the phase plane (Fig. 6)

$$x = -\varphi(y) . \tag{2.22}$$

Through a point $M(x, y)$ we draw the horizontal line intersecting curve (2.22) at the point N. From the point N drop the perpendicular NS onto the x-axis and connect the points S and M of the line. The perpendicular to the interval SM determines the direction of the integral curve at the point M. Indeed,

Fig. 6 Phase trajectory construction by Lienard's method

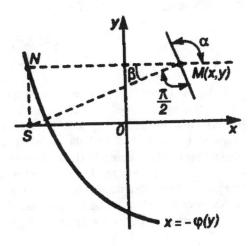

$$\tan \alpha = -\cotan \beta = -\frac{x + \varphi(y)}{y} = \frac{dy}{dx}.$$

Now the phase trajectory passing through the point $M(x, y)$ can be approximately constructed as follows. Having found the direction of the integral curve at the point $M(x, y)$, we replace the phase trajectory in a small neighbourhood of this point by an interval of the tangent line. At the end of this interval we again find the direction of the tangent line by Lienard's method and measure off small interval in this direction, and so on. As a result, one may construct the phase trajectory as a broken line consisting of intervals of sufficiently small length.

We apply Lienard's method to Eq. (2.20), which is valid until the point fall in the stagnation zone. In order to write it in the form (2.21), we change from the time t to the time $\tau = \omega t$. As a result, we have

$$\dot{x} \equiv \frac{dx}{dt} = \omega \frac{dx}{d\tau}, \quad \ddot{x} \equiv \frac{d^2x}{dt^2} = \omega^2 \frac{d^2x}{d\tau^2},$$

and so Eq. (2.20) can be put in the form

$$\frac{d^2x}{d\tau^2} + x + \frac{r}{\omega^2} \operatorname{sign} \frac{dx}{d\tau} = 0.$$

We set $y_1 = dx/d\tau$. The differential equation of the integral curves reads as

$$\frac{dy_1}{dx} = -\frac{x + \frac{r}{\omega^2} \operatorname{sign} y_1}{y_1}.$$

In the case in question, the curve $x = -\varphi(y_1)$ is of the form

$$x = -\frac{r}{\omega^2} \operatorname{sign} y_1$$

and splits into two half-lines (Fig. 7):

$$x = \begin{cases} -r/\omega^2, & y_1 > 0, \\ r/\omega^2, & y_1 < 0. \end{cases}$$

Using the form of the curve $x = -\varphi(y_1)$ one may precisely construct the integral curves by Lienard's method. Indeed, if M is an arbitrary point on the upper half-plane, then the corresponding point N, which lies on the curve $x = -\varphi(y_1)$, is always projected to the point S with the coordinates $(-r/\omega^2, 0)$. But since the arc of the integral curve at the point M should be orthogonal to the direction SM, it may be concluded that in this example all integral curves in the upper half-plane should look like half-circles with common centre at the point S. Similarly, the phase trajectories in the lower half-plane are also composed of half-circles, but their centers lie at the point S_1 with the coordinates $(r/\omega^2, 0)$. So, the phase trajectory consists of eccentric

Fig. 7 Phase trajectory of a
linear oscillator under
Coulomb friction

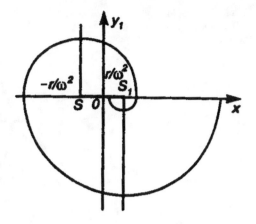

half-circles. The motion terminates when an extremity of a successive half-circle hits
the stagnation zone (Fig. 7).

Self-excited system. As an example of an self-excited system we consider a
system whose motion in the phase variables is described by the differential equations

$$\frac{dx}{dt} = py + \frac{x}{\rho}(1 - \rho^2), \qquad \frac{dy}{dt} = -px + \frac{y}{\rho}(1 - \rho^2), \qquad (2.23)$$

where $p = \text{const}$, $\rho = \sqrt{x^2 + y^2}$ is the distance from the representative point to the
origin. Using system (2.23), we write the differential equation of the form

$$x\frac{dx}{dt} + y\frac{dy}{dt} = \rho(1 - \rho^2)$$

and change in it to the polar coordinates $x = \rho \cos \varphi$, $y = \rho \sin \varphi$. Now the last
equation can be put in the form

$$\frac{d\rho}{dt} = 1 - \rho^2. \qquad (2.24)$$

Deriving the equation

$$y\frac{dx}{dt} - x\frac{dy}{dt} = p\rho^2$$

from system (2.23) and passing in it to the polar coordinates, we get the following
equation for the angle φ:

$$\frac{d\varphi}{dt} = -p. \qquad (2.25)$$

Fig. 8 Example of an
asymptotically stable limit
cycle

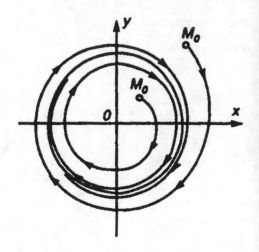

Integrating the differential equations (2.24), (2.25) with the initial conditions $\rho(0) = \rho_0$, $\varphi(0) = \varphi_0$, we find the equation of motion of the representative point in the polar coordinates

$$\rho = \frac{1 + \rho_0 - (1 - \rho_0)e^{-2t}}{1 + \rho_0 + (1 - \rho_0)e^{-2t}}, \quad \varphi = -p\,t + \varphi_0. \tag{2.26}$$

A closed phase trajectory corresponding to a periodic motion of a system is called a *limit cycle*. In practice, self-excited oscillations can be executed only with stable limit cycles. From Eq. (2.24) one may get the solution $\rho = 1$. This closed curve (the circle) is a limit cycle. For $\rho_0 < 1$ it follows from (2.24) that $d\rho/dt > 0$; that is, ρ increases, while ρ decreases for $\rho_0 > 1$. The character of variation of ρ and φ can be traced more clearly from Eqs. (2.26), from which it is seen that regardless of the initial conditions the representative point moves along logarithmic spirals winding around (from inside or outside) the unit circle (Fig. 8). Such a limit cycle is called *asymptotically stable*. To a motion along a limit cycle there correspond self-excited oscillations of a system. A periodic motion is possible also in nonlinear conservative systems, but it depends on the initial conditions. In the case of self-excited oscillations, a periodic motion along a limit cycle is also set when perturbing the initial conditions in certain limits (note that in the case under study the perturbations were assumed to be arbitrary).

We now proceed with consideration of some commonly used methods for approximate solution of equations of nonlinear oscillations.

3 Using Gauss' Principle in Searching Approximate Solutions of Equations of Nonlinear Oscillations. The Bubnov–Galerkin Method

If[2] the motion of a mechanical system is given incompletely, then it is expedient to use Gauss' principle to construct the equations that will complete an incomplete system. In its general form, Gauss' principle reads as (see Sect. 4 of Chap. 9 Vol. I)

$$\delta''(M\mathbf{W} - \mathbf{Y})^2 = 0 \,. \tag{3.1}$$

We employ this principle to find approximate solutions of the nonlinear equation

$$m\ddot{x} = F(t, x, \dot{x}) \,. \tag{3.2}$$

The motion of a material point in the interval $[0, \tau]$ will sought as a series

$$x(t) = \sum_{\nu=1}^{n} a_\nu f^\nu(t) \,, \tag{3.3}$$

where $f^\nu(t)$ are linearly independent functions and a_ν are the parameters to be determined.

The function $x(t)$, as given in the form (3.3), fails in general to satisfy the differential equation, and hence, substituting $x(t)$ into Eq. (3.2) this gives

$$m\ddot{x} - F(t, x, \dot{x}) = R \,, \tag{3.4}$$

where R is the residual. From the mechanical point of view, this residual can be looked upon as the additional force which must be applied to the point for it to move according to the law (3.3).

We shall assume that the motion (3.3) is not completely given, because the parameters a_ν are not specified. To find these parameters we shall vary only accelerations and shall require (as in Gauss' principle) that the averaged value of the squared force R in the interval $[0, \tau]$ would be minimal; that is, the condition

$$\delta'' \int_0^\tau (m\ddot{x} - F(t, x, \dot{x}))^2 dt = 0$$

should be satisfied. In other words, we shall obtain the coefficients a_ν as a consequence of the requirement that the mean-square error in the interval $[0, \tau]$ be minimal.

---

[2] The subsection uses the material of the article: *M. P. Yushkov*. Construction of approximate solutions of equations for nonlinear oscillations on the basis of Gauss' principle Vestn. Leningrad. un-ta. 1984. 13. Pp. 121–123 [in Russian].

Taking into account that in Gauss' principle only accelerations are allowed to vary, we have

$$\int_0^\tau (m\ddot{x} - F(t, x, \dot{x}))\, \delta''\ddot{x}\, dt = 0\,.$$

Substituting this expressions into Eq. (3.3), we get

$$\sum_{\nu=1}^n \delta''a_\nu \int_0^\tau \left(m\sum_{\nu=1}^n a_\nu \ddot{f}^\nu - F\left(t, \sum_{\nu=1}^n a_\nu f^\nu, \sum_{\nu=1}^n a_\nu \dot{f}^\nu\right)\right) \ddot{f}^\nu\, dt = 0\,. \qquad (3.5)$$

The quantities $\delta''a_\nu$ are linearly independent, and hence it follows from Eq. (3.5) that

$$\int_0^\tau \left(m\sum_{\nu=1}^n a_\nu \ddot{f}^\nu - F\left(t, \sum_{\nu=1}^n a_\nu f^\nu, \sum_{\nu=1}^n a_\nu \dot{f}^\nu\right)\right) \ddot{f}^\nu\, dt = 0\,, \nu = \overline{1, n}\,. \qquad (3.6)$$

The conditions under which this system of algebraic equations has nonzero solutions depend on the form of the functions $F(t, x, \dot{x})$ and $f^\nu(t)$, $\nu = \overline{1, n}$.

The quantity R was introduces by formula (3.4) as a force. We now look upon it as an error with with the function $x(t)$, as given in the form (3.3), satisfies Eq. (3.2). Under such an approach from the system of algebraic Eqs. (3.6) with respect to the parameters α_ν one may find a particular approximate solution of Eq. (3.2) in the form (3.3).

Let us use the proposed general method for finding approximate periodic solutions of Eq. (3.2). For simplicity, periodic solutions will be sought in the form

$$x(t) = a_1 \cos kt + a_2 \sin kt\,. \qquad (3.7)$$

We rewrite system (3.6) as

$$\int_0^{2\pi/k} (-mk^2(a_1 \cos kt + a_2 \sin kt) - F(t, a_1 \cos kt +$$
$$+a_2 \sin kt, -a_1 k \sin kt + a_2 k \cos kt)) \cos kt\, dt = 0\,,$$
$$\int_0^{2\pi/k} (-mk^2(a_1 \cos kt + a_2 \sin kt) - F(t, a_1 \cos kt +$$
$$+a_2 \sin kt, -a_1 k \sin kt + a_2 k \cos kt)) \sin kt\, dt = 0\,, \qquad (3.8)$$

in which in this case the upper limit τ should be taken to be equal to the period $2\pi/k$.

These equations are used for approximate construction of the solution of Eq. (3.2) in the form (3.7) in the *Bubnov–Galerkin method*. Usually they are derived from the principle of virtual displacements.

The above general method for finding approximate solutions of Eq. (3.2) can be easily extended to the case of an arbitrary mechanical system with s degrees of freedom. For this purpose, Gauss' principle (3.1) is used in the integral form

$$\delta'' \int_0^T (M\mathbf{W} - \mathbf{Y})^2 dt = 0,$$

and so,

$$\int_0^T (M\mathbf{W} - \mathbf{Y}) \cdot \delta''\mathbf{w} \, dt = 0.$$

Taking into account that

$$M\mathbf{W} - \mathbf{Y} = \left(\frac{d}{dt}\frac{\partial T}{\partial \dot{q}^\sigma} - \frac{\partial T}{\partial q^\sigma} - Q_\sigma \right) \mathbf{e}^\sigma, \quad \delta''\mathbf{w} = \delta''\ddot{q}^\sigma \mathbf{e}_\sigma,$$

we get

$$\int_0^T \left(\frac{d}{dt}\frac{\partial T}{\partial \dot{q}^\sigma} - \frac{\partial T}{\partial q^\sigma} - Q_\sigma \right) \delta''\ddot{q}^\sigma \, dt = 0.$$

From this equation it follows that the functions $q^\sigma(t)$, as given in the form

$$q^\sigma(t) = \sum_{\nu=1}^{n} a_\nu^\sigma f^\nu(t), \quad \sigma = \overline{1, s}, \tag{3.9}$$

can be looked upon as an approximate solution of the Lagrange equations

$$\frac{d}{dt}\frac{\partial T}{\partial \dot{q}^\sigma} - \frac{\partial T}{\partial q^\sigma} = Q_\sigma, \quad \sigma = \overline{1, s},$$

provided that the parameters a_ν^σ satisfy the equations

$$\int_0^T \left(\frac{d}{dt}\frac{\partial T}{\partial \dot{q}^\sigma} - \frac{\partial T}{\partial q^\sigma} - Q_\sigma \right) \ddot{f}^\nu \, dt = 0, \quad \sigma = \overline{1, s}, \quad \nu = \overline{1, n},$$

in which the functions $q^\sigma(t)$ are assumed to be given in the form (3.9).

We now consider several examples showing that this approach of finding approximate periodic solutions of nonlinear equations is constructive.

Application of the Bubnov–Galerkin method for analysis of solutions of the Duffing equation. Let us find in the first place an approximate solution of the equation

$$\ddot{x} + f(x) = 0, \tag{3.10}$$

where $m = 1$, $F(t, x, \dot{x}) = -f(x)$. From system (3.8) it follows that the function

$$x(t) = a \sin kt \tag{3.11}$$

approximately satisfies Eq. (3.10), provided that the parameters a and k are related as

$$\int_0^{2\pi/k} (-ak^2 \sin kt + f(a \sin kt)) \sin kt \, dt = 0 .$$

In particular, for the Duffing equation, in accordance with which $f(x) = \omega^2 x + \mu x^3$, we have, after calculating the integral,

$$k = \sqrt{\omega^2 + (3/4)\mu a^2} .$$

This relation shows that for nonlinear oscillations there is a relation between the amplitude a and the frequency k of free oscillations. So, the free oscillations that are described by the homogeneous Duffing equation are not tautochronic.

In the study of forced oscillations described by inhomogeneous Duffing equation

$$\ddot{x} + \omega^2 x + \mu x^3 = h \sin kt , \tag{3.12}$$

the steady-state motion will again be looked for in the form (3.11), but now k coincides with the frequency of the external force, and the amplitude a is unknown. We note that many problems in mechanics and physics can be brought to Eq. (3.12).

In our setting, system (3.8) is reduced to the equation

$$\int_0^{2\pi/k} (-k^2 a \sin kt + \omega^2 a \sin kt + \mu a^3 \sin^3 kt -$$

$$-h \sin kt) \sin kt \, dt = 0 .$$

Taking the integral, we obtain

$$(3/4)\mu a^3 + (\omega^2 - k^2)a = h . \tag{3.13}$$

Formula (3.13) expresses the dependence of the amplitude a of the steady-state forced oscillations on the frequency k of the external force and is known as the *frequency response function* for the system.

To construct the curve (3.13), we rewrite its equation in the form

$$k^2 = \omega^2 + \frac{3}{4}\mu a^2 - \frac{h}{a} .$$

This curve will be obtained by adding the two curves shown in Fig. 9a by dash lines. One of these curves is the parabola $k^2 = \omega^2 + (3/4)\mu a^2$, and the other one is the hyperbola referred to its asymptotes $k^2 = -h/2$. After adding the values of k^2 corresponding to the these curves, we obtain the requited frequency response function, which is shown in Fig. 9a by the solid line.

Amplitude curves are frequently constructed for the modulus of the amplitude (Fig. 9b), and the change of sign of the latter at the passage through a resonance

Fig. 9 Frequency response
function of the Duffing
equation with no resistance

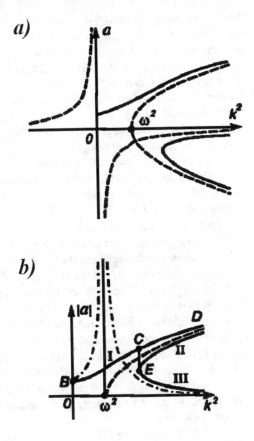

is taken into account by varying the oscillation phase. The dashed line in Fig. 9b
corresponds to free oscillations ($h = 0$) and is called a *skeleton curve*.

For $\mu = 0$ Eq. (3.12) becomes linear. The frequency response function in this case
is shown in Fig. 9b by the dash-dotted line. Comparing the curves corresponding to
the cases $\mu = 0$ and $\mu > 0$ we see that the frequency response function of a nonlinear
system can be obtained as if one would bend to the right the amplitude curves of
a linear system. For a soft characteristic of the elastic force (that is, for $\mu < 0$), the
frequency response function bends in the opposite direction.

From the characteristic thus obtained (Fig. 9b) one may trace the qualitative differences of nonlinear oscillations from the linear ones. For $k = \omega$, when in a linear
system there is a resonance, nonlinear forced oscillations are executed with finite
amplitude, even if the effect of resistance forces is neglected. These amplitudes continue to increase with increasing frequency, and starting from some value of k, to
each frequency there corresponds three possible values of the amplitude a.

It may be shown that the oscillations whose amplitudes have values on the branch
II are unstable, while the oscillations with amplitudes lying on the branch III (that
is, below the point E) are stable. The oscillations corresponding to the branch I in

the area of three solutions (on the segment CD) are stable only if perturbations are sufficiently small. Thus, an increase of the amplitude a with gradual increase of the frequency k of the perturbing force in a system is governed by the branch BC of the frequency response function. If k is further increasing, then a continues to increase along the branch CD. However, if it becomes sufficiently close to the unstable branch II, then for large perturbations there may be a jump from this portion of the curve I to the curve III, which characterizes stable oscillations. Besides, the amplitude a decreases discontinuously, and in what follows, when the frequency is increasing, it becomes monotone decreasing.

With a gradual decrease of the frequency k of the perturbing force, the amplitude a is increasing (the curve III) until it reaches the point E, after which a hopping from the curve III to the curve I takes place, and the amplitude a is increased by the value corresponding to the interval EC. For a further decrease of k, a decrease in a is governed by the branch CB of the curve I.

Thus, one may note yet another peculiarity of forced oscillations of a nonlinear system: the maximal amplitude during an acceleration of the system is greater than the maximal amplitude when the frequency of the perturbing force is decreasing.

Equation (3.12) was composed without consideration of resistance forces, and hence, if there are no perturbations, then in theory there may appear forced oscillations with arbitrarily large amplitude. We recall that in the study of linear systems it was proved to rectify such an inconsistency with experimental data by considering the resistance forces. We also introduce the dissipative force, which takes into account the inelastic resistance of the material.

There are several different conjectures allowing one to approximately take into account the scattering of energy in a material during oscillations. Consider one of the simplest ones.

Usually the resistance forces are out of phase by $\pi/2$ with respect to the elastic forces. Assuming that the resistance does not violate the sine law of oscillations, it is possible from the form of the elastic forces to construct inelastic resistance forces, by replacing $x(t)$ by the quantity $\dot{x}(t)$, which is reflected in the phase shift by $\pi/2$. Besides, we multiply the resulting expression by the coefficient

$$\varphi = \eta/k,$$

where η is the loss coefficient and k is the frequency of the perturbing force. Taking this into account, we assume that in the problem under study the force of inelastic resistance force can be taken in the form

$$\varphi\omega^2\dot{x} + \varphi\mu x^2\dot{x} = \varphi\omega^2\dot{x} + \frac{\varphi}{k^2}\mu\dot{x}^3,$$

and hence, instead of Eq. (3.12), we have

$$\ddot{x} + \omega^2 x + \mu x^3 + \varphi\omega^2\dot{x} + \frac{\varphi}{k^2}\mu\dot{x}^3 - h\sin kt = 0. \tag{3.14}$$

In this case the steady-state forced oscillations will be out of phase with respect to the external force. Hence, the solution of Eq. (3.14) will be sought in the form

$$x(t) = a \sin(kt + \varepsilon).$$

We need to find the amplitude a and the phase ε. Comparing this expression with (3.7), we see that the unknowns a and ε are related to the unknowns a_1 and a_2 as follows

$$a = \sqrt{a_1^2 + a_2^2}; \quad \varepsilon = \operatorname{arccot} \frac{a_2}{a_1}, \quad a_1 > 0;$$

$$\varepsilon = \pi + \operatorname{arccot} \frac{a_2}{a_1}, \quad a_1 < 0;$$

$$\varepsilon = 0, \quad a_1 = 0, \quad a_2 > 0; \quad \varepsilon = \pi, \quad a_1 = 0, \quad a_2 < 0. \tag{3.15}$$

We write system (3.8) as

$$\int_0^{2\pi/k} \psi(a_1, a_2, k, t) \cos kt \, dt = 0,$$

$$\int_0^{2\pi/k} \psi(a_1, a_2, k, t) \sin kt \, dt = 0. \tag{3.16}$$

Here

$$\psi(a_1, a_2, k, t) = -k^2(a_1 \cos kt + a_2 \sin kt) +$$
$$+ \omega^2(a_1 \cos kt + a_2 \sin kt) + \mu(a_1^3 \cos^3 kt +$$
$$+ 3 a_1^2 a_2 \cos^2 kt \sin kt + 3 a_1 a_2^2 \cos kt \sin^2 kt + a_2^3 \sin^3 kt) +$$
$$+ \varphi \omega^2 k (-a_1 \sin kt + a_2 \cos kt) + \varphi \mu k (-a_1^3 \sin^3 kt +$$
$$+ 3 a_1^2 a_2 \sin^2 kt \cos kt - 3 a_1 a_2^2 \sin kt \cos^2 kt + a_2^3 \cos^3 kt) - h \sin kt.$$

Taking the integrals in (3.16), we obtain an algebraic system of nonlinear equations with respect to a_1 and a_2:

$$a_1(\omega^2 - k^2) + \frac{3}{4} \mu a_1^3 + \frac{3}{4} \mu a_1 a_2^2 + \varphi \omega^2 k a_2 +$$

$$+ \frac{3}{4} \varphi \mu k a_1^2 a_2 + \frac{3}{4} \varphi \mu k a_2^3 = 0,$$

$$a_2(\omega^2 - k^2) + \frac{3}{4} \mu a_2^3 + \frac{3}{4} \mu a_2 a_1^2 - \varphi \omega^2 k a_1 -$$

$$- \frac{3}{4} \varphi \mu k a_2^2 a_1 - \frac{3}{4} \varphi \mu k a_1^3 = h.$$

Fig. 10 Frequency response
function of the Duffing
equation with consideration
of resistance

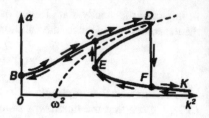

Squaring these equations, adding and collecting terms, this gives

$$\frac{9}{16}\mu^2(1+\varphi^2 k^2)a^6 + \frac{3}{2}\mu(\omega^2 - k^2 + \varphi^2 k^2 \omega^2)a^4 +$$
$$+((\omega^2 - k^2)^2 + \varphi^2 k^2 \omega^4)\,a^2 = h^2\,, \qquad (3.17)$$

where $a^2 = a_1^2 + a_2^2$ in accordance with (3.15).

Thus, we have obtained the equation of the frequency response function with due account of the inelastic resistance force. We note that expression (3.17) with $\varphi = 0$ coincides with the squared expression (3.13). To the frequency response function, as obtained from Eq. (3.17), there corresponds the curve in Fig. 10, which is in a good agreement with experimental data and reflects the variation of the amplitude accompanying an acceleration of the system (the line $BCDFK$) or a decrease in the frequency (the line $KFECB$). Here, one may easily see the hopping of values of the oscillation amplitude, which is characterized by the intervals DF and EC.

4 The Small-Parameter Method

A broad class of nonlinear differential equations can be investigated using the machinery of the small-parameter method by constructing convergent series in terms of powers of the parameter. Let us consider in more detail the Poincaré–Lyapunov's method of searching for periodic solutions of a nonlinear system. Despite of the particular character of the problem it has a great practical value.

Assume that a nonlinear system contains a small parameter μ. For $\mu = 0$ it becomes the so-called *generating* system. Let us consider problems for which the solution of the original nonlinear system tends as $\mu \to 0$ to the solution of the generating system. In this case, the periodic solutions of the generating system (which can be constructed by the assumption) are approximate solutions of the intractable main nonlinear system.

Periodic oscillations of nonautonomous system.

The case of nonzero functional determinant. Assume that the motion equations of a mechanical system are written in the normal form

$$\dot{x}_\sigma = X_\sigma(\mu, t, x_1, \ldots, x_s), \quad \sigma = \overline{1, s}, \tag{4.1}$$

where in some domain the functions $X_\sigma(\mu, t, x)$ are holomorphic functions of the variables $x_1, \ldots, x_s$ and of the parameter μ (provided μ is sufficiently small).

We consider the case when these functions are periodic in time t. Without loss of generality this period may be assumed to be 2π. Expanding the right-hand side of (4.1) in series in powers of the small parameter μ, we may write

$$\dot{x}_\sigma = \sum_{n=0}^{\infty} \mu^n X_\sigma^{(n)}(t, x_1, \ldots, x_s), \quad \sigma = \overline{1, s}, \tag{4.2}$$

where the functions $X_\sigma^{(n)}(t, x)$ are 2π-periodic functions of t that are holomorphic with respect to $x_1, x_2, \ldots, x_s$. To the system of Eqs. (4.2) for $\mu = 0$ there corresponds the following generating system

$$\dot{x}_\sigma^{(0)} = X_\sigma^{(0)}(t, x_1^{(0)}, \ldots, x_s^{(0)}), \quad \sigma = \overline{1, s}. \tag{4.3}$$

Any 2π-periodic solution

$$x_\sigma^{(0)} = x_\sigma^{(0)}(t), \quad \sigma = \overline{1, s}, \tag{4.4}$$

is called a *generating solution*.

We shall look for a 2π-periodic solution of the original system (4.2) that coincides with the generating one for $\mu = 0$.

According to the theory of differential equations, solutions of system (4.2) are functions of the form

$$x_\sigma^{(0)} = x_\sigma(\mu, t, C_1, \ldots, C_s), \quad \sigma = \overline{1, s},$$

where C_σ are arbitrary constants, which may be expressed in terms of the initial data $(x_\sigma)_{t=0}$. However, it is more convenient to express them in terms of the differences

$$(x_\sigma)_{t=0} - (x_\sigma^{(0)})_{t=0} = \alpha_0, \quad \sigma = \overline{1, s}.$$

Now one may write

$$x_\sigma = x_\sigma(\mu, t, \alpha_1, \ldots, \alpha_s), \quad \sigma = \overline{1, s}. \tag{4.5}$$

In their expanded forms, the differences $(x_\sigma)_{t=0} - (x_\sigma^{(0)})_{t=0}$ may be put in the form

$$x_\sigma(\mu, 0, \alpha_1, \ldots, \alpha_s) - x_\sigma^{(0)}(0), \quad \sigma = \overline{1, s}. \tag{4.6}$$

In order to single out from the set of solutions (4.5) a 2π-periodic solution that becomes the generating one for $\mu = 0$, it is required to adjust $\alpha_1, \ldots, \alpha_s$ in a specific way with the help of the periodicity conditions

$$\Phi_\sigma(\mu, \alpha_1, \ldots, \alpha_s) = x_\sigma(\mu, 2\pi, \alpha_1, \ldots, \alpha_s) - \\ -x_\sigma(\mu, 0, \alpha_1, \ldots, \alpha_s) = 0, \quad \sigma = \overline{1, s}. \tag{4.7}$$

According to Poincaré, system (4.7) has a unique holomorphic solution

$$\alpha_\sigma = \alpha_\sigma(\mu), \quad \sigma = \overline{1, s}, \tag{4.8}$$

provided that its functional determinant at the origin is nonzero,

$$\Delta|_{\mu=\alpha_1=\cdots=\alpha_s=0} = \frac{\partial(\Phi_1, \ldots, \Phi_s)}{\partial(\alpha_1, \ldots, \alpha_s)}\bigg|_{\mu=\alpha_1=\cdots=\alpha_s=0} \neq 0. \tag{4.9}$$

Solution (4.8) also vanishes for $\mu = 0$. Besides, Poincaré proved that in this case there exists a unique 2π-periodic solution of system (4.2), which as $\mu \to 0$ tends to the periodic solution of the generating system.

Since the sought-for solution is holomorphic, it will be sough as a series

$$x_\sigma(t) = \sum_{n=1}^{\infty} \mu^n x_\sigma^{(n)}(t), \quad \sigma = \overline{1, s}, \tag{4.10}$$

where the unknown functions $x_\sigma^{(n)}$ are 2π-periodic:

$$x_\sigma^{(n)}(2\pi) = x_\sigma^{(n)}(0), \quad \sigma = \overline{1, s}, \quad n = 0, 1, 2, \ldots.$$

Substituting series (4.10) into Eqs. (4.2) and equating the coefficients with the same powers of the parameter μ, we obtain the systems of differential equations

$$\dot{x}_\sigma^{(n)} = \sum_{i=1}^{s} \frac{\partial X_\sigma^{(0)}}{\partial x_i} x_i^{(n)} + \psi_\sigma^{(n)}(t, x_1^{(0)}, \ldots, x_s^{(0)}, \ldots, x_1^{(n-1)}, \ldots, x_s^{(n-1)}),$$

$$\sigma = \overline{1, s}, \quad n = 0, 1, 2, \ldots, \tag{4.11}$$

where the derivatives are calculated for the generating system (4.4), and $\psi_\sigma^{(n)}$ are some unknown functions of the variables t, $x_\sigma^{(k)}$ ($\sigma = \overline{1, s}$, $k = \overline{0, n-1}$). From these recursive systems one may find in succession the functions $x_\sigma^{(1)}, x_\sigma^{(2)}, \ldots$ ($\sigma = \overline{1, s}$), provided that the generating solution $x_1^{(0)}, \ldots, x_2^{(0)}$ is available.

We note, however, that in general the integration of systems (4.11) involves considerable difficulties, because here we are faced with systems of linear equations with periodic coefficients, whose solution is an independent challenging problem. So, we have established a link between problems of nonlinear oscillations and problems of oscillations of linear systems with variable coefficients.

However, the case when the system under consideration has the form

$$\dot{x}_\sigma(t) = \sum_{i=1}^{s} c_{\sigma i} x_i(t) + f_\sigma(t) + \mu F_\sigma(\mu, t, x_1, \ldots, x_s),$$

$$c_{\sigma i} = \text{const}, \ \sigma = \overline{1, s} \tag{4.12}$$

is much simpler. Here, the generating system and the recursive systems (4.11), which now read as

$$\dot{x}_\sigma = \sum_{i=1}^{s} c_{\sigma i} x_i(t) + \psi_\sigma^{(n)}(\mu, t, x_1^{(0)}, \ldots, x_s^{(0)}, \ldots, x_1^{(n-1)}, \ldots, x_s^{(n-1)}),$$

$$\sigma = \overline{1, s}, \quad n = 1, 2, \ldots,$$

are linear systems with constant coefficients, which can be easily integrated. Equations (4.12) are called *quasi-linear*.

As an example of a quasi-linear system we consider the inhomogeneous Duffing equation

$$\ddot{x} + \omega^2 x = P \sin t + \mu x^3,$$

where ω is noninteger. This equation can be written as a second-order system

$$\dot{x}_1 = x_2, \quad \dot{x}_2 = -\omega^2 x_1 + P \sin t + \mu x_1^3. \tag{4.13}$$

The generating system

$$\dot{x}_1^{(0)} = x_2^{(0)}, \quad \dot{x}_2^{(0)} = -\omega^2 x_1^{(0)} + P \sin t$$

has the general solution

$$x_1^{(0)}(t) = C_1 \sin \omega t + C_2 \cos \omega t + \frac{P}{\omega^2 - 1} \sin t,$$

$$x_2^{(0)}(t) = \omega C_1 \cos \omega t - \omega C_2 \sin \omega t + \frac{P}{\omega^2 - 1} \cos t,$$

which for noninteger ω and $C_1 \neq 0, C_2 \neq 0$ is not 2π-periodic. Setting $C_1 = C_2 = 0$, which can be done by adjusting the initial conditions,

$$x_1^{(0)}(0) = 0, \quad x_2^{(0)}(0) = \frac{P}{\omega^2 - 1}, \tag{4.14}$$

we get the 2π-periodic generating solution

$$x_1^{(0)}(t) = \frac{P}{\omega^2 - 1} \sin t, \quad x_2^{(0)}(t) = \frac{P}{\omega^2 - 1} \cos t. \tag{4.15}$$

Prior to proceeding with finding a solution in the form (4.10), we verify that the functional determinant (4.9) in this example is not zero. To this aim we note that the sought-for solution $x_\sigma(t)$ is holomorphic in $\mu, \alpha_1, \ldots, \alpha_s$; that is, it can be expanded into a series

$$\dot{x}_\sigma(t) = x_\sigma^{(0)}(t) + \mu b_\sigma(t) + \sum_{i=1}^{s} \alpha_n a_{\sigma n}(t) + \ldots, \quad \sigma = \overline{1, s}.$$

Here, the periodicity conditions (4.7) read as

$$\Phi_\sigma = \mu [b_\sigma(2\pi) - b_\sigma(0)] + \sum_{n=1}^{s} \alpha_n [a_{\sigma n}(2\pi) - a_{\sigma n}(0)] + \ldots, \quad \sigma = \overline{1, s}.$$

Calculating the determinant (4.9), we have

$$\Delta|_{\mu=\alpha_1=\cdots=\alpha_s=0} = |a_{\sigma n}(2\pi) - a_{\sigma n}(0)|. \tag{4.16}$$

In this case, the solution of system (4.13) can be written as

$$\begin{aligned} x_1(t) &= x_1^{(0)}(t) + \mu b_1(t) + \alpha_1 a_{11}(t) + \alpha_2 a_{12}(t) + \ldots, \\ x_2(t) &= x_2^{(0)}(t) + \mu b_2(t) + \alpha_1 a_{21}(t) + \alpha_2 a_{22}(t) + \ldots, \end{aligned} \tag{4.17}$$

where α_1, α_2 are the departures of the initial conditions of the sought-for solution from the initial conditions (4.14),

$$x_1(0) - x_1^{(0)}(0) = \alpha_1, \quad x_2(0) - x_2^{(0)}(0) = \alpha_2. \tag{4.18}$$

From the first equation of system (4.13) it follows that

$$b_2(t) = \dot{b}_1(t), \quad a_{21}(t) = \dot{a}_{11}(t), \quad a_{22}(t) = \dot{a}_{12}(t), \ldots. \tag{4.19}$$

Let us find the functions α_{11}, α_{12}. Substituting series (4.17) into system (4.13) and equating the coefficients of α_{11} and α_{12}, we get

$$\ddot{a}_{11} + \omega^2 a_{11} = 0, \quad \ddot{a}_{12} + \omega^2 a_{12} = 0. \tag{4.20}$$

In order to determine the initial conditions we rewrite (4.18) in view of formulas (4.17) and (4.19):

$$\alpha_{11}(0) = 1, \quad \dot{\alpha}_{11}(0) = 0, \quad \alpha_{12}(0) = 0, \quad \dot{\alpha}_{12}(0) = 1.$$

So, to these initial conditions there will correspond the following particular solutions of Eqs. (4.20)

$$\alpha_{11}(t) = \cos \omega t , \quad \alpha_{12}(t) = \frac{1}{\omega} \sin \omega t ,$$

and hence in view of (4.19) the determinant (4.16) will assume the form

$$\begin{vmatrix} \cos 2\omega\pi - 1 & \frac{1}{\omega} \sin 2\omega\pi \\ -\omega \sin 2\omega\pi & \cos 2\omega\pi - 1 \end{vmatrix} = 2(1 - \cos 2\omega\pi) .$$

It follows that it does not vanish for a noninteger ω, and hence, in accordance with Poincaré's theorem, there exists a 2π-periodic solution, which for $\mu = 0$ becomes the solution (4.15). It is worth noting that this conclusion also holds for a more general equation

$$\ddot{x} + \omega^2 x = P \sin t + \mu f(x) ,$$

where $f(x)$ is a nonlinear holomorphic function. This follows from the fact that the functions $a_{11}, a_{12}, a_{21}, a_{22}$ occurring in determinant (4.16) are independent of the form of the function $f(x)$.

So a solution of the inhomogeneous Duffing equation will be sought as a series

$$x(t) = \sum_{i=1}^{\infty} \mu^n x^{(n)}(t) , \tag{4.21}$$

where the sought-for functions $x^{(n)}(t)$, $n = 0, 1, 2, \ldots$, must satisfy the periodicity conditions

$$x^{(n)}(2\pi) = x^{(n)}(0) , \quad \dot{x}^{(n)}(2\pi) = \dot{x}^{(n)}(0) , \quad n = 0, 1, 2, \ldots . \tag{4.22}$$

Substituting series (4.21) into the Duffing equation and equating the coefficients of the same powers of μ, we obtain the system of recursive relations

$$\ddot{x}^{(0)} + \omega^2 x^{(0)} = P \sin t ,$$
$$\ddot{x}^{(1)} + \omega^2 x^{(1)} = (x^{(0)})^3 ,$$
$$\ddot{x}^{(2)} + \omega^2 x^{(2)} = 3(x^{(0)})^2 x^{(1)} ,$$
$$\ddot{x}^{(3)} + \omega^2 x^{(3)} = 3[x^{(0)}(x^{(1)})^2 + (x^{(0)})^2 x^{(2)}] ,$$
$$\cdots\cdots\cdots\cdots\cdots\cdots\cdots\cdots\cdots\cdots$$

The first equation here is a generating one and its 2π-periodic solution should be of form (4.15). Substituting it into the right-hand side of the second equation and taking into account that $\sin t^3 = (1/4)(3 \sin t - \sin 3t)$, we get

$$\ddot{x}^{(1)} + \omega^2 x^{(1)} = \frac{P^3}{4(\omega^2 - 1)^3} (3 \sin t - \sin 3t) .$$

The particular solution of the last equation satisfying conditions (4.22) is as follows

$$x^{(1)}(t) = \frac{3P^3}{4(\omega^2 - 1)^4} \sin t - \frac{P^3}{4(\omega^2 - 1)(\omega^2 - 9)} \sin 3t .$$

The functions $x^{(2)}(t)$, $x^{(3)}(t)$, ... are found similarly.

If we confine ourselves to the first two terms of series (4.21), then we obtain

$$x(t) \approx \frac{P}{\omega^2 - 1} \sin t + \frac{\mu P^3}{4(\omega^2 - 1)^3} \left(\frac{3 \sin t}{\omega^2 - 1} - \frac{\sin 3t}{\omega^2 - 9} \right) .$$

Note that if the functional determinant (4.9) is not zero, then for sufficiently small μ the solution of a quasi-linear system turns out to be close to the solution of the corresponding linear system.

The case of zero functional determinant. Integration proves to be more involved when the functional determinant (4.9) vanishes. To clarify the possibility of such a case, we consider the previous example. For the resonance ($\omega = 1$) the generating equation reads as

$$\ddot{x}^{(0)} + x^{(0)} = P \sin t .$$

According to Sect. 7 of Chap. 4 of Vol. I, the general solution of this equation is as follows:

$$x^{(0)}(t) = C_1 \cos t + C_2 \sin t - \frac{Pt}{2} \cos t .$$

This solution can be 2π-periodic only if $P = 0$; that is, when then the generating equation reads as

$$\ddot{x}^{(0)} + x^{(0)} = 0 .$$

This equation can be looked upon as the limit cycle of the equation

$$\ddot{x} + x = \mu (P \sin t + x^3) , \quad \mu \to 0 .$$

So, for resonance, we need to confine our consideration to the case when both the nonlinear terms and the external force have the same order of smallness. It is essential here that the 2π-periodic generating solution depends on arbitrary constants.

In the general case, the nonlinear system (4.3) may have a generating solution depending on k arbitrary constants $C_1, \ldots, C_k$. It turns out that in this case the functional determinant (4.9) is zero and the constants $C_1, \ldots, C_k$ must satisfy some additional conditions in order that the corresponding solution be 2π-periodic. However, the detailed form of these conditions is fairly cumbersome.[3]

Let us consider the inhomogeneous Duffing equation in the case when ω is 1 or is close to 1; that is, when $\omega^2 = 1 - \mu c$. In this setting, the Duffing equation can be

---

[3] See, for example: *I. G. Malkin.* The method of Lyapunov and Poincaré in the theory of non-linear oscillations. Moscow: Gostekhizdat, 1949 [in Russian].

put into the form

$$\ddot{x} + x = \mu \left(P \sin t + cx + x^3 \right). \tag{4.23}$$

It is quite natural that if in parallel with this equation one writes it in the form of a normal system of differential equations of type (4.13), then to the unknown functions x_1 and x_2 there correspond x and $\dot{x}$.

The generating equation has the 2π-periodic solution

$$x^{(0)}(t) = C_1 \cos t + C_2 \sin t, \tag{4.24}$$

which depends on the parameters C_1 and C_2. In view of (4.6), we have

$$x(\mu, 0, \alpha_1, \alpha_2) - x^{(0)}(0) = \alpha_1,$$
$$\dot{x}(\mu, 0, \alpha_1, \alpha_2) - \dot{x}^{(0)}(0) = \alpha_2. \tag{4.25}$$

The solution of interest can be expanded into a series

$$x(t) = x^{(0)}(t) + \mu b(t) + \sum_{n=1}^{2} \alpha_n a_n(t) + \dots. \tag{4.26}$$

Let us find the functions $b(t)$, $a_1(t)$, $a_2(t)$. To this aim we substitute the series (4.26) into Eq. (4.23). We have

$$\ddot{b} + b = P \sin t + cx^{(0)} + (x^{(0)})^3,$$
$$\ddot{a_1} + a_1 = 0, \quad \ddot{a_1} + a_1 = 0. \tag{4.27}$$

The values of the initial conditions can be calculated if we substitute this series into formulas (4.25),

$$b(0) = 0, \quad \dot{b}(0) = 0, \quad a_1(0) = 1, \quad \dot{a_1}(0) = 0, \quad a_2(0) = 0, \quad \dot{a_2}(0) = 1.$$

Integrating equations (4.27) we have[4]

$$b(t) = \int_0^t P \sin \tau + cx^{(0)}(\tau) + [x^{(0)}]^3 \sin(t - \tau) d\tau,$$

$$a_1(t) = \cos t, \quad a_2(t) = \sin t.$$

---

[4] Here it is worth recalling that the solution of the inhomogeneous equation for harmonic oscillations, as obtained in Sect. 7 of Chap. 4 of Vol. I, can be expressed in terms of the Duhamel integral.

From the condition of 2π-periodicity of solution (4.26) it follows that $b(2\pi) - b(0) = 0$, $\dot{b}(2\pi) - \dot{b}(0) = 0$, and hence for the expression for $b(t)$ we find that

$$\int_0^{2\pi} \left(P \sin \tau + c x^{(0)}(\tau) + [x^{(0)}]^3 \sin (\tau) \right) d\tau = 0,$$

$$\int_0^{2\pi} \left(P \sin \tau + c x^{(0)}(\tau) + [x^{(0)}]^3 \cos (\tau) \right) d\tau = 0, \qquad (4.28)$$

and now, using formula (4.24),

$$P + c\, C_2 + \frac{3}{4}\, C_2 (C_1^2 + C_2^2) = 0, \quad C_1 \left[c + \frac{3}{4}\, (C_1^2 + C_2^2) \right] = 0.$$

Here, the factor $c + (3/4)(C_1^2 + C_2^2)$ may not vanish, because otherwise the first equation will not be satisfied. Hence, a solution of this system is given by $C_1 = 0$, $C_2 = a$, where a is a root of the cubic equation

$$(3/4)\, a^3 + c\, a + P = 0. \qquad (4.29)$$

So, the generating solution reads as

$$x^{(0)}(t) = a \sin t. \qquad (4.30)$$

Now the 2π-periodic solution of Eq. (4.23) can be sought in the form of series (4.21) under the periodicity conditions (4.22). In order to determine the function $x^{(1)}(t)$ we employ the equation

$$\ddot{x}^{(1)} + x^{(1)} = P \sin t + c\, x^{(0)} + (x^{(0)})^3. \qquad (4.31)$$

Substituting here the generating solution (4.30), we have by (4.29)

$$\ddot{x}^{(1)} + x^{(1)} = -\frac{a^3}{4} \sin 3t,$$

whose solution is given by

$$x^{(1)}(t) = C_1^{(1)} \cos t + C_2^{(1)} \sin t + \frac{a^3}{32} \sin 3t.$$

The values of arbitrary constants $C_1^{(1)}$ and $C_2^{(1)}$ can be found from the condition of 2π-periodicity of the function $x^{(2)}(t)$. In this case, the right-hand side of the differential equation for the function $x^{(2)}(t)$ should not contain the terms which are proportional to $\sin t$ and $\cos t$. We note that this method of searching the constants $C_1^{(1)}$ and

$C_2^{(1)}$ is completely analogous to the method of finding the constants C_1 and C_2 in solution (4.24). Equations (4.28), from which we were able to find the constants C_1 and C_2, are exactly the conditions that the right-hand side of Eq. (4.31) should not contain the terms that are proportional to $\sin t$ and $\cos t$. After some calculation aimed at finding the constants $C_1^{(1)}$, $C_2^{(1)}$, we have

$$C_1^{(1)} = 0, \quad C_2^{(1)} = 3a^5 \left(c + \frac{9}{4} a^2 \right)^{-1}.$$

In the same way one may also ascertain the third term of series (4.21), but in this case the labor of computing is much higher.

Periodic free oscillations of autonomous system. We now proceed to find periodic solutions of systems of the form

$$\dot{x}_\sigma = \mu X_\sigma(\mu, x_1, \ldots, x_s), \quad \sigma = \overline{1, s}, \tag{4.32}$$

where functions $X_\sigma, \sigma = \overline{1, s}$ do not not depend explicitly on the time. Such systems are customary called *autonomous*. We shall assume that the right-hand sides of Eqs. (4.32) contain holomorphic functions of their variables.

The study of such systems has a number of peculiarities. First of all, as distinct from the previous cases, here one may not *a priori* indicate the period of oscillations T, which, in general, is a function of μ. Besides, since the differential equations does not explicitly contain t, it follows that, for some known periodic solution $x_\sigma = x_\sigma(t), \sigma = \overline{1, s}$, the solution $x_\sigma = x_\sigma(t - t_1)$, where t_1 is an arbitrary time shift, is also periodic.

Let $x_\sigma^{(0)} = x_\sigma^{(0)}(t), \sigma = \overline{1, s}$, be a T-periodic solution of the generating autonomous system. We shall search for a $(T + \alpha_0)$-periodic solution of the nonlinear system that coincides with the generating one for $\mu = 0$. Necessary and sufficient conditions for its periodicity read as

$$\Phi_\sigma(\mu, \alpha_0, \alpha_1, \ldots, \alpha_s) = x_\sigma(\mu, T + \alpha_0, \alpha_1, \ldots, \alpha_s) -$$
$$-x_\sigma(\mu, 0, \alpha_1, \ldots, \alpha_s) = 0, \quad \sigma = \overline{1, s}, \tag{4.33}$$

where, as before, $\alpha_1, \ldots, \alpha_s$ are the departures of the initial conditions of the sought-for solution from the same conditions of the generating solution. Here, the functions Φ_σ are holomorphic functions of their variables that vanish for $\alpha_\tau = \mu = 0, \tau = \overline{0, s}$. So, we have s equations for $s + 1$ unknown functions $\alpha_\tau(\mu)$, and hence one of these functions may be looked upon as a parameter on which the solution depends.

We also note that the functional determinant $\Delta|_{\mu=\alpha_\tau=0} = 0$ also vanishes, because system (4.33), alongside with the trivial solution with $\mu = \alpha_\tau = 0$, must admit an infinite family of solutions, inasmuch as the generating solution remains periodic when in it the argument t is replaced by the argument $t - t_1$. A. Poincaré, W. MacMillan, and I. G. Malkin examined various cases of vanishing of the functional determinant.

Following Poincaré, we shall again search for a holomorphic solution in the form $x_\sigma(t) = \sum_{n=0}^{\infty} \mu^n x_\sigma^{(n)}(t)$, $\sigma = \overline{1, s}$, where, however, in the case of an autonomous system, as distinct from an nonautonomous system, the functions $x_\sigma^{(n)}(t)$ may fail to satisfy the periodicity conditions $x_\sigma^{(n)}(t + T + \alpha_0) = x_\sigma^{(n)}(t)$, $n = 0, 1, 2, \ldots$. In particular, these functions may contain secular terms, which unboundedly increase in time. A similar picture is seen when expanding the periodic function in a series

$$\sin(\omega + \mu)t = \sin \omega t + \mu t \cos \omega t - \frac{\mu^2 t^2}{2t} \sin \omega t -, \ldots.$$

In this case, for a large value of time, a finite number of terms in the series prove insufficient for characterization of the behaviour of a function.[5] A number of scientists (P. Laplace, J. Lagrange, M. V. Ostrogradskii, A. Lindstedt, A. I. Krylov, etc.) proposed methods to eliminate such undesirable terms from the solutions.

To illustrate one of the methods, we find the periodic solution of the *Rayleigh equation*

$$\ddot{x} + \omega^2 x = \mu (1 - \dot{x}^2) \dot{x} \tag{4.34}$$

with the initial conditions

$$x(0) = A, \quad \dot{x}(0) = 0. \tag{4.35}$$

Such an equation describes, for example, the oscillations of a Froude pendulum, which is a usual physical pendulum attached with friction on the horizontal axis and moving with constant angular velocity. On the pendulum there is the moment of the frictional forces from the axis; it depends on the relative angular velocity.

To find the solution we shall use the method[6] in accordance with which the function x and the squared unknown frequency of oscillations p^2 are sought in the form of series expanded in powers of the small parameter μ:

$$x(t) = x_0(t) + \mu x_1(t) + \mu^2 x_2(t) + \ldots, \tag{4.36}$$

$$p^2 = \omega^2 + \mu h_1 + \mu^2 h_2 + \ldots. \tag{4.37}$$

---

[5] Similar issues were encountered in the 18 century in astronomy when searching for periodic motions of planets using series. This, in turn, suggested the name 'secular terms', when the approximate solution proves unsatisfactory for large values of time (of the order of a century) due to the presence of terms containing this variable as a factor.

[6] A. N. Krylov. Vibration of ships. Leningrad–Moscow. 1936. Pp. 209–214 [in Russian].

The functions $x_0(t)$, $x_1(t)$, $x_2(t)$, ... and the coefficients h_1, h_2, ... are unknowns; the latter will be adjusted to ensure that the functions $x_n(t)$, $n = 0, 1, \ldots$ would contain no secular terms.

Substituting expressions (4.36) and (4.37) into Eq. (4.34) and equating the terms with the same powers of μ, this gives

$$
\begin{aligned}
&\ddot{x}_0 + p^2 x_0 = 0, \\
&\ddot{x}_1 + p^2 x_1 = h_1 x_0 + \left(1 - \dot{x}_0^2\right) \dot{x}_0, \\
&\ddot{x}_2 + p^2 x_2 = h_1 x_1 + h_2 x_0 + \dot{x}_1 - 3\dot{x}_0^2 \dot{x}_1,
\end{aligned} \tag{4.38}
$$

$$\cdots\cdots\cdots\cdots\cdots\cdots\cdots\cdots\cdots$$

In view of (4.35), the following initial conditions should be satisfied

$$x_0(0) = A, \quad \dot{x}_0(0) = 0, \quad x_1(0) = 0, \quad \dot{x}_1(0) = 0,$$
$$x_2(0) = 0, \quad \dot{x}_2(0) = 0, \quad \ldots .$$

Hence, the solution of the generating equation looks like

$$x_0 = A \cos pt.$$

Substituting it into the second equation of system (4.38), we get

$$\ddot{x}_1 + p^2 x_1 = h_1 A \cos pt + Ap \left(\frac{3}{4} A^2 p^2 - 1\right) \sin pt - \frac{1}{4} A^3 p^3 \sin 3pt. \tag{4.39}$$

In order that the function $x_1(t)$ would not contain secular terms, we eliminate the terms with $\cos pt$ and $\sin pt$ from the right-hand side of Eq. (4.39). For this purpose, the conditions $h_1 = 0$, $A^2 = 4/(3p^2)$ should be satisfied, where $p = \omega$.

So, in the zero approximation, Eq. (4.34) has the periodic solution $x = A \cos \omega t$, where the amplitude of oscillations is not arbitrary, but is equal to $A = 2/(\omega\sqrt{3})$. Thus, in a nonconservative system represented by the Rayleigh equation there may appear self-excited oscillations with certain frequency and amplitude.

To find the first approximation, we have the equation

$$\ddot{x}_1 + p^2 x_1 = -\frac{A^3 p^3}{4} \sin 3pt.$$

From its general solution

$$x_1 = M_1 \cos pt + N_1 \sin pt + \frac{A^3 p^3}{32} \sin 3pt$$

one should single out the particular solution $x_1 = (A^3 p/32)(\sin 3pt - 3\sin pt)$ that satisfies the zero initial conditions. So, in the first approximation, the solution of Eq. (4.34) looks like

$$x = A \cos pt + \frac{\mu A^3 p}{32} (\sin 3pt - 3 \sin pt),$$

$$A = 2/(p\sqrt{3}), \quad p = \omega.$$

To find the second approximation, one should solve the equation

$$\ddot{x}_2 + p^2 x_2 = A \left(h_2 - \frac{A^3 p^3}{32} + \frac{21 A^4 p^4}{128} \right) \cos pt +$$

$$+ A^3 p^2 \left(\frac{3}{32} - \frac{21 A^2 p^2}{128} \right) \cos 3pt + \frac{9 A^5 p^4}{128} \cos 5pt. \tag{4.40}$$

There are no secular terms if

$$h_2 = \frac{3 A^2 p^2}{32} - \frac{21 A^4 p^4}{128}; \quad \text{that is} \quad p^2 = \omega^2 + \mu^2 \left(\frac{3 A^2 p^2}{32} - \frac{21 A^2 p^2}{128} \right).$$

Note that it is fairly easy to construct the particular solution of Eq. (4.40) corresponding to the zero initial conditions. A further refinement of the solution and the squared oscillation frequency is made similarly.

The limit cycle of the oscillations thus obtained is close to an ellipse, while the motions themselves resemble the harmonic ones; that is, we have a characteristic case of quasi-linear oscillations. However, if the parameter μ is not small (for example, $\mu = 10$), then the limit cycle is drastically different from an ellipse. This can be verified by constructing the phase trajectory by Lienard's method. In this case, the law of motion is different from the sine law and has characteristic regions of rapid change of variation of the quantity x. Such oscillations are called *relaxation oscillations*.

In passing, we give a relation between the Rayleigh and van der Pol equations. Differentiating equation (4.34) in time and writing $y = \sqrt{3} \dot{x}$, one gets the van der Pol equation, which describes the operation of an electronic generator

$$\ddot{y} + \omega^2 y = \mu (1 - y^2) \dot{y}.$$

5   The Krylov–Bogolyubov Method

Oscillations of autonomous systems. In the previous subsection a holomorphic solution was built in the form of converging series arranged in terms of the powers of the small parameter μ. However, a construction of approximations of larger orders called for a considerable increase in the computational labor. Of equal importance with the above methods are the methods of searching solutions in the form of asymptotic series, in which a finite number of terms tends to the exact solution as $\mu \to 0$. This class of methods involves the method proposed by N. M. Krylov and N. N. Bogolyubov, which will be employed for the study of the motion equations of the autonomous system:

$$\ddot{x} + \omega^2 x = \mu f(x, \dot{x}) . \tag{5.1}$$

For $\mu = 0$ this equation becomes the conventional equation of harmonic oscillations, which has the solution

$$x = a \cos \psi . \tag{5.2}$$

Here, the amplitude a is constant and the phase ψ is a linear time function: $\dot{a} = 0$,
$\dot{\psi} = \omega$, $\psi = \omega t + \theta$, $\theta = $ const.

The approximate solution of Eq. (5.1) for $\mu \neq 0$ will also be sought in the
form (5.2), assuming that a and ψ are the sought-form time functions. This solution is refined by adding a finite number n of terms multiplied by the powers of the
small parameter μ. So, the solution of Eq. (5.1) is sought in the form

$$x = a(t) \cos \psi(t) + \sum_{\nu=1}^{n} \mu^{\nu} u_{\nu}(a, \psi) . \tag{5.3}$$

The functions $u_{\nu}(a, \psi)$ will be sought to be 2π-periodic in the argument ψ and
satisfying the conditions

$$\int_0^{2\pi} u_{\nu}(a, \psi) \cos \psi \, d\psi = 0 , \quad \int_0^{2\pi} u_{\nu}(a, \psi) \sin \psi \, d\psi = 0 , \quad \nu = \overline{1, n} , \tag{5.4}$$

which, on one hand, guarantees the uniqueness of these functions, and on the other
hand, means that in solution (5.3) the term $a(t) \cos \psi(t)$ completely determines the
first harmonics. The unknown functions $a(t)$ and $\psi(t)$ are found from the differential
equations

$$\dot{a} = \sum_{\nu=1}^{n} \mu^{\nu} A_{\nu}(a) , \quad \dot{\psi} = \omega + \sum_{\nu=1}^{n} \mu^{\nu} B_{\nu}(a) . \tag{5.5}$$

The functions $u_{\nu}(a, \psi)$, $A_{\nu}(a)$, $B_{\nu}(a)$, which occur in series (5.3) and (5.5),
should be determined by Eq. (5.1). We first consider its left-hand side. In calculating
the derivative $\ddot{x}$ of the function x, as given in the form of series (5.3), there appear the
derivatives $\dot{a}$ and $\dot{\psi}$, as well as the derivatives $\ddot{a}$ and $\ddot{\psi}$. From series (5.5) it follows
that

$$\ddot{a} = \sum_{\nu=1}^{n} \mu^\nu \frac{dA_\nu}{da} \, \dot{a} = \sum_{\nu=1}^{n} \mu^\nu \frac{dA_\nu}{da} \sum_{\lambda=1}^{n} \mu^\lambda A_\lambda \,,$$

$$\ddot{\psi} = \sum_{\nu=1}^{n} \mu^\nu \frac{dB_\nu}{da} \, \dot{a} = \sum_{\nu=1}^{n} \mu^\nu \frac{dB_\nu}{da} \sum_{\lambda=1}^{n} \mu^\lambda A_\lambda \,.$$

These expressions, as well as expressions (5.3) and (5.5), enable one to represent the left-hand side of Eq. (5.1) in the form of a series in powers of μ (because of their complexity, we write down only the first two expressions):

$$\ddot{x} + \omega^2 x = \mu \left[-2\omega A_1 \sin\psi - 2\omega a B_1 \cos\psi + \omega^2 \left(\frac{\partial^2 u_1}{\partial \psi^2} + u_1 \right) \right] +$$

$$+ \mu^2 \left[\left(A_1 \frac{dA_1}{da} - a B_1^2 - 2\omega a B_2 \right) \cos\psi - \right.$$

$$- \left(2\omega A_2 + 2A_1 B_1 + a A_1 \frac{dB_1}{da} \right) \sin\psi + 2\omega A_1 \frac{\partial^2 u_1}{\partial a \partial \psi} + 2\omega B_1 \frac{\partial^2 u_1}{\partial \psi^2} +$$

$$\left. + \omega^2 \left(\frac{\partial^2 u_2}{\partial \psi^2} + u_2 \right) \right] + \dots . \qquad (5.6)$$

In turn, the expression $\mu f(x, \dot{x})$ from Eq. (5.1) can be expanded in a series

$$\mu f(x, \dot{x}) = \mu f + \mu^2 \left[u_1 \frac{\partial f}{\partial x} + \right.$$

$$\left. + \left(A_1 \cos\psi - a B_1 \sin\psi + \omega \frac{\partial u_1}{\partial \psi} \right) \frac{\partial f}{\partial \dot{x}} \right] + \dots , \qquad (5.7)$$

where in the right-hand side, after calculating all the derivatives of the function f, the arguments x and $\dot{x}$ are replaced, respectively, by the quantities $a \cos\psi$ and $-a\omega \sin\psi$.

Equating in (5.6) and (5.7) the terms with equal powers μ^ν, $\nu = \overline{1, n}$, we obtain

$$\omega^2 \left(\frac{\partial^2 u_\nu}{\partial \psi^2} + u_\nu \right) = f_{\nu-1}(a, \psi) + 2\omega A_\nu \sin\psi + 2\omega a B_\nu \cos\psi \,,$$

$$\nu = \overline{1, n} \,, \qquad (5.8)$$

where we used the notation

$$f_0(a, \psi) = f(a \cos \psi, -a\omega \sin \psi) \,,$$

$$f_1(a, \psi) = u_1 \frac{\partial f}{\partial x} + \frac{\partial f}{\partial \dot{x}} \left(A_1 \cos \psi - a B_1 \sin \psi + \omega \frac{\partial u_1}{\partial \psi} \right) +$$

$$+ \left(a B_1^2 - A_1 \frac{dA_1}{da} \right) \cos \psi + \left(2A_1 B_1 + a A_1 \frac{dB_1}{da} \right) \sin \psi -$$

$$- 2\omega \left(A_1 \frac{\partial^2 u_1}{\partial a \partial \psi} + B_1 \frac{\partial^2 u_1}{\partial \psi^2} \right) \,,$$

$$\dots\dots\dots\dots\dots\dots\dots\dots\dots$$

It is worth noting that Eq. (5.1) is satisfied up to small terms of order μ^{n+1}.

The functions $f_\nu(a, \psi)$ are 2π-periodic in the argument ψ and are expressible in terms of $A_1, B_1, u_1, \dots, A_\nu, B_\nu, u_\nu$.

Let us find A_ν, B_ν, u_ν from the νth equation of system (5.8), assuming that we know all the preceding functions. We expand the functions $f_{\nu-1}(a, \psi)$ and $u_\nu(a, \psi)$ (in view of conditions (5.4) the second one does not contain the first harmonics) in Fourier series:

$$f_{\nu-1}(a, \psi) = g_0^{(\nu-1)}(a) + \sum_{\varkappa=1}^{\infty} \left[g_\varkappa^{(\nu-1)}(a) \cos \varkappa\psi + h_\varkappa^{(\nu-1)}(a) \sin \varkappa\psi \right] \,,$$

$$u_\nu(a, \psi) = v_0^{(\nu)}(a) + \sum_{\varkappa=2}^{\infty} \left[v_\varkappa^{(\nu)}(a) \cos \varkappa\psi + w_\varkappa^{(\nu)}(a) \sin \varkappa\psi \right] \,.$$

Substituting these expressions in the corresponding equation of system (5.8) and equating the coefficients of equal harmonics, we find that

$$A_\nu(a) = -\frac{h_1^{(\nu-1)}(a)}{2\omega} \,, \quad B_\nu(a) = -\frac{g_1^{(\nu-1)}(a)}{2\omega a} \,, \quad v_0^\nu(a) = \frac{g_0^{(\nu-1)}(a)}{\omega^2} \,,$$

$$v_\varkappa^{(\nu)}(a) = \frac{g_\varkappa^{(\nu-1)}(a)}{\omega^2 (1 - \varkappa^2)} \,, \quad w_\varkappa^{(\nu)}(a) = \frac{h_\varkappa^{(\nu-1)}(a)}{\omega^2 (1 - \varkappa^2)} \,, \quad \varkappa = 2, 3, \dots .$$

Thus, we obtain the functions $A_\nu(a)$, $B_\nu(a)$ and the Fourier series of the functions $u_\nu(a, \psi)$, $\nu = \overline{1, n}$; that is, we solve the problem with the required accuracy.

Assume that for a sufficiently large time period of order $1/\mu$ we have the first approximation

$$x = a \cos \psi + \mu u_1(a, \psi) \,, \tag{5.9}$$

where a and ψ satisfy the equations

$$\dot{a} = \mu A_1(a) \,, \quad \dot{\psi} = \omega + \mu B_1(a) \,. \tag{5.10}$$

Hence, by the mean value theorem, the increments of the amplitude and phase over the time t can be written in view of (5.10) in the form

$$a(t) - a(0) = t\mu\overline{A_1(a)}, \quad \psi(t) - \omega t - \psi(0) = t\mu\overline{B_1(a)}, \tag{5.11}$$

where the bar denotes averaging of the corresponding functions in the interval $[0, t]$. From (5.11) it follows that the first term of series (5.9) contains values of order μ; that is, the solution

$$x = a(t)\cos\psi(t), \tag{5.12}$$

where a and ψ satisfy Eqs. (5.10), takes into account the quantities of order μ. On this basis, solution (5.9) is sometimes called *improved first approximation*, as distinct from the *first approximation* (5.12).

In the *van der Pol method* the solution of Eq. (5.1) is sought in the form (5.12). Here, the function $\psi(t)$ is represented as follows:

$$\psi(t) = \omega t + \theta(t).$$

Since one Eq. (5.1) is used to find two unknown functions $a(t)$ and $\theta(t)$, they can be subject to an additional restriction, for example,

$$\dot{a}\cos\psi - a\,\dot{\theta}\sin\psi = 0. \tag{5.13}$$

This condition is convenient because if it is satisfied, then $\dot{x}$ reads as

$$\dot{x} = -a\omega\sin\psi,$$

which is of the same form as for the constant values of a and θ. After subtracting the acceleration $\ddot{x}$ and substituting in into Eq. (5.1), we find that

$$\dot{a}\sin\psi + a\,\dot{\theta}\cos\psi = -\frac{\mu}{\omega}\,f(a\cos\psi, -a\omega\sin\psi). \tag{5.14}$$

From system (5.13), (5.14) we find the derivatives $\dot{a}$ and $\dot{\theta}$:

$$\dot{a} = -\frac{\mu f}{\omega}\sin\psi, \quad \dot{\theta} = \dot{\psi} - \omega = -\frac{\mu f}{\omega a}\cos\psi.$$

Replacing in these equations the function $f\sin\psi$ and $f\cos\psi$ by their mean values over the period 2π, we obtain

$$a = -\frac{\mu}{2\omega\pi} \int_0^{2\pi} f(a\cos\psi, -a\omega\sin\psi)\,\sin\psi\,d\psi,$$

(5.15)

$$\dot\psi = \omega - \frac{\mu}{2\omega a\pi} \int_0^{2\pi} f(a\cos\psi, -a\omega\sin\psi)\,\cos\psi\,d\psi.$$

This is the *transient equations for van der Pol equation*. In the notation used in the Krylov–Bogolyubov method, Eqs. (5.15) can be written in the form (5.10). So, the first approximation (5.12) obtained by the Krylov–Bogolyubov method agrees with the solution obtained by the van der Pol method.

Let us construct the first approximation (5.12) for the *van der Pol equation*

$$\ddot x + \omega^2 x = \mu(1 - x^2)\dot x, \quad f(x,\dot x) = (1 - x^2)\dot x.$$

In accordance with Eqs. (5.15), we have

$$\dot a = -\frac{\mu}{2\omega\pi} \int_0^{2\pi}(1 - a^2\cos^2\psi)(-a\omega\sin\psi)\,\sin\psi\,d\psi = \frac{\mu a}{2}\left(1 - \frac{a^2}{4}\right),$$

$$\dot\psi = \omega - \frac{\mu}{2\omega a\pi} \int_0^{2\pi}(1 - a^2\cos^2\psi)(-a\omega\sin\psi)\,\cos\psi\,d\psi = \omega.$$

Since $2a\dot a = da^2/dt$, this system can be written as

$$\frac{da^2}{dt} = \mu a^2\left(1 - \frac{a^2}{4}\right), \quad \frac{d\psi}{dt} = \omega.$$

(5.16)

Integrating these equations with the initial conditions $a(0) = a_0$, $\psi(0) = \psi_0$, we find that

$$a(t) = a_0 e^{(1/2)\mu t}\left[1 + \frac{1}{4}a_0^2(e^{\mu t} - 1)\right]^{-1/2}, \quad \psi(t) = \omega t + \psi_0.$$

(5.17)

From system (5.16) it follows that the steady-state regime, when the amplitude is constant, is possible with $a^2 = 0$ and $a^2 = 4$. The value $a = 0$ corresponds to a state of rest of the system. However, such an equilibrium position proves unstable, because by formula (5.17) with $a_0 \neq 0$ we have $a \underset{t\to\infty}{\to} 2$, which indicates at stable self-excited oscillations with the amplitude $a = 2$. So, in the system system described by the van der Pol equation, self-excited oscillations appear inadvertently for any nonzero initial values of the amplitude.

Oscillations of nonautonomous systems. We now consider the case when the motion is described by the equation

$$\ddot{x} + \omega^2 x = \mu f(kt, x, \dot{x}),\tag{5.18}$$

where the small perturbing force μf is 2π-periodic in the argument kt. Assume that the function f has the form

$$f(kt, x, \dot{x}) = \sum_{n=0}^{N} f_n(x, \dot{x}) \cos nkt,$$

and that the coefficients $f_n(x, \dot{x})$ are some polynomials of x and $\dot{x}$. In order to find an approximate solution in the form (5.12) we shall expand the functions $f_n(x, \dot{x})$ in Fourier series, replacing in them the quantities x and $\dot{x}$ by, respectively, $a \cos(\omega t + \theta)$ and $-a\omega \sin(\omega t + \theta)$. By this we shall represent the function $f(kt, x, \dot{x})$ by the double Fourier series. The products $\cos nkt \cos m\omega t$ and $\cos nkt \sin m\omega t$ appearing in this series can be written as a sum of harmonics with frequencies $|nk \pm m\omega|$, which are known as *combination frequencies*.

If one of such frequencies becomes close to the eigenfrequency ω, that is,

$$k \approx p\omega/q,$$

where p and q are integer relative primes, then the nonlinear system may undergo a resonance. If $p = q = 1$, then the resonance is called the *main* or *ordinary resonance*. For $p = 1$, $q \neq 1$ a resonance is said to occur on the *external frequency overtone resonance*, otherwise ($q = 1$, $p \neq 1$), it occurs on the *overtone of an eigenfrequency* (the *nth genus resonance*). The last phenomenon was observed already by Helmholtz, who found that human ear is sometimes capable of perceiving a sound of not only a given frequency k, but also of frequencies $k/2$, $k/3$, etc. He thought that this should be primarily explained by nonlinear mechanical properties of the eardrum. The first theoretical investigations in this directions were carried out by N. D. Mandel'shtam and N. D. Papaleksi with the help of the small parameter method.

Let us represent the frequency of the external force k in the form $k = p\omega/q + \Delta$, where Δ is the *detuning with respect to the resonance* with frequency $p\omega/q$. As before, we shall assume that the principal term in the approximate solution is of form (5.12). In this case, the unknown function $\psi(t)$ can be conveniently written as

$$\psi(t) = \frac{q}{p} kt + \theta(t),$$

where $\theta(t)$ is the new unknown function.

In refining the first approximation (5.12) the solution will be sought in the form

$$x = a \cos\left(\frac{q}{p} kt + \theta\right) + \sum_{\nu=1}^{n} \mu^{\nu} u_{\nu}(a, kt, \theta),$$

(5.19)

where the function a and θ are determined from the differential equations

$$\dot{a} = \sum_{\nu=1}^{n} \mu^{\nu} A_{\nu}(a, \theta),$$

$$\dot{\theta} = \omega - \frac{q}{p} k + \sum_{\nu=1}^{n} \mu^{\nu} B_{\nu}(a, \theta),$$

(5.20)

where the difference $(\omega - \frac{q}{p} k)$ is not necessarily small. The sought-for functions A_{ν}, B_{ν}, u_{ν}, $\nu = \overline{1, n}$, will be sought to be 2π-periodic with respect to their variables θ and kt.

To find the solution one should substitute (5.19) into Eq. (5.18) by virtue of (5.20). Next, expanding the right-hand side of Eq. (5.18) in a series in powers of μ and equating the coefficients in the left- and right-hand sides of the same powers of the small parameter, we obtain relations, from which one may determine all unknown functions. However, in the case in question, even for lower approximations the formulas prove to be too cumbersome, and hence we shall not give them in their general form, but rather proceed with a specific example.

Let us consider the oscillations described by Eq. (3.14). We shall seek the solution near the resonance of the principal frequencies, that is, we set $p = q = 1$. As was already mentioned, in the first approximation the Krylov–Bogolyubov method is capable of providing a solution that agrees with the solution derived by the van der Pol method. In accordance with the last mention methods, the solution is sought in the form

$$x = a(t) \cos \psi(t), \quad \psi(t) = kt + \theta(t),$$

(5.21)

where the unknown functions a and θ are related by (5.13). Here it is essential that the derivatives $\dot{x}$ and $\ddot{x}$ are calculated by the relatively simple formulas

$$\dot{x} = -ak \sin \psi,$$
$$\ddot{x} = -ak^2 \cos \psi - \dot{a}k \sin \psi - ak\dot{\theta} \cos \psi.$$

(5.22)

In view of (5.21) and (5.22), Eq. (3.14) can be written as

$$-\dot{a}k \sin \psi - ak\dot{\theta} \cos \psi = h \sin (\psi - \theta) + F(\psi),$$

(5.23)

where $F(\psi) = a(k^2 - \omega^2) \cos \psi - \mu a^3 \cos^3 \psi + \varphi \omega^2 ka \sin \psi + \varphi ka^3 \sin^3 \psi$.
From Eqs. (5.13), (5.23) we find the derivatives $\dot{a}$ and $\dot{\theta}$:

$$\dot{a} = -\frac{h \sin (\psi - \theta) + F(\psi)}{\omega} \sin \psi,$$

$$\dot\theta = -\frac{h\sin(\psi-\theta)+F(\psi)}{a\omega}\cos\psi\,.$$

Averaging the right-hand sides of these equations in terms of ψ over the period 2π, that is, putting,

$$\dot a = -\frac{1}{2\omega\pi}\int_0^{2\pi}(h\sin(\psi-\theta)+F(\psi))\,\sin\psi\,d\psi\,,$$

$$\dot\theta = -\frac{1}{2a\omega\pi}\int_0^{2\pi}(h\sin(\psi-\theta)+F(\psi))\,\cos\psi\,d\psi\,,$$

and calculating the corresponding integrals, we find that

$$\dot a = -\frac{\varphi\omega^2}{2}a - \frac{3\mu\varphi}{8}a^3 - \frac{h}{2k}\cos\theta\,,$$
$$\dot\theta = \frac{h}{2ka}\sin\theta + \frac{3\mu}{8k}a^2 - \frac{\omega^2-k^2}{2k}\,. \tag{5.24}$$

These equations also provide a way of considering stationary oscillations. Making $\dot a = \dot\theta = 0$, we get the system of two equations, which is capable of determining the dependence of the amplitude a and of the phase shift $\varepsilon = \theta - \pi/2$ on the frequency k of the perturbing force. Note that if one substitute the quantities $h\cos\theta$ and $h\sin\theta$, as obtained from (5.24) with $\dot a = \dot\theta = 0$, into the expression $h^2 = h^2\cos^2\theta + h^2\sin^2\theta$, then the resulting expression will coincide with the equation of the frequency response function (3.17), which was earlier obtained by the Bubnov–Galerkin method.

Oscillations in systems with slowly varying parameters. The asymptotic Krylov–Bogolyubov method was extended by A. Yu. Mitropol'skii to systems involving slowly oscillating parameters. This approach is capable, in particular, to study the passage of a mechanical system through resonance when the frequency of the perturbing force varies according to a given law. As an example, we consider the equation

$$\ddot x + \omega^2 x = \mu f(\vartheta, x, \dot x)\,, \tag{5.25}$$

where $f(\vartheta, x, \dot x)$ is a ϑ-periodic function with period 2π. The quantity ϑ is the phase of the perturbing force. We shall assume that the instantaneous frequency $\dot\vartheta$ of this force k slowly varies in time. It is convenient to look upon the quantity k as a function of 'slow' time $\tau = \mu t$. This makes it possible to take into account that the rate of change of the frequency of k,

$$k = \frac{dk}{dt} = \frac{dk}{d\tau}\frac{d\tau}{dt} = \mu\frac{dk}{d\tau} = \mu g(\tau),$$

is a quantity of order μ. Under such an approach, the rate of change of k in time has the same order of smallness as the nonlinear terms. This enables one to take into account both the variability of the frequency and its nonlinearity in the framework of one asymptotic method.

From the above it follows that it is expedient to seek the oscillations in the first approximation in the form $x = \cos(\vartheta + \theta)$, where $a(t)$ and $\theta(t)$ are unknown time functions. Slow variation of the frequency $k = \dot{\vartheta}$ results in excitation of oscillations with the frequency that should be close to that of natural frequencies ω. We therefore assume that $\dot{\psi} = \dot{\vartheta} + \dot{\theta}$ as $\mu \to 0$; that is, $\dot{\theta} \to \omega - k$ as $\mu \to 0$. Hence, when refining the first approximation, the solution of Eq. (5.25) will be sought in the form of a series

$$x = a\cos(\vartheta + \theta) + \sum_{\nu=1}^{n} \mu^{\nu} u_{\nu}(a_{\nu},\ \theta),$$

in which the functions $u_{\nu}(a, \theta)$ are 2π-periodic in the argument θ, and the functions $a(t)$ and $\theta(t)$ satisfy the system of the differential equations

$$\dot{a} = \sum_{\nu=1}^{n} \mu^{\nu} A_{\nu}(a,\ \theta),\quad \dot{\theta} = \omega - k + \sum_{\nu=1}^{n} \mu^{\nu} B_{\nu}(a,\ \theta).$$

As an example, we consider the equation with the following right-hand side

$$\mu f(\vartheta, x, \dot{x}) = h\sin\vartheta - \varphi\omega^2\dot{x} - \frac{\varphi}{k^2}\mu\dot{x}^3 - \mu x^3. \tag{5.26}$$

We construct the first approximation $x = a\cos(\vartheta + \theta)$, in which the functions $a(t)$ and $\theta(t)$ are sought from the equations

$$\dot{a} = \mu A_1(a, \theta),\quad \dot{\theta} = \omega - k - \mu B_1(a,\ \theta). \tag{5.27}$$

Here, the derivatives $\dot{x}$ and $\ddot{x}$ are as follows,

$$\dot{x} = \dot{a}\cos(\vartheta + \theta) - a(\dot{\vartheta} + \dot{\theta})\sin(\vartheta + \theta) = -a\omega\sin(\vartheta + \theta) +$$
$$+\mu(A_1\cos(\vartheta + \theta) - aB_1\sin(\vartheta + \theta)),$$
$$\ddot{x} = -a\omega^2\cos(\vartheta + \theta) + \mu\,[-2A_1\omega\sin(\vartheta + \theta) - 2B_1 a\omega\cos(\vartheta + \theta) +$$
$$+(\omega^{\cdot} - k)\frac{\partial A_1}{\partial\theta}\cos(\vartheta + \theta) - a(\omega - k)\frac{\partial B_1}{\partial\theta}\sin(\vartheta + \theta)\Big],$$

apart from the error term of order μ inclusively. Hence, the left-hand side of Eq. (5.25) can be written as

$$\ddot{x} + \omega^2 x = \mu \left[-2A_1\omega \sin(\vartheta + \theta) - 2B_1 a\omega \cos(\vartheta + \theta) + \right.$$
$$\left. +(\omega - k)\frac{\partial A_1}{\partial \theta} \cos(\vartheta + \theta) - a(\omega - k)\frac{\partial B_1}{\partial \theta} \sin(\vartheta + \theta) \right]. \tag{5.28}$$

The right-hand side of this equations, which is given by (5.26), can be written as

$$\mu f(\vartheta, x, \dot{x}) = \mu \left[\frac{h}{\mu} \sin \vartheta + \frac{\varphi}{\mu} a\omega^3 \sin(\vartheta + \theta) + \right.$$
$$\left. + \frac{\varphi}{k^2} a^3\omega^3 \sin^3(\vartheta + \theta) - a^3 \cos^3(\vartheta + \theta) \right], \tag{5.29}$$

apart from the small error term of order μ inclusively. The quantity $\sin \vartheta$ from this expression can be represented in the form

$$\sin \vartheta = \cos \theta \sin(\vartheta + \theta) - \sin \theta \cos(\vartheta + \theta).$$

Now equating (5.28) and (5.29) and denoting $\psi = \vartheta + \theta$, we find that

$$-2A_1\omega \sin \psi - 2B_1 a\omega \cos \psi + \frac{\partial A_1}{\partial \theta}(\omega - k)\cos \psi -$$
$$-a\frac{\partial B_1}{\partial \theta}(\omega - k)\sin \psi = \frac{h}{\mu}(\cos \theta \sin \psi - \sin \theta \cos \psi) + \tag{5.30}$$
$$+\frac{\varphi}{\mu} a\omega^3 \sin \psi + \frac{\varphi}{k^2} a^3\omega^3 \sin^3 \psi - a^3 \cos^3 \psi.$$

Multiplying Eq. (5.30) in succession by $\cos \psi$ and $\sin \psi$ and integrating in ψ from 0 to 2π, this gives

$$\frac{\partial A_1}{\partial \theta}(\omega - k) - 2B_1 a\omega = -\frac{h}{\mu} \sin \theta - \frac{3}{4}a^3,$$
$$a\frac{\partial B_1}{\partial \theta}(\omega - k) + 2A_1\omega = -\frac{h}{\mu} \cos \theta - \frac{\varphi}{\mu} a\omega^3 - \frac{3}{4}\frac{\varphi}{k^2} a^3\omega^3. \tag{5.31}$$

The solution of this system will be sough in the form

$$A_1 = A_1^0 + A_1^1 \cos \vartheta, \quad B_1 = B_1^0 + B_1^1 \sin \vartheta, \tag{5.32}$$

where A_1^0, A_1^1, B_1^0, B_1^1 are unknowns. We substitute (5.32) into system (5.31) and equate the free terms and the coefficients of $\cos \theta$ and $\sin \theta$. As a result, we get

$$B_1^1(\omega - k)a + 2A_1^1\omega = -\frac{h}{\mu},$$

$$2a\omega B_1^1 + A_1^1(\omega - k) = \frac{h}{\mu},$$

Fig. 11 Amplitude versus
the slowly varying frequency

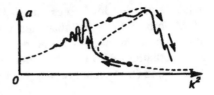

$$2A_1^0 \omega = -\frac{\varphi}{\mu} a\omega^3 - \frac{3}{4}\frac{\varphi}{k^2} a^3\omega^3 ,$$

$$2a\omega B_1^0 = \frac{3}{4} a^3 .$$

Having found from these formulas all the constants in question and substituting (5.32) into system (5.27), we find that

$$\dot{a} = -\frac{\varphi\omega^2}{2} a - \frac{3}{8}\frac{\varphi}{k^2} \mu a^3\omega^2 - \frac{h}{\omega+k}\cos\theta ,$$

$$\dot{\theta} = \omega - k + \frac{3}{8}\frac{\mu}{\omega} a^2 + \frac{h}{a(\omega+k)}\sin\theta . \qquad (5.33)$$

So, the construction of the first approximation was reduced to the integration of system (5.33). However, this integration is unmanageable except numerically. We note that numerical integration of system (5.33) is more expedient than that of the original nonlinear equation, because the latter has a rapidly oscillating solution, and the integration error accumulates rapidly. Now the functions a and θ, as defined from system (5.33), vary slowly, which gives good accuracy of numerical calculations. In other words, Eqs. (5.33) enables one to determine the envelope of the rapidly oscillating curve $x(t)$.

Integrating system (5.33) with given function $k = k(t)$ enables one to find the law of variation of the amplitude a in time. Since the time can be looked upon as a function k, and hence one may also ascertain the law of variation of the amplitude as a function of frequency. A qualitative picture of the amplitude a as a function of k^2, when it increases or decreases, is shown in Fig. 11. The dashed line is the frequency response function, which is shown in Fig. 10.

6 The Method of Direct Separation of Motions

The main idea of the method. Assume that the motion equation of a point reads as

$$m\ddot{x} = F(\dot{x}, x, t) + \Phi(\dot{x}, x, t, kt) , \qquad (6.1)$$

where F is the 'slow' force and Φ is the 'fast' force, which has period 2π in the 'fast' time kt. From the character of the forces acting on a point it is clear that its motion in the plane Otx is composed of the motion along some smooth curve and fast oscillations with k near it. From what has been said it follows that the solution of equation (6.1) should be sought in the form

$$x(t) = X(t) + \xi(t, kt),\qquad(6.2)$$

where ξ corresponds to rapid oscillations near the gradual motion X. The quantity ξ may be looked upon as the difference between x and the mean motion X. Therefore, we require only that the mean value of the function ξ over the period $2\pi/k$ be zero:

$$\frac{1}{2\pi} \int\limits_0^{2\pi} \xi(t, \theta)\,d\theta = 0,\quad \theta = kt.\qquad(6.3)$$

Substituting (6.2) into Eq. (6.1), we obtain

$$m\ddot{X} + m\ddot{\xi} = F(\dot{X} + \dot{\xi}, X + \xi, t) + \Phi(\dot{X} + \dot{\xi}, X + \xi, t, kt).\qquad(6.4)$$

We denote by $\widetilde{F}$ and $\widetilde{\Phi}$, respectively, the mean values of the forces F and Φ over the period $2\pi/k$. They can be approximately recovered by the formulas

$$\widetilde{F}(\dot{X}, X, t) = \frac{1}{2\pi} \int\limits_0^{2\pi} F(\dot{X} + \dot{\xi}, X + \xi, t)\,d\theta,\quad \theta = kt,$$

$$\widetilde{\Phi}(\dot{X}, X, t) = \frac{1}{2\pi} \int\limits_0^{2\pi} \Phi(\dot{X} + \dot{\xi}, X + \xi, t, \theta)\,d\theta.$$

It should be particularly noted that the variables $\dot{X}$, X, t in these formulas are constant parameters. The integration is taken in the fast time $\theta = kt$, which governs only the displacement ξ, velocity $\dot{\xi}$ and force Φ. This approximate calculation of the mean values of the forces F and Φ is the principal idea of the *method of direct separation of motions*, which enables one to single out each type of motions, or in other words, to write down equation for X and ξ separately.[7]

---

[7] This method of solution of nonlinear equations was proposed by Academician P. L. Kapitsa (see: *P. L. Kapitsa*. Dynamic stability of the pendulum with oscillating point of suspension // Zh. Ehksp. Teor. Fiz. 1951. Vol. 21. 5. Pp. 588–598 [in Russian].) This approximate method was further developed by I. I. Blekhman (see: *I. I. Blekhman*. The method of direct separation of motions in problems of vibration effects on nonlinear mechanical systems // Izv. AN SSSR. Mechanics of Solids. 1976. 6. Pp. 13–27 [in Russian]). The very name 'the method of direct separation of motions' became accepted after the publication of this paper.

Equating the mean value of the right-hand side of Eq. (6.4), which equals $\widetilde{F} + \widetilde{\Phi}$, to the mean value of its left-hand side $m\ddot{X}$, we get

$$m\ddot{X} = \widetilde{F}(\dot{X}, X, t) + \widetilde{\Phi}(\dot{X}, X, t). \qquad (6.5)$$

In order to write the differential equation with respect to the displacement ξ, we consider the difference between the force $F + \Phi$ and its mean value $\widetilde{F} + \widetilde{\Phi}$. Equating this difference to $m\ddot{\xi}$, it is found that

$$m\ddot{\xi} = F(\dot{X} + \dot{\xi}, X + \xi, t) + \Phi(\dot{X} + \dot{\xi}, X + \xi, t, kt) - \\ - \widetilde{F}(\dot{X}, X, t) - \widetilde{\Phi}(\dot{X}, X, t). \qquad (6.6)$$

The variables $\dot{X}$, X and t are considered as parameters of this equation.

Equation (6.5) in terms of the function X can be written down only after finding the function ξ. At the same time, in order to write Eq. (6.6) in terms of ξ, one should know the mean motion X. In order to escape from this quandary one needs to circumvent the principal difficulty related to the use of the method of direct separation of motions. The equation with respect to ξ and its solution are found by some other approximate method, which may reside, in particular, on expanding the forces F and Φ in Taylor series and in expanding the function ξ in a series in powers of the small parameter.

We shall show by means of specific examples which important and interesting mechanical phenomena can be investigated by the method of direct separation of motions.

Dynamic stability of the pendulum with oscillating point of suspension.[8] We assume that the point of suspension of a physical pendulum oscillates in the vertical direction according the harmonic law $H \sin kt$. In this case the pendulum equation reads as

$$J\ddot{\varphi} = -mgl \sin\varphi + ml H k^2 \sin kt \sin\varphi. \qquad (6.7)$$

Here, φ is the angular displacement of the pendulum from the lower vertical positions, J is the moment of inertia about the suspension axis, m is the pendulum mass, l is the distance from the axis of rotation to the centre of gravity.

We assume that the frequency k is much larger than the frequency $\omega = \sqrt{mgl/J}$ of small free oscillations of the pendulum. Here, the moment $(-mgl \sin\varphi)$ can be assumed to be slowly varying in comparison with the moment $ml H k^2 \sin kt \sin\varphi$. We write the angle φ as $\varphi = X + \xi$. For $H \ll l$, we may assume that for some initial conditions the quantity ξ is small in comparison with X. Expanding the right-hand side Eq. (6.7) in a Taylor series in ξ and collecting the terms of order ξ, we get

---

[8] This problem was examined by Academician P. L. Kapitsa in the framework of his new method of solution of a nonlinear equation (see the paper mentioned in the previous footnote).

$$-mgl \sin \varphi + ml H k^2 \sin kt \sin \varphi =$$

$$= (ml H k^2 \sin kt - mgl)(\sin X + \xi \cos X) . \tag{6.8}$$

Equating the quantity $J\ddot{X}$ to the mean value of the right-hand side of this expression, this gives

$$J\ddot{X} = -mgl \sin X + ml H k^2 \cos X \frac{1}{2\pi} \int_0^{2\pi} \xi \sin \theta \, d\theta . \tag{6.9}$$

Subtracting the right-hand side of this equation from the right-hand side of (6.8) and equating the result to the quantity $J\ddot{\xi}$, we find that

$$J\ddot{\xi} = ml H k^2 \sin kt \sin X - mgl \xi \cos X +$$

$$+ ml H k^2 \cos X \left(\xi \sin kt - \frac{1}{2\pi} \int_0^{2\pi} \xi \sin \theta \, d\theta \right)$$

or further

$$\ddot{\xi} = \mu k^2 \left[\sin kt (\sin X + \xi \cos X) - \right.$$

$$\left. - \frac{g}{H k^2} \xi \cos X - \cos X \frac{1}{2\pi} \int_0^{2\pi} \xi \sin \theta \, d\theta \right] , \tag{6.10}$$

where $\mu = ml H / J$.

We shall confine our considerations to the case when the parameters H and k are such that the quantity $g/(Hk^2)$ has the order of unity and $\mu \ll 1$. Under these assumptions it readily follows from Eq. (6.10) that the function ξ, which varies with the frequency k, has order μ. Retaining the quantity of order μ on the right of Eq. (6.10), we get

$$\ddot{\xi} = \mu k^2 \sin kt \sin X . \tag{6.11}$$

The partial solution of this equation satisfying condition (6.3) is as follows:

$$\xi = -\mu \sin X \sin kt .$$

We recall that in accordance with the main idea of the method of direct separation of motions the quantity X in Eq. (6.11) is assumed to be constant.

Substituting the so-obtained expression for ξ into Eq. (6.9), we obtain

$$J\ddot{X} = -mgl \sin X - \frac{(mlHk)^2}{4J} \sin 2X. \qquad (6.12)$$

In order to determine under what conditions the vertical position of the pendulum is stable when its point of suspension oscillates in the vertical direction, we employ Eq. (6.12). Putting $X = \pi + X_1$ and assuming that the deviations X_1 are small, we get

$$J\ddot{X}_1 + \left(\frac{(mlHk)^2}{2J} - mgl \right) X_1 = 0.$$

It is seen, therefore, that X_1 varies according to the harmonic law, and hence, the upper vertical position of pendulum is stable if

$$mlH^2k^2 > 2Jg. \qquad (6.13)$$

In the particular case of mathematical pendulum $J = ml^2$, and so condition (6.13) can be put in the form

$$H^2k^2 > 2gl.$$

Loaded spring oscillation with a rapidly oscillating point of suspension and dry friction.[9] In order to dampen various devices whose support executes high-frequency oscillations, they are fastened to their support not rigidly, but elastically. Damping of oscillations of such devices can be achieved with the help of dry friction forces. This involved problem of great technical importance can be considered on a simple mechanical model.

Assume that a load of mass m is connected with its support using a spring of rigidity c. Next, assume that the support is suddenly made to move according to the law (see Sect. 2, formula (2.1))

$$\eta(t) = \eta_1(t) + \eta_2(t), \quad \text{where} \quad \eta_1(t) = \begin{cases} 0, & t < 0, \\ \frac{J_0}{k} t, & t > 0, \end{cases}$$

$$\eta_2(t) = \begin{cases} 0, & t < 0, \\ -\frac{J_0}{k^2} \sin kt, & t > 0. \end{cases}$$

We denote by x the displacement of the load with respect to the base. Here, the absolute displacement of the load is $x + \eta$. We assume that the dry friction force is $(-mJ_c \operatorname{sign} \dot{x})$ for a displacement with relative velocity $\dot{x}$. For simplicity we assume that the effect of gravity is neglected. Then the equation of motion of the load can be written as

---

[9] See S. A. Zegzhda and M. P. Yushkov. Motion in a rapidly oscillating field with dry friction // In the book: *Applied Mechanics*. Issue 1. Leningrad. 1973. Pp. 20–26 [in Russian].

$$\ddot{x} + \omega^2 x = -J_0 \sin kt - J_c \operatorname{sign}\dot{x}, \quad \omega^2 = c/m. \tag{6.14}$$

In accordance with the principal idea of the method we put $x = X + \xi$. Now Eq. (6.14) reads as

$$\ddot{X} + \ddot{\xi} + \omega^2(X + \xi) = -J_0 \sin kt - J_c \operatorname{sign}(\dot{X} + \dot{\xi}). \tag{6.15}$$

It may be approximately assumed that the functions X and ξ vary according to the harmonic law with frequencies ω and k, respectively, and hence in the expression for the velocity $\dot{x} = \dot{X} + \dot{\xi}$ the quantity $\dot{X}$ can be assumed to be small in comparison with $\dot{\xi}$ as $k \gg \omega$. In order to be able to expand the function $\operatorname{sign}(\dot{\xi} + \dot{X})$ in a series in $\dot{X}$ one need, on the basis of physical considerations, to introduce the derivative of the function $\operatorname{sign}\dot{x}$. Here, this function is used to show that the force of dry friction changes the direction with a change in the direction of motion. In reality, the force of dry friction changes continuously, rather then in jumps. Hence, it may be assumed that its variation is described not by the function $\operatorname{sign}\dot{x}$, but rather by the function $f_\varepsilon(\dot{x})$, which we define as

$$f_\varepsilon(\dot{x}) = \begin{cases} -1, & -\infty < \dot{x} < -\varepsilon, \\ \dot{x}/\varepsilon, & -\varepsilon < \dot{x} < \varepsilon, \\ 1, & \varepsilon < \dot{x} < +\infty. \end{cases}$$

In general, the quantity ε is unknown. It is clear that its effect on the motion of the system is insignificant if $\varepsilon \ll \max|\dot{x}|$. The function $\operatorname{sign}\dot{x}$ can be looked upon as the limit of the function $f_\varepsilon(\dot{x})$ as $\varepsilon \to 0$.

Let us calculate the derivative

$$\frac{df_\varepsilon(\dot{x})}{d\dot{x}} = 2\delta_\varepsilon(\dot{x}) = \begin{cases} 0, & -\infty < \dot{x} < -\varepsilon, \\ 1/\varepsilon, & -\varepsilon < \dot{x} < \varepsilon, \\ 0, & \varepsilon < \dot{x} < +\infty. \end{cases}$$

As $\varepsilon \to 0$ the function $\delta_\varepsilon(\dot{x})$ becomes the generalized δ-function. By definition,

$$\int_a^b \delta(\tau) \, d\tau = \begin{cases} 0, & \tau = 0 \notin [a, b], \\ 1, & \tau = 0 \in [a, b]. \end{cases} \tag{6.16}$$

Since

$$\lim_{\varepsilon \to 0} f_\varepsilon(\dot{x}) = \operatorname{sign}\dot{x},$$

$$\lim_{\varepsilon \to 0} \frac{df_\varepsilon(\dot{x})}{d\dot{x}} = 2\delta(\dot{x}),$$

it is natural to assume that

$$\frac{d\,\text{sign}\,\dot{x}}{d\dot{x}} = 2\delta(\dot{x}).\qquad(6.17)$$

Expanding the function $\text{sign}(\dot{\xi} + \dot{X})$ in a Taylor series and considering the first two terms in this expansion, it follows in view of formula (6.17) that

$$\text{sign}(\dot{\xi} + \dot{X}) = \text{sign}\,\dot{\xi} + 2\delta(\dot{\xi})\dot{X}.\qquad(6.18)$$

Using Eq. (6.15), as well as (6.18), in accordance with the basis idea of the method of direct separation of motions one may approximately assume that the functions ξ and X satisfy the equations

$$\ddot{\xi} + \omega^2\xi = -J_0(\sin kt + \mu\,\text{sign}\,\dot{\xi}),\quad \mu = J_c/J_0,\qquad(6.19)$$

$$\ddot{X} + \omega^2 X = -\frac{2J_c}{T}\dot{X}\int_t^{t+T}\delta(\dot{\xi}(\tau))\,d\tau,\quad T = \frac{2\pi}{k}.\qquad(6.20)$$

Equation (6.19) was studied by J. P. Den-Hartog[10] who showed that the periodic solution can be approximately written in the form

$$\xi(t) = A\sin(kt - \alpha),\quad A = \frac{J_0\sqrt{1 - (4\mu/\pi)^2}}{k^2 - \omega^2},\quad \alpha = \arcsin\frac{4\mu}{\pi}.\qquad(6.21)$$

Let us now calculate the mean value of the function $\delta(\dot{\xi})$ over the period $T = 2\pi/k$. We assume that in the interval $[t, t + T]$ the function $\dot{\xi}(t)$ vanishes at times t_1 and $t_2 = t_1 + T/2$. Without loss of generality we may assume that $t_1 > t$, and $t_2 < t + T$. In the neighborhood of the point t_ν, $\nu = 1, 2$, the function $\dot{\xi}(\tau)$ can be written in the form

$$\dot{\xi}(\tau) = \ddot{\xi}(t_\nu)(\tau - t_\nu),\quad \dot{\xi}(t_\nu) = 0,\quad |\ddot{\xi}(t_\nu)| = Ak^2.\qquad(6.22)$$

Since $\dot{\xi}(\tau) \neq 0$ outside the neighborhood of this point, it follows that the integral on the right of Eq. (6.20) can be written in view of (6.16) and (6.22) as

$$\frac{1}{T}\int_t^{t+T}\delta(\dot{\xi}(\tau))\,d\tau = \frac{1}{T}\int_{t_1-\varepsilon}^{t_1+\varepsilon}\delta(\dot{\xi}(\tau))\,d\tau + \frac{1}{T}\int_{t_2-\varepsilon}^{t_2+\varepsilon}\delta(\dot{\xi}(\tau))\,d\tau =$$

---

[10] See: *J. P. Den-Hartog*. Forced Vibrations with combined viscous and Coulomb Damping // Trans. ASME. 1931. Vol. 53. Pp. 107–115.

$$= \frac{1}{T} \int\limits_{-\varepsilon}^{\varepsilon} \delta(a_1 \tau)\, d\tau + \frac{1}{T} \int\limits_{-\varepsilon}^{\varepsilon} \delta(a_2 \tau)\, d\tau\,, \quad a_\nu = \ddot{\xi}(t_\nu)\,.$$

Taking into account that

$$\int\limits_{-\varepsilon}^{\varepsilon} \delta(a_\nu \tau)\, d\tau = \frac{1}{a_\nu} \int\limits_{-\varepsilon a_\nu}^{\varepsilon a_\nu} \delta(\tau_1)\, d\tau_1 = \begin{cases} 1/a_\nu\,, & a_\nu > 0\,, \\ -1/a_\nu\,, & a_\nu < 0\,, \end{cases}$$

$$|a_\nu| = \left|\ddot{\xi}(t_\nu)\right| = Ak^2\,,$$

we finally have

$$\frac{1}{T} \int\limits_{t}^{t+T} \delta(\dot{\xi}(\tau))\, d\tau = \frac{2}{TAk^2}\,.$$

Substituting this value of the integral into Eq. (6.20), one may write

$$\ddot{X} + 2n\dot{X} + \omega^2 X = 0\,, \quad n = \frac{\mu(k^2 - \omega^2)}{\pi k \sqrt{1 - (4\mu/\pi)^2}}\,. \tag{6.23}$$

The initial conditions for integration of this equation are determined using (6.2) and (6.21).

Equation (6.23) means that for high-frequency oscillations with small amplitude the reaction of the system to sufficiently slow motions, as well as to other external actions, is the same as for systems with viscous friction. In this case a system is as if it is floating, that is, it becomes more sensible to external loads. This explains, in particular, inadvertent travels of installations on oscillating supports, unbolting of nuts on vibrating bolts, and so on. It is essential that the method of direct separation of motions is capable of not only explaining the mechanism of transition of dry friction to 'viscous friction', but also of providing the corresponding coefficient for the 'viscous drag'.

Figure 12 shows the results of numerical integration of Eq. (6.14) and the approximate solutions obtained from Eq. (6.23). All dependences are given in dimensionless coordinates. The abscissa is the number of perturbation periods $\eta_2(t)$, the ordinate axis is the dimensionless quantity

$$\frac{k^2 x}{\pi J_0} = \frac{x}{\pi a}\,, \quad a = \frac{J_0}{k^2}\,;$$

that is, the ratio of the relative displacements to the amplitude a of oscillations of the support, as reduced by π times, in order that the line η_2 would have the slope $45°$.

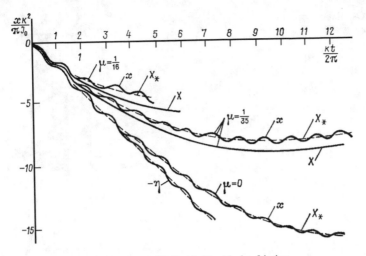

Fig. 12 Motion of a point in a rapidly oscillating field with dry friction

From the exact solution $x(t)$ one may obtain its mean value $X_*(t)$. It is seen from the figure, that the smaller is μ, the closer the solution $X(t)$ to the exact men value $X_*(t)$.

7   The Two-Scale Expansion Method

The two-scale expansion method in nonlinear oscillations is a variant of the Bogolyubov–Mitropol'skii asymptotic method. Such expansions provide a convenient tool for representation of oscillations with slowly varying amplitude. At the initial step of solution, the unknown functions are considered as functions of two arguments $x = x(t, \theta)$, where t is the standard time, $\theta = \mu t$ is the slow time, and μ is the small parameter; in addition,

$$\frac{dx}{dt} = \frac{\partial x}{\partial t} + \mu \frac{\partial x}{\partial \theta}. \tag{7.1}$$

The method will be illustrated on an example of a system with two degrees of freedom that describes the oscillations of a pendulum whose point of support is attached to a spring and which moves along a vertical line (see Fig. 13).

The kinetic and potential energies of the system read as

$$T = \frac{1}{2}m_1 \dot{x}^2 + \frac{1}{2}m_2 \left(\dot{x}^2 + 2l\dot{x}\dot{\varphi}\sin\varphi + l^2\dot{\varphi}^2 \right), \quad \Pi = \frac{1}{2}cx^2 + m_2gl(1 - \cos\varphi),$$

Fig. 13 Oscillations of
a two-mass system

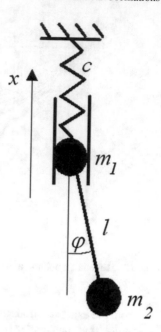

where x is the vertical displacement of the pendulum point of support counted from the equilibrium position, φ is the angular position of the pendulum, m_1, m_2 are the masses, l is the pendulum length, c is the spring stiffness, and g is the acceleration of gravity.

We write the Lagrange equations of the second kind in the form

$$\ddot{x} + \omega^2 x + \eta(\ddot{\varphi}\sin\varphi + \dot{\varphi}^2\cos\varphi) = 0, \quad l\ddot{\varphi} + (g + \ddot{x})\sin\varphi = 0, \qquad (7.2)$$

where

$$\omega^2 = \frac{c}{m_1 + m_2}, \quad \eta = \frac{lm_2}{m_1 + m_2}.$$

We first consider small oscillations of system (7.2) by rejecting the nonlinear term in the first equation and setting approximately $\sin\varphi = \varphi$. Then the solution of the first Eq. (7.2) with the initial condition $x(0) = a$, $\dot{x}(0) = 0$ has the form

$$x(t) = a\cos\omega t.$$

The second equation in (7.2)

$$\ddot{\varphi} + \left(\nu^2 - \frac{a}{l}\omega^2\cos\omega t\right)\varphi = 0, \quad \nu^2 = \frac{g}{l},$$

is the Mathieu equation, which was examined in Sect. 7 of Ch. I. After reducing this equation to the standard form $x'' + (q + \mu\cos 2\tau)x = 0$, we note that the zero solution of this equation can be unstable due to appearing parametric resonances, the principal instability domain on the plane (q, μ) of parameters being described for small μ by the inequality $|q - 1| \leqslant \mu/8$. In our notation, the instability domain reads as

$$\left| \left(\frac{2\nu}{\omega} \right)^2 - 1 \right| \leqslant \frac{a}{2l}, \quad \nu^2 = \frac{g}{l}, \tag{7.3}$$

where ν is the frequency of small oscillations of the pendulum.

This result is meaningful only during the initial stage of motion. With parametric resonance, the oscillation amplitude increases exponentially, which contradicts the energy conservation law ($T + \Pi = \text{const}$) for the conservative system under consideration. To overcome this inconsistency a more accurate analysis of system (7.2) is needed.

From inequalities (7.3) it follows that principal resonance develops only with $\nu \approx \omega/2$. In system (7.2), we introduce the perturbation of frequency δ by the formula

$$\nu^2 = \frac{\omega^2}{4} + \mu\delta.$$

Considering only small oscillations, we shall search the solution to system (7.2) in the form of a series

$$x = l(\mu x_1(t, \theta) + \mu^2 x_2(t, \theta) + \ldots), \quad \varphi = \mu\varphi_1(t, \theta) + \mu^2 f_2(t, \theta) + \ldots, \tag{7.4}$$

where μ is the small parameter, and the functions $x_k(t, \theta)$ and $\varphi_k(t, \theta)$ of two arguments should be consecutively determined when inserted in system (7.2). To evaluate the derivatives, we use (7.1). We take the initial conditions in the form $x(0) = \mu l$, $\dot{x}(0) = 0$, and assume that the initial conditions for $\varphi(t)$ are vanishingly small (but nonzero, since otherwise oscillations of the pendulum are not excited).

Substituting series (7.4) into system (7.2), and taking into account (7.1), we get series with respect to μ and equate to zero the coefficients of μ in different powers. For the first power of μ, we get

$$\frac{\partial^2 x_1}{\partial t^2} + \omega^2 x_1 = 0, \quad \frac{\partial^2 \varphi_1}{\partial t^2} + \frac{\omega^2}{4}\varphi_1 = 0.$$

The solutions to these equations

$$x_1(t, \theta) = x_{1c}(\theta) \cos \omega t + x_{1s}(\theta) \sin \omega t,$$

$$\varphi_1(t, \theta) = \varphi_{1c}(\theta) \cos(\omega t/2) + \varphi_{1s}(\theta) \sin(\omega t/2)$$

contain arbitrary functions $x_{1c}(\theta)$, $x_{1s}(\theta)$, $\varphi_{1c}(\theta)$, $\varphi_{1s}(\theta)$, which should be determined based on the following approximations.

For the coefficients multiplying μ^2 in expansions of Eqs. (7.2), we get

$$\frac{\partial^2 x_2}{\partial t^2} + \omega^2 x_2 + X_2 = 0, \quad X_2(t,\theta) = 2\frac{\partial^2 x_1}{\partial\theta\partial t} + \eta_1\left(\varphi_1\frac{\partial^2\varphi_1}{\partial t^2} + \left(\frac{\partial\varphi_1}{\partial t}\right)^2\right),$$

$$\frac{\partial^2\varphi_2}{\partial t^2} + \frac{\omega^2}{4}\varphi_2 + \Phi_2 = 0, \quad \Phi_2(t,\theta) = 2\frac{\partial^2\varphi_1}{\partial\theta\partial t} + \delta\varphi_1 + \varphi_1\frac{\partial^2 x_1}{\partial t^2},$$

$$\tag{7.5}$$

where $\eta_1 = \eta/l = m_1/(m_1 + m_2)$.

Equations (7.5) have bounded solutions under the conditions

$$\int_0^{2\pi/\omega} X_2(t,\theta)\begin{pmatrix}\sin\omega t \\ \cos\omega t\end{pmatrix}dt = 0, \quad \int_0^{\pi/\omega}\Phi_2(t,\theta)\begin{pmatrix}\sin(\omega t/2) \\ \cos(\omega t/2)\end{pmatrix}dt = 0,$$

which lead to a system of four linear differential equations with respect to the unknown functions $x_{1c}(\theta)$, $x_{1s}(\theta)$, $\varphi_{1c}(\theta)$, $\varphi_{1s}(\theta)$:

$$\frac{dx_{1c}}{d\theta} + \frac{\eta_1\omega}{4}\varphi_{1c}\varphi_{1s} = 0, \quad \frac{dx_{1s}}{d\theta} + \frac{\eta_1\omega}{8}(\varphi_{1s}^2 - \varphi_{1c}^2) = 0,$$

$$\frac{d\varphi_{1c}}{d\theta} - \frac{\delta}{\omega}\varphi_{1s} + \frac{\omega}{2}(\varphi_{1c}x_{1s} - \quad \frac{d\varphi_{1s}}{d\theta} + \frac{\delta}{\omega}\varphi_{1c} - \frac{\omega}{2}(\varphi_{1s}x_{1s} + \tag{7.6}$$

$$-\varphi_{1s}x_{1c}) = 0, \qquad\qquad +\varphi_{1c}x_{1c}) = 0.$$

If the nonlinear terms in the first two equations of system (7.6) are neglected, then from the above initial conditions $x(0) = \mu l$, $\dot{x}(0) = 0$ we get $x_{1c} \equiv 1$, $x_{1s} \equiv 0$. The remaining two Eqs. (7.6) are linear equations with constant coefficients:

$$\frac{d\varphi_{1c}}{d\theta} - \left(\frac{\delta}{\omega} + \frac{\omega}{2}\right)\varphi_{1s} = 0, \quad \frac{d\varphi_{1s}}{d\theta} + \left(\frac{\delta}{\omega} - \frac{\omega}{2}\right)\varphi_{1c} = 0.$$

When searching their solution in the form $\varphi_{1c}(\theta) = \varphi_{1c}^0 e^{\lambda\theta}$, $\varphi_{1s}(\theta) = \varphi_{1c}^0 e^{\lambda\theta}$, we get the characteristic equation

$$\lambda^2 = \frac{\omega^2}{4} - \frac{\delta^2}{\omega^2} \tag{7.7}$$

for the parameter λ. From this equation, we can estimate the instability domain as $|\delta| \leqslant \omega^2/2$; these bounds coincide (with a changed notation) with those determined from formula (7.3).

From Eq. (7.7) we can also find the first approximation for the growth rate of the amplitude when the stability is lost (at the initial motion phase when the angle φ is small). To justify this claim, we consider the nonlinear system (7.6). This system has the integral

Fig. 14 Periodic beats with
consecutive energy
transmissions from one
motion to a different one

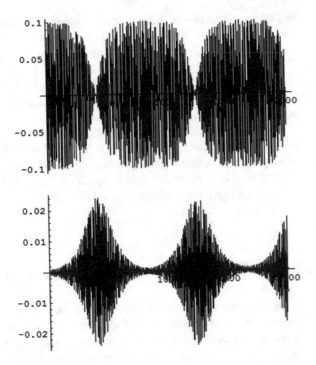

$$4\left(x_{1c}^2(\theta) + x_{1s}^2(\theta)\right) + \eta_1\left(\varphi_{1c}^2(\theta) + \varphi_{1s}^2(\theta)\right) = C. \tag{7.8}$$

Taking into account the above initial conditions, we get $C = 4$, from which the maximum amplitude of oscillation of the pendulum can be estimated as

$$\max_t |\varphi(t)| = \max_t \mu\sqrt{\varphi_{1c}^2(\theta) + \varphi_{1s}^2(\theta)} \leqslant \frac{2\mu}{\sqrt{\eta_1}} = \frac{2a}{l}\sqrt{\frac{m_1 + m_2}{m_2}}, \tag{7.9}$$

where a is the initial amplitude of the suspension point. Calculations show that this estimate is attained if the perturbation of frequency δ is zero. As $|\delta|$ increases, the maximum amplitude of oscillations of the pendulum is decreasing; no oscillations of the pendulum are exited if the instability bounds are overrun.

For illustration, consider the numerical example. Assume that the perturbation of frequency is absent ($\delta = 0$). We take the following parameters: $\omega = 2$, $\nu = 1$, $l = 9.8$. Let $m_1 = m_2$. Then $\eta_1 = 1/2$. Consider the initial conditions: $x(0) = a = 0.1$, $x'(0) = 0$, $\varphi(0) = 0.001$, $\varphi'(0) = 0$. In Fig. 14, for the time interval $0 \leqslant t \leqslant 2000$, we show the results of numerical integration of system (7.2) with the above initial conditions. At the top of the figure we show the variation of the function $x(t)$; the variation of the function $\varphi(t)$ is shown at the bottom.

The motion proceeds as periodic beats with consecutive transmissions of energy of vertical oscillations of the support $x(t)$ into oscillations of the pendulum $\varphi(t)$, and vice versa. At some time instants, vertical oscillations of the support practically vanish and the oscillation amplitude of the pendulum becomes maximal. Since the above system is conservative, we have the integral (7.8). The amplitude of vertical oscillations is majorized by the quantity $a = 0.1$, which is specified by the initial conditions. For the above parameters, from formula (7.9) for evaluation of the pendulum oscillation amplitude we have $\max_t |\varphi(t)| \leqslant 0.028$, which compares favorably with the results of numerical integration.

Beats with transitions of energy also occur in the linear problem with oscillations of two equal pendula connected by a light spring or, in general, in oscillations of weakly coupled systems with equal or close partial frequencies. In the above problem, the coupling is effected in terms of small nonlinear terms, and hence the beats are effected through parametric resonance in which one of the partial frequencies is twice greater than the other one.

8 The Duffing Equation and Strange Attractor

In previous subsections of this chapter we have been concerned with various approximate methods of construction of solutions to nonlinear equations. However, these approaches provide good results only in cases when the nonlinearity is not too large. Otherwise, the methods described above in this chapter are not applicable and the solution can be constructed only via numerical integration. Moreover, the set of solutions to the system of equations behaves unexpectedly: it may happen that several stable solutions can exist, and in addition, it is possible that in the phase space there may appear sets to which the solutions of the system unboundedly approach with increasing time (the so-called attracting sets). Such sets are called *strange attractors*.

In particular, strange attractors can appear under forced oscillations of a mechanical system with one degree of freedom under the action of a periodic perturbing force

$$m\ddot{x} = F(\dot{x}, x, t), \qquad F(\dot{x}, x, T + t) = F(\dot{x}, x, t).$$

Note that strange attractors do not appear in the solution of the equation $\ddot{x} = F(\dot{x}, x)$, but they can appear for nonlinear equations of higher orders, for example, in the equation $\dddot{x} = F(\ddot{x}, \dot{x}, x)$.

Below, we show the results of numerical experiments on construction of stable solutions of the resisted Duffing equation under harmonic excitation. Two types of solutions are shown to exist: first, periodic solutions whose period is a multiple of the excitation period, and second, strange attractors. The dependence of the number of solutions and their type on the excitation level was investigated.

As an example, we consider forced oscillations of a simply supported beam (Fig. 15) subject to a lateral load of intensity $f(x, t)$.[11] The distance between the

---

[11] See: *P. E. Tovstik, T. M. Tovstik*. The Duffing equation and strange attractor // In: Analysis and synthesis of nonlinear mechanical oscillatory systems. St. Petersburg 1998. Vol. 2. Pp. 229–235 [in

Fig. 15 Simply supported
beam with fixed supports

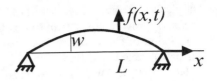

fixed supports is L. For large deflections of the beam, the longitudinal tensile force P
is manifested, and the oscillations are described by the equation

$$EJ\frac{\partial^4 w}{\partial x^4} - P\frac{\partial^2 w}{\partial x^2} + \rho S\frac{\partial^2 w}{\partial t^2} = f(x,t),$$

$$P = \frac{ES}{L}\int_0^L \left(\sqrt{1+(\partial w/\partial x)^2} - 1\right) dx \approx \frac{ES}{2L}\int_0^L (\partial w/\partial x)^2\, dx.$$

Let $f(x,t) = f_0 \sin(\pi x/L)\cos\Omega t$. Then, after separating the variables and scaling,

$$w(x,t) = \alpha x(\tau)\sin\frac{\pi x}{L}, \qquad t = \left(\frac{\pi}{L}\right)^2\sqrt{\frac{\rho S}{EJ}}\,\tau$$

we arrive at the Duffing equation

$$\ddot{x} + c\dot{x} + x + x^3 = b\cos\omega\tau, \quad \Omega = \left(\frac{L}{\pi}\right)^2\sqrt{\frac{\rho S}{EJ}}\,\omega, \quad b = \frac{2L^4}{\pi^4 E\sqrt{JS}}f_0, \quad (8.1)$$

augmented with the additional term $c\dot{x}$, which takes into account the resistance forces
that are proportional to the velocity.

Equation (8.1) contains three parameters: c, b, and ω. We fix two of them ($\omega =
1$, $c = 0.25$), and vary the parameter b in the wide range $0 \leqslant b \leqslant 50$ (here we
do not get unphysical values, because a decrease in the beam thickness results in
a decrease of the momentum of inertia of the transverse section J and an increase
of the parameter b). Putting $\omega = 1$, we consider the perturbation frequency, which
agrees with the frequency of free oscillations of the linear system.

The method of the study of the solution is as follows. Equation (8.1) is numerically
integrated with arbitrary initial conditions $x(0)$, $\dot{x}(0)$. We fix and plot on the phase
plane x, $\dot{x}$ the points

$$x(mT), \ \dot{x}(mT), \quad m = 0, 1, 2, \ldots, \quad T = 2\pi. \quad (8.2)$$

Russian]. On the relation of strange attractors with the classical theory of motion stability, see the
book: *G. A. Leonov.* Strange attractors and classical stability theory. St. Petersburg: St. Petersburg
University Press. 2004. 144 p. [in Russian].

The plane $x, \dot{x}$ with plotted points (8.2) is called the *Poincaré plot*. Poincaré plots (or *sections*) provide a powerful tool capable of ascertaining the qualitative character of variation of the solutions and finding the bifurcations (i.e., transitions from one qualitative state to a different one).

For $c > 0$, a damp is introduced into Eq. (8.1), and as $\tau \to \infty$ the solution $x(\tau)$ tends to some limit solution. Hence one of the three variants of behavior of points (8.2) on the Poincaré plot is realized as $m \to \infty$:

$$(1) \qquad \qquad \{x(mT), \dot{x}(mT)\} \to \{p_1, q_1\},$$
$$(2) \ \{x(mkT + j), \dot{x}(mkT + j)\} \to \{p_j, q_j\}, \quad j = \overline{1, k},$$
$$(3) \qquad \qquad \{x(mT), \dot{x}(mT)\} \in G,$$

where $\{p_j, q_j\}$ are fixed points on the phase plane, G is some set of points on this plane (a *strange attractor*). In case (1), the solution of equation (8.1) with the above initial conditions tends as $\tau \to \infty$ to a periodic solution, whose period T is equal to the period of the perturbation. In case (2), the solution tends to a periodic solution of period kT. However, case (1) can be looked upon as a particular case of (2) with $k = 1$. In case (3), the limit solution is not periodic.

The limit solution depends both on the parameter b and on the initial conditions. For $b \leqslant 50$, 18 intervals of variation of b with different qualitative behavior of solutions were identified. The results are summarized in Table 1, which shows, for the corresponding intervals of variation of b, the quantity n of different limit solutions (which depend on initial conditions); the numbers k which control the periods kT of the limit solutions; strange attractors are indicated in Table 1 by the letter A.

Let us discuss the content of Table 1. For $0 < b \leqslant 2.8$ and for $9.8 \leqslant b \leqslant 22.8$, the solution tends as $\tau \to \infty$ to the same limit solution irrespective of the initial conditions.

For $2.9 \leqslant b \leqslant 9.7$ and for $22.9 \leqslant b \leqslant 35.5$, depending on the initial conditions the solution as $\tau \to \infty$ tends to one of the two stable limit solutions. For example, consider $b = 4$. Two limit solutions are shown in Fig. 16; in Fig. 17 we depict the corresponding domains of initial conditions for $|x(0)| \leqslant 5$, $|\dot{x}(0)| \leqslant 5$.

The influence analysis of initial conditions in the larger domain $|x(0)| \leqslant 100$, $|\dot{x}(0)| \leqslant 100$ shows that the domains corresponding to the first and second limit solutions are interlaced (as in a layer-cake).

Let us continue the discussion of Table 1. For $22.9 \leqslant b \leqslant 39.59$, equation (8.1) has two solutions (corresponding to different initial conditions). First, the period of solutions is T, and then it becomes $2T, 4T, 8T, 16T, \ldots$. Moreover, the intervals of variation of b for which the period is constant become smaller. This process is terminated by the change of two periodic solutions by two strange attractors. In general, such process of period doubling is typical as a strange attractor is approached.

It does not seem possible to describe the entire range of b considered above. We consider only four successive intervals in the range $39.62 \leqslant b \leqslant 48.6$ and consider 6 successive values of b. The results are given in the form of Poincaré plots (Fig. 18), in which points with coordinates $x(mT), \dot{x}(mT)$ for integer $m > 200$ are plotted.

Table 1 Solutions versus the parameter b

b	n	k	b	n	k
0.0 − 2.8	1	1	39.6 − 39.61	3	A, A, 6
2.9 − 9.7	2	1, 1	39.62 − 40.7	2	A, A
9.8 − 22.8	1	1	40.8 − 41.6	1	A
22.9 − 35.5	2	1, 1	41.7 − 44.0	2	A, 3
35.6 − 38.5	2	2, 2	44.3 − 48.6	1	3
38.7 − 39.2	2	4, 4	48.6 − 49.1	2	3, 3
39.3 − 39.35	2	8, 8	49.2	2	6, 6
39.36 − 39.38	2	16, 16	49.3 − 49.4	2	A, A
39.4 − 39.59	2	A, A	49.5 − 50.6	1	A

Fig. 16 Two limit solutions with $b = 4$

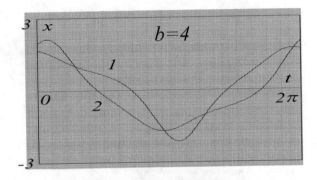

Fig. 17 Domains of initial conditions for $|x(0)| \leqslant 5$, $|\dot{x}(0)| \leqslant 5$

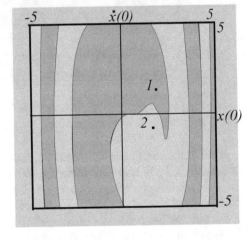

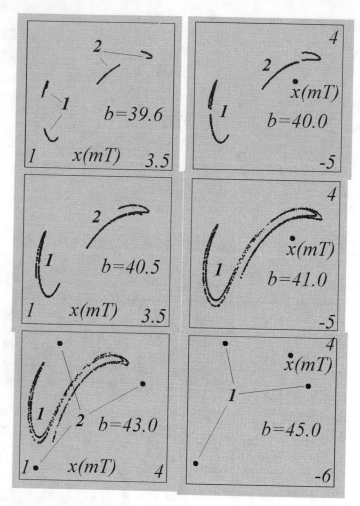

Fig. 18 Strange attractors on Poincaré plots

If the limit solution is kT-periodic, then we have k distinct points on the plot. The number of points in a strange attractor depends on the duration of integration. In Fig. 18, each strange attractor contains 800 points, $201 \leqslant m \leqslant 1000$. The first 200 points are not plotted to exclude the effect of the initial conditions.

Let us consider in succession the plots from Fig. 18. The first plot (for $b = 39.62$) shows two strange attractors *1* and *2*, which can be obtained under some or other initial conditions. Each strange attractor consists of two isolated "pieces". For $b = 40.0$, the pattern is the same, but the pieces approach each other (in comparison with the previous plot). For $b = 40.5$, each attractor now consists of one piece. Next, attractors start to approach each other, and for $b = 41.0$ we already have one attractor *1* for all initial conditions (the solution $x(t)$ for $0 \leqslant t \leqslant 10T$ is shown in Fig. 19).

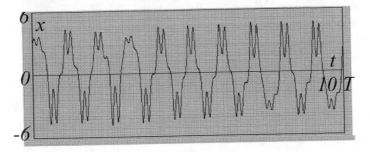

Fig. 19 Chaotic motion with strange attractor

With further increase of b, the strange attractor disappears and there remain only one $3T$-periodic solution 1, which is shown by three points in Fig. 18 for $b = 45.0$.

The motion $x(t)$, as obtained near the strange attractor, is sometimes called the *chaotic motion*. However, the chaos here is not perfect, because some periodicity can be identified (see Fig. 19). When trying to describe a strange attractor by a stationary random process, it turns out that its correlation coefficient $K(\tau)$ assumes values close to 1 with arbitrarily large values of τ.

Chapter 3
Dynamics and Statics of the Stewart Platform

P. E. Tovstik, M. P. Yushkov⊙, S. A. Zegzhda, T. M. Tovstik⊙,
T. P. Tovstik⊙, and V. V. Dodonov⊙

For convenience of the reader, the chapter is subdivided into three parts. In Part I we study the dynamics of a loaded Stewart platform by means of classical methods of theoretical mechanics: the theorem of the center of mass and the law of moment of momentum in relative motion with respect to the mass center. In Part II, the dynamics of a Stewart platform is considered by applying a special form of differential motion equations, which was presented in Sect. 7 of Chap. 8 of Vol. I. In Part III, the statics of a 3-rod Stewart platform is studied using the Lagrange equations of the second kinds.

I) Application of Classical Methods of Rational Mechanics to Studying the Dynamics of a Loaded Stewart Platform[1]

1 Problem Formulation and Kinematics of the Stewart Platform

Formulation of the motion problem of the loaded Stewart platform. In the next forthcoming subsections, as a solution of one applied problem in theoretical mechanics one considers the kinematics and dynamics of the loaded Gaugh–Stewart plat-

---

[1] Part I is an extension of the papers: *G. A. Leonov, S. A. Zegzhda, N. V. Kuznetsov, P. E. Tovstik, T. P. Tovstik, M. P. Yushkov*. Motion of a solid driven by six rods of variable length // Doklady Physics. 2014. Vol. 59. № 3. Pp. 153–157; *G. A. Leonov, S. A. Zegzhda, S. M. Zuev, B. A. Ershov, D. V. Kazunin, D. M. Kostygova, N. V. Kuznetsov, P. E. Tovstik, T. P. Tovstik, M. P. Yushkov*. Dynamics and control of the Stewart platform // Doklady Physics. 2014. Vol. 458. № 1. Pp. 36–41; *G. A. Leonov, P. E. Tovstik, T. M. Tovstik*. Workspaces of the Stewart platform in the 6D space of generalized coordinates // Vestnik St. Petersburg University, Mathematics. 2017. Vol. 50. № 2. Pp. 180–187.

© Springer Nature Switzerland AG 2021
N. N. Polyakhov et al., *Rational and Applied Mechanics*,
Foundations of Engineering Mechanics,
https://doi.org/10.1007/978-3-030-64118-4_3

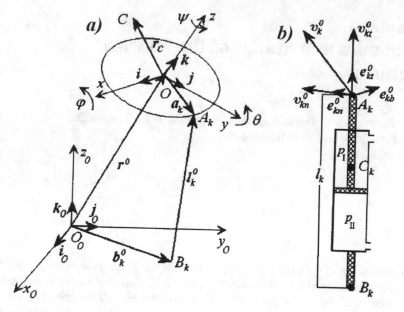

Fig. 1 The Stewart platform (**a**) and the pneumatic cylinder (**b**)

form[2] supported by six hydraulic (or pneumatic) cylinders. One writes down differential motion equations and evaluates the forces required for a prescribed program motion. A feedback is introduced for the stability of this motion. Numerical examples and control delay effects are considered.

A Gaugh–Stewart platform is used in the design of training and drill dynamic benches for pilots and drivers, and also for reliability testing of onboard systems of airplanes and flight vehicles. In addition, Gaugh–Stewart platforms have become extremely widely useful in engineering. The behavior of this involved mechanical system has been extensively studied.

Kinematics of Stewart platforms. We introduce the fixed coordinate system $O_0 x_0 y_0 z_0$ with unit vectors i_0, j_0, k_0 and movable coordinate system $Oxyz$ with vectors i, j, k, rigidly connected with the platform (Fig. 1a). The six rods $B_k A_k$ ($k = \overline{1, 6}$) of variable length simulate the pneumatic cylinders (Fig. 1b) and are attached by means of spherical hinges with the ends B_k to the fixed base and with the ends A_k ($k = \overline{1, 6}$) to the platform. The problem is to achieve a required motion of the platform by varying the rod lengths.

---

[2] First engineering implementation of a platform driven by six rods of variable length was proposed by *V.E. Gaugh* for the performance verification of airplane tyres. Independently, a first article on the theory of motion of such systems was published by *D. Stewart* (A platform with six degrees of freedom // Proc. Inst. Mech. Eng. London. 1965. Vol. 180. № 15. Pp. 371–386).

The orientation of the platform depends on the position of the point O (the pole)

$$\overrightarrow{O_0 O} = r^0(t) = x_0(t)i_0 + y_0(t)j_0 + z_0(t)k_0$$

and on the three successive consecutive angles of rotation around the pole: the yaw (ψ), the pitch θ), and the roll (φ) angles. The rotation tensor $P(\psi, \theta, \varphi)$ reads as

$$P = \begin{pmatrix} p_{11} & p_{12} & p_{13} \\ p_{21} & p_{22} & p_{23} \\ p_{31} & p_{32} & p_{33} \end{pmatrix} =$$

$$= \begin{pmatrix} C_\psi C_\theta & -S_\psi C_\varphi + C_\psi S_\theta S_\varphi & S_\psi S_\varphi + C_\psi S_\theta C_\varphi \\ S_\psi C_\theta & C_\psi C_\varphi + S_\psi S_\theta S_\varphi & -C_\psi S_\varphi + S_\psi S_\theta C_\varphi \\ -S_\theta & C_\theta S_\varphi & C_\theta C_\varphi \end{pmatrix},$$

where for brevity we set $C_\varphi = \cos \varphi$, $S_\theta = \sin \theta$, etc. The Poisson formulas for the derivative with respect to time[3] of the tensor P read as

$$\dot{P} = \omega^0 \times P, \quad \omega^0 = \omega_x^0 i_0 + \omega_y^0 j_0 + \omega_z^0 k_0 = \omega_x i + \omega_y j + \omega_z k,$$
$$\omega_x^0 = \dot{\varphi} \cos \theta \cos \psi - \dot{\theta} \sin \psi, \quad \omega_y^0 = \dot{\varphi} \cos \theta \sin \psi + \dot{\theta} \cos \psi,$$
$$\omega_z^0 = \dot{\psi} - \dot{\varphi} \sin \theta,$$

where ω^0 is the rate of rotation of the platform. We introduce the vector of generalized coordinates that control the position of the platform:

$$q = \{x_0, y_0, z_0, \varphi, \theta, \psi\} = \{q_1, q_2, q_3, q_4, q_5, q_6\}. \tag{1.1}$$

If the quantities (1.1) are given, then the lengths of the pneumatic cylinders (hydraulic cylinders) $l_k = l_k(q_j)$ and their directions e_{kt}^0 can be found from the explicit formulas

$$\overrightarrow{B_k A_k} = l_k^0 = l_k e_{kt}^0 = r^0 + P \cdot a_k - b_k^0, \quad k = \overline{1,6}, \tag{1.2}$$

where the vectors $a_k = \overrightarrow{OA_k}$ and $b_k^0 = \overrightarrow{O_0 B_k}$ are shown in Fig. 1a. On the other hand, if l_k are given, then the quantities q_j are determined from the six nonlinear equations

$$l_k = l_k(q_j), \quad k, j = \overline{1,6}. \tag{1.3}$$

Differentiating (1.2) with respect to time, we get the system of linear equations with respect to $\dot{r}^0$, ω^0. We write this system as

---

[3] See the book: *P.A. Zhilin. Vectors and Tensors of Second Rank in Three-Dimensional Space.* St. Petersburg: Nestor. 2001. 276 p. [in Russian].

$$A \cdot V^0 = i, V^0 = \{\dot{x}_0,\ \dot{y}_0,\ \dot{z}_0,\ \omega_x^0,\ \omega_y^0,\ \omega_z^0\}^T,$$
$$i = \{\dot{l}_1, \dot{l}_2, \dot{l}_3, \dot{l}_4, \dot{l}_5, \dot{l}_6\}^T, \tag{1.4}$$

where the matrix A is composed of the row vectors $A_k = \{e_{kt}^0,\ P \cdot a_k \times e_{kt}^0\}$; the sign ^{T} denotes transposition, and dot denotes differentiation with respect to time.

From the solution of system (1.4), which is considered together with the formulas

$$\dot{\varphi} = \frac{\omega_y^0 \sin\psi + \omega_x^0 \cos\psi}{\cos\theta}, \quad \dot{\theta} = \omega_y^0 \cos\psi - \omega_x^0 \sin\psi, \quad \dot{\psi} = \omega_z^0 + \dot{\varphi}\sin\theta, \tag{1.5}$$

one can represent the derivatives $\dot{q}_i$ as a linear combination of the derivatives $\dot{l}_j$ with coefficients depending on the coordinates q_i. We consider these expressions for $\dot{q}_i$ as a system of differential equations with respect to the functions $q_i(t)$. Integrating this system with respect to the motion, as given in the coordinates l_j, we find the motion in the coordinates q_i.

Vanishing of the determinant of the matrix A indicates that the boundary of the controllability region is attained.

2 Differential Equations of Motion of the Loaded Stewart Platform

Motion equations of the loaded platform. The motion equation of the center of mass C of a loaded platform and the moment equation with respect to the point C read in the moving system of coordinates read as

$$m\left(\ddot{r}^0 + \dot{\omega}^0 \times r_c^0 + \omega^0 \times (\omega^0 \times r_c^0)\right) + mg k_0 = F^0 = \sum_{k=1}^{6} F_k e_{kt}^0,$$
$$J_c \cdot \dot{\omega} + \omega \times (J_c \cdot \omega) = M = \sum_{k=1}^{6} F_k (a_k - r_c) \times e_{kt}, \tag{2.1}$$
$$e_{kt} = P^T \cdot e_{kt}^0, \quad \omega = P^T \cdot \omega^0,$$

where m, and J_c are, respectively, the mass of the loaded platform and its inertia tensor with respect to the point C, g is the acceleration of gravity, $\ddot{r}^0$ is the acceleration of the point O, F_k are the forces acting on the platform from the rods. By F^0 and M we denote the resultant vector and the resultant torque of the forces F_k with respect to the point C. The order of the system equations (2.1) with (1.5) is 12; this system describes the motion of the platform with given forces.

The second equation in system (2.1) is frequently written in the fixed coordinates $O_0 x_0 y_0 z_0$:

$$P \cdot J_c \cdot P^T \dot{\omega}^0 + \omega^0 \times (J_c \cdot P^T \cdot \omega^0) = M^0 =$$
$$= \sum_{k=1}^{6} F_k (a_k^0 - r_c^0) \times e_{kt}^0, \quad r_c^0 = P \cdot r_c.$$

Fig. 2 The Stewart platform
(top view)

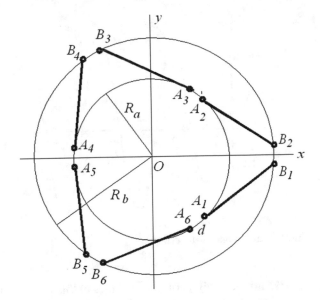

In this case, in particular, if the motion $q(t)$ of the platform is given, then the vectors F^0 and M^0 become known and the forces F_k required for producing this motion can be found from the motion equations, which can be written as

$$A^T \cdot F = \mathcal{F}^0, \quad F = \{F_1, \dots, F_6\}^T, \quad \mathcal{F}^0 = \{F_x^0, F_y^0, F_z^0, M_x^0, M_y^0, M_z^0\}^T,$$

where the matrix A is the same as in (1.4).

Numerical examples. As an example, consider a Stewart platform, in which the hinges A_k and B_k are located symmetrically on the moving and fixed circles which lie on the planes Oxy and $O_0x_0y_0$ and have the radii R_a and R_b, respectively. We assume that the minimal distance between the upper and lower hinges is d and that the angle at the rotation through which we obtain the former arrangement of the hinges is $2\pi/3$ (see Fig. 2). Assume that initially the distance between the planes of the upper and lower hinges is h. We also assume that initially the center of mass C of the platform and the rigid body fastened on it are above the centers of the circles ($z_c > 0$).

Let $R_a = 0.7608$ m, $R_b = 1$ m, $d = 0.2$ m, $h = 1.0196$ m, $z_c = 0.8$ m, $l=1.255$ m, $mg = 10^4$ N. We change to the dimensionless variables in the equations of kinematics and dynamics for this platform. We take R_b for the length unit, and take the gravity force mg of the entire system for the unit force. The dimensionless time is introduced by putting $\tilde{t} = \omega t$ ($\omega^2 = g/R_b$). To differentiate the dimensionless entities from the dimension entities, the former will be given the tilde sign, as this was done for time. In all numerical calculations no account was taken of the inertia force and the weight of the pneumatic cylinders.

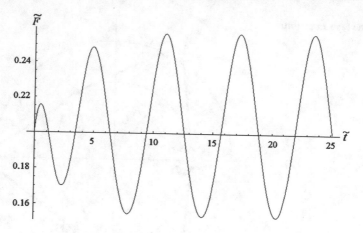

Fig. 3 Loads in pneumatic cylinders

We start the analysis of the dynamics of this platform from calculating the loads in the pneumatic cylinders to produce a prescribed oscillatory vertical motion of the platform according to the law

$$\widetilde{z}_0(\widetilde{t}) = 0.2 \, (1 - e^{-\widetilde{t}/2})^2 \, \sin \widetilde{t}. \qquad (2.2)$$

The vertical motion is given in this way in order that initially both velocity and acceleration of the platform would vanish. Otherwise, at $\widetilde{t} = 0$, a sudden application of the force would be required. Under the above assumptions, the loads created by the pneumatic cylinders for producing motion (2.2) will be equal. The variation of the load $\widetilde{F}(\widetilde{t})$ is shown in Fig. 3.

Let us now discuss the inverse problem in dynamics for vertical motion. This motion is described by one differential equation of second order with respect to $\widetilde{z}_0(\widetilde{t})$. Hence, from the solution of the direct problem one can find analytically the loads $\widetilde{F}(\widetilde{t})$. However, the use of this analytical expression for numerical solution of the inverse problem produces a considerable error on a sufficiently large time interval. Hence we consider a relatively small integration interval, but we add a small perturbation to $\widetilde{F}(\widetilde{t})$ to the load by setting

$$\Delta \widetilde{F}(\widetilde{t}) = 0.0001 \, (1 - e^{-\widetilde{t}/2}) \, \sin 2\widetilde{t}. \qquad (2.3)$$

Integrating, we get the function $\widetilde{z}_0(\widetilde{t})$ (see Fig. 4). The graph shows that on the initial time interval ($0 < \widetilde{t} < 12$) the effect of the above perturbation is insignificant, but later the vertical motion becomes very intense. A stable motion per law (2.2) can be achieved only with the help of the feedback, which will be introduced in Sect. 4.

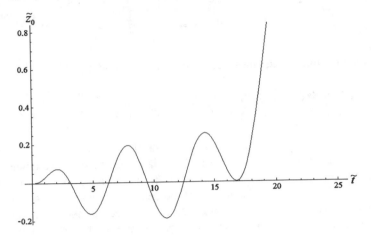

Fig. 4 Vertical motion of the platform

3 The Effect of the Inertia and Weight of the Pneumatic Cylinders

Let us find the corrections that should be introduced into the motion equation (2.1) for taking into account the effect of the forces of inertia and the weight of the pneumatic cylinders. We decompose the velocity $v_k^0 = \dot{r}^0 + \omega^0 \times a_k^0$ of the point A_k as the sum of the direction e_{kt}^0 and the orthogonal direction e_{kn}^0 (see Fig. 1b):

$$v_k^0 = v_{kt}^0 e_{kt}^0 + v_{kn}^0, \quad v_{kt}^0 = v_k^0 \cdot e_{kt}^0 = \dot{l}_k, \quad v_{kn}^0 = v_{kn}^0 e_{kn}^0.$$

Now the angular velocity ω_k^0 of the pneumatic cylinder reads as

$$\omega_k^0 = \frac{v_{kn}^0}{l_k} e_{kb}^0, \quad e_{kb}^0 = e_{kt}^0 \times e_{kn}^0.$$

The vectors e_{kt}^0, e_{kn}^0 and e_{kb}^0 are unit vectors of the moving system of coordinates related to the kth pneumatic cylinder. For the acceleration of the point A_k, we have

$$w_k^0 = \ddot{r}^0 + \dot{\omega}^0 \times a_k^0 + \omega^0 \times (\omega^0 \times a_k^0) = w_{kt}^0 e_{kt}^0 + w_{kn}^0 e_{kn}^0 + w_{kb}^0 e_{kb}^0. \tag{3.1}$$

On the other hand, according to the Coriolis theorem,

$$w_k^0 = \ddot{l}_k e_{kt}^0 + 2\omega_k^0 \times \dot{l}_k e_{kt}^0 + \varepsilon_k^0 \times l_k^0 + \omega_k^0 \times (\omega_k^0 \times l_k^0), \quad l_k^0 = l_k e_{kt}^0, \tag{3.2}$$

where $\varepsilon_k^0 = e_{kn}^0 e_{kn}^0 + e_{kb}^0 e_{kb}^0$ is the sought-for angular acceleration of the pneumatic cylinder. Equating formulas (3.1) and (3.2), we find that

$$w_{kt}^0 = \ddot{l}_k - (\omega_k^0)^2 l_k \,, \quad e_{kn}^0 = -\frac{w_{kb}^0}{l_k} \,, \quad e_{kb}^0 = \frac{w_{kn}^0}{l_k} - \frac{2v_{kn}^0 \dot{l}_k}{l_k^2} \,.$$

We write the motion equations of the pneumatic cylinder and the rod:

$$J_k \varepsilon_k^0 = M_k^0 = l_k^0 \times \widehat{F}_k - S_k g e_{kt}^0 \times k_0 \,, \quad \widehat{F}_k = \widehat{F}_{kn} e_{kn}^0 + \widehat{F}_{kb} e_{kb}^0 \,,$$
$$\widehat{m}(\ddot{l}_k - \omega_k^2 l_{kc}) = \widehat{F}_{kt} - \widehat{m} g e_{kt}^0 \cdot k_0 \,, \quad \widehat{F}_{kt} = P_k - F_{kt}^0 - v \dot{l}_k \,,$$

here $J_k = J_0 + \widehat{J} + \widehat{m} l_{kc}^2$, J_0 is the moment of inertia of the pneumatic cylinder with respect to the point B_k, $\widehat{J}$ is the moment of inertia of the rod with piston with respect to its center of mass C_k (see Fig. 1b), $\widehat{m}$ is the mass of the rod with the piston, $l_{kc} = B_k C_k = l_k - l_0$ is the distance from the center of mass C_k to the fixed point B_k, $l_0 = C_k A_k = \text{const}$, $S_k = S_0 + \widehat{m} l_{kc}$, S_0 is the static moment of the pneumatic cylinder with respect to the point B_k, P_k is the air pressure force on the piston, and v is the coefficient of viscous damping during the motion of the piston.

We can now find the force

$$F_k^0 = F_{kt}^0 e_{kt}^0 + F_{kn}^0 e_{kn}^0 + F_{kb}^0 e_{kb}^0 \,, \tag{3.3}$$

acting on the platform from the pneumatic cylinder, where

$$F_{kt}^0 = P_k - v \dot{l}_k - \widehat{m}(w_{kt}^0 + \omega_k^2 l_0 + g e_{kt}^0 \cdot k_0) \,,$$
$$F_{kn}^0 = -\widehat{F}_{kn} = -\frac{J_k w_{kn}^0}{l_k^2} + \frac{2 J_k v_{kn}^0 \dot{l}_k}{l_k^3} - \frac{S_k g e_{kn}^0 \cdot k_0}{l_k} \,,$$
$$F_{kb}^0 = -\widehat{F}_{kb} = -\frac{J_k w_{kb}^0}{l_k^2} - \frac{S_k g e_{kb}^0 \cdot k_0}{l_k} \,.$$

In the refined variant, system (2.1) can be written as

$$m\left(\ddot{r}^0 + \dot{\omega}^0 \times r_c^0 + \omega^0 \times (\omega^0 \times r_c^0)\right) + m g k_0 = \sum_{k=1}^6 F_k^0 \,,$$

$$J \cdot \dot{\omega} + \omega \times (J \cdot \omega) = \sum_{k=1}^6 a_k \times F_k \,, \quad F_k = P^T \cdot F_k^0 \,, \tag{3.4}$$

where the forces F_k^0 are calculated by formula (3.3).

For $F_{kt} = P_k$, $F_{kn} = F_{kb} = 0$, system (3.4) changes to system (1.1). For system (3.4), the problem of evaluation of forces from a given platform motion is transformed to the problem of evaluation of the quantities P_k.

4 Feedback Construction. Motion Stabilization for a Stewart Platform

Feedback construction. The purpose of the control is to achieve a prescribed motion of the platform. Assume for simplicity that the inertia forces of the pneumatic cylinders are not taken into account. Hence, according to the above, from the given functions $q_i^p(t)$ we find the forces $F_i^p(t)$, which should be applied to the platform from the pneumatic cylinders (the program motion is indicated by the superscript p). However, this method of control is unapplicable, because from Fig. 4 it follows that even a simplest motion of the platform in the form of vertical oscillations is unstable. Hence we propose to introduce the feedbacks and write the applied forces in the form

$$F_k(t) = F_k^p(t) + G(l_k^p(t) - l_k(t)), \quad k = \overline{1,6}, \tag{4.1}$$

where $l_k^p(t)$ are the program lengths of the pneumatic cylinders, as calculated from formulas (1.2), $l_k(t)$ are the lengths measured in the process of motion, G is the feedback coefficient.

For the quantity G we restrict our analysis to linear order. Assume at first that the vector of generalized coordinates $\mathbf{q}$ is 0, that is, we consider the problem of the stabilization of equilibrium of the platform. (Here and in what follows it is assumed that the z_0-coordinate measures the vertical motion of the platform). In this case, in formula (4.1) the functions $F_k^p(t)$ and $l_k^p(t)$ are constants and the system is conservative, because the feedbacks can be considered as springs of stiffness G. The potential energy of the system is as follows:

$$\begin{aligned} \Pi &= -\frac{1}{2} mgz_c(\varphi^2 + \theta^2) + \frac{1}{2} G \sum_{k=1}^{6}(\Delta l_k)^2 = \\ &= -\frac{1}{2} mgz_c(\varphi^2 + \theta^2) + \frac{1}{2} G \, \mathbf{q}^T \cdot \mathbf{C} \cdot \mathbf{q} \, . \end{aligned} \tag{4.2}$$

Here the matrix $\mathbf{C}$ is such that

$$\mathbf{C} = l^2 \mathbf{L} \cdot \mathbf{L}, \quad \mathbf{L} = \begin{pmatrix} \mathbf{L}_1 \\ \dots \\ \mathbf{L}_6 \end{pmatrix}, \quad \mathbf{L}_k = \frac{\mathbf{A}_k}{l}. \tag{4.3}$$

The matrix $\mathbf{L}$ from formula (4.3) differs from the matrix $\mathbf{A}$ from formula (1.4) by the factor l^{-6}, where l is the length of the pneumatic cylinders in equilibrium position. The matrix $\mathbf{C}$ is symmetric and positive definite, and hence the feedback coefficient G can be taken so large that the potential energy (4.2) would be positive definite (which secures the stability of equilibrium of the platform). Calculations show that in the case under consideration this is achieved for $G > 0.5807 mgz_c l^{-2}$.

Stabilization of equilibrium and vertical and horizontal motions. Here a numerical experiment will be employed to show that by introducing feedback (4.1) one can stabilize the vertical and horizontal steady-state oscillations of the platform, which are given in the form

$$\tilde{z}_0^p(\tilde{t}) = 0.2 \sin \tilde{t}, \tag{4.4}$$

$$\tilde{x}_0^p(\tilde{t}) = 0.2 \sin \tilde{t}. \tag{4.5}$$

Above, when the vertical motion instability of the platform was found, we used the exact analytical solution of the direct problem, because the vertical motion was described by one second-order differential equation. Now, when dealing with both horizontal and vertical motions of the platform, we shall follow a unified approach, according to which all six differential motion equations will be used to solve both the direct and inverse problems.

In the case of vertical oscillations (4.4), the dimensionless forces $\tilde{F}^p(\tilde{t})$ in all rods vary according to a periodic law with period 2π and amplitude 0.056 about the mean value 0.212. We perturb these program forces by augmenting them with the forces

$$\Delta \tilde{F}^p(\tilde{t}) = \delta \sin(2\tilde{t} + k\pi/3), \quad k = \overline{1, 6}. \tag{4.6}$$

The stability of motions with these perturbations will be secured by introduction of feedbacks (4.1). Calculations show that, for $\tilde{G} = GR_b/(mg) = 100$ and $\delta = 0.04$, the departures $\tilde{z}_0^p(\tilde{t}) - \tilde{z}_0(\tilde{t})$, $k = \overline{1, 6}$ (here $\tilde{z}_0(\tilde{t})$ is the actual dimensionless motion) are majorized in absolute value by the quantity 10^{-5}, and for $\tilde{G} = 30$ and $\delta = 0.02$ they are majorized by $4 \cdot 10^{-5}$. Note that $\delta = 0.02$ is commensurable with the oscillation amplitude $\tilde{F}^p(\tilde{t})$. It is worth pointing out that if $\tilde{F}^p(\tilde{t})$ is absent in (4.1), then in this case the departures $\tilde{z}_0^p(\tilde{t}) - \tilde{z}_0(\tilde{t})$ are also majorized in absolute value by the quantity $6 \cdot 10^{-3}$ if $\tilde{G} = 100$ and $\delta = 0$.

For horizontal oscillations (4.5), the loads in the pneumatic cylinders 1 and 2, 3 and 6, 5 and 4 are shown in Fig. 5 by the solid, dotted, and dashed lines. Calculations show the following results if the above loads are perturbed by (4.6): for $\tilde{G} = 30$, $\delta = 0.02$ the departures $\tilde{x}_{0k}^p(\tilde{t}) - \tilde{x}_{0k}(\tilde{t})$ are majorized in absolute value by $4 \cdot 10^{-3}$, and for $\tilde{G} = 30$, $\delta = 0.04$, by 0.0075. If $\tilde{G} = 100$, $\delta = 0.02$, then the error is reduced by 10 times. It should be specially pointed out that the quantity $\delta = 0.04$ is commensurable with the amplitude of program loads in the 4th and 5th rods.

5 Linearization of the Platform Motion Equations

Linearization of the platform motion equations. Let us consider small oscillations of the above special platform. As a reference point we take the equilibrium position in which all generalized coordinates are zero and the rod lengths and the loads in them are constant:

$$q = 0, \quad l_k = l_0 = 1.255, \quad F_k = F_0 = 0.205, \quad k = \overline{1, 6}.$$

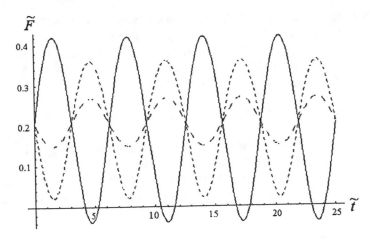

Fig. 5 Loads in the pneumatic cylinders

We introduce the perturbations s_k of the rod lengths (piston displacements) by the formula $l_k = l_0 + s_k$. For small $q = (q_1, \ldots, q_6)^T$ and $s = (s_1, \ldots, s_6)^T$, they are related by the formula

$$s = A_0 \cdot q, \qquad A_0 = \left\{ \frac{\partial l_k}{\partial q_j}\bigg|_{q=0} \right\} = \lim_{q \to 0} A, \qquad (5.1)$$

where the matrix A is the same as in formula (1.4). For the bench with the above sizes, the matrix A_0 is as follows:

$$A_0 = \begin{pmatrix} -0.424 & -0.401 & 0.812 & -0.490 & -0.377 & -0.442 \\ -0.424 & 0.401 & 0.812 & 0.490 & -0.377 & 0.442 \\ 0.559 & -0.166 & 0.812 & 0.571 & -0.236 & -0.442 \\ -0.136 & -0.568 & 0.812 & 0.081 & 0.612 & 0.442 \\ -0.136 & 0.568 & 0.812 & -0.081 & 0.612 & -0.442 \\ 0.559 & 0.166 & 0.812 & -0.571 & -0.236 & 0.442 \end{pmatrix}.$$

For small q, we write down the platform motion equations with separately given linear terms:

$$B \cdot \ddot{q} - C \cdot q + f + f^{\mathrm{err}} = \mathcal{F}^0 = A^T \cdot F. \qquad (5.2)$$

Here, the nonzero elements of the constant matrices B, C and of the vector f are as follows:

$$b_{11} = b_{22} = b_{33} = 1, \quad b_{44} = b_{55} = b_{66} = \rho^2 + z_c^2,$$
$$b_{15} = b_{51} = \rho^2 z_c, \qquad b_{24} = b_{42} = -\rho^2 z_c,$$
$$c_{44} = c_{55} = z_c, \qquad f_3 = 1.$$

In Eq. (5.2), the vector f^{err} includes the nonlinear terms of system (2.1).

Let a program motion $q^p(t)$ be given. By formula (5.1), we find $s^p = A_0 \cdot q^p$. Strictly speaking, the quantity s^p should be determined as above. The resulting departure can be included into f^{err}. The program load $F = F^p$ is found from system (5.2) with $f^{\text{err}} = 0$, $q = q^p$. Now we can write

$$B \cdot \ddot{q}^p - C \cdot q^p + f = A_0^T \cdot F^p . \tag{5.3}$$

Below it will be shown that the resulting motion depends weakly on the accuracy of F^p. Moreover, under certain control constraints it can be assumed that $F^p = 0$.

In order to stabilize the motion, we augment the forces F^p with the control forces F^c by putting $F = F^p + F^c$ in (5.2). Consider the differences of the actual and program motions:

$$\delta q = q - q^p , \qquad \delta s = s - s^p , \qquad \delta s = A_0 \cdot \delta q .$$

We introduce the control force F^c by

$$F^c = -G \delta s - G_f \delta \dot{s} , \tag{5.4}$$

where the constants $G \geqslant 0$ and $G_f \geqslant 0$ should be chosen from the condition of minimality of the vector δs. Subtracting Eqs. (5.2) and (5.3) we find that

$$B \cdot (A_0)^{-1} \cdot \delta \ddot{s} - C \cdot (A_0)^{-1} \cdot \delta s + A_0^T \cdot (G \delta s + G_f \delta \dot{s}) + f^{\text{err}} = 0 . \tag{5.5}$$

Investigation of Eq. (5.5). We rewrite Eq. (5.5) as

$$\delta \ddot{s} - C_* \cdot \delta s + A_* \cdot (G \delta s + G_f \delta \dot{s}) + A_{\text{err}} \cdot f^{\text{err}} = 0 , \tag{5.6}$$

where

$$C_* = A_0 \cdot B^{-1} \cdot C \cdot (A_0)^{-1} , \qquad A_* = A_0 \cdot B^{-1} \cdot A_0^T , \qquad A_{\text{err}} = A_0 \cdot B^{-1} .$$

For $f^{\text{err}} = 0$, the zero solution of equation (5.6) is asymptotically stable if all the roots λ of the characteristic equation

$$\det \left(\lambda^2 E - C_* + A_*(G + \lambda G_f) \right) = 0 \tag{5.7}$$

have negative real parts.

The matrix C_* is symmetric and positive, the matrix A_* is symmetric and positive definite. It follows that under the absence of control ($G = G_f = 0$) the zero solution is unstable, for sufficiently large G it is stable, and for $G_f > 0$ it is asymptotically stable. Let us give numerical results for a platform with the above parameters.

For $G = G_f = 0$, Eq. (5.7) has the zero root (of multiplicity 8) and two pairs of roots $\lambda = \pm 0.887$, indicating instability. Stability occurs for $G > 0.706$, and for $G_f = 0$ self-sustained oscillations occur.

Let $G \gg 1$, $G_f \sim 1$. Then then terms

$$GA_0^T \cdot \delta s + f^{err} = 0.$$

are principal in Eq. (5.5). It follows that in the absence of resonances, we have the estimate

$$\|\delta s\| \sim \frac{\|f^{err}\|}{G}, \tag{5.8}$$

where the norm of a vector is defined by

$$\|s\| = \max_{t,k} |s_k(t)|. \tag{5.9}$$

The effect of control delay. Introducing the constant delay τ in the control, we write Eq. (5.6) as

$$\delta\ddot{s}(t) - C_* \cdot \delta s(t) + A_* \cdot \left(G\delta s(t - \tau) + G_f \delta\dot{s}(t - \tau)\right) + A_{err} \cdot f^{err} = 0.$$

The delay is naturally related to the time required for the pneumatic cylinders to develop the control pressure corresponding to expression (5.4).

Assuming that the delay is small, we introduce the expansion into Eq. (5.9)

$$\delta\ddot{s}(t) - C_* \cdot \delta s(t) + A_* \cdot \left(G\delta s(t - \tau) + G_f \delta\dot{s}(t - \tau)\right) + A_{err} \cdot f^{err} = 0.$$

Then the characteristic equation equation for (5.9) assumes the form

$$\det\left(\lambda^2(E - kA_*) + \lambda A_*(G_f - \tau G) + GA_* - C\right) = 0,$$

where $k = \tau G_f - \tau^2 G/2$. For the asymptotic stability it suffices that the following three conditions be satisfied: the above condition $G > 0.706$, the inequality $G_f > \tau G$, and the positive definiteness of the matrix $E - kA_*$. The last condition is satisfied if

$$k < \frac{1}{\rho_{max}},$$

where ρ_{max} is the largest eigenvalue of the matrix A_*. For the parameters under consideration, we have $\rho_{max} = 3.956$.

For the coefficient G_f, we have the two-side estimate

$$\tau G < G_f < \frac{\tau G}{2} + \frac{1}{\tau \rho_{max}}. \tag{5.10}$$

The left-hand side of (5.10) is smaller than the right-hand side if

$$G < \frac{2}{\tau^2 \rho_{max}}.$$

The last inequality imposes constraints on the coefficient G under the presence of the delay.

Some numerical results. Consider the dynamics of a symmetric controlled platform with the above parameters. The delay effects are not taken into account. Let us model the composite motion of a body which undergoes one translational and two rotational motions. As a program motion, we consider

$$lq_1^p(t) = a \sin \nu t, \quad q_5^p(t) = a \sin \nu t, \quad q^p(t) = a \sin \nu t, \tag{5.11}$$
$$q_2^p(t) = q_3^p(t) = q_4^p(t) = 0, \tag{5.12}$$

where a is the given amplitude of linear and angular oscillations (for simplicity, we assume that the amplitudes of the linear and angular oscillations are equal in the dimensionless variables).

We integrate the system of Eq. (5.2) and calculate the vector program motions of the pistons s^p from the exact formulas (1.2), and evaluating the program loads in the rods F^p from the approximate system (5.3), in which we retain only the linear terms.

The control quality will be estimated from the relative error η, which reads as

$$\eta = \frac{||\delta q||}{||q^p||}, \quad \delta q = q - q^p, \quad ||q^p|| = a,$$

where the norm of a vector is defined according to (5.9).

Let us study the quality of the control depending on the feedback parameters G, G_f, on the amplitude of oscillations of the program motion a, and on the frequency of oscillations $\nu \leqslant 5$ (in the dimensional time, the frequency is at most 2.5 Hz). We first assume that the initial conditions for the actual and program motions are the same.

In Table 1, for two values of the amplitude a and five values of the oscillation frequency ν, we show the values of the control quality function $\eta(\nu)$ depending on the feedback parameters G and G_f. The following conclusions can be made from Table 1.

In accordance with the approximate formula (5.8), the function $\eta(\nu)$ decreases with increasing G approximately as $1/G$. For $G \sim 3$, the adopted control method is inapplicable (a more precise definition of the forces F^p is required).

The function $\eta(\nu)$ increases substantially with the frequency ν.

For relatively small values of $\eta(\nu)$ (say, $\eta(\nu) < 0.5$), comparison of rows 1–4 and 5–8 shows that this function depends weakly on the amplitude a. This dependence is manifested only in the domain in which the above control method is inapplicable.

Table 1 Dependence $\eta(\nu)$

No_	a	G	G_f	$\nu = 1$	2	3	4	5
1	0.1	100	5	0.010	0.016	0.027	0.042	0.061
2	0.1	30	2	0.034	0.055	0.094	0.152	0.231
3	0.1	10	1	0.107	0.183	0.331	0.556	0.911
4	0.1	3	0.3	0.476	1.043	2.660	2.377	1.811
5	0.2	100	5	0.010	0.016	0.027	0.042	0.062
6	0.2	30	2	0.033	0.054	0.094	0.155	0.239
7	0.2	10	1	0.107	0.183	0.341	0.581	0.962
8	0.2	3	0.3	0.482	1.094	3.074	2.417	1.865
9	0.2	30	1	0.033	0.055	0.096	0.163	0.266
10	0.2	30	5	0.033	0.052	0.084	0.126	0.174

The dependence on the parameter G_f can be obtained by comparing row 6 with rows 9 and 10. The function $\eta(\nu)$ is decreasing with increasing G_f (the decrease rate is more substantial with large frequencies ν).

The data in Table 1 were obtained under the assumption that the initial conditions for the actual and the program motions are equal; that is,

$$q_k = 0, \quad k = \overline{1,6}, \quad \dot{q}_1 = \dot{q}_5 = \dot{q}_6 = \nu a, \quad \dot{q}_2 = \dot{q}_3 = \dot{q}_4 = 0 \quad \text{for} \quad t = 0,$$

moreover, the program loads in the rods are evaluated by the above method. However, velocity-inhomogeneous conditions are hardly implementable in practice. Hence, for the actual motion we take the homogeneous conditions

$$q_k = \dot{q}_k = 0, \quad k = \overline{1,6}, \quad \text{for} \quad t = 0. \tag{5.13}$$

The solution thus obtained consists of a transient process for $0 \leqslant t \leqslant t_*$, which eventually tends to a periodic solution with frequency ν. Hence, when evaluating the norm of this solution by formula (5.9) one should assume that $t > t_*$. It is quite natural that the results thus obtained coincide with those given in Table 1. One can speak only about the duration of the transient process, which by the above calculations is shown to be smaller than the halved period (see also Fig. 6).

One approach is to give more simple program forces by setting $F_k^p = 0, k = \overline{1,6}$. In Table 2, we show the results obtained by integration. It is seen that at least for $G = 100$ and $G = 30$ the results are pretty close to those given in Table 1.

To illustrate the closeness of the real and program motions with unequal initial conditions and inaccuracy in specifying the program loads in the rods, we show in Fig. 6 the graphs of the functions $q_1(t)$ and $q_1^p(t)$. We take the parameters $a = 0.2$, $G = 30$, $G_f = 2, \nu = 2$, and consider the case when $F_k^p = 0$ and the functions $q_k(t)$ satisfy the zero initial conditions (5.13).

Table 2 Dependence $\eta(\nu)$ with zero program forces

No_	a	G	G_f	$\nu = 1$	2	3	4	5
5*	0.2	100	5	0.015	0.019	0.029	0.045	0.065
6*	0.2	30	2	0.049	0.065	0.105	0.167	0.252
7*	0.2	10	1	0.153	0.225	0.391	0.643	1.028
8*	0.2	3	0.3	0.781	1.6329	3.316	2.500	1.892

Fig. 6 The actual and the program motion

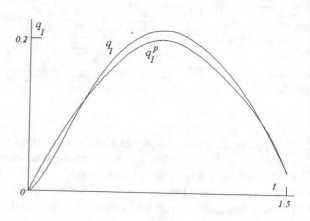

So, due to the instability of the solution to the dynamical inverse problem, feedbacks with respect to translations and rods velocities should be introduced for the realization of the program motion. There are two possibilities to obtain a fairly exact program motion of a Stewart platform: either to precisely define the loads in the rods or to substantially increase the feedback coefficient G from the rods travel. Under relaxed accuracy requirements, consistent results can be obtained for approximately given expression for the program forces in the rods or even if they are equated to zero. However, in the last case the coefficient G should be increased.

6 Workspaces of A Stewart Platform in the Six-Dimensional Space of Generalized Coordinates

Let us now consider the construction of workspaces (the effective domains) in the six-dimensional space of generalized coordinates, which control the position of the platform. In addition, for visualization of the results, we find the projections of these domains onto subspaces of smaller dimension.

1. Workspace. We set, as before,

$$l_k = l_0 + s_k, \quad k = \overline{1, 6},$$

where s_k are variations of the rod lengths with respect to the initial position l_0.

We rewrite (1.3) in the form

$$s = s(q). \tag{6.1}$$

In order to find $q = (q_k, \; k = \overline{1, 6})$ with given $s = (s_k, \; k = \overline{1, 6})$, we solve Eq. (6.1) by Newton's iteration method

$$q^{n+1} = q^n + B(q^n)(s - s^n),$$

$$s = (s_k, \; k = \overline{1, 6}), \quad s^n = (l_k(q^n) - l_0, \; k = \overline{1, 6}),$$

where $B(q) = (A(q))^{-1}$ and $A(q) = \partial s / \partial q$ is the matrix from (1.4). As the zero approximation, we take the initial position of the platform: $q^0 = s^0 = 0$. Calculations show that $q^n \to q$ if $\det(A) \neq 0$.

As an example, we again take the platform with sizes $R_b = 1$ (which is taken for the length unit), $R_a = 0.7608$, $h = 1.0189$, $l_0 = 1.255$, $d = 0.2$. In this case, in the initial position the matrices $A_0 = A(0)$ and $B_0 = B(0) = A_0^{-1}$ are as follows (recall that the matrix A_0 was calculated in §5):

$$A_0 = \begin{pmatrix} -0.424 & -0.401 & 0.812 & -0.490 & -0.377 & -0.442 \\ -0.424 & 0.401 & 0.812 & 0.490 & -0.377 & 0.442 \\ 0.559 & -0.166 & 0.812 & 0.571 & -0.236 & -0.442 \\ -0.136 & -0.568 & 0.812 & 0.081 & 0.612 & 0.442 \\ -0.136 & 0.568 & 0.812 & -0.081 & 0.612 & -0.442 \\ 0.559 & 0.166 & 0.812 & -0.571 & -0.236 & 0.442 \end{pmatrix},$$

$$B_0 = \begin{pmatrix} -0.455 & -0.455 & 0.531 & -0.075 & -0.075 & 0.531 \\ -0.350 & 0.350 & -0.219 & -0.569 & 0.569 & 0.219 \\ 0.205 & 0.205 & 0.205 & 0.205 & 0.205 & 0.205 \\ -0.394 & 0.394 & 0.520 & 0.126 & -0.126 & -0.520 \\ -0.373 & -0.373 & -0.155 & 0.528 & 0.528 & -0.155 \\ -0.377 & 0.377 & -0.377 & 0.377 & -0.377 & 0.377 \end{pmatrix}.$$

In the Euclidean space $\mathbb{R}_6$, consider the cube S_δ with boundary boundary Γ_s:

$$S_\delta: \; |s_k| \leqslant \delta, \; k = \overline{1, 6};$$

we assume (by design considerations) that the rod lengths l_k can vary only in the range $l_0 - \delta \leqslant l_k \leqslant l_0 + \delta$, $k = \overline{1, 6}$. Our purpose here is to describe the domains of possible values of the coordinates q_k (the workspace) $Q_\delta: \; (q_k, \; k = \overline{1, 6})$ with boundary Γ_q in the space of generalized coordinates assuming that the rod lengths vary in this range.

A general approach to the construction of the boundary Γ_q is as follows. We take an arbitrary parametrically given curve

$$q_k = q_k(\tau), \quad \tau \geqslant 0, \quad q_k(0) = 0, \quad k = \overline{1,6}, \tag{6.2}$$

and seek a first point $\tau = \tau_*$ for which

$$s(\tau_*) = s(q(\tau_*)) \in \Gamma_s \quad \text{or} \quad \det(A(q(\tau_*))) = 0. \tag{6.3}$$

Hence $q(\tau_*) \in \Gamma_q$.

Below, under particular assumptions we shall also discuss more general approaches towards construction of the boundary Γ_q.

2. Linear approximation. If the maximal variation δ of the rod lengths is small in comparison with the radius R_b (namely, $\delta \ll 1$), then Eq. (6.1) gives approximate linear relations between the vectors s and q

$$s = A_0 \cdot q, \quad q = B_0 \cdot s,$$

from which one can obtain a number of approximate estimates for the ranges of variation of the generalized coordinates q_k.

2.1. The estimates of the projections of the set Q_δ onto the axis q_k (which are direct consequences of the relations $q = B_0 \cdot s$) read as

$$-q_k^g \delta \leqslant q_k \leqslant q_k^g \delta, \quad q_k^g = \sum_{j=1}^{6} |b_{kj}|, \quad k = \overline{1,6}, \quad B_0 = \{b_{kj}\}. \tag{6.4}$$

2.2. The intersections of the set Q_δ and the axis q_k are estimated as

$$-q_k^r \delta \leqslant q_k \leqslant q_k^r \delta \quad \text{for} \quad q_i = 0, \; i \neq k, \quad k = \overline{1,6}, \tag{6.5}$$

where the coefficients read as

$$q_k^r = \min_j \{1/|a_{jk}|\}, \quad A_0 = \{a_{jk}\}.$$

2.3. The boundary of the two-dimensional projection of the space Q_δ onto any plane $q_k q_n$ can be written in the polar coordinates r, α as follows:

$$q_k = r(\alpha)\delta \cos\alpha, \quad q_n = r(\alpha)\delta \sin\alpha,$$

$$r(\alpha) = \sum_{j=1}^{6} |b_{kj} \cos\alpha + b_{nj} \sin\alpha|, \quad 0 \leqslant \alpha \leqslant 2\pi.$$

2.4. The construction of the intersection of the space Q_δ plane $q_k q_n$ (with $q_i = 0$ for $i \neq k$, $i \neq n$) is more involved. We first solve the auxiliary system of linear equations

Table 3 One-dimensional limits in linear approximation

	$q_1 = x_0$	$q_2 = y_0$	$q_3 = z_0$	$q_4 = \varphi$	$q_5 = \theta$	$q_6 = \psi$
q_k^g	2.125	2.279	1.231	2.079	2.110	2.266
q_k^r	1.789	1.764	1.231	1.751	1.631	2.266

$$\sum_{j=1}^{6} b_{kj}c_j = \hat{q}_k, \quad \sum_{j=1}^{6} b_{nj}c_j = \hat{q}_n, \quad \sum_{j=1}^{6} b_{ij}c_j = 0, \quad i \neq k, \ i \neq n,$$

with the unknowns $\xi_m = (\hat{q}_k, \hat{q}_n, c_i \ (i \neq k, \ i \neq n))$. In this system, c_k and c_n are assumed to be given. The solution reads as $\xi_m = a_{m1}c_k + a_{m2}c_n$. Next, we find the domain on the plane $c_k c_n$ in which $|c_i| \leqslant 1$ for all $i = \overline{1,6}$. Finally, from the formulas $q_k = a_{11}c_k + a_{12}c_n$, $q_n = a_{21}c_k + a_{22}c_n$ we find the domain on the plane $q_k q_n$.

3. Numerical examples. We consider the same platform sizes as before.

3.1. We first discuss the one-dimensional ranges of the generalized coordinates, as given in the linear approximation by formulas (6.4) and (6.5). The coefficients q_k^g and q_k^r are shown in Table 3. In all three cases, we have $q_k^g \geqslant q_k^r$, because the quantities q_k^r are obtained under constraints. However $q_3^g = q_3^r$ and $q_6^g = q_6^r$, because in these cases, the constraints are satisfied automatically due to the symmetry about translation and rotation about the z-axis.

3.2. In Table 4 for $\delta = 0.2$, the linear approximation is compared with the exact boundaries of the projections and intersections. The linear boundaries $q_k^g \delta$ and $q_k^r \delta$ are obtained by formulas (6.4) and (6.5); they are equal (in absolute value) for the upper and lower limits of variation of the generalized coordinates.

To obtain the exact boundaries $q_k^{gm} \leqslant q_k \leqslant q_k^{gp}$ of the projections, we employ the method of iteration to solve the system of 64 equations (6.1) $s = s(q)$ for $s \in S_\delta^0$, $S_\delta^0 = \{s_j = \pm\delta, \ j = \overline{1,6}\}$. Consequently,

$$q_k^{gm} = \min_{s \in S_\delta^0} q_k(s), \quad q_k^{gp} = \max_{s \in S_\delta^0} q_k(s), \quad k = \overline{1,6}.$$

To construct the boundaries of intersection of the space Q_δ and the line q_k we evaluate s_j by formulas (6.1) by varying q_k with $q_i = 0$, $i \neq k$, until $\max_{q_k} |s_j|$ becomes equal to δ. As a result, we find q_k^{rm} and q_k^{rp}. For the coordinates x_0, z_0, θ, the lower and upper exact values of the boundaries are different in sign.

3.3. For the adopted values of the parameters, the quantity δ is bounded by the inequality $\delta < 0.295$, because $\min_{s \in S_\delta} |\det(A)| = 0$ for $\delta = 0.295$.

3.4. The two-dimensional projection of the domain G_δ onto the plane xy in the linear approximation (see paragraph 2.4) is shown as curve *1* in Fig. 7b. Here, for definiteness, we take $\delta = 1$, because in the linear approximation the form of the domain is independent of δ. The boundary of intersection with the plane xy (with $z_0 = \varphi = \theta = \psi = 0$) is shown as curve *2* (in the last case, the platform moves translationally in the plane $z = 0$). Figure 7a depicts the boundary of the domain on the

Table 4 Comparison of the linear approximation and exact values for $\delta = 0.2$

	$q_1 = x_0$	$q_2 = y_0$	$q_3 = z_0$
$|q_k| \leqslant q_k^g \delta$	0.425	0.456	0.246
$q_k^{gm} \leqslant q_k \leqslant q_k^{gp}$	$-0.421/0.431$	± 0.452	$-0.260/0.246$
$|q_k| \leqslant q_k^r \delta$	0.357	0.353	0.246
$q_k^{rm} \leqslant q_k \leqslant q_k^{rp}$	$-0.377/0.318$	± 0.313	$-0.260/0.238$
	$q_4 = \varphi$	$q_5 = \theta$	$q_6 = \psi$
$|q_k| \leqslant q_k^g \delta$	0.416	0.422	0.453
$q_k^{gm} \leqslant q_k \leqslant q_k^{gp}$	± 0.416	$-0.426/0.468$	± 0.470
$|q_k| \leqslant q_k^r \delta$	0.416	0.326	0.453
$q_k^{rm} \leqslant q_k \leqslant q_k^{rp}$	± 0.335	$-0.328/0.340$	± 0.420

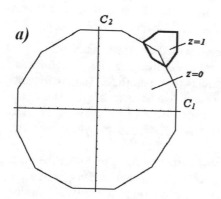 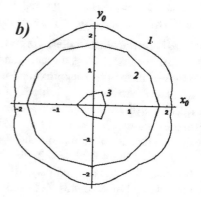

Fig. 7 The auxiliary domains on the plane $c_1 c_2$ (**a**),the boundaries of the projection (*1*) and of the intersections (*2, 3*) in the linear approximation (**b**)

auxiliary plane $c_1 c_2$, which is used in the construction of curve *2* (see paragraph 2.4). In addition, Fig. 7a, b show curves *3*, which correspond to the intersection of the domain G_δ with xy for $z_0 = 1$, $\varphi = \theta = \psi = 0$.

3.5. Let us now consider the construction of the exact curves *1e* and *2e*, which correspond for $\delta = 0.2$ to the approximate curves *1a* and *2a*, and which are shown in Fig. 7 both for the projection onto the plane xy (*1e*) and for the intersection with this plane (*2e*). In the last case, we shall use the polar coordinates and employ formulas (6.2), (6.3) with

$$x_0 = r \cos \alpha, \quad y_0 = r \sin \alpha, \quad z_0 = \varphi = \theta = \psi = 0. \tag{6.6}$$

Using formula (6.1), we find s for the values q given by formulas (6.2). Making the parameter r to be away from zero and having α fixed, we find the first value of $r(\alpha)$ for which $\max_k |s_k| = \delta$. The exact (*2e*) and the approximate (*2a*) boundaries of the intersection of the domain G_δ with the plane xy are shown in Fig. 8b. It is seen that

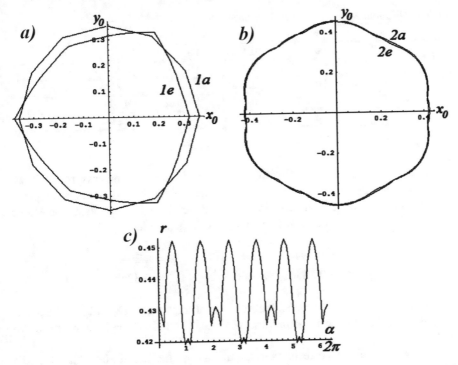

Fig. 8 The approximate (*1a*, *2a*) and versus the exact (*1e*, *2e*) boundaries of the projections (**a**) and the intersections (**b**). The dependence $r(\alpha)$ (**c**)

they are practically equal. The graph of the function $r(\alpha)$ is shown in an expanded scale in Fig. 8c, which shows that the curve $r(\alpha)$ has corners and is not convex.

The construction of the projection of the domain Q_δ onto the plane xy is more laborious. Here it is insufficient to evaluate the solutions x_0 and y_0 of system (4.2) on the set S_δ^0. Instead, one should solve system (4.2) for $s \in S_\delta$ (that is, on the entire cube). The Monte Carlo method shall be used to go through the points of the cube. To enhance the probability of hitting the cube faces, we introduce the random numbers ξ_k (uniformly distributed on the interval $[-N\delta, N\delta]$, $N > 1$) and define

$$s_k = \begin{cases} -\delta, & \xi_k < -\delta, \\ \xi_k, & |\xi_k| \leqslant \delta, \\ \delta, & \xi_k > \delta, \end{cases} \qquad k = \overline{1,6}. \tag{6.7}$$

We took 10^7 tuples of points (6.7) and considered the above domain with boundary *1e* on the plane $x_0 y_0$, as shown in Fig. 8a. It is seen that curves *1e* and *1a* differ considerably. However, the quantity $\delta = 0.2$ should be considered relatively large, because it is 2/3 of the maximally possible value $\delta = 0.295$.

Table 5 Three-dimensional sections of maximal volume

	x_0^*	y_0^*	z_0	φ^*	θ^*	ψ^*
1	0.786δ	0.832δ	0.410δ	0	0	0
2	0.786δ	0.832δ	0	0	0	0.754δ
3	0.786δ	0	0.410δ	0	0.833δ	0
4	0	0.832δ	0.410δ	0.680δ	0	0
5	0	0	0	0.680δ	0.833δ	0.754δ

3.6 The workspace $G_\delta \in R_6$ is fairly involved.[4] Under various assumptions, we give in the linear approximation the solution to the problem of finding a parallelepiped of maximal volume V. In the six-dimensional space $V = \Pi_{k=1}^6 q_k^*$, using the Monte Carlo method, we find the halved rod lengths of this parallelepiped

$$x_0^* = 0.391\delta, \quad y_0^* = 0.417\delta, \quad z_0^* = 0.204\delta,$$
$$\varphi^* = 0.388\delta, \quad \theta^* = 0.440\delta, \quad \psi^* = 0.384\delta.$$

Let us consider the three-dimensional sections of the domain G_δ and find the parallelepipeds of maximal volume in these sections. The halved rod lengths are shown in Table 5.

The rows of Table 5 correspond to the translatory motion of platform (1), to the plane motion of the planes xy (2), xz (3), yz (4), and to the rotational motion around the point O (5).

Thus, we have obtained solvability condition for the system of equations describing the motion of the Stewart platform. In the six-dimensional space of generalized coordinates, the workspace domain of the platform is studied for a given range of rod lengths. This domain is not convex and contains edges. The one- and two-dimensional projections and sections of the workspace domain are considered. The effective domains in the form of parallelepipeds lying in the workspace domain are constructed.

In conclusion of the first part of the chapter note the following. Along with the problems considered above, we could also enumerate a number of supplementary interesting problems arising while studying the dynamics of a loaded Stewart platform. For example, we could turn our attention to the work on the creation of the mathematical apparatus for effective functioning of dynamical simulators which are designed for the training of professional skills of the operators who direct the line-of-sight to the object in the presence of unpredictable displacement of the base. Disturbance is done by the instructor who, by choosing the motions of the base (the Stewart platform), tries to obstruct the work of the operator who is on the platform. Simultaneous actions of the operator and instructor are considered as a zero-sum

---

[4] This is why *F.A. Adkins, E.J. Haug* (Operational envelope of a spatial Stewart platform // Trans. ASME. J. Mech. Des. 1997. Vol. 31. № 368. Pp. 330–332) in solution of control problem used an effective domain $P_\delta \in G_\delta$ in the form of the parallelepiped $|q_k| \leqslant q_k^*$.

game in which each of the participants tries to achieve the best result by choosing his strategy.[5] One seeks the minimax control that provides stabilization of the line-of-sight directed to the object that is fixed with respect to the Earth.[6]

II) Application of a Special Form of Differential Equations to Studying the Motion of a Loaded Stewart Platform[7]

7 Problem Formulation and Coordinate Systems

Problem formulation. In[8] the first part of this chapter, the behavior of a loaded Stewart platform was studied using the classical approach based on the center-of-mass theorem and the theorem on the variation of the kinetic moment of the system in its relative motion about the center of mass. Here, for the same mechanical problem, will shall apply a special form of motion equations outlined in Sect. 7 of Chap. "Dynamics of the Rigid Solid" of Vol. I ("Motion equation of a system of rigid bodies in redundant coordinates").

---

[5] See, for example, the textbook by *L.A. Petrosyan, N.A. Zenkevich, E.V. Shevkoplyas*. Theory of games. BHV-Peterburg. 2012. 432 p. [in Russian].

[6] The main results of investigations are reflected in the following works: *V.V. Aleksandrov, S.S. Lemak, N.A. Parusnikov*. Lectures on the mechanics of controlled systems. Moscow: MAK-SPress. 2012. 240 p. [in Russian]; *V.V. Latonov, V.V. Tikhomirov*. Line-of-sight guidance control using video images // Vestnik Moskovskogo Universiteta. Seriya 1. Matematika. Mekhanika. 2018. № 1. Pp. 53–59 [in Russian] [Moscow Univ. Mech. Bulletin. Vol. 72. № 2. 2017. Pp. 18–26]; *D.S. Burlakov, V.V. Latonov, V.A. Chertopolokhov*. Identification of parameters of a model of a movable motion platform // Fundam. Prikl. Mat. 2018. Vol. 22. Issue 2. Pp. 57–73 [in Russian]; *V.V. Latonov*. Programmed strategies to test the quality of line-of-sight guidance control using video images // Vestnik Moskovskogo Universiteta. Seriya 1. Matematika. Mekhanika. 2018. № 6. Pp. 51–56 [in Russian] [Moscow Univ. Mech. Bulletin. Vol. 72. № 2. 2017. Pp. 135–140]; *V.V. Latonov*. Minimax optimization for a system of line-of-sight stabilization // Vestnik Moskov. Univ. Ser. 1. Mat. Mekh. 2019. № 6. Pp. 64–68 [in Russian] [Moscow Univ. Mech. Bulletin. Vol. 74. Pp. 159–163(2019)]; *V.V. Aleksandrov, V.V. Latonov*. Minimax stabilization of a line-of-sight of inertia object on a movable base // XII All-Russian Congress on fundamental prroblems in theoretical and applied mechanics. Abstracts of reports. Ufa: RITZ BashGU. 2019. P.17 [in Russian].

[7] The second part "Application of a Special Form of Differential Equations to Studying the Motion of a Loaded Stewart Platform" of this chapter was written in collaboration with V.I. Petrova.

[8] The second part of this chapter is based on the papers: *N.N. Polyakhov, S.A. Zegzhda, M.P. Yushkov*. Special form of dynamics equations of a system of rigid bodies // Dokl. AN SSSR. 1989. Vol. 309. № 4. Pp. 805–807 [in Russian]; *S.A. Zegzhda, M.P. Yushkov*. Application of the new form of dynamics equations for motion control of a robotic bench platform using rods of variable length // Vest. St. Petersb. Univ. Ser. 1. 1996. № 3 (№ 15). Pp. 112–114 [in Russian]; *V.I. Petrova*. Relation between coordinate systems describing the dynamics of a loaded Stewart platform // AIP (American Institute of Physics). Conference Proceedings 1959, 030019 (2018), pp. 030019-1–030019-7; *S.A. Zegzhda, V.I. Petrova, M.P. Yushkov*. Application of a special form of differential equations to the study of motion of a loaded Stewart platform // Vest. St. Petersb. Univ. Mathematics. Mechanics. Astronomy. 2020. Vol. 7 (65). Issue 1. Pp. 128–140 [in Russian].

Fig. 9 Model of a loaded
Stewart platform

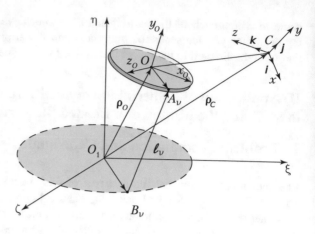

In the study of the motion of the platform of the dynamic simulation bench, it is required to compose the motion equation of the system in the form which would allow one to constructively evaluate the forces appearing in hydraulic (pneumatic) cylinders from a given motion of the platform. In the inverse problem, one needs to determine the motion of the system from the given forces created by legs of hydraulic cylinders. As was already pointed out, for this purpose in this part of the chapter we shall employ a special form of motion equations. However, to be able to work with this form, it is required to know the transformation between coordinate frames used in the study of the problem. Below we shall derive formulas for recalculation of the coordinates in the four coordinate frames applied in the study of the motion of a loaded Stewart platform, which is the central component of the dynamic simulation bench. The notation used below may differ from that adopted in the first part of the chapter.

Coordinate frames. We consider the following coordinate frames (see Fig. 9): $O_1\xi\eta\zeta$ is the fixed frame connected with the bench base; $Ox_oy_oz_o$ is the system rigidly connected with the movable platform of the bench (the Stewart platform); $Cxyz$ is the system with the origin at the common center of gravity of the loaded platform, its axes are directed along the principal axes of inertia of the body under consideration.

The orientation of the platform axes Ox_o, Oy_o, Oz_o with respect to the linearly moving coordinate frame $O\xi'\eta'\zeta'$ is defined by three successive rotations by airplane angles[9] (see Fig. 10): the yaw angle ψ, the pitch angle θ, and the roll angle φ.

---

[9] In this part, unlike the first part of the chapter, we introduce different orientations of Cartesian coordinate systems and different airplane system of coordinates, which were used in USSR (see, for example, the books *Yu.G. Sikharulidze*. Ballistics of Flight Vehicles. Moscow, Nauka, 1982 [in Russian]; *G.S. Byushgens, R.V. Studnev*. Airplane Dynamics. A Spatial Motion. Moscow, Mashinostroenie, 1988 [in Russian]). For more on airplane angles, see chapter 10 "Flight Dynamics".

Fig. 10 Airplane angles
used in USSR

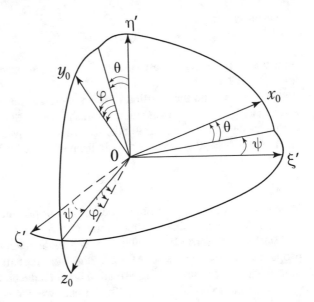

The motion law of the loaded bench platform is considered specified if one knows the motion law of the pole O of this system

$$\xi_o = \xi_o(t), \quad \eta_o = \eta_o(t), \quad \zeta_o = \zeta_o(t) \tag{7.1}$$

and the rotation law of this body around the pole

$$\psi = \psi(t), \quad \theta = \theta(t), \quad \varphi = \varphi(t). \tag{7.2}$$

The functions (7.1) define the vector $\rho_o(t)$

$$\rho_o(t) = \xi_o(t)\varepsilon_1 + \eta_o(t)\varepsilon_2 + \zeta_o(t)\varepsilon_3,$$

where ε_1, ε_2, ε_3 are the unit vectors of the fixed frame $O_1\xi\eta\zeta$.

The vector ρ_c (which specifies the position of the center of mass of a loaded bench platform) and the vectors i, j, k (which are the unit vectors of the $Cxyz$-frame, see Fig. 9), are usually considered as time functions in the study of the dynamics of solids.

8 Formulas for Transitions Between Coordinate Frames

Finding the motion law of the bench platform from the available variations of lengths of hydraulic cylinders. A loaded Stewart platform has 6 degrees of freedom.

The coordinates

$$\xi_o, \ \eta_o, \ \zeta_o, \ \psi, \ \theta, \ \varphi \tag{8.1}$$

were taken above as generalized coordinates. During motion, they vary according to laws (7.1), (7.2).

However, to find the position of the body under investigation, one may consider a different system of curvilinear coordinates. For a dynamic bench, as a new coordinate frame it is convenient to take the lengths of hydraulic (pneumatic) cylinders $B_\nu A_\nu$ ($\nu = \overline{1, 6}$), that support the platform with the cockpit (see Fig. 9):

$$\ell_\nu, \quad \nu = \overline{1, 6}. \tag{8.2}$$

When using such two coordinate frames, it is important to know the transition formulas from one curvilinear frame to the other one.

Usually such formulas involve transcendental functions. In this subsection, we propose a new method (whose foundations were partially alluded to in the first part of the chapter) for establishing a relation between the new and old coordinates in the process of motion of the system. To this end, we compose a system of differential equations with respect to the new coordinates; the inhomogeneity of these differential equations is characterized by the law of variation of the old coordinates.

So, we assume that the law of variation of coordinates (8.2) is known:

$$\ell_\nu = \ell_\nu(t), \quad \nu = \overline{1, 6}. \tag{8.3}$$

It is also assumed that initially (at $t = 0$) the position of the bench platform is known; that is, one knows both the coordinates (8.1) and the coordinates (8.2) at $t = 0$.

We consider the following problem: find the variation of coordinates (8.1) (from $t = 0$ onwards) from the available variations of the lengths of hydraulic cylinders (8.3).

So, from functions (8.3) we need to find the law of variation of coordinates (8.1) from some system of differential equations. To write down the sought-for system of equations, we proceed as follows.

The equality

$$\ell_\nu^2(t) = \ell_\nu{}^2(t), \quad \nu = \overline{1, 6}, \tag{8.4}$$

is clear. Here, $\ell_\nu = \overrightarrow{B_\nu A_\nu}$ (see Fig. 9), B_ν is the attachment point of the νth hydraulic cylinder to the fixed base, A_ν is its attachment point to the moving platform.

Differentiating equalities (8.4) with respect to time and taking into account the relation between the vectors (which follow from Fig. 9), we get

$$\ell_\nu(t)\dot{\ell}_\nu(t) \equiv \boldsymbol{\ell}_\nu \cdot \dot{\boldsymbol{\ell}}_\nu =$$

$$= (\overrightarrow{B_\nu O_1} + \boldsymbol{\rho}_o(t) + \overrightarrow{OA_\nu}) \cdot (\dot{\boldsymbol{\rho}}_o + \overrightarrow{OA_\nu}) = \tag{8.5}$$

$$= (\boldsymbol{\rho}_o(t) + x_{o\nu}\boldsymbol{i}_o + z_{o\nu}\boldsymbol{k}_o - \xi_\nu \boldsymbol{\varepsilon}_1 - \zeta_\nu \boldsymbol{\varepsilon}_3) \cdot$$

$$\cdot (\dot{\boldsymbol{\rho}}_o + x_{o\nu}\dot{\boldsymbol{i}}_o + z_{o\nu}\dot{\boldsymbol{k}}_o), \quad \nu = \overline{1,6}.$$

Here ξ_ν, ζ_ν are the coordinates of the point B_ν in the plane $O_1\xi\zeta$, $x_{o\nu}$, $z_{o\nu}$ are the coordinates of the point A_ν in the plane Ox_oz_o, $\boldsymbol{i}_o$, $\boldsymbol{k}_o$ are the unit vectors of the axes Ox_o, Oz_o.

The vectors from the system of Eq. (8.5) can be written in the form

$$\left.\begin{array}{l} \boldsymbol{\rho}_o(t) = \xi_o(t)\boldsymbol{\varepsilon}_1 + \eta_o(t)\boldsymbol{\varepsilon}_2 + \zeta_o(t)\boldsymbol{\varepsilon}_3, \\ \boldsymbol{i}_o(t) = \alpha_{11}(t)\boldsymbol{\varepsilon}_1 + \alpha_{12}(t)\boldsymbol{\varepsilon}_2 + \alpha_{13}(t)\boldsymbol{\varepsilon}_3, \\ \boldsymbol{k}_o(t) = \alpha_{31}(t)\boldsymbol{\varepsilon}_1 + \alpha_{32}(t)\boldsymbol{\varepsilon}_2 + \alpha_{33}(t)\boldsymbol{\varepsilon}_3, \end{array}\right\} \tag{8.6}$$

where $\alpha_{\sigma\tau}$ are the cosines of the direction angles.

We augment the system of six Eq. (8.5) with additional 6 equations,

$$\left.\begin{array}{l} \boldsymbol{i}_o \cdot \dot{\boldsymbol{i}}_o = 0, \quad \boldsymbol{j}_o \cdot \dot{\boldsymbol{j}}_o = 0, \quad \boldsymbol{k}_o \cdot \dot{\boldsymbol{k}}_o = 0, \\ \boldsymbol{i}_o \cdot \dot{\boldsymbol{j}}_o + \dot{\boldsymbol{i}}_o \cdot \boldsymbol{j}_o = 0, \quad \boldsymbol{j}_o \cdot \dot{\boldsymbol{k}}_o + \dot{\boldsymbol{j}}_o \cdot \boldsymbol{k}_o = 0, \\ \dot{\boldsymbol{k}}_o \cdot \boldsymbol{i}_o + \boldsymbol{k}_o \cdot \dot{\boldsymbol{i}}_o = 0. \end{array}\right\} \tag{8.7}$$

These equations are obtained on differentiating the following clear relations with respect to time:

$$\left.\begin{array}{l} \boldsymbol{i}_o^2 = 1, \quad \boldsymbol{j}_o^2 = 1, \quad \boldsymbol{k}_o^2 = 1, \\ \boldsymbol{i}_o \cdot \boldsymbol{j}_o = 0, \quad \boldsymbol{j}_o \cdot \boldsymbol{k}_o = 0, \quad \boldsymbol{k}_o \cdot \boldsymbol{i}_o = 0. \end{array}\right\} \tag{8.8}$$

The vector $\boldsymbol{j}_o$, which appears in formulas (8.7) and (8.8), can be written as

$$\boldsymbol{j}_o(t) = \alpha_{21}(t)\boldsymbol{\varepsilon}_1 + \alpha_{22}(t)\boldsymbol{\varepsilon}_2 + \alpha_{23}(t)\boldsymbol{\varepsilon}_3.$$

The system of Eqs. (8.5), (8.7) contains 12 unknown time functions

$$\xi_o(t), \quad \eta_o(t), \quad \zeta_o(t), \quad \alpha_{\sigma\tau}(t), \quad \sigma, \tau = \overline{1,3}. \tag{8.9}$$

They are coordinates of the vectors $\boldsymbol{\rho}_o$, $\boldsymbol{i}_o$, $\boldsymbol{j}_o$, $\boldsymbol{k}_o$ with respect to the fixed frame of reference $O_1\xi\eta\zeta$. We shall consider this system as a system of differential equations with respect to the sought-for functions (8.9). We recall that in these equations $\ell_\nu(t)$, $\dot{\ell}_\nu(t)$, $\nu = \overline{1,6}$, are assumed to be given functions of time.

Integrating system (8.5), (8.7) with the initial data

$$\xi_o(0) = \xi_o^0, \quad \eta_o(0) = \eta_o^0, \quad \zeta_o(0) = \zeta_o^0,$$

$$\alpha_{\sigma\tau}(0) = \alpha_{\sigma\tau}^0, \quad \sigma, \tau = \overline{1,3}, \tag{8.10}$$

Table 6 Three-dimensional sections of maximal volume

	$O\xi'$	$O\eta'$	$O\zeta'$
Ox_o	$\cos\psi\cos\theta$	$\sin\theta$	$-\sin\psi\cos\theta$
Oy_o	$-\cos\psi\sin\theta\cos\varphi+$ $+\sin\psi\sin\varphi$	$\cos\theta\cos\varphi$	$\cos\psi\sin\varphi+$ $+\sin\psi\sin\theta\cos\varphi$
Oz_o	$\cos\psi\sin\theta\sin\varphi+$ $+\sin\psi\cos\varphi$	$-\cos\theta\sin\varphi$	$\cos\psi\cos\varphi-$ $-\sin\psi\sin\theta\sin\varphi$

we get the unknown functions (8.9). The values $\alpha_{\sigma\tau}^0$, $\sigma, \tau = \overline{1,3}$, can be determined from the given initial values of the airplane angles

$$\psi(0) = \psi^0, \quad \theta(0) = \theta^0, \quad \varphi(0) = \varphi^0$$

using the direction cosine matrix[10] between the axes of the frames $Ox_oy_oz_o$ and $O_1\xi\eta\zeta$ (Table 6):

So, we have found the functions $\xi_o(t)$, $\eta_o(t)$, $\zeta_o(t)$ involved in the motion law (7.1) of the platform, and now it is required to express the airplane angles ψ, θ, φ in terms of the already obtained functions $\alpha_{\sigma\tau}(t)$, $\sigma, \tau = \overline{1,3}$ (i. e., we need to find also the second part of the motion law (7.2)). To this end, we write down the relations

$$\boldsymbol{i}_o \cdot \boldsymbol{j}_o = r, \quad \dot{\boldsymbol{j}}_o \cdot \boldsymbol{k}_o = p, \quad \dot{\boldsymbol{k}}_o \cdot \boldsymbol{i}_o = q. \tag{8.11}$$

These formulas were obtained from the following considerations. We have

$$\dot{\boldsymbol{i}}_o = \boldsymbol{\omega} \times \boldsymbol{i}_o = \begin{vmatrix} \boldsymbol{i}_o & \boldsymbol{j}_o & \boldsymbol{k}_o \\ p & q & r \\ 1 & 0 & 0 \end{vmatrix} = r\boldsymbol{j}_o - q\boldsymbol{k}_o, \tag{8.12}$$

and hence $\dot{\boldsymbol{i}}_o \cdot \boldsymbol{j}_o = r$. The remaining two formulas in (8.11) are obtained by the same analysis.

So, by solving system (8.5), (8.7) with the help of formulas (8.11), we determine the functions $p = p(t)$, $q = q(t)$, $r = r(t)$. In turn, these functions p, q, r are related to the airplane angles by the kinematic Euler equations[11]:

$$\left.\begin{array}{l} p = \dot{\psi}\sin\theta + \dot{\varphi}, \\ q = \dot{\psi}\cos\theta\cos\varphi + \dot{\theta}\sin\varphi, \\ r = -\dot{\psi}\cos\theta\sin\varphi + \dot{\theta}\cos\varphi. \end{array}\right\} \tag{8.13}$$

---

[10] See, for example, the book: *Yu.G. Sikharulidze*. Ballistics of Flight Vehicles. Moscow, Nauka, 1982 [in Russian].

[11] See the previous footnote.

Hence, using formulas (8.11) we can write Eq. (8.13) as follows:

$$\left.\begin{aligned}
\dot{\psi} \sin\theta + \dot{\varphi} &= \\
&= \dot{\alpha}_{21}(t)\alpha_{31}(t) + \dot{\alpha}_{22}(t)\alpha_{32}(t) + \dot{\alpha}_{23}(t)\alpha_{33}(t)\,, \\
\dot{\psi} \cos\theta \cos\varphi + \dot{\theta} \sin\varphi &= \\
&= \dot{\alpha}_{31}(t)\alpha_{11}(t) + \dot{\alpha}_{32}(t)\alpha_{12}(t) + \dot{\alpha}_{33}(t)\alpha_{13}(t)\,, \\
- \dot{\psi} \cos\theta \sin\varphi + \dot{\theta} \cos\varphi &= \\
&= \dot{\alpha}_{11}(t)\alpha_{21}(t) + \dot{\alpha}_{12}(t)\alpha_{22}(t) + \dot{\alpha}_{13}(t)\alpha_{23}(t)\,.
\end{aligned}\right\} \quad (8.14)$$

The right-hand sides of these equations are the already obtained time functions, and hence system (8.14) can be considered as a system of three differential equations with respect to the airplane angles ψ, θ, φ. Integrating this system with given initial data, we obtain the law of rotation of the loaded platform:

$$\psi = \psi(t)\,, \quad \theta = \theta(t)\,, \quad \varphi = \varphi(t)\,.$$

Finding the law of variation of lengths of hydraulic jacks from the available motion law of the bench platform. It is required to find a transformation from the coordinate frame (8.1) to the coordinate frame (8.2). This transformation can be effected in different ways.

(1) Following the classical approach, we write down the "vectors" of the hydraulic cylinders

$$\boldsymbol{\ell}_\nu = \boldsymbol{p}_o(t) + x_{o\nu}\boldsymbol{i}_o(t) + z_{o\nu}\boldsymbol{k}_o(t) - \xi_\nu\boldsymbol{\epsilon}_1 - \zeta_\nu\boldsymbol{\epsilon}_3\,, \quad \nu = \overline{1,6}\,, \quad (8.15)$$

where the vectors $\boldsymbol{i}_o$, $\boldsymbol{k}_o$ are represented by formulas (8.6) (this system is augmented by the formula for $\boldsymbol{j}_o$):

$$\left.\begin{aligned}
\boldsymbol{i}_o(t) &= \alpha_{11}(t)\boldsymbol{\epsilon}_1 + \alpha_{12}(t)\boldsymbol{\epsilon}_2 + \alpha_{13}(t)\boldsymbol{\epsilon}_3\,, \\
\boldsymbol{j}_o(t) &= \alpha_{21}(t)\boldsymbol{\epsilon}_1 + \alpha_{22}(t)\boldsymbol{\epsilon}_2 + \alpha_{23}(t)\boldsymbol{\epsilon}_3\,, \\
\boldsymbol{k}_o(t) &= \alpha_{31}(t)\boldsymbol{\epsilon}_1 + \alpha_{32}(t)\boldsymbol{\epsilon}_2 + \alpha_{33}(t)\boldsymbol{\epsilon}_3\,.
\end{aligned}\right\}$$

As was pointed out above, the expressions $\alpha_{\sigma\tau}$, $\sigma, \tau = \overline{1,3}$, in terms of the airplane angles are known. Having found the vectors $\boldsymbol{\ell}_\nu$, $\nu = \overline{1,6}$ from formulas (8.15), we can also find their lengths ℓ_ν, $\nu = \overline{1,6}$.

(2) There is another approach to finding the vectors $\boldsymbol{i}_o$, $\boldsymbol{k}_o$ involved in (8.15) as time functions. Using (8.12) and applying cyclic interchanges for the vectors $\boldsymbol{i}_o$, $\boldsymbol{j}_o$, $\boldsymbol{k}_o$ and the quantities p, q, r, we get the following system of differential equations

$$\left.\begin{aligned}
\dot{\boldsymbol{i}}_o &= r\boldsymbol{j}_o - q\boldsymbol{k}_o\,, \\
\dot{\boldsymbol{j}}_o &= p\boldsymbol{k}_o - r\boldsymbol{i}_o\,, \\
\dot{\boldsymbol{k}}_o &= q\boldsymbol{i}_o - p\boldsymbol{j}_o\,.
\end{aligned}\right\} \quad (8.16)$$

Here, with the available motion law (7.1), (7.2) of the bench platform, the functions $p(t)$, $q(t)$, $r(t)$ can be found from the kinematic Euler equations (8.13).

So, under the second approach, one employs formulas (7.1), (7.2), (8.13) to integrate the system of differential equations (8.16), which in an expanded form reads as

$$\dot{\alpha}_{11} = r(t)\alpha_{21} - q(t)\alpha_{31}, \quad \dot{\alpha}_{12} = r(t)\alpha_{22} - q(t)\alpha_{32},$$
$$\dot{\alpha}_{13} = r(t)\alpha_{23} - q(t)\alpha_{33}, \quad \dot{\alpha}_{21} = p(t)\alpha_{31} - r(t)\alpha_{11},$$
$$\dot{\alpha}_{22} = p(t)\alpha_{32} - r(t)\alpha_{12}, \quad \dot{\alpha}_{23} = p(t)\alpha_{33} - r(t)\alpha_{13},$$
$$\dot{\alpha}_{31} = q(t)\alpha_{11} - p(t)\alpha_{21}, \quad \dot{\alpha}_{32} = q(t)\alpha_{12} - p(t)\alpha_{22}, \tag{8.17}$$
$$\dot{\alpha}_{33} = q(t)\alpha_{13} - p(t)\alpha_{23}.' \tag{8.18}$$

Solving system (8.17) with initial data (8.10), we can find the vectors $i_o(t)$, $k_o(t)$. After this, formulas (8.15) can be used to find also the vectors ℓ_ν, $\nu = \overline{1,6}$, and hence their lengths.

(3) When specifying the motion equations (7.1), (7.2), we compose expressions (8.5), and once the vectors $i_o(t)$, $k_o(t)$ are found (as in the previous method), we get

$$\ell_\nu \dot{\ell}_\nu = f_\nu(t), \quad \nu = \overline{1,6}, \tag{8.19}$$

where $f_\nu(t)$ denote the functions from the right-hand sides of formulas (8.5). Formula (8.19) can be looked upon as a system of differential equations with respect to the rod lengths ℓ_ν, $\nu = \overline{1,6}$. This system can be easily integrated. As a result, we find the law of variation of the lengths of hydraulic cylinders. Indeed,

$$f_\nu(t) = \ell_\nu \dot{\ell}_\nu = \frac{1}{2}\frac{d\ell_\nu^2}{dt}, \quad \nu = \overline{1,6},$$

and hence

$$\ell_\nu^2(t) = 2\int_0^t f_\nu(t_1)dt_1, \quad \nu = \overline{1,6}.$$

Finding the vector functions $\rho_c(t)$, $i(t)$, $j(t)$, $k(t)$ from the given motion law of the bench platform. Assume that the motion of the bench platform is given by Eqs. (7.1), (7.2). To describe the motion of the $Cxyz$-frame we shall again construct some differential equations.

Consider the expressions

$$\ell_\nu(t)\dot{\ell}_\nu \equiv \ell_\nu(t) \cdot \dot{\ell}_\nu(t) =$$
$$= (\rho_c(t) + \overrightarrow{CA_\nu} + \overrightarrow{B_\nu O_1}) \cdot (\dot{\rho}_c(t) + \overrightarrow{\dot{CA}_\nu}) =$$
$$= (\rho_c(t) + x_\nu i(t) + y_\nu j(t) + z_\nu k(t) - \xi_\nu \varepsilon_1 - \zeta_\nu \varepsilon_3) \cdot \tag{8.20}$$
$$\cdot (\dot{\rho}_c + x_\nu \dot{i}(t) + y_\nu \dot{j}(t) + z_\nu \dot{k}(t)), \quad \nu = \overline{1,6},$$

where x_ν, y_ν, z_ν are the coordinates of the points $A_\nu (\nu = \overline{1,6})$ in the $Cxyz$-frame.

We note that the position of the center of mass C in the frame $Ox_oy_oz_o$ is considered to be known; the coordinates x_ν, y_ν $z_o\nu$ of the points $A_\nu (\nu = \overline{1,6})$ in the $Cxyz$-frame are also assumed to be known. Now the vectors $\rho_c(t)$, $i(t)$, $j(t)$, $k(t)$ can be expressed in terms of their projections onto the fixed axes $O_1\xi$, $O_1\eta$, $O_1\zeta$ with the unit vectors ε_1, ε_2, ε_3:

$$
\left.
\begin{aligned}
\rho_c(t) &= \xi(t)\varepsilon_1 + \eta(t)\varepsilon_2 + \zeta(t)\varepsilon_3 , \\
i(t) &= \beta_{11}(t)\varepsilon_1 + \beta_{12}(t)\varepsilon_2 + \beta_{13}(t)\varepsilon_3 , \\
j(t) &= \beta_{21}(t)\varepsilon_1 + \beta_{22}(t)\varepsilon_2 + \beta_{23}(t)\varepsilon_3 , \\
k(t) &= \beta_{31}(t)\varepsilon_1 + \beta_{32}(t)\varepsilon_2 + \beta_{33}(t)\varepsilon_3 .
\end{aligned}
\right\}
$$

The initial position of the center of mass and the orientation of the principal axes of inertia are assumed to be known:

$$
\left.
\begin{aligned}
\xi(0) &= \xi^0, \quad \eta(0) = \eta^0, \quad \zeta(0) = \zeta^0 , \\
\beta_{\sigma\tau}(0) &= \beta_{\sigma\tau}^0, \quad \sigma, \tau = \overline{1,3} .
\end{aligned}
\right\}
\tag{8.21}
$$

As before, from the available motion law (7.1), (7.2) of the bench platform one can find the functions

$$
f_\nu(t) = \ell_\nu(t)\dot{\ell}_\nu(t) , \quad \nu = \overline{1,6} .
$$

Now expressions (8.20) can be looked upon as a system of six first-order differential equations with respect to the vectors ρ_c, i, j, k. Proceeding as before, we augment this system by the differential equations

$$
\left.
\begin{aligned}
i \cdot i &= 0, \quad j \cdot j = 0, \quad k \cdot k = 0, \\
i \cdot j + i \cdot j &= 0, \quad j \cdot k + j \cdot k = 0, \\
k \cdot i + k \cdot i &= 0,
\end{aligned}
\right\}
\tag{8.22}
$$

which follow from the abstract constraints

$$
\left.
\begin{aligned}
i^2 &= 1, \quad j^2 = 1, \quad k^2 = 1, \\
i \cdot j &= 0, \quad j \cdot k = 0, \quad k \cdot i = 0,
\end{aligned}
\right\}
$$

which apply to the motion of the loaded platform and which were introduced in Sect. 7 of Chap. 8 of Vol. I.

Integrating the system of differential equations (8.20), (8.22) with the initial data (8.21), we find the vector functions

$$
\rho_c = \rho_c(t) , \quad i = i(t) , \quad j = j(t) , \quad k = k(t) .
$$

Finding the motion law of the bench platform from given functions $\rho_c(t)$, $i(t)$, $j(t)$, $k(t)$. We note that the vector functions

$$\rho_c = \rho_c(t), \quad i = i(t), \quad j = j(t), \quad k = k(t) \tag{8.23}$$

become known as time functions after integration of the differential equations of the bench dynamics written in the special form.

Using (8.23) in formulas (8.20), we find the functions $f_\nu(t) = \ell_\nu \dot{\ell}_\nu(t)$, $\nu = \overline{1,6}$, which we substitute into (8.5). Integrating the so-obtained system of differential equations (8.5) and the adjoint equations (8.7) with the initial data (8.10), we get the vector functions

$$\rho_o = \rho_o(t), \quad i_o = i_o(t), \quad j_o = j_o(t), \quad k_o = k_o(t).$$

Using formulas (8.11), we find the functions $p(t)$, $q(t)$, $r(t)$, and then obtain the law of variation of the angles ψ, θ, φ by integrating system (8.14).

9 Solution of the Direct Dynamics Problem

Evaluation of forces in legs of the hydraulic (pneumatic) cylinders from a given motion law of the bench platform. In Sect. 8 it was shown how from a given motion law of a platform one can find the vector functions $\rho_c(t)$, $i(t)$, $j(t)$, $k(t)$, which are used in the special form of equations of rigid body dynamics.

We recall that $\rho_c = \overrightarrow{O_1 C}$ is the radius vector of the center of mass of a loaded bench platform and i, j, k are the unit vectors of the principal axes of inertia of this mechanical system. The moments of inertia A, B, C with respect to these axes are assumed to be known:

$$A = \int\limits_{(m)} (y^2 + z^2)dm, \quad B = \int\limits_{(m)} (z^2 + x^2)dm, \quad C = \int\limits_{(m)} (x^2 + y^2)dm.$$

When using the special form of dynamics equations (see Sect. 7 of Chap. 8 of Vol. I), we consider the quantities

$$I_x = \int\limits_{(m)} x^2 dm, \quad I_y = \int\limits_{(m)} y^2 dm, \quad I_z = \int\limits_{(m)} z^2 dm,$$

which are related to the classical quantities A, B and C as

$$I_x = \frac{C - A + B}{2}, \quad I_y = \frac{A - B + C}{2}, \quad I_z = \frac{B - C + A}{2}.$$

This mechanical system is subject to the gravity force of the entire system $M\mathbf{g}$ and the six forces $\mathbf{F}_\nu$ acting on the platform from the legs of the hydraulic cylinders. One can neglect the inertia forces due to rotation of the hydraulic cylinders and the moment of the frictional forces in ball-and-socket joints of the hydraulic cylinders, and hence it can be assumed that the forces $\mathbf{F}_\nu$, $\nu = \overline{1,6}$, act along the directions of the vectors $\boldsymbol{\ell}_\nu$, $\nu = \overline{1,6}$. So, the forces $\mathbf{F}_\nu$, $\nu = \overline{1,6}$, can be written in the form

$$\mathbf{F}_\nu = \frac{u_\nu \boldsymbol{\ell}_\nu}{\ell_\nu}, \quad \ell_\nu = |\boldsymbol{\ell}_\nu|, \quad \nu = \overline{1,6}.$$

Here u_ν, $\nu = \overline{1,6}$, are the control parameters responsible for the required platform motion.

According to the center-of-mass theorem,

$$M\ddot{\boldsymbol{\rho}}_c = \sum_{\nu=1}^{6} \mathbf{F}_\nu + M\mathbf{g}, \tag{9.1}$$

where M is the mass of the entire system.

In Sect. 7 of Chap. 8 of Vol. I it was shown that the rotational motion of this system about the center of mass is described by following system of differential equation with respect to the vectors $\mathbf{i}$, $\mathbf{j}$, $\mathbf{k}$:

$$\ddot{\mathbf{i}} = -\left(\dot{\mathbf{i}}\right)^2 \mathbf{i} - \frac{2I_y}{I_x + I_y}(\dot{\mathbf{i}} \cdot \dot{\mathbf{j}})\mathbf{j} -$$
$$- \frac{2I_z}{I_z + I_x}(\dot{\mathbf{k}} \cdot \dot{\mathbf{i}})\mathbf{k} + \frac{L_z}{I_x + I_y}\mathbf{j} - \frac{L_y}{I_z + I_x}\mathbf{k},$$
$$\ddot{\mathbf{j}} = -\left(\dot{\mathbf{j}}\right)^2 \mathbf{j} - \frac{2I_z}{I_y + I_z}(\dot{\mathbf{j}} \cdot \dot{\mathbf{k}})\mathbf{k} -$$
$$- \frac{2I_x}{I_x + I_y}(\dot{\mathbf{i}} \cdot \dot{\mathbf{j}})\mathbf{i} + \frac{L_x}{I_y + I_z}\mathbf{k} - \frac{L_z}{I_x + I_y}\mathbf{i},$$
$$\ddot{\mathbf{k}} = -\left(\dot{\mathbf{k}}\right)^2 \mathbf{k} - \frac{2I_x}{I_z + I_x}(\dot{\mathbf{k}} \cdot \dot{\mathbf{i}})\mathbf{i} -$$
$$- \frac{2I_y}{I_y + I_z}(\dot{\mathbf{j}} \cdot \dot{\mathbf{k}})\mathbf{j} + \frac{L_y}{I_z + I_x}\mathbf{i} - \frac{L_x}{I_y + I_z}\mathbf{j}. \tag{9.2}$$

Here, L_x, L_y, L_z are the projections of the torque of forces $\mathbf{F}_\nu$, $\nu = \overline{1,6}$:

$$\mathbf{L} = \sum_{\nu=1}^{6} (x_\nu \mathbf{i} + y_\nu \mathbf{j} + z_\nu \mathbf{k}) \times \mathbf{F}_\nu.$$

Projecting the first (second, third) of Eq. (9.2) onto the axis Cx (Cy, Cz, respectively), we get

$$\ddot{i} \cdot i = - \left(\dot{i} \right)^2 , \quad \ddot{j} \cdot j = - \left(\dot{j} \right)^2 ,$$
$$\ddot{k} \cdot k = - \left(\dot{k} \right)^2 . \tag{9.3}$$

Projecting the second (third, first) equation onto the axis Cz (Cx, Cy, respectively), we obtain

$$\ddot{j} \cdot k = - \frac{2 I_z}{I_y + I_z} (\dot{j} \cdot k) + \frac{L_x}{I_y + I_z} ,$$
$$\ddot{k} \cdot i = - \frac{2 I_x}{I_z + I_x} (\dot{k} \cdot i) + \frac{L_y}{I_z + I_x} , \tag{9.4}$$
$$\ddot{i} \cdot j = - \frac{2 I_y}{I_x + I_y} (\dot{i} \cdot j) + \frac{L_z}{I_x + I_y} .$$

It is easily checked that these three equations are equivalent to the Euler dynamical equations with respect to the projections p, q, r of the angular velocity vector ω onto the axes Cx, Cy, Cz:

$$A\dot{p} - (B - C)qr = L_x ,$$
$$B\dot{q} - (C - A)rp = L_y ,$$
$$C\dot{r} - (A - B)pq = L_z .$$

So, Eq. (9.4) express the law of moments with respect to the center of mass. Equations (9.3) follow from the fact that i, j, k are unit vectors, i. e.,

$$i^2 = 1 , \quad j^2 = 1 , \quad k^2 = 1 .$$

Differentiating these equations twice with respect to the time, we get Eq. (9.3). By the same operation with the orthogonality conditions of the vectors i, j, k, we get

$$2\dot{i} \cdot \dot{j} + \ddot{i} \cdot j + i \cdot \ddot{j} = 0 ,$$
$$2\dot{j} \cdot \dot{k} + \ddot{j} \cdot k + j \cdot \ddot{k} = 0 , \tag{9.5}$$
$$2\dot{k} \cdot \dot{i} + \ddot{k} \cdot i + k \cdot \ddot{i} = 0 .$$

Inserting in these equations the quantities $\ddot{i} \cdot j$, $\ddot{j} \cdot i$, $\ddot{j} \cdot k$, $\ddot{k} \cdot j$, $\ddot{k} \cdot i$, $\ddot{i} \cdot k$, which were found from Eq. (9.2), we get identities. So, the system of three vector equations (9.2) is equivalent to the system of nine scalar equations (9.3), (9.4), (9.5).

Let us return to the principal problem of determination of the functions $u_\nu(t)$, $\nu = 1, 6$, from the given vector functions $\rho_c(t), i(t), j(t), k(t)$. We recall that in Sect. 8 these functions were found form the given motion law of the platform.

From Eq. (9.1) we get

$$F \equiv \sum_{\nu=1}^{6} F_\nu = M\ddot{\rho}_c - M\mathbf{g}. \tag{9.6}$$

So, from the vector function $\rho_c(t)$ one can find the variation in time of the principal vector F of the system of forces F_ν, $\nu = \overline{1,6}$.

The vector equation (9.6) in scalar form can be written as

$$\sum_{\nu=1}^{6} U_\nu \ell_{\nu x} = F_x(t),$$

$$\sum_{\nu=1}^{6} U_\nu \ell_{\nu y} = F_y(t), \tag{9.7}$$

$$\sum_{\nu=1}^{6} U_\nu \ell_{\nu z} = F_z(t),$$

where

$$U_\nu = \frac{u_\nu}{\ell_\nu}, \quad \nu = \overline{1,6},$$

$$\begin{cases}
F_x(t) = M(\ddot{\rho}_c - \mathbf{g}) \cdot \mathbf{i}(t) = \\
= M[\ddot{\xi}(t)\varepsilon_1 \cdot \mathbf{i}(t) + \ddot{\eta}(t)\varepsilon_2 \cdot \mathbf{i}(t) + \\
+ \ddot{\zeta}(t)\varepsilon_3 \cdot \mathbf{i}(t) - -g\varepsilon_3 \cdot \mathbf{i}(t)] = M[\ddot{\xi}(t)\beta_{11}(t) + \\
+ \ddot{\eta}(t)\beta_{12}(t) + \ddot{\zeta}(t)\beta_{13}(t) - g\beta_{13}(t)], \\
F_y(t) = M[\ddot{\xi}(t)\beta_{21}(t) + \\
+ \ddot{\eta}(t)\beta_{22}(t) + \ddot{\zeta}(t)\beta_{23}(t) - g\beta_{23}(t)], \\
F_z(t) = M[\ddot{\xi}(t)\beta_{31}(t) + \\
+ \ddot{\eta}(t)\beta_{32}(t) + \ddot{\zeta}(t)\beta_{33}(t) - g\beta_{33}(t)], \\
\ell_{\nu x} = \xi(t)\beta_{11}(t) + \eta(t)\beta_{12}(t) + \zeta(t)\beta_{13}(t) + x_\nu - \\
- \xi_\nu \beta_{11}(t) - \zeta_\nu \beta_{13}(t), \\
\ell_{\nu y} = \xi(t)\beta_{21}(t) + \eta(t)\beta_{22}(t) + \zeta(t)\beta_{23}(t) + y_\nu - \\
- \xi_\nu \beta_{21}(t) - \zeta_\nu \beta_{23}(t), \\
\ell_{\nu z} = \xi(t)\beta_{31}(t) + \eta(t)\beta_{32}(t) + \zeta(t)\beta_{33}(t) + z_\nu - \\
- \xi_\nu \beta_{31}(t) - \zeta_\nu \beta_{33}(t), \\
\ell_\nu = \ell_{\nu x}\mathbf{i} + \ell_{\nu y}\mathbf{j} + \ell_{\nu z}\mathbf{k}, \quad \nu = \overline{1,6}.
\end{cases} \tag{9.8}$$

From the system of Eq. (9.4) one can find the projections L_x, L_y, L_z of the principal moment of forces F_ν, $\nu = \overline{1,6}$,

$$L = \sum_{\nu=1}^{6} r_\nu \times F_\nu , \quad r_\nu = x_\nu i + y_\nu j + z_\nu k ,$$

$$F_\nu = U_\nu \ell_\nu , \quad U_\nu = \frac{u_\nu}{\ell_\nu} .$$

We have

$$r_\nu \times F_\nu = \begin{vmatrix} i & j & k \\ x_\nu & y_\nu & z_\nu \\ U_\nu \ell_{\nu x} & U_\nu \ell_{\nu y} & U_\nu \ell_{\nu z} \end{vmatrix} , \tag{9.9}$$

and so

$$\left.\begin{aligned} L_x &= \sum_{\nu=1}^{6} U_\nu (y_\nu \ell_{\nu z} - z_\nu \ell_{\nu y}) , \\ L_y &= \sum_{\nu=1}^{6} U_\nu (z_\nu \ell_{\nu x} - x_\nu \ell_{\nu z}) , \\ L_z &= \sum_{\nu=1}^{6} U_\nu (x_\nu \ell_{\nu y} - y_\nu \ell_{\nu x}) . \end{aligned}\right\} \tag{9.10}$$

Considering formulas (9.7), (9.10) as a system of linear algebraic equations with respect to U_ν, $\nu = \overline{1,6}$, we get the required control parameters

$$u_\nu(t) = \ell_\nu(t) U_\nu(t) , \quad \nu = \overline{1,6} .$$

So, we have solved the problem of determination of the control forces $u_\nu(t)$, $\nu = \overline{1,6}$, providing for the required motion of the system.

10   Solution of the Inverse Dynamics Problem

Determination of the law of motion of a loaded bench platform from the given forces in hydraulic cylinders. The problem is to integrate the system of differential equations (9.1), (9.3), (9.4), (9.5) with initial data

$$\left.\begin{aligned} &\xi(0) = \xi^0 , \eta(0) = \eta^0 , \zeta(0) = \zeta^0 , \\ &\dot\xi(0) = 0 , \dot\eta(0) = 0 , \dot\zeta(0) = 0 , \\ &\beta_{\sigma\tau}(0) = \beta_{\sigma\tau}^0 , \dot\beta_{\sigma\tau}(0) = 0 , \\ &\sigma, \tau = \overline{1,3} . \end{aligned}\right\} \tag{10.1}$$

In projections to the unit vectors i, j, k, the vector equation (9.1) reads as

$$
\left.
\begin{aligned}
M\ddot{\boldsymbol{\rho}}_c \cdot \boldsymbol{i} &\equiv M(\ddot{\xi}(t)\beta_{11}(t) + \ddot{\eta}(t)\beta_{12}(t) + \ddot{\zeta}(t)\beta_{13}(t)) = \\
&= \sum_{\nu=1}^{6} \frac{u_\nu}{\ell_\nu}\ell_{\nu x} + Mg\beta_{13}(t)\,, \\
M\ddot{\boldsymbol{\rho}}_c \cdot \boldsymbol{j} &\equiv M(\ddot{\xi}(t)\beta_{21}(t) + \ddot{\eta}(t)\beta_{22}(t) + \ddot{\zeta}(t)\beta_{23}(t)) = \\
&= \sum_{\nu=1}^{6} \frac{u_\nu}{\ell_\nu}\ell_{\nu y} + Mg\beta_{23}(t)\,, \\
M\ddot{\boldsymbol{\rho}}_c \cdot \boldsymbol{k} &\equiv M(\ddot{\xi}(t)\beta_{31}(t) + \ddot{\eta}(t)\beta_{32}(t) + \ddot{\zeta}(t)\beta_{33}(t)) = \\
&= \sum_{\nu=1}^{6} \frac{u_\nu}{\ell_\nu}\ell_{\nu z} + Mg\beta_{33}(t)\,.
\end{aligned}
\right\} \tag{10.2}
$$

Here the quantities $\ell_{\nu x}, \ell_{\nu y}, \ell_{\nu z}, \ell_\nu = \sqrt{\ell_{\nu x}^2 + \ell_{\nu y}^2 + \ell_{\nu z}^2}$ $(\nu = \overline{1,6})$ can be expressed in terms of the sought-for functions by formulas (9.8). The differential equations (10.2) should be considered together with Eqs. (9.3), (9.4), (9.5).

Note that the quantities L_x, L_y, L_z (which enter Eq. (9.4) as functions of the sought-for variables and given controls u_ν, $\nu = \overline{1,6}$) are given by the expressions:

$$
\left.
\begin{aligned}
L_x &= \sum_{\nu=1}^{6} \frac{u_\nu}{\ell_\nu}(y_\nu \ell_{\nu z} - z_\nu \ell_{\nu y})\,, \\
L_y &= \sum_{\nu=1}^{6} \frac{u_\nu}{\ell_\nu}(z_\nu \ell_{\nu x} - x_\nu \ell_{\nu z})\,, \\
L_z &= \sum_{\nu=1}^{6} \frac{u_\nu}{\ell_\nu}(x_\nu \ell_{\nu y} - y_\nu \ell_{\nu x})\,.
\end{aligned}
\right\}
$$

Integrating the systems of differential equations (10.2), (9.3), (9.4), (9.5) with initial data (10.1), one can find the vector functions $\boldsymbol{\rho}_c(t)$, $\boldsymbol{i}(t)$, $\boldsymbol{j}(t)$, $\boldsymbol{k}(t)$ from the given control $u_\nu(t)$, $\nu = \overline{1,6}$. An algorithm of determination of the platform law motion from these vector functions was described in Sect. 8.

So, with the help of the differential equations obtained in these two subsections one can solve both the direct and the inverse dynamics problems. For better completeness of the study of these two problems, we shall write the system of differential equations in the dimensionless form. As the length unit we take the radius R_b of the circle containing the lower ends of the pneumatic cylinders; as the unit force we take the force of gravity Mg of the entire system. We introduce the dimensionless time $\tau = \omega t$, $(\omega^2 = g/R_b)$. However, in the formulas that follow and in the graphs, the dimensionless time will be denoted by t and ordinary letters will be used for dimensionless coordinates and forces.

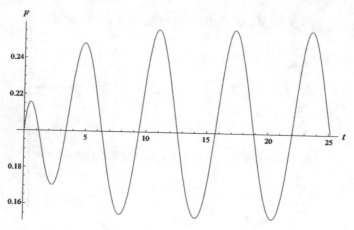

Fig. 11 The force in the cylinder

11 Vertical Oscillations of the Platform[12]

Let us find the forces in the hydraulic cylinders to provide the given oscillatory vertical motion of a symmetrically loaded platform by the law

$$\eta(t) = 0.2\,(\sin t)(1 - e^{-t/2})^2\,. \tag{11.1}$$

As in §2, the vertical displacement $\eta(t)$ is given in this form in order that initially both the velocity and the acceleration of the platform be zero. Otherwise, a sudden application of a force would be required at the initial time. At the same time, vertical sinusoidal oscillations with amplitude 0.2 and frequency 1 are rapidly established in the system.

In the case when the center of mass of the entire system is above the platform center, using the Wolfram Mathematica software package it can be shown that such a motion can be arranged using a force $F(t)$ in the cylinder (the expression of this force is very bulky). Due to the symmetry of the problem, the forces in all cylinders can be shown to be equal. The variation of this force is shown in Fig. 11.

---

[12] Small vertical oscillations of a loaded platform usually appear also in cases when it is in an equilibrium position which is known to be unstable. This can be explained by the fact that the motors providing the required pressures in hydraulic cylinders are usually responsible for small sinusoidal departures of pressures. In engineering, the resulting small oscillations of a loaded bench platform are called "parasitic oscillations". These oscillations, which are partially responsible for the exit of the system from an unstable equilibrium position, have been considered in the paper *S.A. Zegzhda, V.I. Petrova, M.P. Yushkov*. Application of a special form of differential equations to the study of motion of a loaded Stewart platform // Vest. St. Petersb. Univ. Mathematics. Mechanics. Astronomy. 2020. Vol. 7 (65). Issue 1. Pp. 128–140 [in Russian].

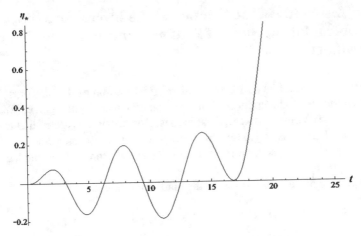

Fig. 12 Motion of the platform with small perturbation of the vertical force

In the inverse problem, when the forces $F(t)$ thus obtained are applied to a loaded platform, we get the required vertical oscillations of the platform according to the law (11.1).

Let us now consider the case of a small perturbation of the control $F(t)$. We denote by $\eta_*(t)$ the displacement of the bench platform corresponding to the small perturbation of the control $F(t)$, as obtained by solving the direct mechanical problem. The small perturbation will be given in the form $0.0001 (\sin 2t)(1 - e^{-t/2})$; i.e., we set

$$F_*(t) = F(t) + 0.0001 (\sin 2t)(1 - e^{-t/2}).$$

The graph of the function $\eta_*(t)$, which is given in Fig. 12, shows that the above perturbation has little effect on the initial time interval $(0 < \tau < 12)$, whereas later the platform starts to ascend steeply. It should be pointed out that in spite of the oscillatory character of the force $F(t)$, the platform ascends rapidly without oscillations.

So, it is seen that the platform starts to ascent steeply, even though the function $F_*(t)$ differs only very insignificantly from the function $F(t)$, which guarantees its motion according to the law (11.1). This is indicative of the instability of the required vertical oscillations (11.1) of the platform.

It is worth specially pointing out that the numerical calculations, as obtained for vertical oscillations of the bench platform, are in full conformity with the motion found in Part I with the use of basic dynamics theorems on the motion of the center of mass and on the variation of the kinetic moment when moving about the center of mass. This can be verified from the complete coincidence of the graphs in Figs. 3 and 4 from Part I and Figs. 11 and 12 from Part II of the present chapter.

12   On Instability of the Solution of the Inverse Dynamics Problem for the Stewart Platform. Introduction of Feedbacks

As in Sect. 11, we shall be concerned with vertical motion of the platform. We also assume that the lower and upper leg joints are located in horizontal planes. We denote by z the variable distance between these planes. Besides, it is assumed that the angles α_k between all six rods (legs) and the vertical direction are the same and equal to $\alpha(z)$, $d\alpha/dz < 0$. We also assume that the longitudinal forces F_k in the legs are the same and equal to F_c.

The equation of vertical motion of the platform reads as

$$m\ddot{z} = -P + F_d, \qquad F_d = F_d(z) = 6F_c \cos\alpha(z), \qquad (12.1)$$

where m and P are, respectively, the mass and weight of the platform, and F_d is the vertical component of the resultant of the force applied to the platform from the legs.

We first consider the equilibrium position of the platform given by the equalities

$$z = z_0, \quad \alpha = \alpha_0 = \alpha(z_0), \quad F_c = \frac{P}{6\cos\alpha_0}. \qquad (12.2)$$

We introduce a small perturbation $x(t)$ and consider the perturbed motion $z(t) = z_0 + x(t)$ near the equilibrium position

$$m\ddot{x}(t) = -P + F_d(z_0 + x(t)).$$

In the linear approximation with respect to x, we have by (12.1) and (12.2)

$$m\ddot{x}(t) = k\,x(t), \qquad k = -6F_c \sin\alpha(z_0)\frac{d\alpha}{dz_0} > 0,$$

which implies that the equilibrium position (12.2) is unstable.

Let us now consider the motion of the platform given by the equalities

$$z = z_0(t), \quad \alpha(t) = \alpha_0(z_0(t)), \quad F_c(t) = \frac{m\ddot{z}_0(t) + P}{6\cos\alpha(z_0(t))}. \qquad (12.3)$$

Here the motion $z_0(t)$ of the platform is specified, and the forces in the legs $F_c(t)$ can be determined from Eq. (12.1) (when solving the direct dynamics problem).

As in the above, we introduce a small perturbation $x(t)$ and consider the perturbed motion $z(t) = z_0(t) + x(t)$ in the vicinity of the unperturbed motion (12.3)

$$m(\ddot{z}_0(t) + \ddot{x}(t)) = -P + F_d(z_0(t) + x(t)).$$

In the linear approximation with respect to $x(t)$, using (12.1) and (12.3) we find that

$$m\ddot{x}(t) = k(t)\,x(t)\,, \qquad k(t) = -6F_c(t)\sin\alpha(z_0(t))\frac{d\alpha}{dz_0}\,. \qquad (12.4)$$

If $F_c(t) > 0$ or $P + \ddot{z}_0(t) > 0$ during the entire motion, then $k(t) > 0$, and the unperturbed motion is unstable.

Indeed, assume that at the initial time $t = 0$ we have $x(0) > 0$, $\dot{x}_0(0) \geqslant 0$. Then, for any t, we have

$$x(t) \geqslant x(0)e^{\beta t}\,, \qquad \beta > 0\,, \qquad \beta^2 = \frac{\min_t k(t)}{m}\,.$$

In Sect. 11, an instability was numerically obtained for constantly acting small perturbations. Here, an instability is established under the perturbation of the initial conditions. From the above analysis it follows that a required motion of the platform cannot be obtained by specifying forces in hydraulic cylinders, which were determined by solving the direct dynamics problem. To this aim, one should introduce feedbacks, which take into account the departure of the resulting motion from the required one. The same conclusion can also be made about the problem of equilibrium of a platform with three rods. This problem will be considered below in Part III of this chapter. The required motion can also be obtained under force action on rods with available feedbacks or with kinematic control (when the rod lengths are specified).

To obtain stable vertical oscillations of the platform according to the law (11.1), we introduce, as in Part I, the classical feedbacks by forming the forces $F_k(t)$, $k = \overline{1, 6}$, created by hydraulic cylinders in the form

$$F_k(t) = F_k^p(t) + G(l_k^p(t) - l_k(t))\,, \qquad k = \overline{1, 6}\,.$$

Here $F_k^p(t)$ are the programmed control forces, $l_k^p(t)$ are the programmed lengths of the hydraulic cylinders, $l_k(t)$ are the measured actual lengths of hydraulic cylinders, G is the feedback coefficient. As in Part I, numerical experiments show that under certain choice of G one can always assure that the true motion departs from the programmed motion (11.1) with a given accuracy. Moreover, for sufficiently large feedback coefficient G, one can always execute the programmed motion (11.1) even with $F_k^p(t) \equiv 0$.

III) Application of the Lagrange Equations of the Second Kind to Stabilization of the Equilibrium State of a Three-Rod Stewart Platform[13]

13 Kinematics of a Three-Rod Stewart Platform

Introduction. Above we have considered loaded Stewart platforms controlled by six rods (legs) of variable length. Platforms based on three rods are also frequently used in parallel with such problems. Such designs are convenient when it is enough to control only three coordinates, for example, when designing position-sensing mechanisms of active surfaces of radiotelescope mirrors. In this case, it is enough to achieve an accurate angular orientation of the mirrors, and the position of their center of mass does not matter.[14]

In this part of the chapter, we shall study the statics of a loaded Stewart platform with the help of the machinery of the Lagrange equations of the second kind. Later, we shall linearize the Lagrange equations of the second kind, which describe the behavior of the system near the platform equilibrium and derive parameters of the control forces under which the equilibrium position is stable. The following notation may differ from that adopted earlier.

Description of kinematics. Consider the kinematics of a platform based on three rods of adjustable length.

A movable platform is modelled by a flat plate in the form of a regular triangle. The motion is controlled by three rods $A_i B_i$ $(i = \overline{1, 3})$ connected with the movable platform by ball-and-socket joints at the points B_i and connected with the base at the points A_i by cylindrical hinges (see Fig. 13). The position and orientation of the movable platform is controlled by changes in the lengths of the rods on which it rests with corresponding changes in the angles of their inclination to the base. The points B_i $(i = \overline{1, 3})$ form a regular triangle with circumscribed circle of radius R_b; the points A_i $(i = \overline{1, 3})$ in the base also form a regular triangle inscribed in a circle of radius R_a and centered at the point O'. The axis a_i of a cylindrical hinge at each point A_i lies in the base plane and $a_i \perp \overrightarrow{O'A_i}$.

We introduce the fixed Cartesian frame $O'xyz$ and the frame $O\xi\eta\zeta$ attached to the movable platform. The points A_i in the fixed frame of reference have the coordinates

---

[13] This part of chapter "Dynamics and statics of the Stewart Platform" is based on the papers: *V. V. Alexandrov, B. Ya. Lokshin, L. Gomez Esparza, and H. A. Salazar Ibargüen.* Stabilization of a platform under wind loads // J. Math. Sci. (2007) 146: 5863; *S. M. Zuev.* Stabilization of the equilibrium Stewart platform with three degrees of freedom // Vest. St. Petersb. Univ. 2013. Ser. 1. Vyp. 4. Pp. 84–92 [in Russian].

[14] The first telescopes of this type began to work in Hawaii and are currently being built in Mexicos.

Fig. 13 Platform scheme

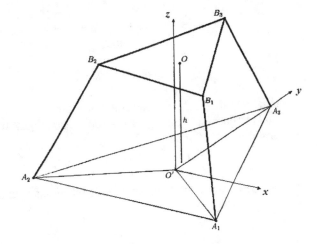

$$A_1 = \left(\frac{\sqrt{3}R_a}{2}, -\frac{R_a}{2}, 0 \right), \quad A_2 = \left(-\frac{\sqrt{3}R_a}{2}, -\frac{R_a}{2}, 0 \right),$$
$$A_3 = (0, R_a, 0).$$

The coordinates of the point B_i in the moving frame $O\xi\eta\zeta$ are as follows:

$$B_1 = \left(\frac{\sqrt{3}R_b}{2}, -\frac{R_b}{2}, 0 \right), \quad B_2 = \left(-\frac{\sqrt{3}R_b}{2}, -\frac{R_b}{2}, 0 \right),$$
$$B_3 = (0, R_b, 0).$$

The radius vectors of the points A_i and B_i in the fixed frame of reference can be looked upon as the columns

$$\mathbf{r}_a^i = \begin{pmatrix} x_a^i \\ y_a^i \\ z_a^i \end{pmatrix}, \quad \mathbf{r}_b^i = \begin{pmatrix} x_b^i \\ y_b^i \\ z_b^i \end{pmatrix}, \quad i = \overline{1,3}.$$

If a point is specified by a vector ρ_ν in the moving frame, then in the fixed frame of reference it is defined by the vector

$$\mathbf{r}^\nu = \mathbf{r}^0 + K\rho_\nu. \tag{13.1}$$

Here $\mathbf{r}^0$ is the radius vector of the point O (the origin of the moving frame), K is the matrix of rotation of the moving frame with respect to the fixed frame of reference:

$$K^{-1} = K^T, \quad \det K = 1,$$

$$K = \begin{pmatrix} c_2 c_3 & -c_2 s_3 & s_2 \\ c_1 s_3 + c_3 s_2 s_1 & s_3 c_1 - s_2 s_3 s_1 & -c_2 s_1 \\ s_1 s_3 - c_1 s_2 c_3 & c_1 s_2 s_3 + c_1 s_3 & c_2 c_1 \end{pmatrix}, \tag{13.2}$$

$$c_i = \cos \psi_i, \quad s_i = \sin \psi_i, \quad i = \overline{1,3}.$$

From formula (13.1) one can find the radius vectors $\mathbf{r}_b^i$ of the points B_i at which each rod is attached to the upper platform. The lengths of the rods are evaluated by the formulas

$$l_i = \sqrt{\left(r_b^i - r_a^i\right)^T \left(r_b^i - r_a^i\right)}, \quad i = \overline{1,3}.$$

We next consider the unit vectors $\mathbf{e}_i$ directed along each of the rods from A_i to B_i:

$$\mathbf{e}_i = \frac{\mathbf{r}_b^i - \mathbf{r}_a^i}{l_i}. \tag{13.3}$$

The platform position is uniquely specified by the vector $\mathbf{q}^T = (x_0, y_0, z_0, \psi_1, \psi_2, \psi_3)$ (with six generalized coordinates), which describes the position and orientation of the platform with respect to the fixed bases, x_0, y_0, z_0 are the Cartesian coordinates of the point O in the fixed frame $O'xyz$, and ψ_1, ψ_2, ψ_3 are, respectively, the roll, pitch, and yaw angles.

14 Dynamics Equations for a Three-Rod Platform

Assume that the center of mass lies at the center of the triangle, the rods are assumed to be of zero mass. As generalized coordinates we choose $\mathbf{q}^T = (q^1, q^2, q^3, q^4, q^5, q^6)$ and write the Lagrange equations of the second kind.

If V_0 is the velocity of the point O, then

$$V_0^2 = \dot{x}_0^2 + \dot{y}_0^2 + \dot{z}_0^2 = \sum_{i=1}^{3} \dot{q}_i^2.$$

We denote by J^i, $i = \overline{1,3}$, the principal central moments of inertia of the loaded platform. Let ω_i be the components of the instantaneous angular velocity vector. We have

$$\omega = \begin{pmatrix} \cos \psi_2 \cos \psi_1 & \sin \psi_1 & 0 \\ -\cos \psi_2 \sin \psi_1 & \sin \psi_1 & 0 \\ \sin \psi_2 & 0 & 1 \end{pmatrix} \begin{pmatrix} \dot{\psi}_1 \\ \dot{\psi}_2 \\ \dot{\psi}_3 \end{pmatrix}.$$

Now the expression for the kinetic energy can be written in the form

Fig. 14 Platform base

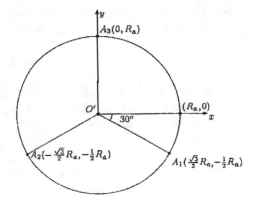

$$T(\mathbf{q}) = \frac{M V_0^2(\mathbf{q})}{2} + \sum_{i=1}^{3} \frac{1}{2} J^i \omega_i^2(\mathbf{q}) . \tag{14.1}$$

Note that the total number of coordinates in the vector $\mathbf{q}$ is superfluous for specifying the platform position. At the points B_i, the rods are connected with the lower platform using cylindrical hinges so that the projection of any of the points A_i to the plane $O'xy$ lies on the line $O'B_i$ (see Fig. 14). To find the dependence between the coordinates, we derive the constraint equations corresponding to the platform dynamics. We have

$$y_b^1 = -\frac{x_b^1}{\sqrt{3}}, \quad y_b^2 = \frac{x_b^2}{\sqrt{3}}, \quad x_b^3 = 0 .$$

As independent coordinates, we take the coordinate q_3 of the center of the movable platform along the axis $O'z$ and the angles of rotation q_4, q_5 about the axes $O'x$ and $O'y$. The remaining coordinates are expressed in terms of the independent ones as follows:

$$\begin{cases} q_1 = -\dfrac{R_b \cos q_5 \sin q_4 \sin q_5}{\cos q_4 \cos q_5 + 1}, \\[2mm] q_2 = \dfrac{R_b}{2} \dfrac{\sin^2 q_5 - \cos^2 q_5 \sin^2 q_4}{\cos q_4 \cos q_5 + 1}, \\[2mm] q_6 = -\arctan\left(\dfrac{\sin q_4 \sin q_5}{\sin q_4 + \sin q_5}\right). \end{cases} \tag{14.2}$$

Here $-\pi/2 < q_4 < \pi/2$, $-\pi/2 < q_5 < \pi/2$.

We relabel the chosen independent coordinates as $p_1 \equiv q_3$, $p_2 \equiv q_4$, $p_3 \equiv q_5$, and take them as new generalized coordinates, which uniquely specify the position of the platform. Now the system of the Lagrange equations of the second kind reads as

$$\frac{d}{dt}\left(\frac{\partial T}{\partial \dot{p}_i}\right) - \frac{\partial T}{\partial p_i} = Q_i, \quad i = \overline{1,3}, \tag{14.3}$$

where Q_i are the generalized forces corresponding to the coordinates p_i $(i = \overline{1,3})$.

Using (14.2) one can express the vector $\mathbf{q}$ in terms of the vector $\mathbf{p}$:

$$\mathbf{q} = \mathbf{q}(\mathbf{p}) \,. \tag{14.4}$$

Substituting (14.4) into (14.1), we find the kinetic energy in terms of the generalized coordinates $\mathbf{p}$ and the generalized velocities $\dot{\mathbf{p}}$.

To find the generalized forces Q_i, we write down the forces and the radius vector of the points of application of the forces in projections onto the axes of the fixed frame of reference. The platform is subject to the gravity force $\mathbf{F}_O$, which is applied at the point O with the radius vector $\mathbf{r}_b^o$ and directed along the axis $O'z$, and to three forces $\mathbf{F}_i$ $(i = \overline{1,3})$, which are applied at the points B_i $(i = \overline{1,3})$ with radius vectors $\mathbf{r}_b^i$ and which are directed along the vectors $\mathbf{e}_i$. From formulas (14.4) and (13.3) we can express $\mathbf{e}_i$ in terms of the generalized coordinates $\mathbf{p}$. As a result, we get

$$\mathbf{F}_i(\mathbf{p}) = F_i \mathbf{e}_i(\mathbf{p}) \,, \quad \mathbf{F}_O = (0, 0, -Mg)^T \,. \tag{14.5}$$

The elementary work can be written as

$$\delta A = \mathbf{F}_O(\mathbf{p})\delta \mathbf{r}^o(\mathbf{p}) + \sum_{i=1}^{3} \mathbf{F}_i(\mathbf{p})\delta \mathbf{r}_b^i(\mathbf{p}) =$$

$$= -Mg\delta p_1 + \sum_{i=1}^{3}\sum_{j=1}^{3} \mathbf{F}_i(\mathbf{p})\frac{\partial \mathbf{r}_b^i(\mathbf{p})}{\partial p_j}\delta p_j \,.$$

Denoting by the components F_{jx}, F_{jy}, F_{jz} of the vector $\mathbf{F}_j$, we find the generalized forces, which are equal to the coefficients multiplying the independent variations δp_i:

$$Q_i = \sum_{j=0}^{3}\left(F_{jx}\frac{\partial x^j}{\partial p_i} + F_{jy}\frac{\partial y^j}{\partial p_i} + F_{jz}\frac{\partial z^j}{\partial p_i} \right) \,. \tag{14.6}$$

We recall that the index $j = 0$ denotes the coordinates of the center of mass of the movable platform and the gravity force. Note that with the help of formulas (14.5) the projections F_{ix}, F_{iy}, F_{iz} can be expressed in terms of the generalized coordinates p_i and the external control forces $\mathbf{F}_i$, $i = \overline{1,3}$, which can be given as functions of time or as functions of the generalized coordinates p_i (or if each function can depend only on "its own" length l_i, which also eventually depends on the generalized coordinates).

Substituting (14.6) into the Lagrange equations (14.3), we get a system of differential equations with respect to the generalized coordinates.

The other way around, with a given programmed motion (i.e., if $p_i = p_i(t)$ are given as functions of time), from (14.3) one can find the control forces F_i $(i = \overline{1,3})$ also as functions of time.

15 Stabilization of Equilibrium of the Horizontal Position of the Platform

Assume that in the system the steady-state regime is implemented,

$$p_1 = h, \quad p_2 = p_3 = 0. \tag{15.1}$$

From the Lagrange equations (14.3), we find the stationary values of the forces F_i^* that secure this equilibrium state:

$$F_i^* = \frac{1}{3p} Mg\sqrt{(R_a - R_b)^2 + h^2}, \quad i = \overline{1,3}.$$

To investigate the behavior of the system near position (15.1), we introduce small increments of coordinates and additional small control forces ΔF_i:

$$p_1 = h + \Delta p_1,$$
$$p_2 = \Delta p_2,$$
$$p_3 = \Delta p_3,$$
$$F_i = F_i^* + \Delta F_i, \quad i = \overline{1,3}.$$

We also introduce the dimensionless control forces:

$$u_i = \frac{\Delta F_i}{F_i^*}, \quad i = \overline{1,3}.$$

From the Lagrange equations (14.3) we get the equations of first approximation, which we write in the matrix form:

$$\ddot{\mathbf{p}} = H\mathbf{p} + G\mathbf{u}. \tag{15.2}$$

Here H and G are constant matrices of size 3×3, $\mathbf{u} = (u_1, u_2, u_3)^T$.

Putting $u_i = 0$, $i = \overline{1,3}$, in Eq. (15.2), we get the system

$$\ddot{\mathbf{p}} = H\mathbf{p},$$

whose solution is unstable. To verify this, it suffices to write the matrix H as

$$H = \begin{pmatrix} H_{11} & 0 & 0 \\ 0 & H_{22} & 0 \\ 0 & 0 & H_{33} \end{pmatrix},$$

where

$$H_{11} = \frac{gs^2}{h(s^2 + h^2)}, \qquad H_{22} = \frac{MgR_b(s^2 R_a + h^2(R_a - R_b))}{2h J_1(s^2 + h^2)},$$

$$H_{33} = \frac{MgR_b(s^2 R_a + h^2(R_a - R_b))}{2h J_2(s^2 + h^2)}, \qquad s^2 = (R_a - R_b)^2.$$

From the form of the matrix H it follows that the oscillations in terms of the generalized coordinates are separated, and so the trivial solution is exponentially unstable. This means that an additional control action is required for implementation of the stationary position (15.1) of the platform.

We write the system of differential equations in the Cauchy form:

$$\dot{\mathbf{z}} = A\mathbf{z} + B\mathbf{u},$$
$$\mathbf{z} = (p_1, \dot{p}_1, p_2, \dot{p}_2, p_3, \dot{p}_3)^T; \qquad (15.3)$$

here

$$A = \begin{pmatrix} 0 & 1 & 0 & 0 & 0 & 0 \\ A_{21} & 0 & 0 & 0 & 0 & 0 \\ 0 & 0 & 0 & 1 & 0 & 0 \\ 0 & 0 & A_{43} & 0 & 0 & 1 \\ 0 & 0 & 0 & 0 & A_{55} & 0 \end{pmatrix},$$

$$A_{21} = \frac{gs^2}{h(s^2 + h^2)}, \qquad A_{43} = \frac{MgR_b(s^2 R_a + h^2(R_a - R_b))}{2h J_1(s^2 + h^2)},$$

$$A_{55} = \frac{MgR_b(s^2 R_a + h^2(R_a - R_b))}{2h J_2(s^2 + h^2)},$$

$$B = \begin{pmatrix} 0 & 0 & 0 \\ \frac{g}{3} & \frac{g}{3} & \frac{g}{3} \\ 0 & 0 & 0 \\ -\frac{MgR_b}{6J_1} & -\frac{MgR_b}{6J_1} & -\frac{MgR_b}{3J_1} \\ 0 & 0 & 0 \\ -\frac{MgR_b}{2\sqrt{(3)}J_2} & \frac{MgR_b}{2\sqrt{(3)}J_2} & 0 \end{pmatrix}.$$

We construct the control in the form of linear feedbacks

$$\mathbf{u} = K\mathbf{z}, \qquad (15.4)$$

where $K = |k_{ij}|_{(3,6)}$ are the coefficients of the constant matrix which need to be determined. We choose the coefficients of the feedback control matrix so as to split the system into three independent subsystems. We test each of these subsystems for stability. Substituting (15.4) into (15.3), we get the closed system

$$\dot{z} = (A + BK)z + Cz, \tag{15.5}$$

where the matrix C reads as

$$C = \begin{pmatrix} 0 & 1 & 0 & 0 & 0 & 0 \\ c_{1,1} & c_{1,2} & c_{1,3} & c_{1,4} & c_{1,5} & c_{1,6} \\ 0 & 0 & 0 & 1 & 0 & 0 \\ c_{2,1} & c_{2,2} & c_{2,3} & c_{2,4} & c_{2,5} & c_{2,6} \\ 0 & 0 & 0 & 0 & 0 & 1 \\ c_{3,1} & c_{3,2} & c_{3,3} & c_{3,4} & c_{3,5} & c_{3,6} \end{pmatrix}.$$

We split the system into three independent systems by equating to zero the succeeding coefficients c_{ij} (in order that the matrix C would become of block diagonal form):

$$\begin{cases} c_{1,3} = c_{1,4} = c_{1,5} = c_{1,6} = 0, \\ c_{2,1} = c_{2,2} = c_{2,5} = c_{2,6} = 0, \\ c_{3,1} = c_{3,2} = c_{3,3} = c_{3,4} = 0, \end{cases} \tag{15.6}$$

$$C = \begin{pmatrix} C_1 & 0 & 0 \\ 0 & C_2 & 0 \\ 0 & 0 & C_3 \end{pmatrix}.$$

Here all the matrices C_k, $k = \overline{1,3}$, are of size 2×2.

Consider conditions (15.6) as a system of 12 algebraic equations with respect to 18 unknowns $k_{i,j}$. As independent 6 coefficients we take $k_{1,i}$, $i = \overline{1,6}$; the remaining coefficients of the matrix K can be expressed in terms of $k_{1,i}$, $i = \overline{1,6}$ using system (15.6).

So, the system of differential equations (15.5) splits into three subsystems. Consider their characteristic equations

$$\det(C_k - E\lambda) = 0, \quad k = \overline{1,3}. \tag{15.7}$$

For the stability of the solution it is necessary and sufficient[15] that the real parts of the roots of the characteristic equations be negative. A sufficient condition for this is that all coefficients for each of the resulting characteristic equations be of the same sign, because all equations are of second degree.

The characteristic equations read as

$$\lambda^2 + d_{1i}\lambda + d_{2i} = 0, \quad i = \overline{1,3}.$$

Therefore, a sufficient condition for the stability is that all the coefficients d_{1i} and d_{2i}, $i = \overline{1,3}$, be positive.

---

[15] See, for example, the book: *D.R. Merkin*. Introduction to the theory of stability (Texts in Applied Mathematics, Vol. 24). Springer Science and Business Media. 2012. 320 p.

As a result, we get the following constraints on the coefficients $k_{1,i}$ sufficient for the asymptotic stability of the trivial solution of system (15.5):

$$
\begin{cases}
k_{1,1} < -\dfrac{s^2}{h(s^2 + h^2)}, \\[2mm]
k_{1,2} < 0, \\[2mm]
k_{1,3} > \dfrac{s^2 R_a + h^2(R_a - R_b)}{2h(s^2 + h^2)}, \\[2mm]
k_{1,4} > 0, \\[2mm]
k_{1,5} > \dfrac{\sqrt{3}(s^2 R_a + h^2(R_a - R_b))}{2h(s^2 + h^2)}, \\[2mm]
k_{1,6} > 0.
\end{cases}
\tag{15.8}
$$

So, we have found the parameters of the feedback constraint equation securing the asymptotic stability against small deviations from the stationary position.

16 A Numerical Example

Consider the following platform parameter:

$$
M = 200\,(\text{kg}), \quad R_a = 3\,(\text{m}), \quad R_b = 2\,(\text{m}), \quad h = 2\,(\text{m}).
$$

We have

$$
J_x = J_y = \frac{M R_b^2}{4} = 200\,(\text{kg}\,\text{m}^2), \quad J_z = \frac{M R_b^2}{2} = 400\,(\text{kg}\,\text{m}^2),
$$

$$
F_i^* = 731.194\,(\text{N}),
$$

$$
A = \begin{pmatrix}
0 & 1 & 0 & 0 & 0 & 0 \\
0.981 & 0 & 0 & 0 & 0 & 0 \\
0 & 0 & 0 & 1 & 0 & 0 \\
0 & 0 & 6.867 & 0 & 0 & 1 \\
0 & 0 & 0 & 0 & 6.867 & 0
\end{pmatrix},
$$

$$
B = \begin{pmatrix}
0 & 0 & 0 \\
3.27 & 3.27 & 3.27 \\
0 & 0 & 0 \\
-3.27 & -3.27 & -6.54 \\
0 & 0 & 0 \\
-5.664 & 5.664 & 0
\end{pmatrix}.
$$

By the above constraints (15.8), we have the following requirements on the coefficients of the feedback matrix:

$$k_{1,1} < -0.1 , \quad k_{1,2} < 0 , \quad k_{1,3} > 0.35 ,$$
$$k_{1,4} > 0 , \quad k_{1,5} > 0.606 , \quad k_{1,6} > 0 .$$

Let

$$k_{1,1} = -1 , \quad k_{1,2} = -1 , \quad k_{1,3} = 1 , \quad k_{1,4} = 1 , \quad k_{1,5} = 1 , \quad k_{1,6} = 1 .$$

Then the characteristic equations (15.7) have roots with negative real parts. For the first equation

$$\lambda^2 + 3\lambda + 0.826 = 0 ,$$

which is responsible for the oscillations of the center of mass, the roots are as follows: $-0.307, -2.69$. For the second and third equations,

$$\lambda^2 + 6\lambda + 1.19 = 0 ,$$

$$\lambda^2 + 3.464\lambda + 0.416 = 0$$

(these equations are responsible for the oscillations of the platform near its center of mass), the roots are as follows: $-0.205, -5.79$ and $-0.125, -3.34$, respectively.

So, we have evaluated the parameters of the feedback for which the horizontal equilibrium position of the platform is stable. Note that the above chosen coefficients of the feedback matrix are only one of various possible tuples of coefficients which provide the stability.

Chapter 4
Vibrations and Autobalancing of Rotor Systems

V. G. Bykov⊙, A. S. Kovachev⊙, and P. E. Tovstik

This chapter discusses the simplest models of rotor systems with a finite number of degrees of freedom. Various types of rotor vibrations due to their unbalance, the unequal elastic characteristics of the shaft or bearings, as well as the influence of internal friction and structural damping are studied. Problems of balancing of rotors equipped with a passive automatic ball balancers are investigated.

1 Forced and Self-excited Oscillations of a Rotor with an Isotropic Viscoelastic Shaft[1]

Basic concepts. The rotor is called the rotating element of a mechanism or machine, held in the supporting roller or slide bearings. We will consider the mechanical model of a rotor in the form of a massive solid body mounted on a weightless flexible or rigid shaft rotating in rigid or elastically fixed supporting bearings. The straight line connecting the bearing centers is called the *axis of rotation* of the rotor. A rotor is called *balanced* if one of its main central axes of inertia (hereinafter, for brevity, we call it *the polar axis*) coincides with the axis of rotation. A balanced rotor with isotropic (that is, identical in all directions) viscoelastic characteristics of the shaft and supports does not experience variable inertial loads during rotation that cause shaft bending or deformation of elastic supports.

If the polar axis of the rotor does not coincide with the axis of rotation, then there are three main types of unbalance:

(1) *static unbalance* occurs in the case when the polar axis of the rotor is parallel to the axis of rotation (Fig. 1a);

---

[1] This section contains materials partially published in the article: *V.G. Bykov, P.E. Tovstik.* Synchronous Whirlings and Self-oscillations of a Statically Unbalanced Rotor in Limited Excitation // Mechanics of Solids. 2018. Vol. 53. Suppl. 2. Pp. S60–S70.

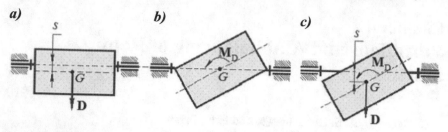

Fig. 1 The types of rotor unbalances

(2) *torque unbalance* occurs if the polar axis intersects the axis of rotation at the center of gravity of the rotor G (Fig. 1b);
(3) *dynamic unbalance* implies a case when the polar axis is not parallel to and does not intersect the shaft axis at the center of gravity (Fig. 1c).

A measure of unbalance of some mass is the *unbalance vector* directed along *the eccentricity*, which is the radius-vector of the center of this mass relative to the rotor axis. The magnitude of the unbalance vector is equal to the product of the unbalanced mass by the eccentricity value. All rotor unbalances are reduced to two vectors: the main vector **D** and the main moment of unbalances $\mathbf{M}_D$ regardless of the reasons that caused the displacement s of the center of mass G from the axis of rotation.

Rotor unbalance is responsible for the appearance the inertial forces and moments that cause forced lateral vibrations of the shaft, which, in turn, can lead to increased loads on the bearings and dangerous vibrations of the rotor machine. In practice, the structural unbalance of the rotor is eliminated by balancing (i.e., fixing corrective masses of certain places). However, during the exploitation of certain types of high-speed rotary machines (for example, centrifuges, separators, etc.), in the process of their operations, mode changes of unbalance may occur that cannot be eliminated by the rotor spot balancing. In these cases, the most promising and sometimes the only possible way to eliminate the unbalance is the use of autobalancing devices of various types.[2] However, even a fully balanced rotor can cause self-excited vibrations due to a loss of stability of the main rotation mode. The reasons for their appearance are associated with parametric excitation due to the uneven stiffness of elastic shaft or elastic supports, as well as the influence of internal friction in the shaft and structural damping between several elements of the rotor system.

Jeffcott rotor model. Equations of motion in fixed, rotating and polar coordinate systems. To study the dynamics and stability of rotor systems, the simplest mechanical model of the rotor in the form of a massive stiff disk fixed in the middle of a weightless flexible shaft is often used. In context of this model, which in the literature is called a Jeffcott rotor,[3] only flat motion of the disk is considered and the influence of gyroscopic forces and moments is neglected.

---

[2] *A.A. Gusarov*. Self-balancing Devices of Direct Action. Moscow: Nauka. 2002. 120 p. [in Russian].
[3] *G. Genta*. Dynamics of Rotating Systems. Springer. 2005. 658 p.

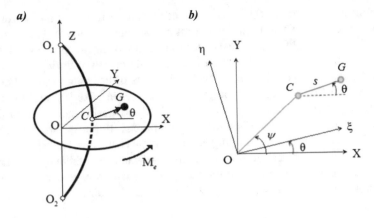

Fig. 2 Jeffcott rotor model

Consider a model of the Jeffcott rotor mounted in vertical hinge supports, as shown in Fig. 2a. We assume that the motion of the rotor is subject an external torque M_e applied to the disk. Supposing that the center of mass of the disk G does not coincide with the point of attachment of the disk to the shaft C, we denote by $s = |\overrightarrow{CG}|$ the magnitude of the rotor static eccentricity vector. We introduce fixed frame $OXYZ$ as follows: the Z-axis is directed vertically up along the line connecting the centers of the bearings O_1 and O_2, and we place the X- and Y-axes in the horizontal plane passing through the points C and G (static eccentricity planes). The spin angle of the rotor θ is the angle between the OX axis and the eccentricity vector $\overrightarrow{CG}$. In the plane of static eccentricity, we additionally introduce the rotating frame $O\xi\eta$, as shown in Fig. 2b.

The above rotor model has three degrees of freedom. Let us choose the absolute coordinates x,y of the point C and the spin angle of the rotor θ as generalized coordinates . Expressing the coordinates of the center of mass of the disk through the generalized coordinates, the kinetic energy of the rotor reads as

$$T = \frac{1}{2}m\big((\dot{x} - s\dot{\theta}\sin\theta)^2 + (\dot{y} + s\dot{\theta}\cos\theta)^2\big) + \frac{1}{2}J_G\dot{\theta}^2 , \qquad (1.1)$$

where m is the mass of the disk, and J_G is its moment of inertia about the perpendicular axis passing through the point G.

Suppose that the rotor shaft is isotropic, that is, the coefficient of elasticity of the shaft k is the same for all directions perpendicular to its axis. Hence the potential energy of the elastic shaft can be written as

$$V = \frac{1}{2}k(x^2 + y^2) . \qquad (1.2)$$

We will assume further that four types of resistance force act on the rotor: (1) friction forces from a fixed part of the rotor machine (friction in bearings, fixed dampers); (2) external forces of resistance to lateral movements of the shaft; (3) the forces of internal friction in the shaft due to its deformation during bending; (4) structural damping forces between several elements of the rotor system. Forces of resistance of the first and second types, when considered in a fixed frame, refer to *non-rotational damping*, and damping forces of the third and fourth type, when considered in a rotating frame, are referred as *rotational damping*. Suppose for simplicity that non-rotational and rotational damping is viscous, that is, the resistance forces depend linearly on absolute or relative velocities. Then we can define them by means of the Rayleigh dissipative function R, considering it as the sum of two components R_n and R_r representing the non-rotational and rotational damping

$$R = R_n + R_r, \quad R_n = \frac{1}{2} c_n(\dot{x}^2 + \dot{y}^2) + \frac{1}{2} c_\theta \dot{\theta}^2, \quad R_r = \frac{1}{2} c_r(\dot{\xi}^2 + \dot{\eta}^2). \quad (1.3)$$

Here c_n denotes the damping coefficient corresponding to the forces of external resistance to the lateral movement of the shaft, c_θ is the damping coefficient corresponding to the forces of external resistance to rotational motion, c_r is the coefficient of rotational damping, ξ and η are coordinates of the point C in the rotating frame. Considering the relation between the fixed and rotating frames

$$\xi = x \cos\theta + y \sin\theta, \quad \eta = -x \sin\theta + y \cos\theta, \quad (1.4)$$

the rotation component of the Rayleigh function can be represented as

$$R_r = \frac{1}{2} c_r((\dot{x} + \dot{\theta}y)^2 + (\dot{y} - \dot{\theta}x)^2). \quad (1.5)$$

To find the generalized forces, we write the expression for the elementary work of external forces

$$\delta A = Q_x \delta x + Q_y \delta y + Q_\theta \delta\theta = M_e \delta\theta,$$

from which it follows that

$$Q_x = Q_y = 0, \quad Q_\theta = M_e. \quad (1.6)$$

With regard to expressions (1.1)–(1.6) the Lagrange equations of the second kind, describing the dynamics of the presented rotor model in a fixed frame, have the form

$$\begin{cases} m\ddot{x} + (c_n + c_r)\dot{x} + c_r\dot{\theta}y + kx = ms(\dot{\theta}^2 \cos\theta + \ddot{\theta} \sin\theta), \\ m\ddot{y} + (c_n + c_r)\dot{y} - c_r\dot{\theta}x + ky = ms(\dot{\theta}^2 \sin\theta - \ddot{\theta} \cos\theta), \\ J_c\ddot{\theta} + c_\theta\dot{\theta} + c_r(\dot{x}y - x\dot{y} + (x^2 + y^2)\dot{\theta}) = M_e + ms(\ddot{x} \sin\theta - \ddot{y} \cos\theta), \end{cases} \quad (1.7)$$

where $J_C = J_G + ms^2 = mr^2$, r is the disk inertia radius.

To reduce the number of parameters, we proceed to dimensionless coordinates and time

$$\bar{x} = \frac{x}{r}, \quad \bar{y} = \frac{y}{r}, \quad \bar{t} = \frac{t}{\Omega}, \quad \text{where } \Omega = \sqrt{k/m}$$

and write the system of equations (1.7) in the dimensionless form

$$\begin{cases} \ddot{\bar{x}} + (\delta_n + \delta_r)\dot{\bar{x}} + \delta_r\dot{\theta}\bar{y} + \bar{x} = \varepsilon(\dot{\theta}^2\cos\theta + \ddot{\theta}\sin\theta)\,, \\ \ddot{\bar{y}} + (\delta_n + \delta_r)\dot{\bar{y}} - \delta_r\dot{\theta}\bar{x} + \bar{y} = \varepsilon(\dot{\theta}^2\sin\theta - \ddot{\theta}\cos\theta)\,, \\ \ddot{\theta} + \delta_\theta\dot{\theta} + \delta_r(\dot{\bar{x}}\bar{y} - \bar{x}\dot{\bar{y}} + (\bar{x}^2 + \bar{y}^2)\dot{\theta}) = \mu + \varepsilon(\ddot{\bar{x}}\sin\theta - \ddot{\bar{y}}\cos\theta)\,, \end{cases} \tag{1.8}$$

where

$$\delta_n = \frac{c_n}{m\Omega}, \quad \delta_r = \frac{c_r}{m\Omega}, \quad \delta_\theta = \frac{c_\theta}{mr^2\Omega}, \quad \mu = \frac{M_e}{mr^2\Omega^2}, \quad \varepsilon = \frac{s}{r}.$$

In system (1.8), the points above the variables mean differentiation with respect to the dimensionless time $\bar{t}$, and the dimensionless parameters δ_n, δ_r, δ_θ characterize viscous damping, μ is the external torque and ε is the static eccentricity. In what follows to simplify notation, we omit the bar over dimensionless variables.

To study whirling motions of the rotor, it is convenient to use equations in the rotating frame $O\xi\eta$, which can be obtained by substituting the relations between the fixed and rotating dimensionless coordinates into the Eqs. (1.8)

$$x = \xi\cos\theta - \eta\sin\theta, \quad y = \xi\sin\theta + \eta\cos\theta. \tag{1.9}$$

The sum and difference of first two equations of the resulting system, give, after elementary transformations,

$$\begin{cases} \ddot{\xi} + (\delta_n + \delta_r)\dot{\xi} + (1 - \nu^2)\xi - 2\nu\dot{\eta} - (\dot{\nu} + \delta_n\nu)\eta = \varepsilon\nu^2\,, \\ \ddot{\eta} + (\delta_n + \delta_r)\dot{\eta} + (1 - \nu^2)\eta + 2\nu\dot{\xi} + (\dot{\nu} + \delta_n\nu)\xi = -\varepsilon\dot{\nu}\,, \\ \dot{\nu} + \delta_\theta\nu + \delta_r(\dot{\xi}\eta - \dot{\eta}\xi) = \mu - \varepsilon(\ddot{\eta} + 2\nu\dot{\xi} + \xi\dot{\nu} - \eta\nu^2)\,, \end{cases} \tag{1.10}$$

where $\nu = \dot{\theta}$ is the dimensionless spin speed of the shaft.

Along with the rotating frame, we will also use the polar coordinates: the dimensionless amplitude $a = |\overrightarrow{OC}|/r$ of whirling motion and the total phase angle ψ of the radius vector $\overrightarrow{OC}$ in a fixed frame (see Fig. 2b). We substitute $x = a\cos\psi$, $y = a\sin\psi$ into the system (1.8) and perform the following transformations: (1) multiply the first equation by $\cos\psi$ and add it with the second one multiplied by $\sin\psi$; (2) multiply the first equation by $\sin\psi$ and subtract the second equation multiplied by $\cos\psi$. t As a result, we obtain

$$\begin{cases} \ddot{a} + (\delta_n + \delta_r)\dot{a} + (1 - \dot{\psi}^2)a = \varepsilon(\dot{\theta}^2 \cos(\theta - \psi) + \ddot{\theta}\sin(\theta - \psi)) , \\ (\ddot{\psi} + \delta_n\dot{\psi} - \delta_r(\dot{\theta} - \dot{\psi}))a + 2\dot{a}\dot{\psi} = \varepsilon(\dot{\theta}^2\sin(\theta - \psi) - \ddot{\theta}\cos(\theta - \psi)) , \\ \ddot{\theta} + \delta_\theta\dot{\theta} + \delta_r a^2(\dot{\theta} - \dot{\psi}) = \mu - \varepsilon((\ddot{a} - a\dot{\psi}^2)\sin(\theta - \psi) + (a\ddot{\psi} + 2\dot{a}\dot{\psi})\cos(\theta - \psi)) . \end{cases}$$

$$(1.11)$$

The order of system (1.11) can be reduced, if we take the phase angle $\phi = \theta - \psi$, the dimensionless whirling speed $\dot{\psi} = \omega$ and the spin speed $\dot{\theta} = \nu$ as the new variables:

$$\begin{cases} \ddot{a} + (\delta_n + \delta_r)\dot{a} + (1 - \omega^2)a = \varepsilon(\nu^2 \cos\phi + \dot{\nu}\sin\phi) , \\ (\dot{\omega} + \delta_n\omega - \delta_r\dot{\phi})a + 2\dot{a}\omega = \varepsilon(\nu^2\sin\phi - \dot{\nu}\cos\phi) , \\ \dot{\nu} + \delta_\theta\nu + \delta_r\dot{\phi}a^2 = \mu - \varepsilon((\ddot{a} - a\omega^2)\sin\phi + (a\dot{\omega} + 2\dot{a}\omega)\cos\phi) , \\ \dot{\phi} = \nu - \omega. \end{cases}$$

$$(1.12)$$

System (1.12) is fully equivalent to systems (1.8) and (1.10), but is more convenient for the study of asynchronous rotor whirling.

Note that all the presented systems of equations are non-linear, so their analysis involves known difficulties. At the same time, a number of phenomena in the dynamics of rotor systems can be studied in terms of a simpler model, in which the spin speed is considered to be constant.

Forced and self-excited oscillations of a rotor rotating at a constant angular speed. Let us suppose that the rotor shaft rotates at a constant dimensionless angular speed $\dot{\theta} = \nu = \text{const}$. In this case, the rotor rotation angle is considered to be given, and the rotor has only two degrees of freedom. The equations of motion of the rotor with respect to the coordinates x and y can be obtained by substituting $\theta = \nu t$ into the first two equations of system (1.8):

$$\begin{cases} \ddot{x} + (\delta_n + \delta_r)\dot{x} + \delta_r\nu y + x = \varepsilon\nu^2 \cos\nu t , \\ \ddot{y} + (\delta_n + \delta_r)\dot{y} - \delta_r\nu x + y = \varepsilon\nu^2 \sin\nu t . \end{cases}$$

$$(1.13)$$

Introducing the complex variable $z = x + iy$, we write the system (1.13) as one complex equation

$$\ddot{z} + (\delta_n + \delta_r)\dot{z} + (1 - i\delta_r\nu)z = \varepsilon\nu^2 e^{i\nu t} .$$

$$(1.14)$$

Since the Eq. (1.14) is linear in z, its general solution can be represented as the sum of the particular solution of the inhomogeneous equation and the general solution of the corresponding homogeneous equation

$$z = z_0 e^{i\nu t} + C_1 e^{\lambda_1 t} + C_2 e^{\lambda_2 t} ,$$

$$(1.15)$$

where C_1 and C_2 are arbitrary constants, and λ_1 and λ_2 are the roots of the characteristic equation

$$\lambda^2 + (\delta_n + \delta_r)\lambda + 1 - i\delta_r\nu = 0 .$$

$$(1.16)$$

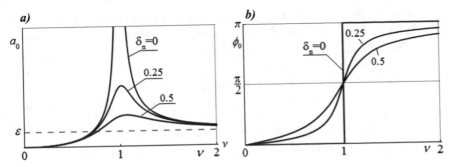

Fig. 3 Amplitude-frequency (**a**) and phase-frequency (**b**) responses of forced vibrations

Substituting into Eq. (1.14) a particular solution of the form $z = z_0 e^{i\nu t}$, we find

$$z_0 = \frac{\varepsilon \nu^2}{1 - \nu^2 + i\delta_n \nu} \, . \tag{1.17}$$

Returning to the real coordinates, we obtain a particular solution of system (1.13)

$$x_0 = a_0 \cos(\nu t + \phi_0), \qquad y_0 = a_0 \sin(\nu t + \phi_0), \tag{1.18}$$

where

$$a_0 = |z_0| = \frac{\varepsilon \nu^2}{\sqrt{(1 - \nu^2)^2 + (\delta_n \nu)^2}}, \qquad \tan \phi_0 = \frac{\delta_n \nu}{\nu^2 - 1} \, .$$

A particular solution (1.18), due to the unbalance of the rotor, describes the forced vibrations in the form of circular synchronous whirling of the disk center. The amplitude a_0 of whirling of the point C and the phase angle ϕ_0 between the vectors $\overrightarrow{OC}$ and $\overrightarrow{CG}$ depend on the frequency ν, which coincides with the angular speed of rotation of the disk. Note that expression (1.18) does not include the rotational damping coefficient δ_r. This is due to the fact that the bent rotor shaft at synchronous regular whirling during one revolution does not experience tensile and compression deformations, which are responsible for internal friction.

Figure 3 shows the amplitude-frequency and phase-frequency responses of the forced vibrations of the rotor, as calculated for various values of the external damping coefficient δ_n. The resonance nature of the whirling amplitude is manifested when the angular speed of the rotor ν approaches the critical value ν_c corresponding to the natural frequency of the lateral oscillations of the shaft:

$$\nu_c = \frac{1}{\sqrt{1 - \dfrac{\delta_n^2}{2}}} \, . \tag{1.19}$$

Analysis of the phase-frequency response shows that when the angular speed of the rotor is below the critical value ($\nu < 1$), the phase angle $\phi_0 < \pi/2$, i.e., vector $\overrightarrow{CG}$ is directed away from the axis of rotation. When $\nu = 1$, the angle ϕ_0 becomes $\pi/2$, and in the supercritical region $\nu > 1$, the vector $\overrightarrow{CG}$ rotates in the direction of the axis of rotation. With increase of the angular speed, the center of mass of the disk G approaches the point O, therefore this effect is called *rotor self-centering* in the supercritical region.

We now consider the general solution of the homogeneous equation (1.14) and write the expression for the roots of the characteristic equation (1.16)

$$\lambda_{1,2} = -\frac{1}{2}\left(\delta_n + \delta_r \pm \sqrt{(\delta_n + \delta_r)^2 - 4 + i4\delta_r\nu}\right). \qquad (1.20)$$

The nature of the general solution can be visualized by considering the signs of the real and imaginary parts of λ_1 and λ_2. Using the well-known formula

$$\sqrt{a \pm ib} = \frac{1}{\sqrt{2}}\left(\sqrt{\sqrt{a^2 + b^2} + a} \pm i\sqrt{\sqrt{a^2 + b^2} - a}\right), \qquad (1.21)$$

we find that $\qquad \text{Re}\,\lambda_{1,2} = -\dfrac{\delta_n + \delta_r}{2} \mp \mathfrak{R}^+, \quad \text{Im}\,\lambda_{1,2} = \mp\mathfrak{R}^-,$

where $\qquad \mathfrak{R}^\pm = \dfrac{1}{\sqrt{2}}\sqrt{\sqrt{((\delta_n+\delta_r)^2-4)^2+(4\delta_r\nu)^2}\pm((\delta_n+\delta_r)^2-4)}.$

Figure 4 depicts the real and imaginary parts of the roots of the characteristic equation versus the frequency ν, as calculated with $\delta_n = \delta_r = 0.06$. Dashed lines correspond to the case $\delta_r = 0$. It can be seen from the graphs that the inequalities $\text{Im}\,\lambda_1 < 0$, $\text{Im}\,\lambda_2 > 0$ and $\text{Re}\,\lambda_1 < 0$ are valid in the whole range of angular speeds, while the inequality $\text{Re}\,\lambda_2 < 0$ is valid only under the condition $\nu < \nu_*$, where ν_* is the threshold angular speed defined by the equation $\text{Re}\,\lambda_2 = 0$, which has a single positive root

$$\nu = \nu_* = 1 + \frac{\delta_n}{\delta_r}. \qquad (1.22)$$

A comparison of formulas (1.19) and (1.22) shows that when $\delta_n = 0$, i. e., in the absence of external friction, the threshold frequency coincides with the critical one, and if external friction is present, then the threshold frequency is greater than the critical threshold.

Analysis of the roots of the characteristic equation allows us to conclude that the free vibrations of the rotor, as described by the general solution of the homogeneous equation (1.14), are the sum of the forward and backward whirlings. In this case, the amplitude of the backward whirling decays faster than the amplitude of the forward one, since $|\text{Re}\,\lambda_1| > |\text{Re}\,\lambda_2|$. If the angular speed of the rotor ν is less than the threshold value ν_*, then the proper whirling motions decay exponentially, and, after a certain period of time, a purely forced synchronous whirling mode (1.18) is

Fig. 4 Real and imaginary parts of the roots of the characteristic equation (1.20)

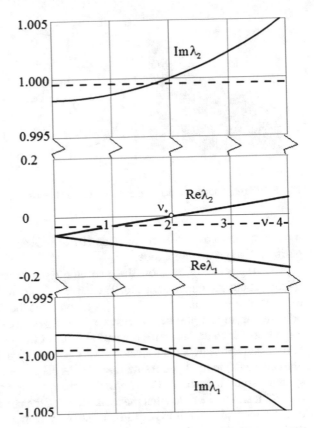

established in the system. In the case when the angular speed exceeds the supercritical threshold value, the movement of the rotor will be unstable due to the exponential growth of the amplitude of its own forward whirling.

Let us consider in more detail the case of $\nu = \nu_*$, when the angular speed is exactly equal to the threshold one. Considering that the roots of the characteristic equation (1.20) in this case are of the form

$$\lambda_1 = -(\delta_n + \delta_r + i), \qquad \lambda_2 = i,$$

we represent the general solution (1.15) as the sum of forward and backward whirlings with variable amplitudes

$$z = (z_0 e^{i(\delta_n/\delta_r)t} + C_2)e^{it} + C_1 e^{-(\delta_n + \delta_r)t} e^{-it}. \tag{1.23}$$

The amplitude of the backward whirling decreases exponentially with time, and since the amplitude of the forward whirling changes periodically, the steady-state

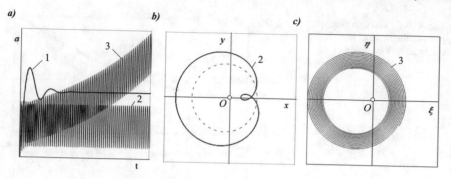

Fig. 5 Whirling motions types of the rotor rotating at constant angular speed

motion of the rotor will have the character of forward asynchronous whirling with periodically varying amplitude.

The graphs in Fig. 5a show three specific types of whirling motion of the rotor rotating at constant angular speed. The curve 1 (synchronous whirling) corresponds to the case $\nu < \nu_*$; the curve 2 ($\nu = \nu_*$) represents the forward whirling with periodically varying amplitude; the curve 3 ($\nu > \nu_*$) demonstrates an unstable whirling motion, in which the mean for the period value of the amplitude increases exponentially. Figure 5b shows the trajectories of the point C (solid curve) and the point G (dashed curve) in the fixed frame Oxy for the case $\nu = \nu_*$. Figure 5c represents the spinning spiral trajectory of the point C in the rotating frame $O\xi\eta$ in the case when $\nu > \nu_*$. It should be noted that the assumption of the constancy of the angular speed of the rotor under conditions of loss of stability of the rotational mode leads to an unlimited increase in the amplitude of the whirling. However, such an increase in the amplitude can occur only due to the transfer of the energy of the rotor's spin. Therefore, maintaining a constant angular speed requires an energy source of unlimited power. In reality, such a situation is impossible, therefore, to study the rotor dynamics at frequencies exceeding the threshold value, a mechanical model is necessary capable of taking into account the factor of limited excitation.

Stability of spin rotation of a balanced rotor with limited excitation. Consider the question of the stability of the rotational motion of a fully balanced rotor under the action of an external torque. Assuming that $\varepsilon = 0$ in system (1.10), we obtain equations describing the motion of a balanced rotor in a rotating frame

$$\begin{cases} \ddot{\xi} + (\delta_n + \delta_r)\dot{\xi} + (1 - \nu^2)\xi - 2\nu\dot{\eta} - (\dot{\nu} + \delta_n\nu)\eta = 0, \\ \ddot{\eta} + (\delta_n + \delta_r)\dot{\eta} + (1 - \nu^2)\eta + 2\nu\dot{\xi} + (\dot{\nu} + \delta_n\nu)\xi = 0, \\ \dot{\nu} + \delta_\theta\nu + \delta_r(\dot{\xi}\eta - \dot{\eta}\xi)\eta = \mu. \end{cases} \quad (1.24)$$

Assuming that the parameter μ, which characterizes the external torque, to be constant, we substitute in (1.24) a stationary solution of the form $\nu = \nu_0 = \text{const}$, $\xi = \xi_0 = \text{const}$, $\eta = \eta_0 = \text{const}$. As a result, we obtain an algebraic system of equations for ν_0, ξ_0 η_0

$$\begin{cases} (1 - \nu_0^2)\xi_0 - \delta_n \nu_0 \eta_0 = 0, \\ \delta_n \nu_0 \xi_0 + (1 - \nu_0^2)\eta_0 = 0, \\ \delta_\theta \nu_0 = \mu. \end{cases} \quad (1.25)$$

Considering the first two Eqs. (1.25) as a linear system with respect to ξ_0 and η_0, we write its determinant

$$D = (1 - \nu_0^2)^2 + (\delta_n \nu_0)^2. \quad (1.26)$$

Since $D > 0$, system (1.25) has a unique trivial solution

$$\xi_0 = \eta_0 = 0, \qquad \nu_0 = \mu/\delta_\theta, \quad (1.27)$$

which corresponds to the rotation of the disk at constant angular speed in the absence of bending of the shaft. Further, solution (1.27) will be called the steady-state spin mode.

To study the stability of this mode, we substitute $\xi = \Delta\xi$, $\eta = \Delta\eta$, $\nu = \nu_0 + \Delta\nu$ into Eqs. (1.24), where $\Delta\xi$, $\Delta\eta$ $\Delta\nu$ are small variations of coordinates and angular speed. Expanding the obtained expressions in series and neglecting small quantities of second order and higher, we obtain a linearized system of equations in variations

$$\begin{cases} (\ddot{\Delta\xi}) + (\delta_n + \delta_r)(\dot{\Delta\xi}) - 2\nu_0(\dot{\Delta\eta}) + (1 - \nu_0^2)\Delta\xi - \delta_n \nu_0 \Delta\eta = 0, \\ (\ddot{\Delta\eta}) + (\delta_n + \delta_r)(\dot{\Delta\eta}) + 2\nu_0(\dot{\Delta\xi}) + (1 - \nu_0^2)\Delta\eta + \delta_n \nu_0 \Delta\xi = 0, \\ (\dot{\Delta\nu}) + \delta_\theta \Delta\nu = 0. \end{cases} \quad (1.28)$$

The third equation of system (1.28) is independent and easily integrated; therefore, to study the stability, it suffices to find a solution to a linear system of two second-order equations with respect to the variations of the coordinates $\Delta\xi$ and $\Delta\eta$. Introduction of the complex variation $\Delta\zeta = \Delta\xi + i\Delta\eta$ leads this system to a single second-order equation

$$(\ddot{\Delta\zeta}) + (\delta_n + \delta_r + 2i\nu_0)(\dot{\Delta\zeta}) + (1 - \nu_0^2 + i\delta_n \nu_0)\Delta\zeta = 0, \quad (1.29)$$

which corresponds to the characteristic equation

$$\lambda^2 + (\delta_n + \delta_r + 2i\nu_0)\lambda + 1 - \nu_0^2 + i\delta_n \nu_0 = 0, \quad (1.30)$$

whose the roots

$$\lambda_{1,2} = -\frac{1}{2}\left(\delta_n + \delta_r + 2i\nu_0 \pm \sqrt{(\delta_n + \delta_r)^2 - 4 + i4\delta_r \nu_0)}\right). \quad (1.31)$$

Note that the real parts of the complex roots (1.31) coincide with the real parts of the roots (1.20), whence Re $\lambda_1 < 0$ for any values of the parameters, and Re $\lambda_2 < 0$

if the condition (1.22) is met. Hence, taking into account the third Eq. (1.25), we obtain the following condition for asymptotic stability

$$\mu < \mu_* = \delta_\theta \left(1 + \frac{\delta_n}{\delta_r} \right) . \tag{1.32}$$

Thus, the steady-state spin mode is asymptotically stable when the value of the parameter μ is strictly less than the threshold value μ_*, which depends on the coefficients of external and rotational damping.

Self-oscillations of a balanced rotor. It should be noted that using the equations of motion of a balanced rotor in a rotating frame (1.24), we were able to detect only one steady-state solution—the spin mode. We will show that equations in the polar frame allow us to find another stationary solution that corresponds to the steady-state mode of asynchronous self-oscillations. Putting $\varepsilon = 0$ in system (1.12), we get

$$\begin{cases} \ddot{a} + (\delta_n + \delta_r)\dot{a} + (1 - \omega^2)a = 0 , \\ (\dot{\omega} + \delta_n\omega - \delta_r(\nu - \omega))a + 2\dot{a}\omega = 0 , \\ \dot{\nu} + \delta_\theta\nu + \delta_r(\nu - \omega)a^2 = \mu . \end{cases} \tag{1.33}$$

The stationary solution of system (1.33) $a = a_0 = \text{const}, \omega = \omega_0 = \text{const}, \nu = \nu_0 = \text{const}$ must satisfy the following system of algebraic equations:

$$\begin{cases} (1 - \omega_0^2)a_0 = 0 , \\ (\delta_n\omega_0 - \delta_r(\nu_0 - \omega_0))a_0 = 0 , \\ \delta_\theta\nu_0 + \delta_r(\nu_0 - \omega_0)a_0^2 = \mu . \end{cases} \tag{1.34}$$

As the system of equations (1.25), system (1.34) has an obvious solution $a_0 = 0$, $\nu_0 = \mu/\delta_\theta$ representing the steady-state spin mode. But if we assume $a_0 \neq 0$, then it is easy to find another steady-state solution of system (1.34)

$$\omega_0 = 1, \quad \nu_0 = \nu_* = 1 + \frac{\delta_n}{\delta_r}, \quad a_0 = \sqrt{\frac{1}{\delta_n}(\mu - \delta_\theta\nu_*)} . \tag{1.35}$$

Solution (1.35) has the character of regular asynchronous whirling, with the angular speed of the rotor spin exactly coinciding with the threshold frequency (1.22), which was previously defined for the rotor model rotating at constant angular speed. Since the amplitude and frequency of asynchronous whirling are completely determined by the parameters of the system, this steady-state mode is self-oscillatory, and the condition for its existence is as follows:

$$\mu > \delta_\theta\nu_* = \delta_\theta \left(1 + \frac{\delta_n}{\delta_r} \right) = \mu_* . \tag{1.36}$$

Thus, it can be stated that the threshold value μ_*, which defines the region of existence of the self-oscillation mode, coincides with the threshold value (1.32), which limits the region of asymptotic stability of the steady-state spin mode.

We will study the stability of the self-oscillation mode in the first approximation. We substitute $a = a_0 + \Delta a$, $\omega = 1 + \Delta\omega$, $\nu = \nu_* + \Delta\nu$ into system (1.33), where Δa, $\Delta\omega$ and $\Delta\nu$ are small variations of variables. Expanding the expressions obtained in series in small variations and neglecting small quantities of second order and higher, we obtain a linearized system of equations in variations

$$
\begin{cases}
(\ddot{\Delta a}) + (\delta_n + \delta_r)(\dot{\Delta a}) - 2a_0 \Delta\omega = 0, \\
(\dot{\Delta\omega}) + 2\dfrac{(\dot{\Delta a})}{a_0} + (\delta_n + \delta_r)\Delta\omega - \delta_r \Delta\nu = 0, \\
(\dot{\Delta\nu}) + 2\delta_n a_0 \Delta a - \delta_r a_0^2 \Delta\omega + (\delta_\theta + \delta_r a_0^2)\Delta\nu = 0.
\end{cases} \tag{1.37}
$$

We represent system (1.37) in the matrix form

$$
\dot{\mathbf{Z}} = \mathbf{A}\mathbf{Z}, \tag{1.38}
$$

where

$$
\mathbf{Z} = \begin{Vmatrix} \Delta\dot{a} \\ \Delta a \\ \Delta\omega \\ \Delta\nu \end{Vmatrix}, \quad
\mathbf{A} = \begin{Vmatrix} -(\delta_n+\delta_r) & 0 & 2a_0 & 0 \\ 1 & 0 & 0 & 0 \\ -2/a_0 & 0 & -(\delta_n+\delta_r) & \delta_r \\ 0 & -2\delta_n a_0 & \delta_r a_0^2 & -(\delta_\theta+\delta_r a_0^2) \end{Vmatrix}.
$$

The characteristic polynomial of the system (1.38)

$$
|\mathbf{A} - \lambda\mathbf{E}| = b_0\lambda^4 + b_1\lambda^3 + b_2\lambda^2 + b_3\lambda + b_4 = 0 \tag{1.39}
$$

has the fourth order, where

$$
b_0 = 1,
$$
$$
b_1 = 2(\delta_n + \delta_r) + \delta_\theta + \delta_r a_0^2,
$$
$$
b_2 = 4 + (\delta_n + \delta_r)^2 + 2\delta_\theta(\delta_n + \delta_r) + \delta_r(2\delta_n + \delta_r)a_0^2,
$$
$$
b_3 = \delta_\theta(4 + (\delta_n + \delta_r)^2) + \delta_r(4 + \delta_n\delta_r + \delta_n^2)a_0^2,
$$
$$
b_4 = 4\delta_n\delta_r a_0^2.
$$

Note that all the coefficients of b_i in the case of $a_0 \neq 0$ are positive, therefore the necessary and sufficient condition for asymptotic stability, in accordance with the Lienard–Chipart criterion,[4] has the form

---

[4] F.R. Gantmaher. The Theory of Matrices. AMS Chelsea Publishing: Reprinted by American Mathematical Society. Vol. 2. 2000. 660 p.

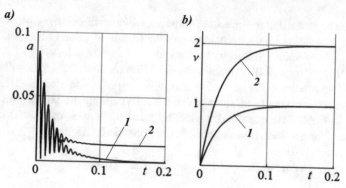

Fig. 6 Free vibrations of a balanced rotor at $\mu \leqslant \mu_*$

$$\Delta_3 = \begin{vmatrix} b_1 & b_3 & 0 \\ b_0 & b_2 & b_4 \\ 0 & b_1 & b_3 \end{vmatrix} > 0, \tag{1.40}$$

where Δ_3 is a minor of the third-order Hurwitz matrix. Simple algebra shows that

$$\begin{aligned}
\Delta_3 &= b_1 b_2 b_3 - b_1^2 b_4 - b_0 b_3^2 = (\delta_n + \delta_r)(2\delta_\theta (4 + (\delta_n + \delta_r)^2)(4 + (\delta_n + \delta_r + \delta_\theta)^2) + \\
&\quad + a_0^2 \delta_r (\delta_r^3 (2\delta_n + 3\delta_\theta) + \delta_r^2 (8 + 6\delta_n^2 + 14\delta_n\delta_\theta + 3\delta_\theta^2) + \delta_r (8\delta_n + 6\delta_n^3 + \\
&\quad + (28 + 19\delta_n^2)\delta_\theta + 9\delta_n\delta_\theta^2) + (16 + \delta_n^4 + 4\delta_n(2 + \delta_n^2)\delta_\theta + (10 + 3\delta_n^2)\delta_\theta^2)) + \\
&\quad + a_0^4 \delta_r^2 (4\delta_n^3 + 2(8 + 3\delta_n^2)\delta_\theta + \delta_r^2 (3\delta_n + \delta_\theta + \delta_r (12 + 7\delta_n^2 + 6\delta_n\delta_\theta)) + \\
&\quad + a_0^6 \delta_r^3 (4 + 2\delta_n^2 + \delta_n\delta_r).
\end{aligned}$$

Since the expression for Δ_3 does not contain negative terms, it can be argued that the steady-state self-oscillatory mode will be asymptotically stable in its entire domain (that is, if condition (1.36) is satisfied).

To illustrate the results obtained, system (1.33) was numerically integrated with the following values of dimensionless parameters: $\delta_n = 0.11$, $\delta_r = 0.11$, $\delta_\theta = 0.08$. Substituting these values into formula (1.36), we obtain the threshold value of the parameter characterizing the external rotational moment: $\mu_* = 0.16$.

Figure 6a shows the graphs of the amplitude a of the whirling motion of the rotor and the angular speed ν depending on time in the cases $\mu = 0.08 < \mu_*$ (curve 1) and $\mu = \mu_*$ (curve 2). The free vibrations of the rotor in these cases are due to initial disturbances. The curve 1 demonstrates the asymptotic stability of the spin rotation mode, and the curve 2 corresponds to the critical steady-state mode. Note that in the critical case the magnitude of the steady-state amplitude of the whirling motion depends on the initial integration conditions. Figure 6b presents similar graphs for the angular speeds of the rotor. In both cases, the steady-state angular speed of the rotor is equal to μ/δ_θ.

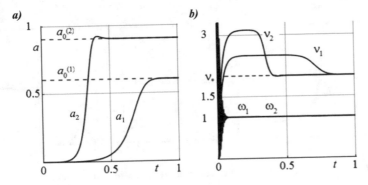

Fig. 7 Asynchronous self-oscillations of a balanced rotor at $\mu > \mu_*$

Figure 7a, b depict the graphs of the amplitudes of self-oscillations and angular speeds of ν and ω in cases when the external torque exceeds the threshold value, i.e. $\mu > \mu_*$. The curves a_1, ν_1 and ω_1 are drawn for the case $\mu = 0.2$, and the curves a_2, ν_2 and ω_2 are drawn for the case $\mu = 0.25$. Dashed lines correspond to stationary solutions defined by formulas (1.35). The results of the calculations confirm that the steady-state amplitudes of self-oscillations depend on the external torque, and the steady-state angular speeds of spin and whirling motion are independent and are determined only by the parameters of the system.

Forced oscillations of an unbalanced rotor with limited excitation. Let us return to the above-described model of a statically unbalanced Jeffcott rotor rotating under the action of an applied external torque. To study the forced vibrations of the rotor, we use the equations in the rotating frame (1.10). The parameter μ, which characterizes the external torque, as before, will be considered constant.

We substitute a stationary solution of the form $\xi = \xi_0 = $ const, $\eta = \eta_0 = $ const, $\nu = \nu_0 = $ const into system (1.10). As a result, we obtain a system of algebraic equations

$$\begin{cases} (1 - \nu_0^2)\xi_0 - \delta_n\nu_0\eta_0 = \varepsilon\nu_0^2, \\ \delta_n\nu_0\xi_0 + (1 - \nu_0^2)\eta_0 = 0, \\ \delta_\theta\nu_0 - \varepsilon\nu_0^2\eta_0 = \mu. \end{cases} \quad (1.41)$$

The absence in the equations of the rotational damping coefficient indicates that internal friction does not work, that is, the steady-state motion of the rotor is a regular synchronous whirling. From the first two equations of (1.41) we have

$$\xi_0 = \frac{\varepsilon\nu_0^2(1 - \nu_0^2)}{(1 - \nu_0^2)^2 + \delta_n^2\nu_0^2}, \quad \eta_0 = \frac{-\varepsilon\delta_n\nu_0^3}{(1 - \nu_0^2)^2 + \delta_n^2\nu_0^2}, \quad (1.42)$$

which gives us the expression for the amplitude of the forced vibrations

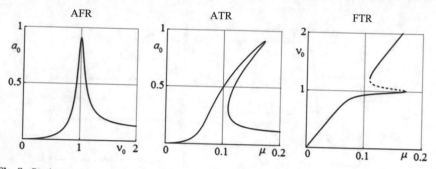

Fig. 8 Stationary responses of forced vibrations of the rotor with limited excitation

$$a_0 = \sqrt{\xi_0^2 + \eta_0^2} = \frac{\varepsilon \nu_0^2}{\sqrt{(1 - \nu_0^2)^2 + \delta_n^2 \nu_0^2}}\,. \tag{1.43}$$

In view of (1.42) and (1.43), the last equation of (1.41) can be written as

$$(\delta_\theta + \delta_n a_0^2)\nu_0 = \mu\,. \tag{1.44}$$

Figure 8 presents the stationary amplitude-frequency (AFR), amplitude-torque (ATR) and frequency-torque (FTR) responses of forced vibrations calculated by formulas (1.43) and (1.44) with $\varepsilon = 0.1$, $\delta_n = 0.11$, $\delta_r = 0.11$, $\delta_\theta = 0.08$.

The maximum amplitude of the amplitude-frequency response corresponds to the critical frequency ν_c, defined by formula (1.19). We note that ATR and FTR have a clear nonlinear character, which consists in the existence of a region where three steady-state modes correspond to the same value of the parameter μ.

To study the stability of synchronous whirling by the first approximation, we put in system (1.10) $\xi = \xi_0 + \Delta\xi$, $\eta = \eta_0 + \Delta\eta$, $\nu = \nu_0 + \Delta\nu$, where $\Delta\xi$, $\Delta\eta$ И $\Delta\nu$ are small deviations from the stationary values of ξ_0, η_0, which can be found from system (1.41).

Expanding the expressions thus obtained in a series and neglecting small quantities of the second order and higher, we obtain a linear system of equations in variations

$$\mathbf{A}\dot{\zeta} + \mathbf{B}\zeta = \mathbf{0}\,, \tag{1.45}$$

where

$$\zeta = \begin{Vmatrix} \Delta\dot{\xi} \\ \Delta\dot{\eta} \\ \Delta\nu \\ \Delta\xi \\ \Delta\eta \end{Vmatrix}, \quad \mathbf{A} = \begin{Vmatrix} 1 & 0 & -\eta_0 & 0 & 0 \\ 0 & 1 & \varepsilon + \xi_0 & 0 & 0 \\ 0 & \varepsilon & 1 + \varepsilon\xi_0 & 0 & 0 \\ 0 & 0 & 0 & 1 & 0 \\ 0 & 0 & 0 & 0 & 1 \end{Vmatrix},$$

$$\mathbf{B} = \begin{Vmatrix} \delta_n + \delta_r & -2\nu_0 & -\delta_n\eta_0 - 2\nu_0(\varepsilon + \xi_0) & 1 - \nu_0^2 & -\delta_n\nu_0 \\ 2\nu_0 & \delta_n + \delta_r & -2\eta_0\nu_0 + \delta_n\xi_0 & \delta\nu_0 & 1 - \nu_0^2 \\ \delta_r\eta_0 + 2\varepsilon\nu_0 & -\delta_r\xi_0 & \delta_\theta - 2\varepsilon\eta_0\nu_0 & 0 & -\varepsilon\nu_0^2 \\ -1 & 0 & 0 & 0 & 0 \\ 0 & -1 & 0 & 0 & 0 \end{Vmatrix}.$$

Coefficients of the characteristic polynomial of system (1.45)

$$|\lambda\mathbf{A} + \mathbf{B}| = \sum_{k=0}^{5} b_k\lambda^{5-k} = 0, \tag{1.46}$$

read as

$b_0 = 1 - \varepsilon^2$,

$b_1 = 2(\delta_n + \delta_r) + \delta_\theta + \delta_r a_0^2 + 2\varepsilon\delta_r\xi_0 - \varepsilon^2(\delta_n + \delta_r)$,

$b_2 = 2(1 + \nu_0^2) + (\delta_n + \delta_r)^2 + 2\delta_\theta(\delta_n + \delta_r) + \delta_r(2\delta_n + \delta_r)a_0^2 +$
$\quad + \varepsilon((2\delta_r^2 + 2\delta_n\delta_r + 1)\xi_0 - \delta_n\nu_0\eta_0) - \varepsilon^2(1 + 2\nu_0^2)$,

$b_3 = 2(\delta_n + \delta_r) + \delta_\theta(2 + (\delta_n + \delta_r)^2) + 2(\delta_n - \delta_r + \delta_\theta)\nu_0^2 +$
$\quad + \delta_r(1 + \delta_n\delta_r + \delta_n^2 + 3\nu_0^2)a_0^2 + \varepsilon((\delta_n + 3\delta_r + (\delta_n + 6\delta_r)\nu_0^2)\xi_0 -$
$\quad - \delta_n(\delta_n + \delta_r)\nu_0\eta_0) + \varepsilon^2(\delta_n + 5\delta_r)\nu_0^2$,

$b_4 = (1 - \nu_0^2)^2 + 2\delta_\theta(\delta_n + \delta_r) + (\delta_n^2 + 2\delta_\theta(\delta_n - \delta_r))\nu_0^2 +$
$\quad + \delta_n\delta_r(1 + \nu_0^2)a_0^2 + \varepsilon((1 + 3(1 + \delta_n\delta_r)\nu_0^2)\xi_0 - (2(\delta_n + \delta_r) + \delta_n\nu_0^2)\nu_0\eta_0 +$
$\quad + (1 + 3(1 + \delta_n\delta_r)\nu_0^2)\xi_0) + \varepsilon^2\nu_0^2(5 - \nu_0^2)$,

$b_5 = \delta_\theta((1 - \nu_0^2)^2 + \delta_n^2\nu_0^2) + \varepsilon(\delta_n\nu_0^2(1 + \nu_0^2)\xi_0 - \nu_0(2 - (2 - \delta_n^2)\nu_0^2)\eta_0) + 2\varepsilon^2\delta_n\nu_0^4$.

Necessary and sufficient conditions for the asymptotic stability of regular synchronous whirling are given by the Lienard–Chipart criterion

$$b_k > 0, \quad k = \overline{0,5}, \quad \Delta_2 > 0, \quad \Delta_4 > 0,$$

where Δ_2 and Δ_4 are the corresponding minors of the Hurwitz matrix.

Figure 9, where graphs of the dependences of the coefficients b_k and minors Δ_2, Δ_4 on ν_0 are presented, demonstrates that the condition $b_5 > 0$ specifies the stability regions of synchronous whirling in the form $\nu_0 < \tilde{\nu}_1$ or $\nu_0 > \tilde{\nu}_2$, and the condition $\Delta_4 > 0$ sets the stability region as $\nu_0 < \nu_*$, where ν_* is the supercritical threshold frequency given by the equation $\Delta_4 = 0$.

Figure 10 demonstrates two-parameter stability diagrams of synchronous whirling in the (ν_0, δ_r), (ν_0, δ_n) and (δ_n, δ_r) parameter planes. The dark color in the diagrams indicates the areas of instability in which at least one of the inequalities holds: $b_5 < 0$ or $\Delta_4 < 0$. The left diagram, which is built for the case $\delta_n = 0.11$, and the middle one, built with $\delta_r = 0.11$, clearly indicates the instability zone in the form

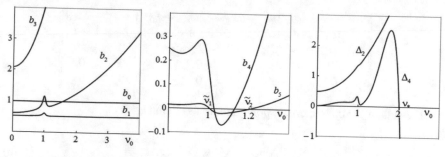

Fig. 9 The coefficients b_k and minors Δ_2, Δ_4 versus ν_0

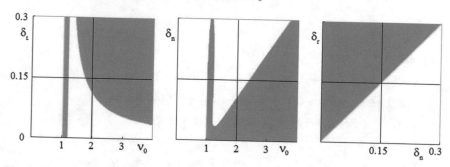

Fig. 10 Two-parameter stability diagrams

of a narrow vertical band, which corresponds to the falling section of the frequency-torque response, which is marked by dashes in Fig. 8c. The right-hand diagram bulit for the case of $\nu_0 = 2$ shows that synchronous whirling will be unstable under the condition $\delta_n < \delta_r$.

The interaction of forced vibrations and self-oscillations. We study the rotor motion in the case when the μ parameter, which characterizes the torque, exceeds the threshold value μ_*, which limits the stability region of synchronous whirling. Considering the results obtained in the previous paragraphs, it can be assumed that the nature of the rotor movement in this case will be asynchronous self-oscillations.

The value of the parameter μ_* is found by substituting the threshold value of the frequency ν_* into the expression (1.44). Then for the design parameters used in the paragraph above, the threshold values of frequency and torque will be equal to $\nu_* = 2.00364$ and $\mu_* = 0.16126$ respectively.

Figure 11 demonstrates the results of the numerical integration of the system (1.12) for a balanced and unbalanced rotor in the case $\mu = 0.2$. The curves 1 are calculated with $\varepsilon = 0$, and the curves 2 are calculated with $\varepsilon = 0.05$. The graphs in Fig. 11a show the change in the amplitude of the whirling motion of the point C with time for balanced (curve 1) and unbalanced (curve 2) rotors. A comparison of curves 1 and 2 shows that the unbalance of a rotor affects both the time of the establishment of the self-oscillatory mode and its character. The figure shows that the steady-state

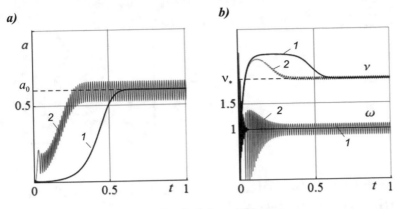

Fig. 11 Self-oscillations of balanced (**a**) and unbalanced (**b**) rotors

amplitude of the whirling motion of a balanced rotor is constant and equal to the stationary value a_0, as calculated by formula (1.35), whereas for an unbalanced rotor the self-oscillation amplitude is a time function that quickly oscillates around the value of a_0. In the case of an unbalanced rotor, the angular speeds of spin ν and whirling motion ω, the graphs of which are shown in Fig. 11b, has the same rapidly oscillating character. At the same time, self-oscillations are asynchronous, since oscillations ν occur near the threshold frequency ν_*, and oscillations ω occur near the critical frequency ν_c.

Main conclusions. Let us formulate the main results of the analytical and numerical study of the model of an unbalanced Jeffcott rotor with isotropic viscoelastic shaft.

1. The rotation of a static unbalanced rotor with angular speed close to the natural frequency of lateral oscillations (the critical speed) leads to resonant oscillations that have the character of regular synchronous whirling. In the supercritical frequency range, the effect of rotor self-centering is manifested.

2. The balanced Jeffcott rotor has supercritical threshold frequency due to the influence of forces associated with rotational damping. In case of rotation with angular speed that exceeds the threshold frequency, a loss of stability occurs in the spin mode. The assumption of the constancy of the angular speed of the rotor in conditions of loss of stability of the rotational mode leads to an unlimited increase in the amplitude of the whirling motion, which requires an energy source of unlimited power.

3. For a balanced rotor with limited excitation, there is a threshold torque depending on the coefficients of the external and internal (rotational) damping, upon reaching which the spin mode loses stability. If the torque exceeds the threshold value, then an asymptotically stable asynchronous self-oscillation mode with constant amplitude and constant angular speeds of spin and whirling motion is established in the system.

4. In the case of an unbalanced rotor, the magnitude of the threshold torque remains the same, but the nature of the self-oscillation mode changes: the amplitude and angular speeds are functions of time that oscillate rapidly around stationary values.

Fig. 12 Jeffcott rotor with
an orthotropic shaft

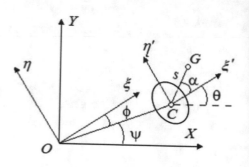

2 Forced and Self-excited Oscillations of a Rotor with an Orthotropic Viscoelastic Shaft[5]

Mechanical model of the rotor and the equation of motion. Let us consider a mechanical model of a statically unbalanced Jeffcott rotor under the assumption that its shaft has different elastic properties in different directions perpendicular to its polar axis. For definiteness, we will take into account the orthotropy of the viscoelastic characteristics of the rotor shaft, assuming that its stiffness diagram has the form of an ellipse.

Let us direct the axes of the frame $C\xi'\eta'$ connected with the rotor along the axes of the rigidity ellipse, and the corresponding elastic coefficients of the shaft will be denoted by k_ξ and k_η. We also introduce the rotating frame $O\xi\eta$, whose axes are collinear with the corresponding axes of the associated frame. The position of the center of mass G of the disk in the associated frame is determined by two parameters: the eccentricity value $\overrightarrow{CG}$ and the phase angle of the unbalance vector α, as shown in Fig. 12. Then the expression for the kinetic energy of the rotor takes the following form:

$$T = \frac{1}{2}m((\dot{x} - s\dot\theta\sin(\theta+\alpha))^2 + (\dot{y} + s\dot\theta\cos(\theta+\alpha))^2 + \frac{1}{2}J_G\dot\theta^2 . \qquad (2.1)$$

The dissipative function of non-rotational damping for a rotor with orthotropic shaft remains the same as in the case of an isotropic shaft (see expression (1.3)), and the potential energy of the shaft and the dissipative function of rotational damping are written as

$$V = \frac{1}{2}(k_\xi\xi^2 + k_\eta\eta^2) , \quad R_r = \frac{1}{2}(c_\xi\dot\xi^2 + c_\eta\dot\eta^2) , \qquad (2.2)$$

where c_ξ and c_η are the coefficients of rotational damping.

Substituting the formulas connecting the fixed and rotating coordinates

[5] The section contains material published in the article: *V.G. Bykov*. Synchronous and asynchronous whirling of the balanced rotor with an orthotropic elastic shaft AIP Conference Proceedings. 2018. Vol. 1959. № 080010. https://doi.org/10.1063/1.5034727.

$$x = \xi \cos\theta - \eta \sin\theta, \quad y = \xi \sin\theta + \eta \cos\theta \tag{2.3}$$

into expressions (1.3) and (2.1) we write the Lagrange equations of the second kind in the rotating frame

$$
\begin{cases}
m\ddot{\xi} + (c_n + c_\xi)\dot{\xi} - 2m\dot{\theta}\dot{\eta} + (k_\xi - m\dot{\theta}^2)\xi - (m\ddot{\theta} + c_n\dot{\theta})\eta = \\
\qquad\qquad = ms(\dot{\theta}^2 \cos\alpha + \ddot{\theta} \sin\alpha), \\
m\ddot{\eta} + (c_n + c_\eta)\dot{\eta} + 2m\dot{\theta}\dot{\xi} + (m\ddot{\theta} + c_n\dot{\theta})\xi + (k_\eta - m\dot{\theta}^2)\eta = \\
\qquad\qquad = ms(\dot{\theta}^2 \sin\alpha - \ddot{\theta} \cos\alpha), \\
mr^2\ddot{\theta} + c_\theta\dot{\theta} + (c_n\dot{\theta} + m\ddot{\theta})(\xi^2 + \eta^2) + \xi(c_n\dot{\eta} + 2m\dot{\theta}\dot{\xi} + m\ddot{\eta}) - \\
\quad - \eta(c_n\dot{\xi} - 2m\dot{\theta}\dot{\eta} + m\ddot{\xi}) = M_e + ms(\ddot{\xi} - 2\dot{\theta}\dot{\eta} - 2\ddot{\theta}\eta)\sin\alpha - \\
\quad - ms(\ddot{\eta} + 2\dot{\theta}\dot{\xi} + 2\ddot{\theta}\xi)\cos\alpha,
\end{cases} \tag{2.4}
$$

where $mr^2 = J_G + ms^2$, and M_e is the external torque.

The last equation of system (2.4) can be simplified if we subtract from it the second equation multiplied by ξ and add to it the first equation multiplied by η; as a result, we have

$$
\begin{aligned}
mr^2\ddot{\theta} + c_\theta\dot{\theta} + c_\xi\dot{\xi}\eta - c_\eta\dot{\eta}\xi + (k_\xi - k_\eta)\xi\eta = M_e + \\
+ ms\big((\ddot{\xi} - 2\dot{\theta}\dot{\eta} - \ddot{\theta}\eta - \dot{\theta}^2\xi)\sin\alpha - (\ddot{\eta} + 2\dot{\theta}\dot{\xi} + \ddot{\theta}\xi - \dot{\theta}^2\eta)\cos\alpha\big).
\end{aligned}
$$

Changing to the dimensionless variables $\bar{\xi} = \xi/r$, $\bar{\eta} = \eta/r$ and the dimensionless time $\bar{t} = \Omega t$, where $\Omega = \sqrt{(k_\xi + k_\eta)/2m}$, we get

$$
\begin{cases}
\ddot{\bar{\xi}} + (\delta_n + \delta_\xi)\dot{\bar{\xi}} + (\kappa_\xi - \nu^2)\bar{\xi} - 2\nu\dot{\bar{\eta}} - (\dot{\nu} + \delta_n\nu)\bar{\eta} = \varepsilon(\nu^2\cos\alpha + \dot{\nu}\sin\alpha), \\
\ddot{\bar{\eta}} + (\delta_n + \delta_\eta)\dot{\bar{\eta}} + (\kappa_\eta - \nu^2)\bar{\eta} + 2\nu\dot{\bar{\xi}} + (\dot{\nu} + \delta_n\nu)\bar{\xi} = \varepsilon(\nu^2\sin\alpha - \dot{\nu}\cos\alpha), \\
\dot{\nu} + \delta_\theta\nu + \delta_\xi\dot{\bar{\xi}}\bar{\eta} - \delta_\eta\dot{\bar{\eta}}\bar{\xi} + (\kappa_\xi - \kappa_\eta)\bar{\xi}\bar{\eta} = \mu + \\
\quad + \varepsilon\big((\ddot{\bar{\xi}} - 2\nu\dot{\bar{\eta}} - \dot{\nu}\bar{\eta} - \nu^2\bar{\xi})\sin\alpha - (\ddot{\bar{\eta}} + 2\nu\dot{\bar{\xi}} + \dot{\nu}\bar{\xi} - \nu^2\bar{\eta})\cos\alpha\big).
\end{cases} \tag{2.5}
$$

where

$\nu = \dot{\theta}$ is the dimensionless angular speed of the rotor,

$$\delta_n = \frac{c_n}{m\Omega}, \quad \delta_\xi = \frac{c_\xi}{m\Omega}, \quad \delta_\eta = \frac{c_\eta}{m\Omega}, \quad \delta_\theta = \frac{c_\theta}{mr^2\Omega},$$

$$\kappa_\xi = \frac{k_\xi}{m\Omega^2}, \quad \kappa_\eta = \frac{k_\eta}{m\Omega^2}, \quad \mu = \frac{M_e}{mr^2\Omega^2}, \quad \varepsilon = \frac{s}{r}.$$

We will use the system of equations (2.5) in the study of the stability of the steady-state spin mode, the analysis of forced rotor vibrations and the stability of forced synchronous whirling. As before, we omit the line above the dimensionless variables.

Let us switch from the rotating frame $O\xi\eta$ to the polar coordinates $\xi = a\cos\phi$, $\eta = a\sin\phi$, where $a = |\overrightarrow{OC}|/r$ is the value of the radius-vector of the point C, and ϕ is its phase angle (see Fig. 12). After substituting polar coordinates into (2.5), we perform the following transformations: (1) multiply the first equation by $\cos\phi$ and add it to the second one multiplied by $\sin\phi$; (2) multiply the first equation by $\sin\phi$ and subtract the second one multiplied by $\cos\phi$; (3) multiply the second equation by a and add it to the third equation. As a result, after elementary transformations, we obtain the following system of equations:

$$
\begin{cases}
\ddot{a} + (\delta_n + \delta_r)\dot{a} + (1 - \omega^2)a + (\delta\dot{a} + \kappa a)\cos 2\phi - \delta\dot{\phi}a\sin 2\phi = \\
\qquad = \varepsilon(\nu^2\cos(\phi + \alpha) + \dot{\nu}\sin(\phi + \alpha)), \\
(\dot{\omega} + \delta_n\omega - \delta_r\dot{\phi})a + 2\dot{a}\omega + (\delta\dot{a} + \kappa a)\sin 2\phi + \delta\dot{\phi}a\cos 2\phi = \\
\qquad = \varepsilon(\nu^2\sin(\phi + \alpha) - \dot{\nu}\cos(\phi + \alpha)), \\
\dot{\nu} + \delta_\theta\nu + (\dot{\omega} + \delta_n\omega)a^2 + 2\omega\dot{a}a = \\
\qquad = \mu - \varepsilon((\ddot{a} - a\omega^2)\sin(\phi + \alpha)) + (a\dot{\omega} + 2\dot{a}\omega)\cos(\phi + \alpha))), \\
\dot{\phi} = \nu - \omega,
\end{cases}
\tag{2.6}
$$

where $\omega = \dot{\psi} = \dot{\theta} - \dot{\phi}$ is the angular speed of the whirling motion in a fixed frame. For convenience, the following notation is introduced in system (2.6):

$$
\kappa = (\kappa_\xi - \kappa_\eta)/2, \quad \delta = (\delta_\xi - \delta_\eta)/2, \quad \delta_r = (\delta_\xi + \delta_\eta)/2.
\tag{2.7}
$$

The parameters κ and δ can be called, respectively, the orthotropy coefficients for stiffness and damping.

Next, we will use system (2.6) to study parametric and self-excited vibrations of rotors.

Steady state spin mode stability of a balanced rotor. Let us consider the case of a fully balanced rotor with an orthotropic shaft, on which a constant torque acts. We put $\varepsilon = 0$, $\mu = \text{const}$ in system (2.5) and substitute into it a stationary solution $\xi = \xi_0 = \text{const}$, $\eta = \eta_0 = \text{const}$, $\nu = \nu_0 = \text{const}$. As the result, in notation (2.7), we obtain an algebraic system of equations for ξ_0, η_0 and ν_0

$$
\begin{cases}
(1 + \kappa - \nu_0^2)\xi_0 - \delta_n\nu_0\eta_0 = 0, \\
\delta_n\nu_0\xi_0 + (1 - \kappa - \nu_0^2)\eta_0 = 0, \\
\delta_\theta\nu_0 + 2\kappa\xi_0\eta_0 = \mu.
\end{cases}
\tag{2.8}
$$

Assuming that ν_0 is a parameter, we consider the first two Eqs. (2.8) as a linear system with respect to ξ_0, η_0 with the determinant

$$
D = (1 - \nu_0^2)^2 + \delta_n^2\nu_0^2 - \kappa^2.
\tag{2.9}
$$

If $D \neq 0$, then system (2.8) has a unique solution $\xi_0 = \eta_0 = 0$, $\nu_0 = \mu/\delta_\theta$ corresponding to the steady state spin mode. Since this case always takes place for sufficiently small values of the orthotropy coefficient κ, we will call it the case of a "slightly orthotropic" shaft. If the orthotropy coefficient satisfies the condition

$$|\kappa| > \delta_n \sqrt{1 - \frac{\delta_n^2}{4}}, \tag{2.10}$$

then the equation $D = 0$ has two real roots

$$\nu_{0i} = \sqrt{1 - \frac{\delta_n^2}{2} \pm \sqrt{\kappa^2 - \delta_n^2 + \frac{\delta_n^4}{4}}}, \quad i = 1, 2, \tag{2.11}$$

representing the critical angular speeds of the rotor. The nontrivial solutions of system (2.8), corresponding to these roots

$$\xi_{0i}^2 = \frac{\mu - \delta_\theta \nu_{0i}}{2\kappa} \cdot \frac{\delta_n \nu_{0i}}{1 + \kappa - \nu_{0i}^2}, \quad \eta_{0i}^2 = \frac{\mu - \delta_\theta \nu_{0i}}{2\kappa} \cdot \frac{1 + \kappa - \nu_{0i}^2}{\delta_n \nu_{0i}}, \quad i = 1, 2, \tag{2.12}$$

describe parametric rotor vibrations, at which the point C performs a circular whirling motion with a frequency ν_{01} and ν_{02}. The condition for the existence of two critical frequencies can also be written as

$$\delta_n < \delta_n^* = \sqrt{2(1 - \sqrt{1 - \kappa^2})}, \tag{2.13}$$

where δ_n^* is the limiting value of the external damping coefficient for a fixed value of the orthotropy parameter κ.

Figure 13a shows the graphs of critical frequencies as a function of the parameter $|\kappa|$; they were calculated from formula (2.11) for three values of the external damping coefficient δ_n. Similar graphs, but depending on the coefficient δ_n with fixed values of $|\kappa|$, are shown in Fig. 13b. Analysis of the graphs allows us to conclude that the interval between the critical frequencies (ν_{01}, ν_{02}) expands with increasing the orthotropy parameter κ and shrinks with increasing the external damping coefficient

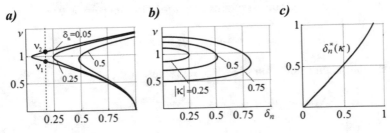

Fig. 13 Critical frequencies (**a**, **b**) and damping coefficient δ_n^* (**c**)

δ_n. The graph of the dependence of the limiting damping coefficient $\delta_n{}^*$ on κ is presented in Fig. 13c.

Let us study the stability of the steady-state spin mode in the first approximation. To do this, we substitute $\xi = \Delta\xi$, $\eta = \Delta\eta$, $\nu = \nu_0 + \Delta\nu$ into system (2.5), where $\Delta\xi$, $\Delta\eta$ and $\Delta\nu$ are small variations of coordinates and angular speed. Expanding in series and discarding small quantities of second order and higher, we obtain the linearized system of equations in variations

$$\begin{cases} (\ddot{\Delta\xi}) + (\delta_n + \delta_r + \delta)(\dot{\Delta\xi}) - 2\nu_0(\dot{\Delta\eta}) + (1 - \nu_0^2 + \kappa)\Delta\xi - \delta_n\nu_0\Delta\eta = 0, \\ (\ddot{\Delta\eta}) + (\delta_n + \delta_r - \delta)(\dot{\Delta\eta}) + 2\nu_0(\dot{\Delta\xi}) + (1 - \nu_0^2 - \kappa)\Delta\eta + \delta_n\nu_0\Delta\xi = 0, \quad (2.14) \\ (\dot{\Delta\nu}) + \delta_\theta\Delta\nu = 0. \end{cases}$$

The first two equations in (2.14) can be considered as an independent system of second-order linear differential equations for variations $\Delta\xi$ and $\Delta\eta$ of the coordinates. The characteristic equation of this system is a polynomial of degree 4

$$\lambda^4 + b_1\lambda^3 + b_2\lambda^2 + b_3\lambda + b_4 = 0, \tag{2.15}$$

with the coefficients

$$b_1 = 2(\delta_n + \delta_r),$$
$$b_2 = (\delta_n + \delta_r)^2 + 2(1 + \nu_0^2) - \delta^2,$$
$$b_3 = 2(\delta_n + \delta_r - \kappa\delta + (\delta_n - \delta_r)\nu_0^2),$$
$$b_4 = (1 - \nu_0^2)^2 + \delta_n^2\nu_0^2 - \kappa^2.$$

Obviously, $b_1 > 0$ and $b_2 > 0$, therefore, using the Lienard–Chipart criterion, the necessary and sufficient conditions for asymptotic stability can be written as

$$b_4 > 0, \qquad \Delta_3 > 0, \tag{2.16}$$

where Δ_3 is the minor of the third order Hurwitz matrix.

Note that $b_4 = D$ and the roots ν_1 and ν_2 of the equation $D = 0$ are defined by (2.11). It follows that the first condition in (2.16) is fulfilled in the domains $\nu_0 < \nu_{01}$ and $\nu_0 > \nu_{02}$. Using the third equation of system (2.8) and taking into account that $\xi_0 = \eta_0 = 0$, we find the corresponding conditions for the parameter μ

$$\mu < \mu_1 = \delta_\theta\nu_{01}, \qquad \mu > \mu_2 = \delta_\theta\nu_{02}. \tag{2.17}$$

To test the second condition of (2.16), we represent Δ_3 as

$$\Delta_3 = b_1(b_2b_3 - b_1b_4) - b_3^2 = A\nu_0^4 + B\nu_0^2 + C,$$

where

$$A = -16\delta_r^2, \quad B = 4(4 - \delta_r^2)(\delta_n^2 + \delta_r^2) - 4\delta^2(\delta_n^2 - \delta_r^2) - 16\delta\kappa\delta_r,$$
$$C = 4((\delta_n + \delta_r)^2 - \delta^2)((\delta_n + \delta_r)(\delta_n + \delta_r - \delta\kappa) + \kappa^2).$$

Since $A < 0$ and $C > 0$, the biquadratic equation $A\nu_0^4 + B\nu_0^2 + C = 0$ has a single positive root

$$\nu_* = \sqrt{\frac{-B - \sqrt{B^2 - 4AC}}{2A}} \tag{2.18}$$

corresponding to the threshold value of the angular speed. If we assume that $\delta = 0$, i.e., we consider the anisotropy of only the elastic characteristics of the shaft, then for the threshold speed we obtain the following formula

$$\nu_* = \left(1 + \frac{\delta_n}{\delta_r}\right)\sqrt{\frac{1}{2}\left(1 - \frac{\delta_r^2}{4}\right) + \sqrt{\left(1 + \frac{\delta_r^2}{4}\right)^2 + \left(\frac{\kappa\delta_r}{\delta_n + \delta_r}\right)^2}}. \tag{2.19}$$

The threshold value of the parameter μ characterizing the external torque, at which the steady-state spin mode loses stability, has the form

$$\mu_* = \delta_\theta \nu_*. \tag{2.20}$$

Figure 14 shows the graphs of the dependences of the coefficient b_4 and the minor Δ_3 on the parameter μ, which were calculated with $\delta_n = 0.11$, $\delta_r = 0.1$, $\delta = 0$, $\delta_\theta = 0.1$. Case (a) corresponds to a slightly orthotropic shaft ($\kappa = 0.05$), when condition (2.10) is not met, and case (b) ($\kappa = 0.3$) corresponds to a highly orthotropic shaft. In the areas where graphs lie above the abscissa, the asymptotic stability of the spin mode is guaranteed. It can be seen from the figure that in the case of a slightly orthotropic shaft, the steady-state spin mode is asymptotically stable in the region $\mu < \mu_*$. In the case of a highly orthotropic shaft, an additional "critical" instability region (μ_1, μ_2) appears.

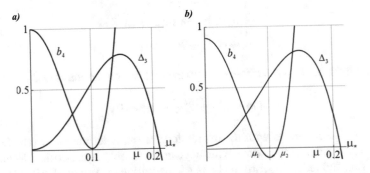

Fig. 14 Areas of asymptotic stability of the steady-state spin mode

Further, it will be shown that if $\mu \in (\mu_1, \mu_2)$, then a rotor with a highly orthotropic shaft produces parametric vibrations in the form of synchronous whirling of the point C at the frequency of the first critical speed ν_{01}.

Synchronous whirling and asynchronous self-oscillations of a balanced rotor. To study the steady-state synchronous motion modes of a balanced rotor, we use Eqs. (2.6), which in the case $\varepsilon = 0$ have the form

$$\begin{cases} \ddot{a} + (\delta_n + \delta_r)\dot{a} + (1 - \omega^2)a + (\delta\dot{a} + \kappa a)\cos 2\phi - \delta a\dot{\phi}\sin 2\phi = 0, \\ (\dot{\omega} + \delta_n\omega - \delta_r\dot{\phi})a + 2\omega\dot{a} + \delta a\dot{\phi}\cos 2\phi + (\delta\dot{a} + \kappa a)\sin 2\phi = 0, \\ \dot{\nu} + \delta_\theta\nu + (\dot{\omega} + \delta_n\omega)a^2 + 2\omega\dot{a}a = \mu, \\ \dot{\phi} = \nu - \omega. \end{cases} \qquad (2.21)$$

Substituting $a = a_0 = \mathrm{const}$, $\omega = \omega_0 = \mathrm{const}$, $\nu = \nu_0 = \mathrm{const}$ into the system (2.21), we get the system of transcendental equations for a_0, ω_0, ν_0

$$\begin{cases} (1 - \omega_0^2 + \kappa\cos 2\phi - \delta(\nu_0 - \omega_0)\sin 2\phi)a_0 = 0, \\ (\delta_n\omega_0 + \kappa\sin 2\phi - (\delta_r - \delta\cos 2\phi)(\nu_0 - \omega_0))a_0 = 0, \\ \delta_\theta\nu_0 + \delta_n\omega_0 a_0^2 = \mu. \end{cases} \qquad (2.22)$$

System (2.22) has an obvious trivial solution $a_0 = 0$, $\nu_0 = \mu/\delta_\theta$ corresponding to the steady-state spin mode. If we assume that $a_0 \neq 0$, then system (2.22) takes the form

$$\begin{cases} 1 - \omega_0^2 + \kappa\cos 2\phi = \delta(\nu_0 - \omega_0)\sin 2\phi, \\ \delta_n\omega_0 + \kappa\sin 2\phi = (\nu_0 - \omega_0)(\delta_r - \delta\cos 2\phi), \\ \delta_\theta\nu_0 + \delta_n\omega_0 a_0^2 = \mu. \end{cases} \qquad (2.23)$$

Let us show that system (2.23) has a stationary solution of the type of regular synchronous whirling. Putting $\omega_0 = \nu_0$ and $\phi = \phi_0$, we get

$$\begin{cases} 1 - \nu_0^2 = -\kappa\cos 2\phi_0, \\ \delta_n\nu_0 = -\kappa\sin 2\phi_0, \\ \delta_n\nu_0 a_0^2 = \mu - \delta_\theta\nu_0. \end{cases} \qquad (2.24)$$

Eliminating the unknown variable ϕ_0 from the first two equations of (2.24), we obtain the equation

$$(1 - \nu_0^2)^2 + \delta_n^2\nu_0^2 - \kappa^2 = 0, \qquad (2.25)$$

whose left side coincides with the expression for the determinant D (see (2.9)). It follows that regular synchronous whirling can occur only at the critical frequencies ν_{01} or ν_{02}, as defined by the formula (2.11). Substituting the values of critical frequen-

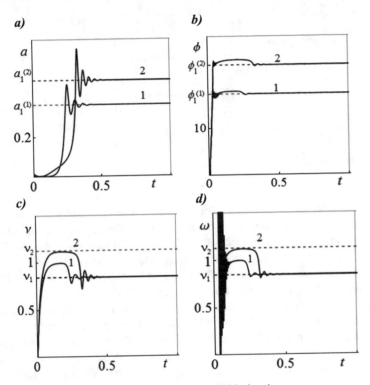

Fig. 15 Synchronous whirling of balanced rotor in the critical region

cies into the last equation of (2.24), we find the expressions for the corresponding amplitudes and phase angles

$$a_{0i}^2 = \frac{\mu - \delta_\theta \nu_{0i}}{\delta_n \nu_{0i}}, \qquad \phi_{0i} = \frac{1}{2} \arctan \frac{\delta_n \nu_{0i}}{1 - \nu_{0i}^2}, \qquad i = 1, 2. \qquad (2.26)$$

If the parameter $\mu \in (\delta_\theta \nu_{01}, \delta_\theta \nu_{02})$, then from the first formula of (2.26) we have $a_{01}^2 > 0$ and $a_{02}^2 < 0$, whence it follows that synchronous whirling can occur only at the frequency ν_{01}.

Figure 15a, b, c, d represents the graphs of amplitudes, phase shift angles, angular speeds of spin and whirling motion, respectively, which are plotted as a result of numerical integration of system (2.21) with the following values of dimensionless parameters: $\delta_n = \delta_r = 0.11$, $\delta_\theta = 0.08$, $\alpha = 0.3$, $\delta = 0$, $\kappa = 0.3$. The critical values $\mu_1 = 0.06764$ and $\mu_2 = 0.09026$ are calculated using formulas (2.11) and (2.17). Curves 1 and 2 are constructed for the cases $\mu = 0.08$ and $\mu = 0.09$, which belong to the critical region (μ_1, μ_2). The dashed lines correspond to the steady-state values of the variables and are calculated using formulas (2.26). The graphs confirm that in cases where $\mu \in (\mu_1, \mu_2)$ the steady-state rotor motions are regular synchronous whirling with frequency equal to the first critical speed ν_{01}.

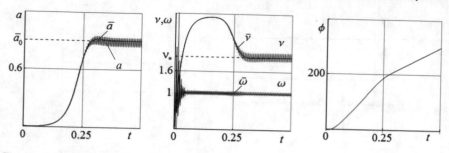

Fig. 16 Asynchronous self-oscillations

Let us consider the motion of a rotor in the case when the external torque exceeds the threshold value that defines the boundary of the stability region of the spin mode (2.20). For the selected parameter values, the threshold frequency, as calculated by formula (2.19), is $\nu_* = 2.0056$, from which we find $\mu_* = \delta_\theta \nu_* = 0.1605$.

Figure 16 presents the results of the numerical integration of system (2.21) for $\mu = 0.25 > \mu_*$. The graphs show how the amplitude a of the whirling motion of the rotor, the angular speeds of the whirling motion ω and spin ν, and the phase angle ϕ change with time. The figure also shows that the orthotropy of the elastic characteristics of the shaft has the same effect on the rotor self-oscillation mode as the unbalance (see Fig. 11), generating small rapid oscillations of amplitude and angular speeds superimposed on slowly varying functions of time. In this case, the phase angle is a rapidly changing function of time, which makes a large number of full revolutions during one revolution of the rotor.

Taking into account the nature of the rotor self-oscillations, we will look for a particular solution of the system of equations (2.21) in the form of a superposition of slow and fast motions

$$a(t, \phi) = \tilde{a}(t) + \epsilon A(\phi), \quad \omega(t, \phi) = \tilde{\omega}(t) + \epsilon W(\phi), \quad \nu(t, \phi) = \tilde{\nu} + \epsilon V(\phi), \quad (2.27)$$

where $\tilde{a}(t)$, $\tilde{\omega}(t)$ и $\tilde{\nu}(t)$ are slowly varying functions of time, $A(\phi)$, $W(\phi)$ and $V(\phi)$ are periodic functions of a rapidly changing angle ϕ, which have zero mean values over the period, and $\epsilon \ll 1$ is a small parameter.

Of greatest interest in the solution of (2.27) are the slow components, reflecting the general trend of the nature of the self-oscillation mode. These components can be found using the averaging method. Substituting (2.27) into (2.21), averaging the obtained relations over ϕ over the period, taking into account the smallness of the parameters ϵ, κ and δ, we obtain in the first approximation the system of equations for the "slow" variables $\tilde{a}$, $\tilde{\omega}$ and $\tilde{\nu}$

$$\begin{cases} \ddot{\tilde{a}} + (\delta_n + \delta_r)\dot{\tilde{a}} + (1 - \tilde{\omega}^2)\tilde{a} = 0, \\ (\dot{\tilde{\omega}} + \delta_n\tilde{\omega} - \delta_r(\tilde{\nu} - \tilde{\omega}))\tilde{a} + 2\tilde{\omega}\dot{\tilde{a}} = 0, \\ \dot{\tilde{\nu}} + \delta_\theta\tilde{\nu} + \delta_r(\tilde{\nu} - \tilde{\omega})\tilde{a}^2 = \mu. \end{cases} \quad (2.28)$$

The approximate system (2.28) coincides with the exact system (1.33) for a rotor with an isotropic shaft, therefore it has a stationary solution in the form of regular asynchronous whirling, similar to the solution (1.35)

$$\tilde{\omega}_0 = 1, \quad \tilde{\nu}_0 = \nu_* = 1 + \frac{\delta_n}{\delta_r}, \quad \tilde{a}_0 = \sqrt{\frac{1}{\delta_n}\left(\mu - \delta_\theta(1 + \frac{\delta_n}{\delta_r})\right)}, \quad (2.29)$$

and the condition for the existence and asymptotic stability of this solution read as

$$\mu > \delta_\theta\left(1 + \frac{\delta_n}{\delta_r}\right) = \delta_\theta \nu_* = \mu_* . \quad (2.30)$$

For comparison with the solution of the exact system of equations (2.21) Fig. 16 in parallel also shows the results of the numerical integration of the approximate system (2.28). Note that the exact solutions differ from the approximate equations by the appearance of high-frequency oscillations of amplitude and angular speeds due to the orthotropy of the rotor shaft.

Forced synchronous whirling and self-oscillations of an unbalanced rotor. Forced synchronous whirling of a statically unbalanced rotor under the action of inertial load can be conveniently investigated using equations in a rotating frame. Assuming that the external torque is constant, we substitute in system (2.5) the following stationary solution: $\xi = \xi_0 = \text{const}, \eta = \eta_0 = \text{const}, \nu = \nu_0 = \text{const}$. As a result, taking into account (2.7), we obtain an algebraic system of equations for the coordinates ζ_0, η_0 and the angular speed ν_0

$$\begin{cases} (1 - \nu_0^2 + \kappa)\xi_0 - \delta_n\nu_0\eta_0 = \varepsilon\nu_0^2\cos\alpha, \\ \delta_n\nu_0\xi_0 + (1 - \nu_0^2 - \kappa)\eta_0 = \varepsilon\nu_0^2\sin\alpha, \\ \delta_\theta\nu_0 + 2\kappa\xi_0\eta_0 = \mu - \varepsilon\nu_0^2(\xi_0\sin\alpha - \eta_0\cos\alpha). \end{cases} \quad (2.31)$$

From the first two equations of (2.31) we find

$$\begin{aligned} \xi_0 &= \varepsilon\nu_0^2\frac{(1-\nu_0^2-\kappa)\cos\alpha + \delta_n\nu_0\sin\alpha}{(1-\nu_0^2)^2 + \delta_n^2\nu_0^2 - \kappa^2}, \\ \eta_0 &= \varepsilon\nu_0^2\frac{(1-\nu_0^2+\kappa)\sin\alpha - \delta_n\nu_0\cos\alpha}{(1-\nu_0^2)^2 + \delta_n^2\nu_0^2 - \kappa^2}, \end{aligned} \quad (2.32)$$

which yields the expression for calculating the amplitude and phase angle of the forced vibration depending on the angular speed of the rotor

$$a_0 = \sqrt{\xi_0^2 + \eta_0^2}, \quad \phi_0 = \arctan\frac{\eta_0}{\xi_0}. \quad (2.33)$$

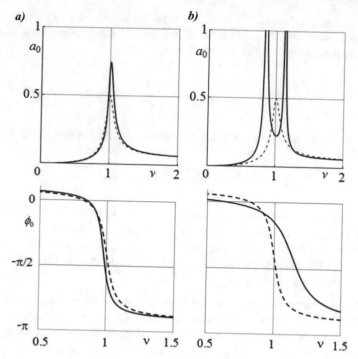

Fig. 17 AFR and PFR of the rotor with a slightly orthotropic shaft (**a**) and with a highly orthotropic shaft (**b**)

Expressions (2.32) and (2.33) allow us to construct stationary amplitude-frequency and phase-frequency responses of forced vibrations of the rotor.

Figure 17a presents the AFR and PFR for a slightly orthotropic rotor calculated at $\varepsilon = 0.05$, $\delta_n = 0.1$, $\delta_\theta = 0.1$, $\kappa_\eta = 0.9\kappa_\xi$ or $|\kappa| = 0.05$, $\delta_n^* = 0.05$, and Fig. 17b shows them for a highly orthotropic rotor ($\kappa_\eta = 2\kappa_\xi$ ИЛИ$|\kappa| = 0.333$, $\delta_n^* = 0.338$). For comparison, the AFR and PFR of an isotropic rotor are shown dashed ($\kappa = \alpha = 0$).

The graphs show that in the case of a slightly orthotropic shaft, when the parameters of the system do not satisfy conditions (2.10) and (2.13), the rotor has one critical frequency. In this case, the maximum amplitude of the disk center deviation is finite. In the case of a strong shaft orthotropy, when conditions (2.10), (2.13) are fulfilled, the splitting of the critical frequency and discontinuity of the frequency response occur. Note that despite the presence of external damping, the denominator in expressions (2.32) at the critical frequencies (2.11) vanishes, which makes the amplitude tend to infinity. Therefore, a more real picture of the forced vibrations of the rotor is given by the amplitude-torque (ATR) and frequency-torque (FTR) responses constructed using the third equation of system (2.31), which takes into account the limitations of external excitation.

Figure 18 represents the ATR and FTR calculated for a rotor with slightly orthotropic shaft, and Fig. 19 shows similar characteristics for a rotor with highly

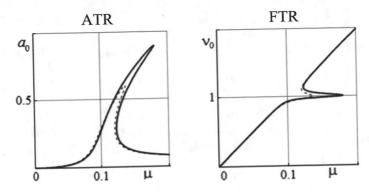

Fig. 18 ATR and FTR of forced rotor vibrations in case of a slightly orthotropic shaft

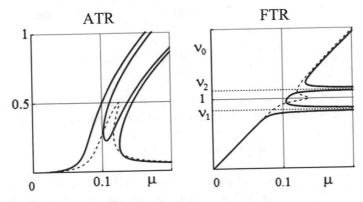

Fig. 19 ATR and FTR of forced rotor vibrations in case of a highly orthotropic shaft

orthotropic shaft. In both cases, the pronounced nonlinear nature of ATR and FTR is manifested which leads to the appearance of a region in which several stationary solutions correspond to the same value of the parameter μ; the practical feasibility of such solutions will be assessed on the basis of stability conditions.

To study the stability of stationary solutions, we substitute into system (2.5) $\xi = \xi_0 + \Delta\xi$, $\eta = \eta_0 + \Delta\eta$, $\nu = \nu_0 + \Delta\nu$, where $\Delta\xi$, $\Delta\eta$ and $\Delta\nu$ are small variations. Expanding the expressions thus obtained in a series and neglecting small quantities of second order and higher, we obtain a linear system of equations in variations

$$\mathbf{A}\dot{\boldsymbol{\zeta}} + \mathbf{B}\boldsymbol{\zeta} = 0, \tag{2.34}$$

where

$$\boldsymbol{\zeta} = \{\Delta\dot{\xi}, \Delta\dot{\eta}, \Delta\nu, \Delta\xi, \Delta\eta\}^T,$$

$$\mathbf{A} = \begin{Vmatrix} 1 & 0 & -\eta_0 - \varepsilon S_\alpha & \delta_n + \delta_\varepsilon & -2\nu_0 \\ 0 & 1 & \xi_0 + \varepsilon C_\alpha & 2\nu_0 & \delta_n + \delta_\eta \\ -\varepsilon S_\alpha & \varepsilon C_\alpha & 1 + \varepsilon(C_\alpha\xi_0 + S_\alpha\eta_0) & \delta_\varepsilon\eta_0 + 2\varepsilon C_\alpha\nu_0 & -\delta_\eta\xi_0 + 2\varepsilon C_\alpha\nu_0 \\ 0 & 0 & 0 & 1 & 0 \\ 0 & 0 & 0 & 0 & 1 \end{Vmatrix},$$

$$\mathbf{B} = \begin{Vmatrix} 0 & 0 & -\delta_n\eta_0 - 2\nu_0(\varepsilon C_\alpha + \xi_0) & 1 + \kappa - \nu_0^2 & -\delta_n\nu_0 \\ 0 & 0 & \delta_n\xi_0 - 2\nu_0(\varepsilon S_\alpha + \eta_0) & \delta_n\nu_0 & 1 - \kappa - \nu_0^2 \\ 0 & 0 & \delta_\theta - 2\varepsilon\nu_0(C_\alpha\eta_0 - S_\alpha\xi_0) & 2\kappa\eta_0 + \varepsilon S_\alpha\nu_0^2 & 2\kappa\xi_0 - \varepsilon C_\alpha\nu_0^2 \\ -1 & 0 & 0 & 0 & 0 \\ 0 & -1 & 0 & 0 & 0 \end{Vmatrix}.$$

Here, to simplify notation we write: $S_\alpha = \sin\alpha$, $C_\alpha = \cos\alpha$.

The characteristic polynomial of system (2.34) is of the fifth order,

$$|\lambda\mathbf{A} + \mathbf{B}| = \sum_{k=0}^{5} b_k\lambda^{5-k} = 0, \tag{2.35}$$

and the necessary and sufficient conditions for the asymptotic stability of stationary solutions based on the Lienard–Chipart criterion are as follows:

$$b_k > 0, \quad k = \overline{0,5}, \quad \Delta_2 > 0, \quad \Delta_4 > 0;$$

here Δ_2, and Δ_4 are the corresponding minors of the Hurwitz matrix.

Figures 20 and 21 shows the graphs of the dependences of the coefficients b_k and minors Δ_2, Δ_4 on ν_0 for a slightly and highly orthotropic rotor, respectively. The frequency intervals in which all the graphs lie above the abscissa correspond to the asymptotically stable forced synchronous whirling of the rotor. The graphs show that in both cases the boundaries of the critical region of instability are defined by two positive real roots ν_1 and ν_2 of the equation $b_5 = 0$, and the threshold value ν_* of the frequency of synchronous whirling is the maximum real root of the equation $\Delta_4 = 0$.

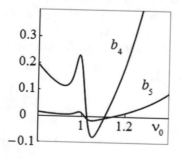 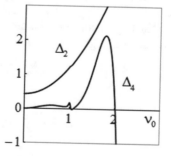

Fig. 20 Stability areas for a slightly orthotropic rotor

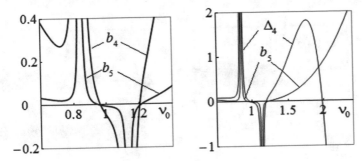

Fig. 21 Stability areas for a highly orthotropic rotor

Table 1 The margins of the "critical region" and the threshold torques

κ	α	ν_1	ν_2	μ_*
0	0	1.00916	1.09362	0.20122
0.1	0.1	0.99751	1.10100	0.20151
0.2	0.2	0.99795	1.15168	0.20247
0.3	0.3	0.94301	1.20615	0.20405
0.4	0.4	0.86862	1.26059	0.20616
0.5	0.5	0.78862	1.31403	0.20870

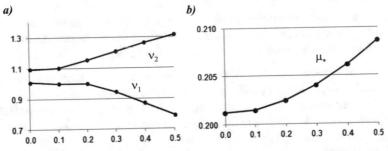

Fig. 22 Margins of the critical region of instability (**a**) and threshold value of the parameter μ (**b**)

Now the threshold value of the parameter μ_* can be found from the last equation of (2.31).

Table 1 shows the calculation results of the margins of an instability "critical region" ν_1, ν_2 and the magnitude of the threshold torque μ_* depending on the value of the orthotropy coefficient κ. The data from the table presented in Fig. 22 shows the expansion of the critical region of instability and an increase in the threshold value μ_* with increasing parameter κ.

Let us consider the rotor motion in the case when $\mu > \mu_*$; in this case, the external torque exceeds the threshold value that controls the boundary of the stability region of the forced synchronous whirling. Figure 23. presents the results of numerical

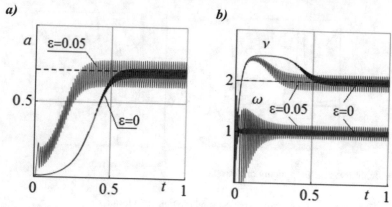

Fig. 23 Asynchronous self-oscillations of an unbalanced orthotropic rotor

integration of system (2.6) in the mode of asynchronous self-oscillations with the following values of the dimensionless parameters $\delta_n = \delta_\xi = \delta_\eta = \delta_\theta = 0.1$, $\delta = 0$, $\kappa = 0.3$, $\alpha = 0.3$, $\mu = 0.25 > \mu_* = 0.20405$ (see Table 1).

The graphs show how the amplitude a of the whirling motion of the rotor (a) and the angular speed of the whirling motion ω and spin ν (b) change with time. For comparison, the figures show the similar dark-colored graphs for a balanced rotor $(\varepsilon = 0)$. It follows from the figures that the unbalance of the rotor does not affect average steady-state amplitudes of self-oscillations and the angular speeds of spin and whirling motion of the rotor, but rapid oscillations of the amplitude and angular speeds of an unbalanced rotor are more intense than those of a balanced one.

Main conclusions. Based on the results of analytical and numerical studies of the rotor model with an orthotropically elastic shaft, the following can be stated:

1. A balanced rotor with orthotropic elastic shaft can have three types of steady-state motion modes: spin mode in the absence of shaft bending, regular synchronous whirling, and asynchronous self-oscillations.

2. In the case of a slightly orthotropic shaft (if condition (2.10) is violated), the spin mode is asymptotically stable in the subcritical and supercritical frequency range, provided that the external torque is less than the threshold value, as defined by (2.20).

3. In the case of sufficiently high orthotropy of the shaft, the critical frequency splits into two, resulting in a "critical region" of instability of the pure rotation mode, for which asymptotically stable regular synchronous rotor whirling occurs. The steady-state angular speed of the rotor in the critical region does not depend on the applied torque and is equal to the first critical speed.

4. In the case when the external torque exceeds the threshold value, the system settles down to the asymptotically stable mode of asynchronous self-oscillations.

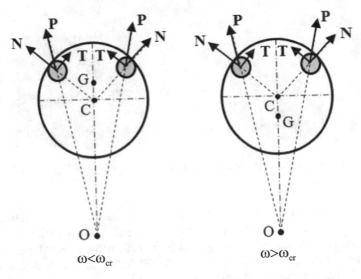

Fig. 24 The ball autobalancer operating principle

3 Automatic Balancing of a Statically Unbalanced Rotor[6]

Operating principle of an automatic ball balancer. In some types of high-speed rotary machines such as centrifuges, separators, centrifugal pumps, etc., the rotor unbalance may change during their operation. In such cases, continuous balancing of the rotor is necessary to reduce the level of vibration of the machine. To solve this problem, technologies have been developed based on the use of active and passive auto-balancing devices of various types. In practice, passive automatic ball balancers (ABB) are widely useful and found applications not only in the above-mentioned rotary machines, but also in high-precision mechatronic devices such as drives, disk drives, micro-electric motors, etc.

ABB is designed to compensate the unbalance of an elastically suspended rotor at angular speeds exceeding its first critical speed. It has a race containing one or more circular cavities in which massive metal balls can freely move. The ABB is fixed on the rotor so that the axis of symmetry of the race coincides with the polar axis of the rotor. The principle of operation of an ABB, which was first patented[7] at the end of the nineteenth century, is shown in Fig. 24.

When the angular speed of a rotor is below the critical one, the balls will converge under the action of the tangential components T of the centrifugal forces P, moving

---

[6] The section contains materials published in the articles: *V.G. Bykov*. Steady-state modes of motion of an unbalanced rotor with an auto-balancing mechanism // Vestn. S.-Peterb. Univ. Ser. 1. 2006. Vol. 2. Pp. 90–101 [in Russian]; *V.G. Bykov*. Unsteady modes of motion of a statically unbalanced rotor with an auto-balancing mechanism // Vestn. S.-Peterb. Univ. Ser. 1. 2010. Vol. 3. Pp. 89–96 [in Russian].

[7] *G.M. Herrick*. Self-adjusting counter-balance: patent No 414642 US, 1889.

a)

b)

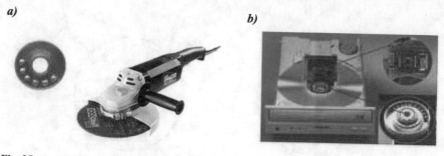

Fig. 25 Autobalancer technology for hand grinders (**a**); DVD player LG Electronics (**b**)

to the "heavy" side of the rotor, which leads to an increase in the overall unbalance of the system. If the angular speed is higher than the critical one then the balls also converge under the action of centrifugal forces, but due to the effect of the rotor self-centering, their movement will be directed to the "light" side and, with sufficient weight of balancing balls, will lead to a complete balancing of the system. In this case, the points C and O will coincide, the forces of T will vanish, and the motion of the balls will stop. Thus, it can be stated that the balancing effect with the use of ABB is manifested only in the supercritical frequency range.

Despite the fact that the first theoretical study on autobalancing of rotors with the help of a ball balancer appeared already in 1932,[8] the interest in this problem continues unabated. This is evidenced by both numerous articles and patents, as well as proprietary technologies that have appeared over the past few years on the use of ABB in mechanical engineering and household appliances (Fig. 25).

Mechanical model of a rotor equipped with an automatic ball balancer. We consider a model of a statically unbalanced rotor in the form of a massive stiff disk symmetrically fixed in the middle of a vertical weightless elastic shaft rotating in the hinged bearings O_1 and O_2 under the action of an applied external torque M_e (Fig. 26).

To compensate for the unbalance, the rotor is equipped with a passive ABB in the form of a circular cavity fixed on the same axis as the disk and filled with a viscous liquid in which balancing balls of the same mass can move freely. In accordance with the Jeffcott rotor model, we assume that the disk moves only in the horizontal plane perpendicular to the shaft axis.

Let points G and C denote, respectively, the center of mass of the disk and the point of attachment of the disk to the shaft. To describe the motion of the rotor and of the balancing balls, we introduce two frames: the stationary frame $OXYZ$ and the rotating frame $C\xi\eta\zeta$, as shown in Fig. 26. We denote by s the static eccentricity of the rotor, m is the mass of the rotor disk and the ABB's hull, which is fixed on the rotor (excluding the mass of the balls), J_G is the moment of inertia of the rotor

---

[8] *E.L. Thearle*. A New Type of Dynamic Balancing Machine // Transactions of ASME. **54**(12). 1932. Pp. 131–141.

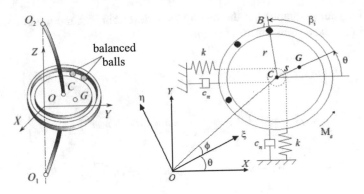

Fig. 26 Mechanical model of a rotor with ABB

about the polar axis, k is coefficient the elasticity of the shaft, m_b is the mass of one balancing ball and r is the radius of the race of the ABB.

If the ABB contains n balls, then due to the assumptions made the described mechanical system has $(n+3)$ degrees of freedom. We choose absolute coordinates x and y of the point C, angle θ of rotation of the disk in the plane XOY, angles β_i $(i = \overline{1, n})$ of the balls deflection relative to the circular race of the ABB as the generalized coordinates, and write the expression for the kinetic energy of the system and the potential energy of the elastic shaft as follows:

$$T - \frac{1}{2}m(\dot{x}_G^2 + \dot{y}_G^2) + \frac{1}{2}J_G\dot{\theta}^2 + \frac{1}{2}m_b \sum_{i=1}^{n}(\dot{x}_{B_i}^2 + \dot{y}_{B_i}^2),$$

$$V = \frac{1}{2}k(x^2 + y^2);$$

here

$$x_G = x + s\cos\theta, \qquad x_{B_i} = x + r\cos(\theta + \beta_i),$$
$$y_G = y + s\sin\theta, \qquad y_{B_i} = y + r\sin(\theta + \beta_i).$$

Assuming that the internal friction and the structural damping are negligible, we will take into account in our model only the forces of external viscous resistance acting on the rotor and the forces of viscous resistance acting on the balancing balls in the ABB's race. Denoting the damping coefficients by c_n, c_θ and c_b (that characterize the energy losses during the lateral movements of the disk, the shaft rotation in the bearings and the movement of the balancing balls, respectively) we write the expression for the Rayleigh dissipative function as

$$R = \frac{1}{2}c_n(\dot{x}^2 + \dot{y}^2) + \frac{1}{2}c_\theta\dot{\theta}^2 + \frac{1}{2}c_b \sum_{i=1}^{n}(\dot{\psi}_i)^2.$$

The Lagrange equations of the second kind, describing the motion of a rotor with an ABB in a fixed frame, have the form

$$
\begin{cases}
(m + nm_b)\ddot{x} + c_n\dot{x} + kx = -\dfrac{d^2}{dt^2}\left[ms\cos\theta + m_b r \sum_{j=1}^{n} \cos\varphi_j \right], \\[2ex]
(m + nm_b)\ddot{y} + c_n\dot{y} + ky = -\dfrac{d^2}{dt^2}\left[ms\sin\theta + m_b r \sum_{j=1}^{n} \sin\varphi_j \right], \\[2ex]
J_C\ddot{\theta} + c_\theta\dot{\theta} = M_e + ms(\ddot{x}\sin\theta - \ddot{y}\cos\theta) + c_b \sum_{j=1}^{n}(\dot{\varphi}_j - \dot{\theta}), \\[2ex]
m_b r^2\ddot{\varphi}_j + c_b(\dot{\varphi}_j - \dot{\theta}) = m_b r(\ddot{x}\sin\varphi_j - \ddot{x}\cos\varphi_j), \quad j = \overline{1,n}.
\end{cases}
\tag{3.1}
$$

where $\varphi_i = \theta + \beta_i$, $J_C = J_G + ms^2$.

Moving to the dimensionless coordinates $\bar{x} = x/r$, $\bar{y} = y/r$ and the dimensionless time $\bar{t} = \Omega t$ ($\Omega = \sqrt{k/m}$), we represent Eqs. (3.1) in the dimensionless form

$$
\begin{cases}
(1+n\chi)\ddot{x} + \delta_n\dot{x} + x = \varepsilon(\dot{\theta}^2\cos\theta + \ddot{\theta}\sin\theta) + \chi\sum_{i=1}^{n}(\dot{\varphi}_i^2\cos\varphi_i + \ddot{\varphi}_i\sin\varphi_i), \\[2ex]
(1+n\chi)\ddot{y} + \delta_n\dot{y} + y = \varepsilon(\dot{\theta}^2\sin\theta - \ddot{\theta}\cos\theta) + \chi\sum_{i=1}^{n}(\dot{\varphi}_i^2\sin\varphi_i - \ddot{\varphi}_i\cos\varphi_i), \\[2ex]
\rho\ddot{\theta} + \delta_\theta\dot{\theta} = \mu + \varepsilon(\ddot{x}\sin\theta - \ddot{y}\cos\theta) + \chi\delta_b\sum_{i=1}^{n}(\dot{\varphi}_i - \dot{\theta}), \\[2ex]
\ddot{\varphi}_i + \delta_b(\dot{\varphi}_i - \dot{\theta}) = \ddot{x}\sin\varphi_i - \ddot{y}\cos\varphi_i, \quad i = \overline{1,n},
\end{cases}
\tag{3.2}
$$

where

$$
\chi = \frac{m_b}{m}, \quad \varepsilon = \frac{s}{r}, \quad \rho = \frac{J_C}{mr^2\Omega}, \quad \mu = \frac{M_e}{mr^2\Omega^2},
$$
$$
\delta_n = \frac{c_n}{m\Omega}, \quad \delta_\theta = \frac{c_\theta}{mr^2\Omega}, \quad \delta_b = \frac{c_b}{m_b r^2\Omega}.
$$

In the system of equations (3.2) the dot denotes derivative with respect to dimensionless time, and the bar over dimensionless variables is omitted for simplicity.

Let us suppose that the shaft is spinning at constant angular speed $\dot{\theta} = \nu$. We denote by ξ and η the coordinates of the point C in the rotating frame and express the fixed coordinates in term of them

$$
x = \xi\cos\nu t - \eta\sin\nu t, \quad y = \xi\sin\nu t + \eta\cos\nu t.
\tag{3.3}
$$

We substitute expressions (3.3) into system (3.2) and perform the following transformations. Multiply the first equation by $\cos\nu t$ and add it to the second one multiplied by $\sin\nu t$; next we multiply the second equation by $\cos\nu t$ and subtract from it the first one multiplied by $\sin \nu t$. As a result, we obtain the system of equations for the coordinates ξ, η and β_i

$$
\begin{cases}
(1+n\chi)(\ddot{\xi}-2\nu\dot{\eta}-\nu^2\xi)+\delta_n(\dot{\xi}-\nu\eta)+\xi=\varepsilon\nu^2+\chi\sum_{i=1}^{n}((\nu+\dot{\beta}_i)^2\cos\beta_i+\ddot{\beta}_i\sin\beta_i), \\
(1+n\chi)(\ddot{\eta}+2\nu\dot{\xi}-\nu^2\eta)+\delta_n(\dot{\eta}+\nu\xi)+\eta=\chi\sum_{i=1}^{n}((\nu+\dot{\beta}_i)^2\sin\beta_i-\ddot{\beta}_i\cos\beta_i), \\
\ddot{\beta}_i+\delta_b\dot{\beta}_i=(\ddot{\xi}-2\nu\dot{\eta}-\nu^2\xi)\sin\beta_i-(\ddot{\eta}+2\nu\dot{\xi}-\nu^2\eta)\cos\beta_i, \quad i=\overline{1,n}.
\end{cases}
\tag{3.4}
$$

Modes of steady rotation of a rotor, where in the coordinates ξ, η and β_i remain constant, will be called steady-state. Equations describing steady-state modes will be obtained by putting $\xi=\xi_0=$ const, $\eta=\eta_0=$ const and $\beta_i=\beta_{i0}=$ const into Eqs. (3.4)

$$
\begin{cases}
(1-(1+n\chi)\nu^2)\xi_0-\delta_n\nu\eta_0=\varepsilon\nu^2+\chi\nu^2\sum_{i=1}^{n}\cos\beta_{i0}, \\
(1-(1+n\chi)\nu^2)\eta_0+\delta_n\nu\xi_0=\chi\nu^2\sum_{i=1}^{n}\sin\beta_{i0}, \\
\xi_0\sin\beta_{i0}-\eta_0\cos\beta_{i0}=0, \quad i=\overline{1,n}.
\end{cases}
\tag{3.5}
$$

The last Eq. (3.5) implies that, for all $i=\overline{1,n}$,

$$
\beta_{i0}=\arctan\left(\frac{\eta_0}{\xi_0}\right)+\pi k, \quad k=0,1,\dots.
$$

From which it follows that in the case of unbalanced steady-state mode, the balancing balls can take up a position only on a straight line passing through the points O and C, located either on one or on opposite sides of the circular race.

Balanced steady-state mode. Existence and stability conditions. The stationary mode of a rotor, in which the geometrical center of the disk C lies on the axis $O_1 O_2$, will be called a balanced steady-state mode. Assuming in Eqs. (3.5) $\xi_0=\eta_0=0$, we obtain a system of two equations for the deflection angles of the balancing balls β_{i0}

$$
\chi\sum_{i=1}^{n}\cos\beta_{i0}=-\varepsilon, \quad \sum_{i=1}^{n}\sin\beta_{i0}=0.
\tag{3.6}
$$

For $n=1$, system (3.6) has a unique solution $\beta_1=\pi$, which exists only if the condition $\chi=\varepsilon$ is fulfilled. It follows that one ball is not enough to balance the rotor with a variable unbalance. For $n=2$, the solution of system (3.6) exist if $2\chi\geqslant\varepsilon$ and looks like

$$\beta_{10} = \beta, \quad \beta_{20} = -\beta, \quad \text{ГДе} \quad \beta = \pm \arccos\left(-\frac{\varepsilon}{2\chi}\right). \tag{3.7}$$

In the case $n > 2$, i.e. when the ABB contains more than two balancing balls, system (3.6) has an infinite set of solutions that exist if the condition $n\chi \geqslant \varepsilon$ is fulfilled, or, in the dimensional parameters

$$nm_b r \geqslant ms. \tag{3.8}$$

Thus, for the existence of a balanced steady-state mode it is necessary that the maximum unbalance of all the balancing balls be not less than the unbalance of the rotor.

We will study the stability of a balanced steady-state mode for a rotor equipped with an ABB with two balls. We substitute $n = 2$, $\xi = \Delta\xi$, $\eta = \Delta\eta$, $\beta_1 = \beta + \Delta\beta_1$ and $\beta_2 = -\beta + \Delta\beta_2$ into Eqs. (3.4), where $\Delta\xi$, $\Delta\eta$ and $\Delta\beta_i$, $i = 1, 2$ are small variations of the generalized coordinates. Expanding the expressions thus obtained in series and neglecting small quantities of second order and higher, we obtain the linear system of equations in variations

$$\mathbf{A}\ddot{\mathbf{Z}} + \mathbf{B}\dot{\mathbf{Z}} + \mathbf{C} = \mathbf{0}, \tag{3.9}$$

where

$$\mathbf{Z} = \begin{Vmatrix} \Delta\xi \\ \Delta\eta \\ \Delta\beta_1 \\ \Delta\beta_2 \end{Vmatrix}, \quad \mathbf{A} = \begin{Vmatrix} 1+2\chi & 0 & -\chi\sin\beta & \chi\sin\beta \\ 0 & 1+2\chi & \chi\cos\beta & \chi\cos\beta \\ -\sin\beta & \cos\beta & 1 & 0 \\ \sin\beta & \cos\beta & 0 & 1 \end{Vmatrix},$$

$$\mathbf{B} = \begin{Vmatrix} \delta_n & -2(1+2\chi)\nu & -2\chi\nu\cos\beta & -2\chi\nu\cos\beta \\ 2(1+2\chi)\nu & \delta_n & -2\chi\nu\sin\beta & 2\chi\nu\sin\beta \\ 2\nu\cos\beta & 2\nu\sin\beta & \delta_b & 0 \\ 2\nu\cos\beta & -2\nu\sin\beta & 0 & \delta_b \end{Vmatrix},$$

$$\mathbf{C} = \begin{Vmatrix} 1-(1+2\chi)\nu^2 & -\delta_n\nu & \chi\nu^2\sin\beta & -\chi\nu^2\sin\beta \\ \delta_n\nu & 1-(1+2\chi)\nu^2 & -\chi\nu^2\cos\beta & -\chi\nu^2\cos\beta \\ \nu^2\sin\beta & -\nu^2\cos\beta & 0 & 0 \\ -\nu^2\sin\beta & -\nu^2\cos\beta & 0 & 0 \end{Vmatrix}.$$

The characteristic polynomial of system (3.9) is of the eighth order

$$|\lambda^2\mathbf{A} + \lambda\mathbf{B} + \mathbf{C}| = \sum_{i=0}^{8} b_i \lambda^{8-i} = 0, \tag{3.10}$$

and its coefficients are as follows:

$$b_0 = 1 + 2\chi + \chi^2 \sin^2 2\beta, \qquad b_1 = 2(1 + \chi)(\delta_n + \delta_b(1 + 2\chi)),$$

$$b_2 = 2(1 + \chi) + \delta_n^2 + 2\delta_n\delta_b(2 + 3\chi) + \delta_b^2(1 + 2\chi)^2 +$$
$$+ 2\nu^2(1 + 3\chi + \chi^2(3 - \cos 4\beta)),$$

$$b_3 = 2(\delta_n + \delta_b(2 + 3\chi) + \delta_n\delta_b(\delta_n + \delta_b(1 + 2\chi)) +$$
$$+ \nu^2(\delta_n(1 + 4\chi) + \delta_b(2 + 7\chi + 6\chi^2))),$$

$$b_4 = (1 + \delta_n\delta_b)^2 + 2\delta_b(\delta_n + \delta_b(1 + 2\chi)) + \nu^4(1 + 6\chi + \chi^2(11 - 3\cos 4\beta)) +$$
$$+ \nu^2(-2 + 8\chi + \delta_n^2 + 2\delta_b(2\delta_n(1 + 3\chi) + \delta_b(1 + 2\chi)^2),$$

$$b_5 = 2\delta_b(1 + \delta_n\delta_b) + 2\nu^2(\delta_n\delta_b(\delta_n + \delta_b(1 + 2\chi) + 2\delta_b(\chi - 1)) +$$
$$+ 2\nu^4(3\chi\delta_n + \delta_b(1 + 5\chi + 6\chi^2)),$$

$$b_6 = \delta_b^2 + \nu^2\delta_b^2(\delta_n^2 - 2(1 + 2\chi)) + \nu^4(\delta_b^2(1 + 2\chi)^2 + 6\delta_n\delta_b\chi - 2\chi) +$$
$$+ 2\delta_b\chi\nu^6(\chi + \chi^2(3 - \cos 4\beta)),$$

$$b_7 = 2\delta_b\nu^4\chi(\nu^2(1 + 2\chi) - 1), \qquad b_8 = \nu^8\chi^2 \sin^2 2\beta.$$

A necessary condition for the stability of the steady-state mode is the positivity of all coefficients b_i. In particular, from the condition $b_7 > 0$ we have

$$\nu > \frac{1}{\sqrt{1 + 2\chi}}, \tag{3.11}$$

whence it follows that the balanced steady-state mode is unstable in the subcritical frequency range.

The stability study in the supercritical region is conveniently carried out using the Routh criterion, according to which a necessary and sufficient condition for the asymptotic stability of a balanced mode is the positivity of the Routh coefficients $c_{i,1}$, $i = \overline{1,9}$, as given by recurrent formula[9]

$$c_{i,j} = c_{i-2,j+1} - \frac{c_{i-1,j+1}c_{i-2,1}}{c_{i-1,1}}, \tag{3.12}$$

where
$$c_{1,1} = b_0; \quad c_{1,2} = b_2; \quad c_{1,3} = b_4; \quad c_{1,4} = b_6; \quad c_{1,5} = b_8;$$
$$c_{2,1} = b_1; \quad c_{2,2} = b_3; \quad c_{2,3} = b_5; \quad c_{2,4} = b_7; \quad c_{2,5} = 0.$$

The expression for the coefficients of the characteristic polynomial includes four dimensionless parameters: ε, χ, δ_n, δ_b. If three parameters are fixed (i.e., if they are given specific values) then in the plane corresponding to the remaining parameter and the dimensionless frequency ν we can construct, on the basis of the Routh criterion, a region of stability and instability of a balanced steady-state mode, which we will call two-parameter stability diagrams. As an example, in Fig. 27 stability diagrams

---

[9] *F.R. Gantmaher.* The Theory of Matrices. AMS Chelsea Publishing: Reprinted by American Mathematical Society. Vol. 2. 2000. 660 p.

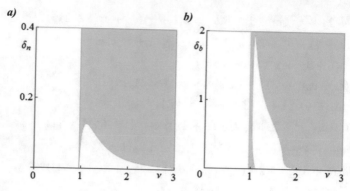

Fig. 27 Stability diagrams of the balanced steady-state mode

are presented in the case when $\varepsilon = 0.05$ and $\chi = 0.04$, that is, if the condition for the existence of a balanced steady-state mode (3.8) is fulfilled.

The diagram (a) in the parameters plane (ν, δ_n) is calculated for $\delta_b = 1$; An analogous diagram (b) in the parameter plane (ν, δ_b) is calculated for $\delta_n = 0.1$. The areas corresponding to the asymptotic stability of a balanced steady-state mode are shown in dark. Analysis of the diagrams shows that, with a sufficient mass of balancing balls, a balanced steady-state mode is stable in the supercritical region for certain values of the damping coefficients.

The graphs of the rotor whirling motion and the balancing balls movements as calculated by numerical integration of system (3.4) with $n = 2$, $\varepsilon = 0.05$, $\chi = 0.04$, $\delta_n = 0.1$ and $\delta_b = 1$, are presented in Fig. 28a (the case $\nu = 0.8$), in Fig. 28b (the case $\nu = 1.2$) and in Fig. 28c (the case $\nu = 2$).

In the first case, the subcritical unbalanced steady-state mode of whirling motion of the point C with constant amplitude is established in the system. In this case, the balancing balls touch, and their common center of mass lies on the line passing through the points O and C. The second case corresponds to a supercritical angular speed, while the parameters of the system correspond to unstable (bright) regions of the stability diagrams shown in Fig. 27. The graphs indicate the presence of a self-oscillating whirling mode in the system, accompanied by rapid rotational movements of the balancing balls relative to the rotor disk. In the third case, the system parameters correspond to the stable (dark) area of the diagrams in Fig. 27 and we see that a balanced spin steady-state mode is established in the system, the balancing balls occupying positions in accordance with the values of the angles β_{10} and β_{20}, as calculated by the formula (3.7).

Unbalanced steady-state modes. Existence conditions. As previously demonstrated, the study of whirling (unbalanced) motion modes of the rotor can be conveniently carried out in the polar coordinates a and ϕ, where $a = |\overrightarrow{OC}|/R$ is the radius vector of C, and ϕ is its phase angle in a rotating frame (Fig. 26). We substitute the relations $\xi = a \cos \phi$, $\eta = a \sin \phi$ into Eqs. (3.4) and perform the following transformations: multiply the first equation by $\cos \phi$ and add to it the second equation

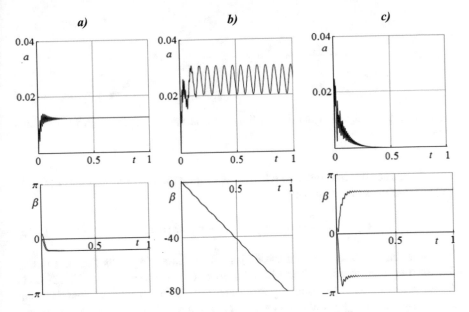

Fig. 28 Specific types of the rotor whirling motion and the balancing balls movements

multiplied by $\sin\phi$; next we multiply the second equation by $\cos\phi$ and subtract from it the first equation multiplied by $\sin\phi$. As a result, we get

$$
\begin{cases}
(1+2\chi)(\ddot{a} - (\nu + \dot{\phi})^2 a) + \delta_n\dot{a} + a = \\
\quad = \varepsilon\nu^2\cos\phi + \chi\sum_{i=1}^{n}((\nu + \dot{\beta}_i)^2\cos(\phi - \beta_i) - \ddot{\beta}_i\sin(\phi - \beta_i)), \\
(1+2\chi)(a\ddot{\phi} + 2\dot{a}(\nu + \dot{\phi})) + \delta_n a(\nu + \dot{\phi}) = \\
\quad = -\varepsilon\nu^2\sin\phi - \chi\sum_{i=1}^{n}((\nu + \dot{\beta}_i)^2\sin(\phi - \beta_i) + \ddot{\beta}_i\cos(\phi - \beta_i)), \\
\ddot{\beta}_i + \delta_b\dot{\beta}_i = (a(\nu + \dot{\phi})^2 - \ddot{a})\sin(\phi - \beta_i) - (a\ddot{\phi} + 2\dot{a}(\nu + \dot{\phi}))\cos(\phi - \beta_i), \\
i = \overline{1,n}.
\end{cases}
\tag{3.13}
$$

Let us consider an ABB with two balls and substitute into Eqs. (3.13) constant values $a = a_0$, $\phi = \phi_0$, $\beta_1 = \beta_{10}$ and $\beta_2 = \beta_{20}$; as a result, we get a system of transcendental equations describing an unbalanced stationary regime of rotor motion

$$
\begin{cases}
a_0(1 - (1 + 2\chi)\nu^2) - \chi\nu^2(\cos(\phi_0 - \beta_{10}) + \cos(\phi_0 - \beta_{20})) = \varepsilon\nu^2\cos\phi_0, \\
a_0\delta_n\nu + \chi\nu^2(\sin(\phi_0 - \beta_{10}) + \sin(\phi_0 - \beta_{20})] = -\varepsilon\nu^2\sin\phi_0, \\
a_0\sin(\phi_0 - \beta_{10}) = 0, \\
a_0\sin(\phi_0 - \beta_{20}) = 0.
\end{cases}
\tag{3.14}
$$

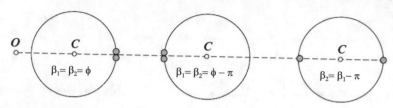

Fig. 29 Possible configuration of balancing balls in case of unbalanced steady-state modes of rotor motion

Since we are considering the case when $a_0 \neq 0$, from the last two equations of (3.14) we have two possible solutions

$$\beta_{i0}^{(1)} = \phi_0\,, \quad \beta_{i0}^{(2)} = \phi_0 - \pi\,, \quad i = 1, 2\,, \tag{3.15}$$

that show that the common center of mass of the balancing balls lies on the line OC. In this case, there are three possible locations of the balls (see Fig. 29): (1) $\beta_1 = \beta_2 = \phi$—the balls are touching at a point located farther from O than the point C; (2) $\beta_1 = \beta_2 = \phi - \pi$—the balls are touching at a point that is between the points O and C; (3) $\beta_1 = \beta_2 - \pi$—the balls are located on opposite sides of the race on opposite sides of the point C.

Obviously, due to the effect of centrifugal forces on the balls, the second and third configurations will be unstable, therefore only the first configuration of the balancing balls is practically realized. Substituting $\beta_{10} = \beta_{20} = \phi_0$ into Eqs. (3.14), we get

$$\begin{cases} a_0(1 - (1 + 2\chi)\nu^2) - 2\chi\nu^2 = \varepsilon\nu^2 \cos \phi_0\,, \\ a_0 \delta_n \nu = -\varepsilon\nu^2 \sin \phi_0\,. \end{cases} \tag{3.16}$$

To eliminate ϕ_0, we will square both sides of (3.16) and add them thereby obtaining a square equation for a_0,

$$((1 - (1 + 2\chi)\nu^2)^2 + \delta_n^2\nu^2)a_0^2 - 4\chi\nu^2(1 - (1 + 2\chi)\nu^2)a_0 + (4\chi^2 - \varepsilon^2)\nu^4 = 0\,. \tag{3.17}$$

Consider the dimensionless parameter

$$\sigma = \frac{2\chi}{\varepsilon} = \frac{2m_b r}{ms}\,,$$

expressing the ratio of the maximum unbalance introduced by the balancing balls to the unbalance magnitude of the rotor, we call it *the balancing ratio*. Let us find out how the value of the parameter σ affects the condition for the existence of unbalanced steady-state modes of the rotor motion. To do this, we represent the quadratic equation (3.17) in the form

$$pa_0^2 - qa_0 + r = 0\,, \tag{3.18}$$

where
$$p = (1 - (1 + 2\chi)\nu^2)^2 + \delta_n^2\nu^2 > 0 ,$$
$$q = 4\chi\nu^2(1 - (1 + 2\chi)\nu^2) , \tag{3.19}$$
$$r = 4\nu^4(\chi^2 - \varepsilon^2) = \varepsilon^2\nu^4(\sigma^2 - 1) ,$$

and write the expression for its roots

$$a_0^{(1,2)} = \frac{q \mp \sqrt{q^2 - 4pr}}{2p} . \tag{3.20}$$

Since a_0 is the length of the radius vector $\overrightarrow{OC}$, only real and non-negative roots have physical meaning. Analysis of expressions (3.19) and (3.20) gives the following conditions for the existence of unbalanced steady-state modes, depending on the value of the balancing ratio.

If $\sigma < 1$, then $r < 0$. In this case, for any value of ν, Eq. (3.18) has one positive root $a_0^{(2)}$ corresponding to the unbalanced steady-state mode.

If $\sigma = 1$, then $r = 0$. Here, Eq. (3.18) has one zero root or one positive root $a_0^{(2)}$ provided that $q > 0$ that is equivalent to

$$\nu < \frac{1}{\sqrt{1 + 2\chi}} . \tag{3.21}$$

If $\sigma > 1$, then $r > 0$. Expression (3.20) gives two positive roots provided that $q^2 - 4pr > 0$ or

$$\nu < \frac{1}{1 + 2\chi}\left(\sqrt{1 + 2\chi + \frac{\delta_n^2}{4}(\sigma^2 - 1)} - \frac{\delta_n}{2}\sqrt{\sigma^2 - 1}\right) . \tag{3.22}$$

Having determined the magnitude of the stationary amplitude a_0, the corresponding stationary value of the phase angle ϕ_0 can be found from the first equation of (3.16)

$$\phi_0^{(i)} = \arccos\left(a_0^{(i)}\left(\frac{1 - \nu^2}{\varepsilon\nu^2} - \sigma\right) - \sigma\right) , \quad i = 1, 2 . \tag{3.23}$$

Figure 30 presents the amplitude-frequency responses (AFR) and phase-frequency responses (PFR) of steady-state unbalanced modes calculated by (3.20) and (3.23) with $\varepsilon = 0.05$, $\delta_n = 0.1$ for three values of the balancing ratio σ. The solid curves correspond to the values $a_0^{(2)}$ and $\phi_0^{(2)}$ and the dashed lines correspond to $a_0^{(1)}$ and $\phi_0^{(1)}$.

Stability of unbalanced steady-state modes. We substitute $n = 2$, $a = a_0 + \Delta a$, $\phi = \phi_0 + \Delta\phi$, $\beta_1 = \phi_0 + \Delta\beta_1$ И $\beta_2 = \phi_0 + \Delta\beta_2$ into Eqs. (3.13), where Δa, $\Delta\phi$, $\Delta\beta_1$ and $\Delta\beta_2$ are small variations of the generalized coordinates from the stationary values corresponding to the unbalanced mode. Expanding the obtained expressions

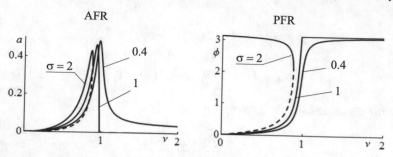

Fig. 30 AFR and PFR of unbalanced steady-state modes

in a series of small variations and omitting the second-order terms of smallness, we obtain, in view of (3.16), the linear system of equations

$$\mathbf{A}\ddot{\mathbf{Z}} + \mathbf{B}\dot{\mathbf{Z}} + \mathbf{C} = \mathbf{0},\tag{3.24}$$

where

$$\mathbf{Z} = \left\|\begin{array}{c} \Delta a \\ \Delta\phi \\ \Delta\beta_1 \\ \Delta\beta_2 \end{array}\right\|, \quad \mathbf{A} = \left\|\begin{array}{cccc} 1+2\chi & 0 & 0 & 0 \\ 0 & (1+2\chi)a_0 & \chi & \chi \\ 0 & a_0 & 1 & 0 \\ 0 & a_0 & 0 & 1 \end{array}\right\|,$$

$$\mathbf{B} = \left\|\begin{array}{cccc} \delta_n & -2a_0(1+2\chi)\nu & -2\chi\nu & -2\chi\nu \\ 2(1+2\chi)\nu & a_0\delta_n & 0 & 0 \\ 2\nu & 0 & \delta_b & 0 \\ 2\nu & 0 & 0 & \delta_b \end{array}\right\|,$$

$$\mathbf{C} = \left\|\begin{array}{cccc} 1-(1+2\chi)\nu^2 & -a_0\delta_n\nu & 0 & 0 \\ \delta_n\nu & a_0(1-(1+2\chi)\nu^2 & -\chi\nu^2 & -\chi\nu^2 \\ 0 & -a_0\nu^2 & a_0\nu^2 & 0 \\ 0 & -a_0\nu^2 & 0 & a_0\nu^2 \end{array}\right\|.$$

Conclusions about the stability of steady-state unbalanced modes can be made by analysing the coefficients of the characteristic polynomial and the Routh criterion. We write expressions for the coefficients of the characteristic polynomial, which has the form similar to (3.10)

$$b_0 = 1 + 2\chi, \quad b_1 = 2(1+\chi)(\delta_n + \delta_b(1+2\chi)),$$
$$b_2 = 2(1+\chi)+\delta_n^2+2\delta_n\delta_b(2+3\chi)+\delta_b^2(1+2\chi)^2+2\nu^2(1+3\chi+2\chi^2)(1+a_0),$$
$$b_3 = 2(\delta_n+\delta_b(2+3\chi)+\delta_n\delta_b(\delta_n+\delta_b(1+2\chi))+$$
$$\quad + \nu^2(\delta_n(1+4\chi)+\delta_b(2+7\chi+6\chi^2)) + a_0(\delta_n(2+3\chi)+\delta_b(1+2\chi))),$$
$$b_4 = (1+\delta_n\delta_b)^2 + 2\delta_b(\delta_n+\delta_b(1+2\chi))+\nu^4(1+6\chi+8\chi^2)+$$

$$+ \nu^2(-2 + 8\chi + \delta_n^2 + 2\delta_b(2\delta_n(1+3\chi) + \delta_b(1+2\chi)^2) + a_0^2\nu^4(1+2\chi)^2 +$$
$$+ 2a_0\nu^2(2 + 3\chi + 2\delta_n\delta_b(1+2\chi) + \delta_n^2 + \nu^2(2 + 7\chi + 6\chi^2)),$$
$$b_5 = 2\delta_b(1 + \delta_n\delta_b) + 2\nu^2(\delta_n\delta_b(\delta_n + \delta_b(1+2\chi) + 2\delta_b(\chi - 1)) +$$
$$+ 2\nu^4(3\chi\delta_n + \delta_b(1 + 5\chi + 6\chi^2)) + 2\delta_n a_0^2\nu^4(1+2\chi) +$$
$$+ 2a_0\nu^2(\delta_n^2\delta_b + 2\delta_n(1 + (1+3\chi)\nu^2)),$$
$$b_6 = \delta_b^2 + \nu^2\delta_b^2(\delta_n^2 - 2(1+2\chi)) + \nu^4(\delta_b^2(1+2\chi)^2 + 6\delta_n\delta_b\chi - 2\chi) +$$
$$+ a_0^2\nu^4(2(1 + \nu^2) + 4\chi(1 + 2(1+\chi)\nu^2 + \delta_n^2)) +$$
$$+ a_0\nu^2(2(1-2\nu^2) + 2\nu^2(\nu^2 + 2\chi(1 + (5+6\chi)\nu^2) + 2\nu^2\delta_n^2 +$$
$$+ 4\delta_n\delta_b(1 + 4(1+2\chi)\nu^2)),$$
$$b_7 = 2\delta_b\chi\nu^4((1+2\chi)\nu^2 - 1) + 2a_0^2\nu^4\delta_n(1 + (1+2\chi)\nu_2) +$$
$$+ 2a_0\nu^2(\delta_n^2\delta_b\nu^2 + 3\chi\delta_n\nu^4 + \delta_b((1 - \nu^2)^2 - 4\nu^2\chi(1 - (1+\chi)\nu^2)),$$
$$b_8 = a_0\nu^4(a_0((1 - (1+2\chi)\nu^2)^2 + \delta_n^2\nu^2) - 2\chi\nu^2(1 - (1+2\chi)\nu^2).$$

Note that in view of (3.19) the coefficient b_8 can be written as

$$b_8 = a_0\nu^4\left(pa_0 - \frac{q}{2}\right).$$

Obviously, the sign of the coefficient b_8 is determined by the expression in brackets, which, after substituting the stationary values of a_0 defined by formula (3.20), takes the form

$$pa_0^{(1,2)} - \frac{q}{2} = \mp\frac{1}{2}\sqrt{q^2 - 4pr}.$$

The "+" sign corresponds to the upper branches of the frequency response and phase response shown in Fig. 30 with solid curves, and the "−" sign corresponds to the lower branches marked with strokes. It follows that the lower branches of the frequency responses of unbalanced steady-state modes do not satisfy the necessary stability condition for any values of the system parameters.

The stability of steady-state modes corresponding to the upper branches of the frequency responses can be established, as before, by using the Routh criterion. The results of numerical calculations with $\varepsilon = 0.05$, $\chi = 0.02$ and $\sigma = 2\chi/\varepsilon = 0.8$ are presented in Fig. 31a, b in the form of two-parameter stability diagrams constructed in the planes of parameters (ν, δ_n) and (ν, δ_b), respectively. The areas of asymptotic stability of unbalanced steady-state modes are shown in dark. Since the balancing ratio is $\sigma < 1$, the mass of the balancing balls is insufficient to balance the unbalance. The left diagram is calculated with $\delta_b = 0.5$, and the right one, with $\delta_n = 0.1$. Analysis of the diagrams shows that in the subcritical region (that is, when $\nu < 1$), the unbalanced steady-state mode is asymptotically stable for any parameter values,

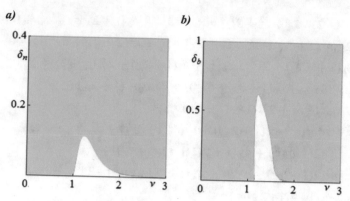

Fig. 31 Two-parameter stability diagrams of unbalanced steady-state modes

while in the supercritical region stability is guaranteed only for certain values of the damping coefficients δ_n and δ_b.

These results illustrated in Fig. 32, as the results of the numerical integration of the system of equations (3.14), carried out with the above values of δ_n and δ_b, are presented. The graphs (*a*) show the change with time of the amplitude a, the phase angle ϕ and the deflection angles of the balancing balls β_1 and β_2 in the case where the dimensionless angular speed of the rotor $\nu = 0.8 < 1$, i.e., is below the critical speed. Similar graphs (*b*) and (*c*) are constructed for the case when the angular speed of the rotor is higher than the critical one. In this case, the graphs (*b*) were calculated for $\nu = 1.3$ and correspond to the unstable region of the diagram 31, and the graphs (*c*) were obtained for the case $\nu = 2$ and correspond to the stable region of the diagram. It can be seen from the figure that steady-state unbalanced modes in the form of regular whirling motion of the rotor correspond to stable regions of the diagram.

Non-stationary passage through resonance. It is of interest to investigate the behavior of a rotor equipped with an ABB with non-stationary passage of critical frequency. This can be done by numerical integration of system (3.2) assuming that the dimensionless angular speed of the rotor varies with linear time

$$\dot{\theta}(t) = \nu(t) = \alpha t, \qquad \theta(t) = \frac{1}{2}\alpha t^2 + \theta_0, \qquad \alpha = \text{const.}$$

In Fig. 33 the graphs of amplitude curves (*a*) and angles of deflection of the balancing balls (*b*) were plotted with $\varepsilon = 0.05$, $\chi = 0.02$, $\delta_n = 0.1$, $\delta_b = 10$ for two values of the angular acceleration α.

The stationary frequency response is marked by a dashed curve. Since in this case $\sigma = 0.8 < 1$, i.e. the mass of the balancing balls is insufficient to compensate for the unbalance, an unbalanced steady-state mode is established in the supercritical region. An analysis of the behavior of the amplitude graphs reveals an interesting feature: the maximum amplitude of the whirling motion of the rotor during a non-stationary passage through the resonance exceeds the maximum amplitude of the stationary

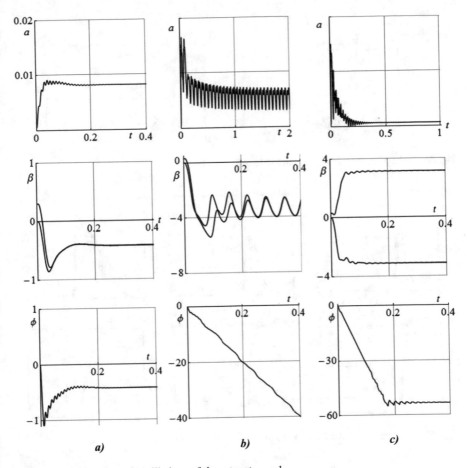

Fig. 32 Specific types of oscillations of the rotor at $\sigma < 1$

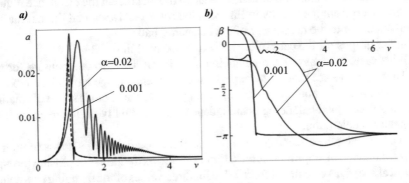

Fig. 33 Passage through resonance in the case of $\sigma = 0.8$

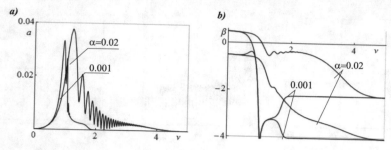

Fig. 34 Passage through resonance in the case of $\sigma = 1.6$

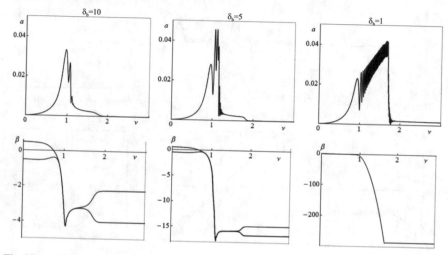

Fig. 35 Influence of viscous friction in ABB

frequency response. This phenomenon is due to the fact that in the resonance region, as can be seen from the graphs in Fig. 33b, the rotor receives an additional inertial perturbation due to the movement of the balancing balls.

Similar graphs for the case of $\sigma = 1.6$ are presented in Fig. 34.

In this case, the mass of the balancing balls is sufficient to compensate for the unbalance, and we see that in the supercritical frequency range a balanced mode is established. It can be noted that the minimum value of the angular speed of rotation of the rotor, at which balancing occurs, depends substantially on the magnitude of the angular acceleration.

Let us consider the effect of the magnitude of the coefficient of viscous friction in ABB on the shape of amplitude curves. Figure 35 shows the curves of passing through the critical speed, as calculated for $\sigma = 1.6$ for three values of the damping coefficient δ_b. We see that for $\delta_b = 10$ the amplitude of the whirling motion of the rotor when passing the resonance region changes quite smoothly. At the same time, for $\delta_b = 5$ and $\delta_b = 1$, the graphs show that after passing through the critical frequency, high-frequency oscillations of increasing amplitude occur and are followed by a sharp "breakdown". This effect is due to the fact that insufficient damping in the ABB leads to slipping of the balancing balls in the circular race of the ABB; therefore, their deflection angles relative to the rotor disk change rapidly, making a large number of full revolutions. As a result, high-frequency oscillations due to the movements of the balls are superimposed on the whirling motion of the rotor.

Main conclusions.

1. A statically unbalanced Jeffcott rotor can be fully balanced in the supercritical region with a single automatic ball balancer, provided that the maximum total unbalance of the balls exceeds the unbalance of the rotor and the asymptotic stability conditions of the balanced steady-state mode are met.

2. An important role in the process of auto-balancing is played by structural damping in the ABB. If the value of the damping coefficient is not sufficient, than self-oscillations of rotor may occur in the supercritical region they can be accompanied by rapid movements of the balancing balls in the ABB.

3. If the total mass of the balancing balls is insufficient, then in the subcritical region, for any values of the system parameters, an unbalanced steady-state mode is established in the form of regular synchronous whirling. In the supercritical region, this mode will be stable only for certain ratios between the parameters of the system.

4. With non-stationary passage of the critical region the amplitude of the forced vibrations of a rotor equipped with an ABB, often depends not only on the angular acceleration and external damping, but also on the mass of the balancing balls. In this case, the maximum amplitude with a non-stationary passage of the resonance region may exceed the maximum amplitude of the stationary frequency response, and the rotor auto-balancing process significantly depends on its angular acceleration.

4 Automatic Balancing of a Jeffcott Rotor with an Orthotropic Elastic Shaft[10]

Mechanical model. Consider a mechanical model of a statically unbalanced Jeffcott rotor with an ABB, assuming that the rotor shaft is orthotropic, that is, it has different viscoelastic characteristics in two mutually perpendicular directions. We introduce three systems of coordinates: the fixed frame OXY, the rotating frame $O\xi\eta$, and the frame $C\xi'\eta'$ rigidly connected with the rotor (Fig. 36).

Let the axes ξ and η be collinear with the axes ξ' and η', which are directed along the axes of the rigidity ellipse. We denote by k_ξ and k_η respectively the elastic coefficients of the shaft corresponding to these axes. The position of the center of mass of the disk is given by the eccentricity $s = |\overrightarrow{CG}|$ and the phase angle α between

---

[10] The section contains material published in the article: *B.G. Bykov*. Auto-balancing of a rotor with an orthotropic elastic shaft // Journal of Applied Mathematics and Mechanics. 2013. Vol. 77. Pp. 369–379.

Fig. 36 System of
coordinates

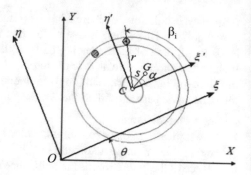

the unbalance vector and the ξ' axis. As the generalized coordinates we choose the
coordinates x and y of point C in a fixed frame, the angle θ of rotation of the rotor
and the angles of deflection β_i $(i = \overline{1, n})$ of the balancing balls relative to the circular
race of the ABB.

The expressions for the kinetic energy and potential energy of the system have
the form

$$T = \frac{1}{2}m(\dot{x}_G^2 + \dot{y}_G^2) + \frac{1}{2}J_G\dot{\theta}^2 + \frac{1}{2}m_b \sum_{i=1}^{n}(\dot{x}_{B_i}^2 + \dot{y}_{B_i}^2),$$

$$V = \frac{1}{2}k_\xi(x\cos\theta + y\sin\theta)^2 + \frac{1}{2}k_\eta(-x\sin\theta + y\cos\theta)^2,$$

where

$$x_G = x + s\cos(\theta + \alpha), \qquad x_{B_i} = x + r\cos(\theta + \beta_i),$$
$$y_G = y + s\sin(\theta + \alpha), \qquad y_{B_i} = y + r\sin(\theta + \beta_i),$$

m is the mass of the rotor, m_b is the mass of the balancing ball, I_G is the moment of
inertia of the rotor, and r is the radius of the circular race of the ABB.

Assuming that the rotor is subject to external viscous damping forces only, and
the circular race contains a viscous fluid, we write the expression for the Rayleigh
dissipative function

$$R = \frac{1}{2}c_n(\dot{x}^2 + \dot{y}^2) + \frac{1}{2}c_\theta\dot{\theta}^2 + \frac{1}{2}c_b \sum_{i=1}^{n}\dot{\beta}_i^2,$$

where c_n and c_θ are the external damping factor and c_b is the coefficient of resistance
to the motion of balls in the race of the ABB.

Considering that the angle $\theta = \theta(t)$ is a specified function of time, we introduce
the new variable $\varphi_i = \theta + \beta_i$ and write the Lagrange equations of the second kind
in the generalized coordinates x, y and φ_i

$$\begin{cases} (m+nm_b)\ddot{x}+c_n\dot{x}+(k_\xi \cos^2\theta+k_\eta \sin^2\theta)x+\dfrac{1}{2}(k_\xi-k_\eta)y\sin 2\theta = \\ \qquad = -\dfrac{d^2}{dt^2}\left[ms\cos(\theta+\alpha)+m_b r\sum\cos\varphi_i\right], \\ (m+nm_b)\ddot{y}+c_0\dot{y}+(k_\xi \sin^2\theta+k_\eta \cos^2\theta)y+\dfrac{1}{2}(k_\xi-k_\eta)x\sin 2\theta = \\ \qquad = -\dfrac{d^2}{dt^2}\left[ms\sin(\theta+\alpha)+m_b r\sum\sin\varphi_i\right], \\ m_b r^2\ddot{\varphi}_i+c_\psi(\dot{\varphi}-\dot{\theta})_i=m_b r(\ddot{x}\sin\varphi_i-\ddot{y}\cos\varphi_i), \quad i=\overline{1,n}. \end{cases} \tag{4.1}$$

In the case when the rotor rotates with constant angular speed $\dot{\theta}=\omega$, it is convenient to write the equations of motion in the rotating frame $O\xi\eta$. We substitute the expressions

$$x=\xi\cos\omega t-\eta\sin\omega t, \qquad y=\xi\sin\omega t+\eta\cos\omega t$$

into system (4.1) and perform the following transformations. We multiply the first equation by $\cos\omega t$ and add it to the second equation multiplied by $\sin\omega t$. We then multiply the second equation by $\cos\omega t$ and subtract the first one multiplied by $\sin\omega t$. As a result, we obtain autonomous equations of motion in the variables ξ, η and β_i. Changing to the dimensionless coordinates and time, we represent them in the form

$$\begin{cases} (1+n\chi)(\ddot{\bar{\xi}}-2\nu\dot{\bar{\eta}}-\nu^2\bar{\xi})+\delta_n(\dot{\bar{\xi}}-\nu\bar{\eta})+\kappa_\xi\bar{\xi} = \\ \qquad = \varepsilon\nu^2\cos\alpha+\chi\sum_{i=1}^n\left((\nu+\dot{\beta}_i)^2\cos\beta_i+\ddot{\beta}_i\sin\beta_i\right), \\ (1+n\chi)(\ddot{\bar{\eta}}+2\nu\dot{\bar{\xi}}-\nu^2\bar{\eta})+\delta_n(\dot{\bar{\eta}}+\nu\bar{\xi})+\kappa_\eta\bar{\eta} = \\ \qquad = \varepsilon\nu^2\sin\alpha+\chi\sum_{i=1}^n\left((\nu+\dot{\beta}_i)^2\sin\beta_i-\ddot{\beta}_i\cos\beta_i\right), \\ \ddot{\beta}_i+\delta_\beta\dot{\beta}_i=(\ddot{\bar{\xi}}-2\nu\dot{\bar{\eta}}-\nu^2\bar{\xi})\sin\beta_i-(\ddot{\bar{\eta}}+2\nu\dot{\bar{\xi}}-\nu^2\bar{\eta})\cos\beta_i, \quad i=\overline{1,n}, \end{cases} \tag{4.2}$$

where

$$\bar{\xi}=\frac{\xi}{r}, \quad \bar{\eta}=\frac{\eta}{r}, \quad \bar{t}=\Omega t, \quad \Omega=\sqrt{\frac{k_\xi+k_\eta}{2m}}, \quad \nu=\frac{\omega}{\Omega},$$

$$\chi=\frac{m_b}{m}, \quad \varepsilon=\frac{s}{r}, \quad \delta_n=\frac{c_n}{m\Omega}, \quad \delta_b=\frac{c_b}{m_b\Omega r^2},$$

$$\kappa_\xi=\frac{k_\xi}{m\Omega^2}=\frac{2k_\xi}{k_\xi+k_\eta}, \quad \kappa_\eta=\frac{k_\eta}{m\Omega^2}=\frac{2k_\eta}{k_\xi+k_\eta}.$$

For simplicity, we will henceforth omit the tilde above the dimensionless variables.

Unbalanced steady-state modes. To find partial solutions of the form of $\xi=\xi_0-const$, $\eta=\eta_0=const$, $\psi_i=\beta_{i0}=const$ in system (4.2) we set to zero the values of all derivatives of the generalized coordinates. As a result, we obtain a system of transcendental equations that describes the steady modes of motion of the rotor

$$
\begin{cases}
\left(\kappa_\xi - (1 + n\chi)\nu^2\right)\xi_0 - \delta_n\nu\eta_0 = \varepsilon\nu^2\cos\alpha + \chi\nu^2\sum_{i=1}^{n}\cos\beta_{i0}\,, \\[2mm]
\left(\kappa_\eta - (1 + n\chi)\nu^2\right)\eta_0 + \delta_0\nu\xi_0 = \varepsilon\nu^2\sin\alpha + \chi\nu^2\sum_{i=1}^{n}\sin\beta_{i0}\,, \\[2mm]
\xi_0\sin\beta_{i0} - \eta_0\cos\beta_{i0} = 0\,, \qquad i = \overline{1, n}\,.
\end{cases}
\tag{4.3}
$$

It is convenient to investigate the steady modes of motion of a rotor in the polar frame. We choose the length a_0 and the angle of rotation ϕ_0 of the radius vector $\overrightarrow{OC}$ about the $O\xi$ axis as required parameters. Substituting the expressions $\xi_0 = a_0\cos\phi_0$ and $\eta_0 = a_0\sin\phi_0$ into the last equation of (4.3), we obtain

$$
a_0\sin(\beta_{i0} - \phi_0) = 0\,, \qquad i = \overline{1, n}\,.
\tag{4.4}
$$

If $a_0 \neq 0$, then from (4.4) we get

$$
\beta_{i0} = \phi_0 + \pi k\,, \quad i = \overline{1, n}\,;\ k \in \mathbb{Z}.
\tag{4.5}
$$

We will next consider an ABB with two balancing balls. Then in the case of unbalanced steady-state mode, that is the case of synchronous whirling, only two variants of their mutual arrangement are possible: with $\beta_{20} = \beta_{10}$, or with $\beta_{20} = \beta_{10} + \pi$. We transform the first two equations of system (4.3), multiplying the first equation by $\cos\phi_0$ and add it to the second multiplied by $\sin\phi_0$. Next we multiply the first equation by $\sin\phi_0$ and subtract from it the second multiplied by $\cos\phi_0$. This gives us

$$
\begin{cases}
a_0\left(\kappa_\xi\cos^2\phi_0 + \kappa_\eta\sin^2\phi_0 - (1 + 2\chi)\nu^2\right) = \varepsilon\nu^2\cos(\phi_0 - \alpha) + \chi\nu^2\gamma\,, \\[2mm]
a_0\left((\kappa_\xi - \kappa_\eta)\cos\phi_0\sin\phi_0 - \delta_n\nu\right) = \varepsilon\nu^2\sin(\phi_0 - \alpha)\,,
\end{cases}
\tag{4.6}
$$

where

$$
\gamma = \cos(\beta_{10} - \phi_0) + \cos(\beta_{20} - \phi_0)\,.
$$

In the case of regular whirling, the coefficient γ can take only one of three possible values: 2, -2 or 0. Each of them corresponds to one of three variants of arrangement of the balancing balls on the circular race of the ABB (Fig. 29). Note that the second and third variants are unstable for any values of the parameters, since under the action of centrifugal forces the balls tend to occupy the position most distant from the point O on the disk. Therefore we will only explore the first variant of arrangement of balls (for which $\gamma = 2$). Expressing in system (4.6) $\cos^2\phi_0$ and $\sin^2\phi_0$ in terms $\cos 2\phi_0$, and taking into account that $\kappa_\xi + \kappa_\eta = 2$ and $\kappa_\xi - \kappa_\eta = 2\kappa$, we get

$$\begin{cases} a_0\kappa \cos 2\phi_0 + a_0\left(1 - (1+2\chi)\nu^2\right) - 2\chi\nu^2 = \varepsilon\nu^2 \cos(\phi_0 - \alpha), \\ a_0\kappa \sin 2\phi_0 - a_0\delta_n\nu = \varepsilon\nu^2 \sin(\phi_0 - \alpha). \end{cases} \quad (4.7)$$

Multiplying the second equation in (4.7) by the imaginary unit i and adding it to the first one, we come to a complex equation for a_0, ϕ_0 and ν

$$a_0\kappa e^{i2\phi_0} + a_0(1 - (1+2\chi)\nu^2 - i\delta_n\nu) - 2\chi\nu^2 = \varepsilon\nu^2 e^{i(\phi_0 - \alpha)}. \quad (4.8)$$

Considering Eq. (4.8) as a quadratic equation with respect to $e^{i\phi_0}$, we find

$$e^{i\phi_0} = \frac{\varepsilon\nu^2 e^{-i\alpha} \pm \sqrt{D(a_0, \nu)}}{2a_0\kappa}, \quad (4.9)$$

where

$$D(a_0, \nu) = (\varepsilon\nu^2 e^{-i\alpha})^2 - 4a_0\kappa(a_0(1 - (1+2\mu)\nu^2 - i\delta_n\nu) - 2\chi\nu^2).$$

Equating the modules of the left- and right-hand sides of relation (4.9), we obtain an equation for a_0 and ν

$$|\varepsilon\nu^2 e^{-i\alpha} \pm \sqrt{D(a_0, \nu)}| = 2a_0|\kappa|. \quad (4.10)$$

This equations suitable for calculating the rotor frequency response. Eliminating the variable a_0 from system (4.7), we obtain the equation for calculating the phase response

$$(1-(1+2\chi)\nu^2)\sin(\phi_0-\alpha)+\delta_n\nu\cos(\phi_0-\alpha)-\kappa\sin(\phi_0+\alpha) =$$
$$= \frac{2\chi}{\varepsilon}(\kappa\sin 2\phi_0 - \delta_n\nu), \quad (4.11)$$

where $\sigma = 2\mu/\varepsilon$ is the balancing ratio.

Figure 37 shows the amplitude-frequency and phase-frequency responses for a rotor with slightly orthotropic shaft ($|\kappa| = 0.05$), as calculated by formulas (4.10) and (4.11) with the following values of dimensionless parameters $\varepsilon = 0.05$, $\delta_n = 0.1$, $\kappa_\xi = 1.05$, $\kappa_\eta = 0.95$, $\alpha = 0.25$. Variant (a) corresponds to the case when the balancing ratio is $\sigma = 0.8$, i.e. the mass of the balancing balls is insufficient to compensate for the unbalance. We see that in this case unbalanced mode exists in the entire frequency range. Variant (b), as calculated with $\sigma = 1.2$, shows that the unbalanced mode exists only in the subcritical frequency range. For comparison, the dashed curves show the frequency response and phase response of a rotor with isotropic shaft.

Figure 38 presents the similar responses for a rotor with highly orthotropic shaft ($|\kappa| = 0.333$), which were calculated for the following values of the dimensionless parameters $\kappa_\xi = 0.666$ and $\kappa_\eta = 1.333$. In this case, there are two regions of the

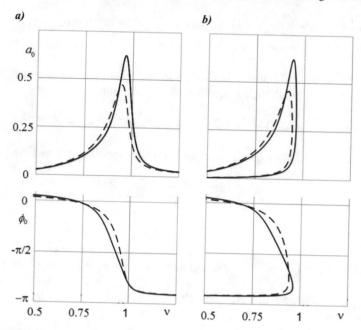

Fig. 37 AFR and PFR of the rotor with slightly orthotropic shaft

resonant oscillations of the rotor, the amplitude of which at two critical frequencies tends to infinity. The figure (a) shows that in the case $\sigma = 0.8$, the unbalanced mode of an orthotropic rotor exists in the entire frequency range, except for critical frequencies, and the figure (b) shows that in the case $\sigma = 1.2$, the unbalanced mode of an orthotropic rotor exists in the region of not very high supercritical frequencies, in contrast to the isotropic rotor.

Balanced steady-state mode. Setting $\xi_0 = \eta_0 = 0$ in (4.3), we obtain the equations

$$\chi \sum_{i=1}^{n} \cos \beta_{i0} = -\varepsilon \cos \alpha, \qquad \chi \sum_{i=1}^{n} \sin \beta_{i0} = -\varepsilon \sin \alpha, \qquad (4.12)$$

which describe a balanced steady-state mode.

They are distinguished from the similar equations for an isotropic rotor (3.6) only by the presence of the phase angle α. If $n = 2$ and the condition $\sigma \geqslant 1$ holds, system (4.12) has a unique solution, which corresponds to a balanced mode:

$$\beta_{10} = \arccos(-\varepsilon/2\chi) + \alpha, \qquad \beta_{20} = -\arccos(-\varepsilon/2\chi) + \alpha. \qquad (4.13)$$

We will investigate the stability of the balanced steady-state mode in first approximation. Suppose that $\Delta\xi$, $\Delta\eta$ and $\Delta\beta_i$ ($i = 1, 2$) are small deviations of the generalized coordinates from the stationary values corresponding to the balanced mode. Substi-

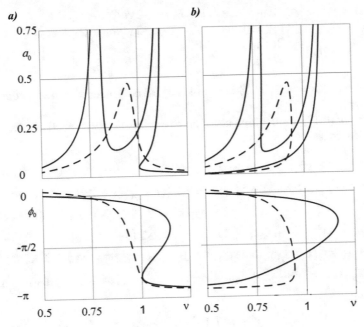

Fig. 38 AFR and PFR of the rotor with a highly orthotropic shaft

tuting the expressions

$$\xi = \Delta\xi, \quad \eta = \Delta\eta, \quad \beta_1 = \beta_{10} + \Delta\beta_1, \quad \beta_2 = \beta_{20} + \Delta\beta_2,$$

into Eqs. (4.2), expanding in series in small deviations and neglecting the small quantities of second order and higher, we obtain a linear system of variations equations, which we write in the matrix form as a linear eighth-order system

$$\mathbf{A}\dot{\mathbf{Z}} + \mathbf{B}\mathbf{Z} = \mathbf{0}, \tag{4.14}$$

where

$$\mathbf{Z} = \{\Delta\xi, \Delta\eta, \Delta\beta_1, \Delta\beta_2, \Delta\dot{\xi}, \Delta\dot{\eta}, \Delta\dot{\beta}_1, \Delta\dot{\beta}_2\}^T,$$

$$\mathbf{A} = \begin{Vmatrix} \mathbf{E} & \mathbf{O} \\ \mathbf{O} & \tilde{\mathbf{A}} \end{Vmatrix}, \quad \tilde{\mathbf{A}} = \begin{Vmatrix} 1+2\chi & 0 & -\chi S_{10} & -\chi S_{20} \\ 0 & 1+2\chi & \chi C_{10} & \chi C_{20} \\ -S_{10} & C_{10} & 1 & 0 \\ -S_{20} & C_{20} & 0 & 1 \end{Vmatrix}, \quad \mathbf{B} = \begin{Vmatrix} \mathbf{O} & -\mathbf{E} \\ \tilde{\mathbf{B}} & \tilde{\tilde{\mathbf{B}}} \end{Vmatrix},$$

$$\tilde{\mathbf{B}} = \begin{Vmatrix} (k_1 - (1+2\chi)\nu^2) & -\delta_n\nu & \chi\nu^2 S_{10} & \chi\nu^2 S_{20} \\ \delta_n\nu & (\kappa_\eta - (1+2\chi)\nu^2) & -\chi\nu^2 C_{10} & -\chi\nu^2 C_{20} \\ \nu^2 S_{10} & -\nu^2 C_{10} & 0 & 0 \\ \nu^2 S_{20} & -\nu^2 C_{20} & 0 & 0 \end{Vmatrix},$$

$$\tilde{\mathbf{B}} = \begin{Vmatrix} \delta_0 & -2(1+2\chi)\nu & -2\chi\nu C_{10} & -2\chi\nu C_{20} \\ 2(1+2\chi)\nu & \delta_n & -2\chi\nu S_{10} & -2\chi\nu S_{20} \\ 2\nu C_{10} & 2\nu S_{10} & \delta_b & 0 \\ 2\nu C_{20} & 2\nu S\psi_{20} & 0 & \delta_b \end{Vmatrix},$$

$S_{i0} = \sin \beta_{i0}$, $C_{i0} = \cos \beta_{i0}$, $\mathbf{E}$ is the (4×4) identity matrix, and $\mathbf{O}$ is the (4×4) zero matrix.

In view of (4.13), the coefficients of the characteristic polynomial of system (4.14) can be written as

$$a_0 = 1 + 2\chi + \varepsilon^2 \gamma_1, \qquad a_1 = 2(1 + \chi)\gamma_2,$$

$$a_2 = 2(1 + \chi)(1 + \delta_n \delta_b) + \gamma_2^2 + \chi\gamma_3 + 2\left((1 + \chi)(1 + 2\chi) + 2\varepsilon^2 \gamma_1\right)\nu^2,$$

$$a_3 = 2\gamma_2(1 + \delta_n \delta_b) + (2(1 + \chi) + \chi\gamma_3)\delta_b + 2\left(\delta_n(1 + 4\chi) + 2\delta_b(1 + 2\chi)\right)\nu^2,$$

$$a_4 = (\nu^2 - 1)^2 - \lambda^2 + \delta_0^2 \delta_1^2 + 2\delta_1(\delta_0 + \gamma_2) + 2\left(3\mu + 4\mu^2 + 3\varepsilon^2 \gamma_1\right)\nu^4 +$$
$$+ \left(\delta_n^2 + 2\delta_b^2(1 + 2\chi)^2 + 4\delta_n \delta_b(1 + 3\chi) + 8\chi + 2\chi\gamma_3\right)\nu^2,$$

$$a_5 = 2\delta_b\left((\nu^2 - 1)^2 - \kappa^2 + \delta_n \delta_b + \chi\gamma_3 + \delta_n \gamma_2 + 2\chi\nu^2 + \chi(5 + 6\chi)\nu^4\right) +$$
$$+ 6\chi\delta_n \nu^4,$$

$$a_6 = \delta_b^2\left(((1 + 2\chi)\nu^2 - 1)^2 - \kappa^2\right) + \delta_n^2 \delta_b^2 \nu^2 + \chi\nu^4(\gamma_3 + 6\delta_n \delta_b + 2(\nu^2 - 1)) +$$
$$+ 4(\chi^2 + 2\varepsilon^2 \gamma_1)\nu^6,$$

$$a_7 = \chi\delta_b\left(2(1 + 2\chi)\nu^2 - 2 + \gamma_3\right)\nu^4,$$

$$a_8 = \varepsilon^2 \gamma_1 \nu^8,$$

where

$$\gamma_1 = 1 - \varepsilon^2/(4\chi^2), \quad \gamma_2 = \delta_n + (1 + 2\chi)\delta_b, \quad \gamma_3 = 2\kappa(1 - \varepsilon^2/(2\chi^2))\cos 2\alpha.$$

A necessary condition for stability of the balanced steady-state mode, which follows from the condition that the coefficient a_7 is positive, has the form

$$\nu^2 > \left(1 - 2\kappa(1 - 2/\sigma^2)\cos 2\alpha\right)/(1 + 2\chi). \tag{4.15}$$

For an isotropic rotor (that is, in the case when $\kappa = 0$), condition (4.15) is identical to the previously obtained condition (3.11).

A necessary and sufficient condition for the real parts of the roots of the characteristic equation to be negative is that the Routh coefficients $c_{i,1}$, $i = \overline{1, 9}$, be positive when they are calculated from the recursion formulas $c_{i,j} = c_{i-2,j+1} - c_{i-1,j+1}c_{i-2,1}/c_{i-1,1}$, where

$$c_{1,1} = a_0; \quad c_{1,2} = a_2; \quad c_{1,3} = a_4; \quad c_{1,4} = a_6; \quad c_{1,5} = a_8;$$
$$c_{2,1} = a_1; \quad c_{2,2} = a_3; \quad c_{2,3} = a_5; \quad c_{2,4} = a_7; \quad c_{2,5} = 0.$$

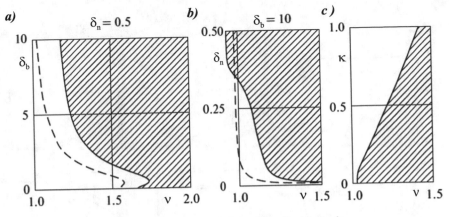

Fig. 39 Two-parameter stability diagrams of a balanced steady-state mode

The calculations, as performed with $\varepsilon = 0.05$, $\chi = 0.03$, $\delta_n = 0.05$, $\delta_b = 10$, specify the following regions of asymptotic stability of the balanced steady-state mode:

for rotor with isotropic shaft ($\kappa = 0$, $\alpha = 0$) $\nu > 1.024$,

for rotor with orthotropic shaft ($|\kappa| = 0.333$, $\alpha = 0.25$) $\nu > 1.177$.

It is convenient to evaluate the influence of the individual parameters of the system using two-parameter stability diagrams. Figure 39 presents stability diagrams in the (ν, δ_b) and (ν, δ_n) parameter planes, which were calculated using the Routh criterion for isotropic and orthotropic rotors. The diagram (a) corresponds to $\delta_n = 0.05$, and the diagram (b) corresponds to $\delta_b = 10$. The stability regions of the balanced steady-state mode for the orthotropic rotor are shown hatched. The boundaries of the stability region for the isotropic rotor are marked by dashed lines. The diagram (c) demonstrates the narrowing of the stability region in the (ν, κ) parameter plane as the anisotropy parameter κ increases.

Transitional modes. The behavior of an anisotropic rotor during the transition to the steady-state mode was investigated by numerical integration of system (4.2) for constant value of the angular speed ν. The results of the calculations are shown in Fig. 40, where the solid curves correspond to an orthotropic rotor ($|\kappa| = 0.333$) and the dashed curves correspond to an isotropic rotor.

Figure 40a presents graphs of the time dependence of the displacement of the centre of the disk $a = \sqrt{\xi^2 + \eta^2}$ and of the angles of deflection of the balancing balls β_1 and β_2 calculated in the subcritical region ($\nu < 1$). The figure (b) and (c) show similar results in the supercritical region ($\nu > 1$). It is seen from the graphs that for $\nu = 0.75$ the amplitude of circular whirling of the anisotropic rotor is greater than that of the isotropic rotor. For $\nu = 1.2$, the anisotropy of the shaft results in an increase in the duration of the transitional process, and for $\nu = 1.5$, it has virtually no effect on the motion of the rotor and the balancing balls.

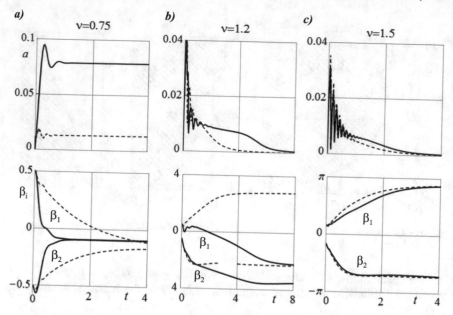

Fig. 40 The effect of shaft orthotropy on transitional processes

Of special practical interest are modes with unsteady passage through the critical region, which appear when the rotor accelerates from the state of rest or in the case of stopping, when its angular velocity is above the critical value. Figure 41 demonstrated the results of numerical integration of system (4.2) for a rotor spinning with constant angular acceleration $\ddot{\theta}(t) = u = \text{const}$. The graph ($a$) shows the variation of the amplitudes of oscillations of the point C versus the instantaneous angular velocity in a mode of "slow" passage ($u = 0.02$) through the critical region; the graphs (b) fits to "fast" passage ($u = 0.08$).

It is seen that in the case of slow passage through the critical region the maximum deviation of the center of the orthotropic rotor (the solid curve) is several times greater than the maximum deviation of the rotor with isotropic shaft (the dashed curve). With rapid passage through the critical region, no strong discrepancy between the deviations for the isotropic and orthotropic rotors is observed. Figure (c) and (d) shows the effect of rotor acceleration on the movement of the balancing balls. When the angular speed change is slow, the balls (which move in one direction) join together, and in the supercritical region they diverge in different directions, occupying a position corresponding to the balanced mode. With a rapid change in the angular speed of the rotor, the balls do not touch each other and their movement to stationary points can occur in one or in different directions.

Main conclusions. Based on the above, the following conclusions can be made regarding the influence of the orthotropy of the elastic properties of a flexible shaft on the motion of a rotor equipped with ABB.

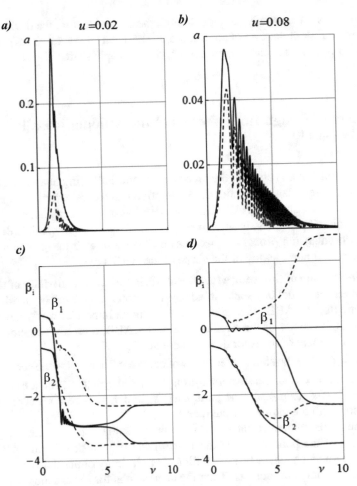

Fig. 41 Unstationary passage of the critical frequency

1. The orthotropy of the flexible shaft results in the appearance of a second critical frequency and also causes (when the external damping is less than the critical damping) an unrestricted increase in the amplitude of the whirling motion.

2. The conditions of existence of a balanced steady-state mode for an orthotropic rotor have the same form as those for an isotropic rotor, but the stability region for a given mode narrows as the anisotropy parameter increases.

3. Numerical investigation of transitional modes of motion shows that when rotor rotates with constant angular velocity, the anisotropy has an appreciable effect on the motion of the rotor in the subcritical region and near the critical frequencies. In the supercritical region (sufficiently far from the critical frequencies) the orthotropy of the shaft does not have a significant effect on the auto-balancing process.

4. During slow unsteady passage through the critical region, the deviation of the center of the disk in the case of an orthotropic shaft can be several times greater than the similar deviation in the case of a rotor with isotropic shaft.

5 Effect of Design Imperfection of the Automatic Ball Balancer[11]

In Sects. 3 and 4 of this chapter, the processes of auto-balancing of a statically unbalanced rotor were considered under the assumption that the trajectory of movement of the centers of the balancing balls relative to ABB has the shape of an ideal circle with center at point C. However, in some cases, for an adequate mathematical description of the auto-balancing process, it becomes necessary to take into account possible eccentricity and non-ideality of the shape of the ABB's race.

Autobalancing of the rotor with consideration of the eccentricity of the ABB. We consider the model of a static unbalanced Jeffcott rotor described in §4, but here we assume that the ABB is installed on a disk "in an imperfect way", i.e. the point E, which is the center of the circular race of the ABB, does not coincide with the point of attachment of the rotor disk to the shaft C.

Let $s_1 = |\vec{CG}|$ be the static eccentricity of the rotor. To describe the eccentricity of ABB, we introduce two parameters: the displacement $s_2 = |\vec{CE}|$ and the eccentricity angle $\gamma = \angle GCE$ (Fig. 42.). The parameters s_1 and s_2 will be considered small in comparison with the radius r of the race.

Let the ABB contain n balancing balls of the same mass m_b. Let x, y be the coordinates of the point C, m be the mass of the rotor disk with the ABB's hull (without balancing balls), k be the coefficient of elasticity of the shaft and c_n, c_θ and c_b are the damping coefficients. Then the expressions for the kinetic energy of the system, the potential energy of the elastic shaft, and the Rayleigh dissipative function can be represented as

$$T = \frac{1}{2}m(\dot{x}_G^2 + \dot{y}_G^2) + \frac{1}{2}J_G\dot{\theta}^2 + \frac{1}{2}m_b\sum_{j=1}^{n}(\dot{x}_{B_j}^2 + \dot{y}_{B_j}^2),$$

$$V = \frac{1}{2}k(x^2 + y^2), \qquad R = \frac{1}{2}c_n(\dot{x}^2 + \dot{y}^2) + \frac{1}{2}c_b\sum_{j=1}^{n}\dot{\beta}_j^2 + \frac{1}{2}c_\theta\dot{\theta}^2, \tag{5.1}$$

where

---

[11] The section contains materials published in articles: *V.G. Bykov, A.S. Kovachev*. Dynamic of a statically unbalanced rotor with eccentric ball autobalancer // Vestn. S.-Peterb. Univ. Ser. 1. 2014. Vol. 1. № 4. Pp. 579–588 [in Russian]; *V.G. Bykov, A.S. Kovachev*. Dynamics of a statically unbalanced rotor with an elliptic automatic ball balancer // Vestnik St. Petesrburg University, Mathematics. 2019. Vol. 52. No. 3. Pp. 301–308.

Fig. 42 Rotor with an imperfectly installed ABB

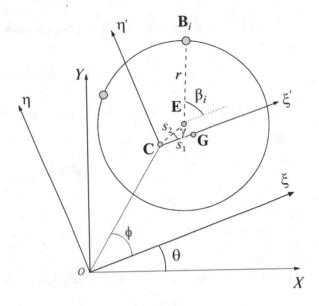

$$x_G = x + s_1 \cos\theta, \qquad y_G = y + s_1 \sin\theta,$$
$$x_{B_j} = x + s_2 \cos(\theta + \gamma) + r\cos(\theta + \beta_j),$$
$$y_{B_j} = y + s_2 \sin(\theta + \gamma) + r\sin(\theta + \beta_j).$$

Assuming that the angle of rotation $\theta = \theta(t)$ is a specified function of time, we write the Lagrange equations of the second kind

$$
\begin{cases}
(m + nm_b)\ddot{x} + c_n\dot{x} + kx = \\
\quad = -\dfrac{d^2}{dt^2}\left[ms_1\cos\theta + nm_b s_2\cos(\theta + \gamma) + m_b r\displaystyle\sum_{j=1}^{n}\cos(\theta + \beta_j)\right], \\[2mm]
(m + nm_b)\ddot{y} + c_n\dot{y} + ky = \\
\quad = -\dfrac{d^2}{dt^2}\left[ms_1\sin\theta + nm_b s_2\sin(\theta + \gamma) + m_b r\displaystyle\sum_{j=1}^{n}\sin(\theta + \beta_j)\right], \\[2mm]
m_b r^2(\ddot{\beta}_j + \ddot{\theta}) + c_b\dot{\beta}_j = m_b r\left(\ddot{x}\sin(\theta + \beta_j) - \ddot{y}\cos(\theta + \beta_j)\right) + \\
\quad + m_b r s_2\left(\dot{\theta}^2\sin(\gamma - \beta_j) - \ddot{\theta}\cos(\gamma - \beta_j)\right), \quad j = \overline{1, n}.
\end{cases}
\tag{5.2}
$$

Changing to the dimensionless variables in the system (5.2)

$$\bar{x} = \frac{x}{r}, \quad \bar{y} = \frac{y}{r}, \quad \bar{t} = \Omega t, \quad \Omega = \sqrt{\frac{k}{m}},$$

we obtain

$$\begin{cases} (1+n\chi)\ddot{\bar{x}}+\delta_n\dot{\bar{x}}+\bar{x}=-\dfrac{d^2}{d\bar{t}^2}\left[\varepsilon_1\cos\theta+n\varepsilon_2\chi\cos(\theta+\gamma)+\chi\sum_{j=1}^{n}\cos(\theta+\beta_j)\right], \\[2mm] (1+n\chi)\ddot{\bar{y}}+\delta_n\dot{\bar{y}}+\bar{y}=-\dfrac{d^2}{d\bar{t}^2}\left[\varepsilon_1\sin\theta+n\varepsilon_2\chi\sin(\theta+\gamma)+\chi\sum_{j=1}^{n}\sin(\theta+\beta_j)\right], \\[2mm] \ddot{\beta}_j+\ddot{\theta}+\delta_b\dot{\beta}_j=\ddot{\bar{x}}\sin(\theta+\beta_j)-\ddot{\bar{y}}\cos(\theta+\beta_j)+\varepsilon_2\left(\dot{\theta}^2\sin(\gamma-\beta_j)-\ddot{\theta}\cos(\gamma-\beta_j)\right), \end{cases}$$
$$(5.3)$$

where

$$j=\overline{1,n}, \quad \chi=\frac{m_b}{m}, \quad \varepsilon_1=\frac{s_1}{r}, \quad \varepsilon_2=\frac{s_2}{r}, \quad \delta_n=\frac{c_n}{m\Omega}, \quad \delta_b=\frac{c_b}{m_br^2\Omega}.$$

Next we assume that rotor angular speed is $\nu=\dot{\theta}=\text{const}$. Substituting $\theta=\nu t$ into system (5.3), and omitting the bar above dimensionless variables for simplicity, we have

$$\begin{cases} (1+n\chi)\ddot{x}+\delta_n\dot{x}+x=(\varepsilon_1\cos\nu t+n\varepsilon_2\chi\cos(\nu t+\gamma))\nu^2+ \\[2mm] \qquad\qquad +\chi\sum_{j=1}^{n}\left((\nu+\dot{\beta}_j)^2\cos(\nu t+\beta_j)+\ddot{\beta}_j\sin(\nu t+\beta_j)\right), \\[2mm] (1+n\chi)\ddot{y}+\delta_n\dot{y}+y=(\varepsilon_1\sin\nu t+n\varepsilon_2\chi_2\sin(\nu t+\gamma))\nu^2+ \\[2mm] \qquad\qquad +\chi\sum_{j=1}^{n}\left((\nu+\dot{\beta}_j)^2\sin(\nu t+\beta_j)-\ddot{\beta}_j\cos(\nu t+\beta_j)\right), \\[2mm] \ddot{\beta}_j+\delta_b\dot{\beta}_j=\ddot{x}\sin(\nu t+\beta_j)-\ddot{y}\cos(\nu t+\beta_j)+\varepsilon_2\nu^2\sin(\gamma-\beta_j), \quad j=\overline{1,n}. \end{cases}$$
$$(5.4)$$

Introducing the complex variable $z=x+iy$, we represent system (5.4) in the complex form

$$\begin{cases} (1+n\chi)\ddot{z}+\delta_n\dot{z}+z=\left((\varepsilon_1+n\varepsilon_2\chi e^{i\gamma})\nu^2+\chi\sum_{j=1}^{n}((\nu+\dot{\beta}_j)^2-i\ddot{\beta}_j)e^{i\beta_j}\right)e^{i\nu t}, \\[2mm] \ddot{\beta}_j+\delta_b\dot{\beta}_j=-\mathrm{Im}\left[\ddot{z}e^{-i(\nu t+\beta_j)}\right]+\varepsilon_2\nu^2\sin(\gamma-\beta_j), \quad j=\overline{1,n}. \end{cases}$$
$$(5.5)$$

Let us proceed with the consideration of the motion equations in the rotating frame $O\xi\eta$. To do this, we introduce the complex variable $\zeta=\xi+i\eta$ and substitute the relations

$$z=\zeta e^{i\nu t}, \quad \dot{z}=(\dot{\zeta}+i\nu\zeta)e^{i\nu t}, \quad \ddot{z}=(\ddot{\zeta}+2i\nu\dot{\zeta}-\nu^2\zeta)e^{i\nu t}, \qquad (5.6)$$

into system (5.5). As a result we obtain

$$\begin{cases} (1+n\chi)\ddot{\zeta} + (\delta_n + 2i\nu(1+n\chi))\dot{\zeta} + (1-(1+n\chi)\nu^2 + i\nu\delta_n)\zeta = \\ \qquad = (\varepsilon_1 + n\varepsilon_2\chi e^{i\gamma})\nu^2 + \chi\sum_{j=1}^{n}((\nu+\dot{\beta}_j)^2 - i\ddot{\beta}_j)e^{i\beta_j}, \qquad (5.7) \\ \ddot{\beta}_j + \delta_b\dot{\beta}_j = -\mathrm{Im}[(\ddot{\zeta}+2i\nu\dot{\zeta}-\nu^2\zeta)e^{-i\beta_j}] + \varepsilon_2\nu^2\sin(\gamma-\beta_j), \quad j = \overline{1,n}. \end{cases}$$

In contrast to (5.5), the system of equations (5.7) is autonomous and convenient for studying steady-state modes of motion of the rotor.

Stationary solutions of system (5.7) will be sought in the form

$$\zeta = \zeta_0 = \xi_0 + i\eta_0, \qquad \beta_j = \beta_{0j} = const, \quad (j = \overline{1,n}), \qquad (5.8)$$

where $a_0 = \sqrt{\xi_0^2 + \eta_0^2} = const$ is the amplitude of the whirling movement of the disk center. Substituting expressions (5.8) into Eqs. (5.7), we obtain the system of transcendental equations for the unknowns ξ_0, η_0 and β_{0j}

$$\begin{cases} (1-(1+n\chi)\nu^2 + i\nu\delta_n)(\xi_0+i\eta_0) = \nu^2\left(\varepsilon_1 + n\varepsilon_2\chi e^{i\gamma} + \chi\sum_{j=1}^{n}e^{i\beta_{0j}}\right), \\ \qquad\qquad\qquad\qquad\qquad\qquad\qquad\qquad\qquad\qquad\qquad\qquad (5.9) \\ \eta_0\cos\beta_{0j} - \xi_0\sin\beta_{0j} + \varepsilon_2\sin(\gamma-\beta_{0j}) = 0, \quad j = \overline{1,n}. \end{cases}$$

Let us check the possibility of the existence of a balanced steady-state mode in which the geometric center of the disk C lies on the axis O_1O_2. To do this, we substitute $\xi_0 = \eta_0 = 0$ into system (5.9) and separate the real and imaginary parts in the first equation, as a result we will have

$$\begin{cases} \sum_{j=1}^{n}\cos\beta_{0j} = -\dfrac{\varepsilon_1}{\chi} - n\varepsilon_2\cos\gamma, \\ \sum_{j=1}^{n}\sin\beta_{0j} = -n\varepsilon_2\sin\gamma, \qquad\qquad\qquad (5.10) \\ \sin(\gamma-\beta_{0j}) = 0, \quad j = \overline{1,n}. \end{cases}$$

Multiplying the first equation of system (5.10) by $\sin\gamma$ and subtracting the second equation multiplied by $\cos\gamma$ from it, we obtain

$$\sum_{j=1}^{n}\sin(\gamma-\beta_{0j}) = -\varepsilon_1\sin\gamma = 0.$$

It follows that $\gamma = 0$ or $\gamma = \pi$, i.e., a fully balanced steady-state mode for a rotor with non-ideal positioning of the ABB is possible only in cases when the points E, C and G lie on the same straight line. In view of this, system (5.10) takes the following form

$$\begin{cases} \sum_{j=1}^{n} \cos \beta_{0j} = -\dfrac{\varepsilon_1}{\chi} \pm n\varepsilon_2, \\ \sin \beta_{0j} = 0, \quad j = \overline{1,n}, \end{cases} \tag{5.11}$$

whence it follows that all angles of deflection of the balancing balls can take only two values $\beta_{0j} = \{0, \pi\}$ ($j = \overline{1,n}$), i.e. the centers of all balls must also lie on the same line with the points C and G. In addition, the system parameters must be such that the first equation in (5.11) is satisfied. For example, for a two-ball ABB ($n = 2$), in case $\gamma = 0$, a balanced mode will take place under the condition $\chi = \varepsilon_1/2(1 - \varepsilon_2)$, and in case $\gamma = \pi$ a balanced mode will take place under the condition $\chi = \varepsilon_1/2(1 + \varepsilon_2)$. Considering the above, it can be stated that due to the presence of eccentricity in the ABB it is incapable of providing complete balancing of the rotor with arbitrary unbalance.

Let us find the conditions for the existence of unbalanced steady-state modes of whirling motion of the rotor for the case when the ABB contains two balancing balls. Let us transform the equations describing the motion of the balancing balls

$$\begin{cases} \eta_0 \cos \beta_{01} - \xi_0 \sin \beta_{01} + \varepsilon_2 \sin(\gamma - \beta_{01}) = 0, \\ \eta_0 \cos \beta_{02} - \xi_0 \sin \beta_{02} + \varepsilon_2 \sin(\gamma - \beta_{02}) = 0, \end{cases} \tag{5.12}$$

as follows. Multiply the first equation in (5.12) by $\sin \beta_{02}$ and subtract from it the second equation multiplied by $\sin \beta_{01}$. Similarly, we subtract the second equation multiplied by $\cos \beta_{01}$ from the first equation multiplied by $\cos \beta_{02}$. As a result, we get

$$\begin{cases} \sin(\beta_{02} - \beta_{01})(\eta_0 + \varepsilon_2 \sin \gamma) = 0, \\ \sin(\beta_{02} - \beta_{01})(\xi_0 + \varepsilon_2 \cos \gamma) = 0. \end{cases} \tag{5.13}$$

System (5.13) has two types of solutions. Solutions of the first type satisfy the equation

$$\sin(\beta_{01} - \beta_{02}) = 0,$$

from which it follows that $\beta_{01} = \beta_{02} + \pi k$ ($k = 0, 1$), i.e. the balancing balls either touch each other or are located on opposite sides of the circular race of the ABB. Solutions of the second type are as follows:

$$\eta_0 = -\varepsilon_2 \sin \gamma, \quad \xi_0 = -\varepsilon_2 \cos \gamma, \quad \text{or} \quad a_0 = \varepsilon_2, \quad \phi_0 = \gamma \pm \pi(2k+1). \tag{5.14}$$

Thus, a characteristic feature of the solutions of the second type is the presence of residual vibration of the rotor, the amplitude of which is exactly equal to the eccentricity of the ABB. Since parameter ε_2 is small, unbalanced steady-state modes corresponding to the solutions of the second type can be called almost balanced.

Let us find the angles of deflection of the balancing balls corresponding to the almost balanced steady-state mode. Substituting solution (5.14) into the first equation of (5.9), we obtain the complex equation

$$\chi \nu^2 (e^{i\beta_{01}} + e^{i\beta_{02}}) = -\varepsilon_1 \nu^2 + \varepsilon_2 (1 - (1 + 4\chi)\nu^2 + i\nu\delta_n)e^{i\gamma}, \qquad (5.15)$$

which is equivalent to the system of two real equations

$$\begin{cases} \cos \beta_{01} + \cos \beta_{02} = -\dfrac{\varepsilon_1}{\chi} - \dfrac{\varepsilon_2 ((1 - \nu^2) \cos \gamma - \nu \delta_n \sin \gamma)}{\chi \nu^2}, \\[3mm] \sin \beta_{01} + \sin \beta_{02} = -\dfrac{\varepsilon_2 ((1 - \nu^2) \sin \gamma + \nu \delta_n \cos \gamma)}{\chi \nu^2}. \end{cases} \qquad (5.16)$$

Denoting the right-hand sides of Eqs. (5.16) as p_1 and p_2, respectively, we represent them as

$$\begin{cases} 2 \cos \dfrac{\beta_{01} + \beta_{02}}{2} \cos \dfrac{\beta_{01} - \beta_{02}}{2} = p_1, \\[3mm] 2 \sin \dfrac{\beta_{01} + \beta_{02}}{2} \cos \dfrac{\beta_{01} - \beta_{02}}{2} = p_2, \end{cases} \qquad (5.17)$$

whence it follows that

$$\begin{cases} \tan \dfrac{\beta_{01} + \beta_{02}}{2} = \dfrac{p_2}{p_1}, \\[3mm] \cos \dfrac{\beta_{01} - \beta_{02}}{2} = \pm \dfrac{1}{2} \sqrt{p_1^2 + p_2^2}. \end{cases} \qquad (5.18)$$

Since ε_2 is small, the solution of system (5.18) reads as

$$\begin{aligned} \beta_{01} &= \pm \arccos \left(- \frac{1}{2} \sqrt{p_1^2 + p_2^2} \right) + \arctan \frac{p_2}{p_1}, \\[3mm] \beta_{02} &= \mp \arccos \left(- \frac{1}{2} \sqrt{p_1^2 + p_2^2} \right) + \arctan \frac{p_2}{p_1}. \end{aligned} \qquad (5.19)$$

Formulas (5.19) imply the condition for the existence of an almost balanced mode

$$p_1^2 + p_2^2 \leqslant 4. \qquad (5.20)$$

Note that in the case of an ideally fixed ABB (that is, for $\varepsilon_2 = 0$ and $\gamma = 0$) we will have $p_1 = -\varepsilon_1/\chi$, $p_2 = 0$. Moreover, formulas (5.19) will take the form similar to the previously obtained formulas (3.7).

Figure 43a presents $p_1^2 + p_2^2$ versus the dimensionless angular speed ν as calculated for the three values of the balancing ratio $\sigma = 2m_b r/(ms_1)$. It can be seen from the graphs that condition (5.20) is fulfilled if the balancing ratio is $\sigma \geqslant 1$ (i.e., in the supercritical frequency range). The curves in Fig. 43b show the stationary values of

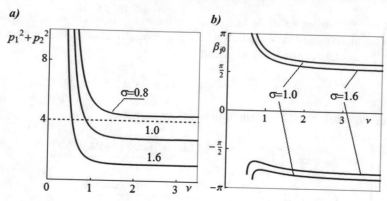

Fig. 43 Almost balanced steady-state mode: **a** existence conditions; **b** the deflection angles of the balancing balls

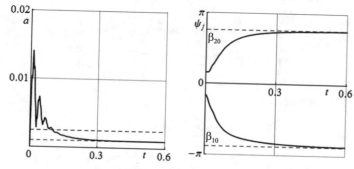

Fig. 44 Transition to the almost balanced mode with $\nu = 1.5$ and $\sigma = 1$

the angles of deflection of the balancing balls versus the angular speed of the rotor, as calculated using the formulas (5.19).

Results of the numerical integration of system (5.4) at constant angular speed of the rotor $\nu = 1.5$ and the values of the dimensionless parameters $\chi = 0.025$, $\varepsilon_1 = 0.05$, $\varepsilon_2 = 0.02$, $\gamma = 1$, $\delta_n = 0.21$, $\delta_b = 4.26$, corresponding to the almost balanced steady-state mode, are shown in Fig. 44. The graphs demonstrate that after a short transitional process, the dimensional amplitude of the whirling movement of the point C $a = \sqrt{x^2 + y^2}$ becomes equal to the value of the ABB's eccentricity s_2, and the balancing balls occupy the position corresponding to solution (5.19) (marked by dashed lines).

We now investigate the conditions for the existence of unbalanced steady-state motion modes of the rotor, corresponding to the solution of system (5.13) of the form $\beta_{01} = \beta_{02} = \beta_0$. Substituting this solution into system (5.9) and separating the real and imaginary parts of the first equation, we get

$$\begin{cases} (1 - (1 + 2\chi)\nu^2)\xi_0 - \delta_n\nu\eta_0 - (\varepsilon_1 + 2\varepsilon_2\chi\cos\gamma)\nu^2 = 2\chi\nu^2\cos\beta_0\,, \\ \delta_n\nu\xi_0 + (1 - (1 + 2\chi)\nu^2)\eta_0 - 2\varepsilon_2\chi\nu^2\sin\gamma = 2\chi\nu^2\sin\beta_0\,, \\ (\xi_0 + \varepsilon_2\cos\gamma)\sin\beta_0 = (\eta_0 + \varepsilon_2\sin\gamma)\cos\beta_0\,. \end{cases} \quad (5.21)$$

Denoting by B_1 and B_2 the left-hand sides of the first two equations of system (5.21) and eliminating the angle β_0, we obtain the two equations

$$\begin{cases} B_1{}^2 + B_2{}^2 = 4\chi^2\nu^4\,, \\ (\xi_0 + \varepsilon_2\cos\gamma)B_2 = (\eta_0 + \varepsilon_2\sin\gamma)B_1\,, \end{cases} \quad (5.22)$$

which represent the quadratic system of the form

$$\begin{cases} \kappa_{10}(\xi_0^2 + \eta_0^2) + \kappa_{11}\xi_0 + \kappa_{12}\eta_0 + \kappa_{13} = 0\,, \\ \kappa_{20}(\xi_0^2 + \eta_0^2) + \kappa_{21}\xi_0 + \kappa_{22}\eta_0 + \kappa_{23} = 0\,, \end{cases} \quad (5.23)$$

where

$$\kappa_{10} = (1 - (1 + 2\chi)\nu^2)^2 + \delta_n^2\nu^2\,,$$
$$\kappa_{11} = -2\varepsilon_1(1 - (1 + 2\chi)\nu^2)\nu^2 - 4\chi\varepsilon_2(1 - (1 + 2\chi)\nu^2)\cos\gamma + \delta_n\nu\sin\gamma)\nu^2\,,$$
$$\kappa_{12} = 2\varepsilon_1\delta_n\nu^3 - 2\chi\varepsilon_2((1 - (1 + 4\chi)\nu^2)\sin\gamma - \delta_n\nu\cos\gamma)\nu^2\,,$$
$$\kappa_{13} = (\varepsilon_1^2 + 4\chi\varepsilon_1\varepsilon_2\cos\gamma + 4\chi^2(\varepsilon_2^2 - 1))\nu^4\,, \quad \kappa_{20} = \delta_n\nu\,,$$
$$\kappa_{21} = \varepsilon_2(\delta_n\nu\cos\gamma - (1 - \nu^2)\sin\gamma)\,,$$
$$\kappa_{22} = \varepsilon_1\nu^2 + \varepsilon_2(\delta_n\nu\sin\gamma + (1 - \nu^2)\cos\gamma)\,, \quad \kappa_{23} = \varepsilon_1\varepsilon_2\nu^2\sin\gamma\,.$$

The numerical solution of system (5.23) can be obtained in the form of amplitude-frequency responses (AFR) of unbalanced steady-state modes presented in Fig. 45.

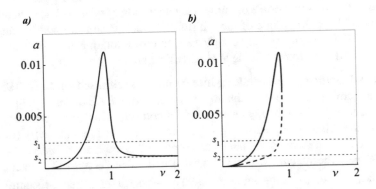

Fig. 45 AFR of unbalanced steady-state modes

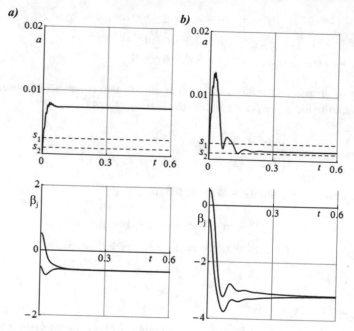

Fig. 46 Transition to unbalanced steady-state mode: **a** $\nu = 0.8$, **b** $\nu = 1.2$

Calculations were carried out for the following values of dimensionless parameters: $\varepsilon_1 = 0.05$, $\varepsilon_2 = 0.02$, $\gamma = 1$, $\delta_n = 0.21$, $\delta_b = 1$. Figure (*a*) corresponds to the case when the balancing ratio is $\sigma = 0.8$, i.e., the mass of the balancing balls is insufficient to compensate for the unbalance, and figure (*b*) corresponds to the case $\sigma = 1.6$. The character of the frequency response shows that in the case $\sigma < 1$ unbalanced steady-state modes exist in the whole range of rotor angular speeds, while at $\sigma > 1$ unbalanced modes occur only in the subcritical frequency range.

Figure 46 shows the results of numerical integration of system (5.4) with $\sigma = 0.8$. Time plots of the amplitude of the whirling motion of the rotor and the angles of deflection of the balancing balls are plotted for a rotor rotating at subcritical angular speed (*a*), and for a rotor rotating at supercritical angular speed (*b*).

Autobalancing of a rotor taking into account the ellipticity of the ABB's race.
Here we examine the impact of the imperfect shape of the ABB's race on the process of autobalancing of a statically unbalanced Jeffcott rotor. To do this, we consider the model of the ABB presented in Fig. 47, assuming that its internal cavity (race) has the form not of an ideal circle, but of an ellipse with semi-axes r_1 and r_2.

Let C mark the attachment point of the disk to the shaft, G mark the center of mass of the disk, $s = |\overrightarrow{CG}|$ be the value and α be the phase angle of static eccentricity. To determine the position of the rotor and balancing balls, let us introduce three frames: the inertial frame $OXYZ$, the frame rigidly connected with the rotor $C\xi\eta$ and the rotating frame $O\tilde{\xi}\tilde{\eta}$. The axes of the fixed frame are directed along the axes of the

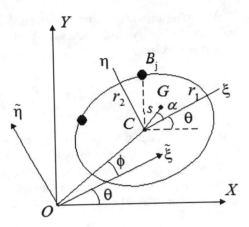

Fig. 47 ABB with an elliptical race

ellipse of the ABB's race, and the axes of the rotating frame are directed parallel to the corresponding axes of the fixed frame, as shown in Fig. 47. The angle of rotation between the axes OX and $C\xi$ is denoted by θ.

If we assume that the angular speed of the rotor $\dot{\theta}(t)$ changes with time according to a given law, and the disk moves only in the plane of static eccentricity, then the described mechanical model has $n+2$ degrees of freedom, where n is the number of balancing balls considered as material points. Let us choose as generalized coordinates the absolute coordinates x and y of a point C in the inertial frame and the relative coordinates of the balancing balls ξ_j and η_j that satisfy the equations of holonomic constraints

$$f_j = \frac{\xi_j^2}{r_1^2} + \frac{\eta_j^2}{r_2^2} - 1 = 0, \quad j = \overline{1, n}. \tag{5.24}$$

We express the absolute coordinates of the points G and B_j in terms of the above generalized coordinates:

$$
\begin{aligned}
x_G &= x + s\cos(\theta + \alpha), & x_{B_j} &= x + \xi_j \cos\theta - \eta_j \sin\theta, \\
y_G &= y + s\sin(\theta + \alpha), & y_{B_j} &= y + \xi_j \sin\theta + \eta_j \cos\theta,
\end{aligned}
\tag{5.25}
$$

and write the expressions for the kinetic and potential energy of the system:

$$T = \frac{1}{2}m(\dot{x}_G^2 + \dot{y}_G^2) + \frac{1}{2}I_G\dot{\theta}^2 + \frac{1}{2}m_b \sum_{j=1}^{n}(\dot{x}_{B_j}^2 + \dot{y}_{B_j}^2), \quad V = \frac{1}{2}k(x^2 + y^2); \tag{5.26}$$

where m and I_G are the mass and moment of inertia of the disk, m_b is the balancing ball mass, and k is shaft coefficient of elasticity.

Assuming that only external viscous damping forces act on the rotor and the balancing balls move in a viscous medium, we write the expression for the Rayleigh dissipative function

$$R = \frac{1}{2}c_n(\dot{x}^2 + \dot{y}^2) + \frac{1}{2}c_b \sum_{j=1}^{n}(\dot{\xi}_j^2 + \dot{\eta}_j^2), \qquad (5.27)$$

where c_n is the coefficient of resistance to transverse motion of the disk, and c_b is the coefficient of resistance to motion of the balancing balls in the ABB's race.

To construct a mathematical model of the system, let us use the Lagrange equations of the second kind with indefinite multipliers λ_j corresponding to the equations of additional constraints (5.24)

$$\frac{d}{dt}\left(\frac{\partial T}{\partial \dot{q}_i}\right) - \frac{\partial T}{\partial q_i} = -\frac{\partial V}{\partial q_i} - \frac{\partial R}{\partial \dot{q}_i} + \sum_{j=1}^{n}\lambda_j \frac{\partial f_j}{\partial q_i}, \quad i = \overline{1, n+2}, \qquad (5.28)$$

where $q_1 = x$, $q_2 = y$, $q_{2j+1} = \xi_j$, $q_{2j+2} = \eta_j$, $j = \overline{1, n}$.

Taking into account expression (5.24)–(5.26), Eq. (5.28) can be written as

$$\begin{cases} (m + nm_b)\ddot{x} + c_n\dot{x} + kx = ms(\dot{\theta}^2\cos(\theta + \alpha) + \ddot{\theta}\sin(\theta + \alpha)) - \\ -m_b \sum_{j=1}^{n}((\ddot{\xi}_j - \dot{\theta}^2\xi_j - 2\dot{\theta}\dot{\eta}_j - \ddot{\theta}\eta_j)\cos\theta - (\ddot{\eta}_j - \dot{\theta}^2\eta_j + 2\dot{\theta}\dot{\xi}_j + \ddot{\theta}\xi_j)\sin\theta), \\ (m + nm_b)\ddot{y} + c_n\dot{y} + ky = ms(\dot{\theta}^2\sin(\theta + \alpha) - \ddot{\theta}\cos(\theta + \alpha)) - \\ -m_b \sum_{j=1}^{n}((\ddot{\xi}_j - \dot{\theta}^2\xi_j - 2\dot{\theta}\dot{\eta}_j - \ddot{\theta}\eta_j)\sin\theta + (\ddot{\eta}_j - \dot{\theta}^2\eta_j + 2\dot{\theta}\dot{\xi}_j + \ddot{\theta}\xi_j)\cos\theta), \\ m_b\ddot{\xi}_j + c_b\dot{\xi}_j - m_b(2\dot{\theta}\dot{\eta}_j + \dot{\theta}^2\xi_j + \ddot{\theta}\eta_j) = -m_b(\ddot{x}\cos\theta + \ddot{y}\sin\theta) + \lambda_j\frac{2\xi_j}{r_1^2}, \\ m_b\ddot{\eta}_j + c_b\dot{\eta}_j + m_b(2\dot{\theta}\dot{\xi}_j + \dot{\theta}^2\eta_j + \ddot{\theta}\xi_j) = m_b(\ddot{x}\sin\theta - \ddot{y}\cos\theta) + \lambda_j\frac{2\eta_j}{r_2^2}, \quad j = \overline{1, n}. \end{cases}$$

$$(5.29)$$

Equation (5.26) together with the constraint Eqs. (5.24) form a closed system of $3n+2$ equations for $2n+2$ generalized coordinates and n Lagrange multipliers.

Eliminating the indefinite factors λ_j from the last $2n$ equation of system (5.29), which describe the balancing balls motion, we get the system of $n+2$ equations, which we write down in dimensionless form:

$$
\begin{cases}
(1+n\chi)(\ddot{\bar{x}} + \delta_n \dot{\bar{x}} + \bar{x}) = \varepsilon(\dot{\theta}^2 \cos(\theta + \alpha) + \ddot{\theta}\sin(\theta + \alpha)) - \\
-\chi \sum_{j=1}^{n}((\ddot{\bar{\xi}}_j - \dot{\theta}^2 \bar{\xi}_j - 2\dot{\theta}\dot{\bar{\eta}}_j - \ddot{\theta}\bar{\eta}_j)\cos\theta - (\ddot{\bar{\eta}}_j - \dot{\theta}^2 \bar{\eta}_j + 2\dot{\theta}\dot{\bar{\xi}}_j + \ddot{\theta}\bar{\xi}_j)\sin\theta), \\
(1+n\chi)(\ddot{\bar{y}} + \delta_n \dot{\bar{y}} + \bar{y}) = \varepsilon(\dot{\theta}^2 \sin(\theta + \alpha) - \ddot{\theta}\cos(\theta + \alpha)) - \\
-\chi \sum_{j=1}^{n}((\ddot{\bar{\xi}}_j - \dot{\theta}^2 \bar{\xi}_j - 2\dot{\theta}\dot{\bar{\eta}}_j - \ddot{\theta}\bar{\eta}_j)\sin\theta + (\ddot{\bar{\eta}}_j + \dot{\theta}^2 \bar{\eta}_j + 2\dot{\theta}\dot{\bar{\xi}}_j + \ddot{\theta}\bar{\xi}_j)\cos\theta), \\
\rho_1^2 \bar{\eta}_j(\ddot{\bar{\xi}}_j + \delta_b \dot{\bar{\xi}}_j - 2\dot{\theta}\dot{\bar{\eta}}_j - \dot{\theta}^2 \bar{\xi}_j - \ddot{\theta}\bar{\eta}_j + \ddot{\bar{x}}\cos\theta + \ddot{\bar{y}}\sin\theta) = \\
= \rho_2^2 \bar{\xi}_j(\ddot{\bar{\eta}}_j + \delta_b \dot{\bar{\eta}}_j + 2\dot{\theta}\dot{\bar{\xi}}_j - \dot{\theta}^2 \bar{\eta}_j + \ddot{\theta}\bar{\xi}_j + \ddot{\bar{y}}\cos\theta - \ddot{\bar{x}}\sin\theta), \quad j = \overline{1,n},
\end{cases}
\tag{5.30}
$$

where

$$
\bar{x} = \frac{x}{r}, \quad \bar{y} = \frac{y}{r}, \quad \bar{\xi}_j = \frac{\xi_j}{r}, \quad \bar{\eta}_j = \frac{\eta_j}{r}, \quad \bar{t} = \Omega t, \quad \Omega = \sqrt{\frac{k}{m + nm_b}},
$$

$$
\chi = \frac{m_b}{m}, \quad \varepsilon = \frac{s}{r}, \quad \delta_n = \frac{c_n}{(m + nm_b)\Omega}, \quad \delta_b = \frac{c_b}{m_b \Omega}, \quad \rho_1 = \frac{r_1}{r}, \quad \rho_2 = \frac{r_2}{r},
$$

$$
r = \frac{r_1 + r_2}{2} \quad \text{is the scale factor}.
$$

We obtain a closed and suitable for numerical integration system of $2n + 2$ second-order differential equations by adding to the system (5.30) n double differentiated equations of constraint

$$
\rho_2^2(\dot{\bar{\xi}}_j^2 + \bar{\xi}_j \ddot{\bar{\xi}}_j) + \rho_1^2(\dot{\bar{\eta}}_j^2 + \bar{\eta}_j \ddot{\bar{\eta}}_j) = 0, \quad j = \overline{1,n}.
\tag{5.31}
$$

In what follows the bar over dimensionless variables will be omitted for simplicity.

Steady-state motion modes. Consider a rotor rotating with constant dimensionless angular speed $\dot{\theta} = \nu = const$. Introducing complex variables $z = x + iy$ and $\zeta_j = \xi_j + i\eta_j$, $j = \overline{1,n}$, let us write the first two equations of system (5.30) as a single equation in complex form:

$$
(1 + n\chi)(\ddot{z} + \delta_n \dot{z} + z) = \varepsilon \nu^2 e^{i(\nu t + \alpha)} - \chi \sum_{j=1}^{n}(\ddot{\zeta}_j - \nu^2 \zeta_j + 2i\nu \dot{\zeta}_j)e^{i\nu t}.
\tag{5.32}
$$

In Eq. (5.32) the variable z defines the position of the point C in the inertial frame. Using the exponential form, we write it as

$$
z = ae^{i(\nu t + \phi)} = \tilde{z}e^{i\nu t},
$$

where $\tilde{z} = ae^{i\phi} = \tilde{\xi} + i\tilde{\eta}$ is the complex variable describing the position of the point C in the rotating frame $O\tilde{\xi}\tilde{\eta}$. Proceeding to the new dimensionless variables

$\tilde{\xi}$ and $\tilde{\eta}$ in system (5.30), we obtain an autonomous system of equations, which is convenient for the study of the stationary modes of rotor motion:

$$
\begin{cases}
\ddot{\tilde{\xi}}+\delta_n\dot{\tilde{\xi}}-2\nu\dot{\tilde{\eta}}+(1-\nu^2)\tilde{\xi}-\nu\delta_n\tilde{\eta}=\dfrac{1}{1+n\chi}\left(\varepsilon\nu^2\cos\alpha-\chi\sum_{j=1}^{n}(\ddot{\xi}_j-\nu^2\xi_j-2\nu\dot{\eta}_j)\right), \\[2mm]
\ddot{\tilde{\eta}}+\delta_n\dot{\tilde{\eta}}+2\nu\dot{\tilde{\xi}}+(1-\nu^2)\tilde{\eta}+\nu\delta_n\tilde{\xi}=\dfrac{1}{1+n\chi}\left(\varepsilon\nu^2\sin\alpha-\chi\sum_{j=1}^{n}(\ddot{\eta}_j-\nu^2\eta_j+2\nu\dot{\xi}_j)\right), \\[2mm]
\rho_1^2\eta_j(\ddot{\xi}_j+\delta_b\dot{\xi}_j-\nu(2\dot{\eta}_j+\nu\xi_j))+\ddot{\tilde{\xi}}-2\nu\dot{\tilde{\eta}}-\nu^2\tilde{\xi}= \\[2mm]
\quad = \rho_2^2\xi_j(\ddot{\eta}_j+\delta_b\dot{\eta}_j+\nu(2\dot{\xi}_j-\nu\eta_j))+\ddot{\tilde{\eta}}-2\nu\dot{\tilde{\xi}}-\nu^2\tilde{\eta}), \quad j=\overline{1,n}.
\end{cases}
$$
$$(5.33)$$

Substituting stationary solutions of the form $\tilde{\xi}=\tilde{\xi}_c=\text{const}$, $\tilde{\eta}=\tilde{\eta}_c=\text{const}$, $\xi_j=\xi_{j0}=\text{const}$ and $\eta_j=\eta_{j0}=\text{const}$ into system (5.33) and the constraint equations (5.24), we obtain a closed system of $2n+2$ algebraic equations

$$
\begin{cases}
(1+2\chi)((1-\nu^2)\tilde{\xi}_c-\nu\delta_n\tilde{\eta}_c)=\nu^2\left(\varepsilon\cos\alpha+\chi\sum_{j=1}^{n}\xi_{j0}\right), \\[2mm]
(1+2\chi)(\nu\delta_n\tilde{\xi}_c+(1-\nu^2)\tilde{\eta}_c)=\nu^2\left(\varepsilon\sin\alpha+\chi\sum_{j=1}^{n}\eta_{j0}\right), \\[2mm]
\rho_1^2\eta_{j0}\tilde{\xi}_c-\rho_2^2\xi_{j0}\tilde{\eta}_c=(\rho_2^2-\rho_1^2)\xi_{j0}\eta_{j0}, \\[2mm]
\rho_2^2\xi_{j0}^2+\rho_1^2\eta_{j0}^2=\rho_1^2\rho_2^2, \quad j=\overline{1,n}.
\end{cases}
$$
$$(5.34)$$

For $n=2$ system (5.34) consists of six equations describing the stationary motion modes of motion of the rotor in the case of an ABB with two balancing balls. Let us investigate it more detail. First of all, consider the question of existence of a fully balanced mode. Substituting $\tilde{\xi}_c=\tilde{\eta}_c=0$ into the first four equations, we get

$$
\begin{cases}
\varepsilon\cos\alpha+\chi(\xi_{10}+\xi_{20})=0, \\[1mm]
\varepsilon\sin\alpha+\chi(\eta_{10}+\eta_{20})=0, \\[1mm]
(\rho_2^2-\rho_1^2)\xi_{10}\eta_{10}=0, \\[1mm]
(\rho_2^2-\rho_1^2)\xi_{20}\eta_{20}=0.
\end{cases}
$$
$$(5.35)$$

In the case of a "perfect" ABB, when $\rho_1=\rho_2=1$, the third and fourth equations in (5.35) are satisfied identically, and the first and second, together with the equations of constraints, form a closed system of equations. In this case, since the ξ and η axes of the fixed frame can be directed arbitrarily, it can be assumed without loss of generality that $\alpha=0$. As a result, we get the following system

$$\begin{cases} \xi_{10} + \xi_{20} = -\varepsilon/\chi \,, \\ \eta_{10} + \eta_{20} = 0 \,, \\ \xi_{10}^2 + \eta_{10}^2 = 1 \,, \\ \xi_{20}^2 + \eta_{20}^2 = 1 \,, \end{cases} \tag{5.36}$$

which has a unique solution

$$\xi_{10} = \xi_{20} = -\frac{\varepsilon}{2\chi} \,, \quad \eta_{10} = \sqrt{1 - \frac{\varepsilon^2}{4\chi^2}} = -\eta_{20} \,, \tag{5.37}$$

corresponding to fully balanced rotor rotation.

In the case of an elliptic ABB we have $\rho_1 \neq \rho_2$. Therefore, combining Eqs. (5.36) with the equations of constraints, we obtain a system of six equations for four unknowns

$$\begin{cases} \xi_{10} + \xi_{20} = -\dfrac{\varepsilon}{\chi} \cos\alpha \,, \\[2mm] \eta_{10} + \eta_{20} = -\dfrac{\varepsilon}{\chi} \sin\alpha \,, \\[2mm] \xi_{10}\eta_{10} = 0 \,, \\[2mm] \xi_{20}\eta_{20} = 0 \,, \\[2mm] \rho_2^2\xi_{10}^2 + \rho_1^2\eta_{10}^2 = \rho_1^2\rho_2^2 \,, \\[2mm] \rho_2^2\xi_{20}^2 + \rho_1^2\eta_{20}^2 = \rho_1^2\rho_2^2 \,. \end{cases} \tag{5.38}$$

System (5.38) is overdetermined and has solutions of the form

$$\begin{aligned} & \{\xi_{10} = 0, \ \eta_{10} = \pm\rho_2, \ \xi_{20} = \pm\rho_1, \ \eta_{20} = 0\} \,, \\ & \{\xi_{10} = \pm\rho_1, \ \eta_{10} = 0, \ \xi_{20} = 0, \ \eta_{20} = \pm\rho_2\} \end{aligned} \tag{5.39}$$

only if the additional conditions are met

$$\begin{cases} |\alpha| = \arctan\dfrac{\rho_2}{\rho_1} \,, \\[3mm] \chi = \dfrac{\varepsilon}{\rho_1^2 + \rho_2^2} \,. \end{cases} \tag{5.40}$$

Solutions (5.39) have a simple physical explanation. In the case of a fully balanced rotor disk, the point C (the center of the elliptical race of the ABB) lies on the axis of rotation. But then the equilibrium position of the balancing balls can only be at the vertices of the ellipse, because otherwise the balls will move under the action of the tangential components of the centrifugal forces. However, solutions (5.39) have no particular meaning, since conditions (5.40) cannot be satisfied for an arbitrary unbalance. Therefore, we can state that a fully balanced mode in the case of an elliptic ABB is practically unrealizable.

Let us now investigate the problem of existence of unbalanced stationary modes of rotor motion. Assuming that the amplitude $a = \sqrt{\tilde{\xi}_c^2 + \tilde{\eta}_c^2}$ of the steady-state whirling motion of the point C is nonzero, we transform the last four equations of system (5.35) as follows: multiply the third equation by $\xi_{20}\eta_{20}$ and subtract from it the fourth equation multiplied by $\xi_{10}\eta_{10}$, and also subtract the sixth equation from the fifth. As a result, we get two equations

$$\rho_1^2 \tilde{\xi}_c (\xi_{20} - \xi_{10}) \eta_{10}\eta_{20} = \rho_2^2 \tilde{\eta}_c (\eta_{20} - \eta_{10}) \xi_{10}\xi_{20} \,,$$
$$\rho_2^2(\xi_{10}^2 - \xi_{20}^2) + \rho_1^2(\eta_{10}^2 - \eta_{20}^2) = 0 \,, \tag{5.41}$$

that are satisfied identically under the condition

$$\xi_{10} = \xi_{20} = \xi_0 \,, \quad \eta_{10} = \eta_{20} = \eta_0 \,. \tag{5.42}$$

Since the considered ABB model neglects the dimensions of the balancing balls, condition (5.42) means that the two balls are in contact with each other.

Substituting expressions (5.42) into system (5.35), we write it as

$$\begin{cases} p_1 \tilde{\xi}_c - p_2 \tilde{\eta}_c = \varepsilon \cos\alpha + 2\chi\xi_0 \,, \\ p_2 \tilde{\xi}_c + p_1 \tilde{\eta}_c = \varepsilon \sin\alpha + 2\chi\eta_0 \,, \\ \rho_1^2 \eta_0 \tilde{\xi}_c - \rho_2^2 \xi_0 \tilde{\eta}_c = (\rho_2^2 - \rho_1^2)\xi_0\eta_0 \,, \\ \rho_2^2 \xi_0^2 + \rho_1^2 \eta_0^2 = \rho_1^2 \rho_2^2 \,, \end{cases} \tag{5.43}$$

where

$$p_1 = (1 + 2\chi)\frac{1 - \nu^2}{\nu^2} \,, \quad p_2 = (1 + 2\chi)\frac{\delta_n}{\nu} \,.$$

Expressing from the first two equations of (5.43) the variables $\tilde{\xi}_c$ and $\tilde{\eta}_c$ in terms of ξ_0 and η_0

$$\tilde{\xi}_c = \frac{\varepsilon(p_1 \cos\alpha + p_2 \sin\alpha) + 2\chi(p_1\xi_0 + p_2\eta_0)}{p_1^2 + p_2^2} \,,$$
$$\tilde{\eta}_c = \frac{\varepsilon(p_1 \sin\alpha - p_2 \cos\alpha) + 2\chi(p_1\eta_0 - p_2\xi_0)}{p_1^2 + p_2^2} \,, \tag{5.44}$$

and substituting formulas (5.44) into the third equation of (5.43), we obtain a system of two second-order equations for the variables ξ_0 and η_0

$$\begin{cases} A_1\xi_0\eta_0 + A_2\xi_0 + A_3\eta_0 + A_4 = 0 \,, \\ \dfrac{\xi_0^2}{\rho_1^2} + \dfrac{\eta_0^2}{\rho_2^2} = 1 \,, \end{cases} \tag{5.45}$$

where

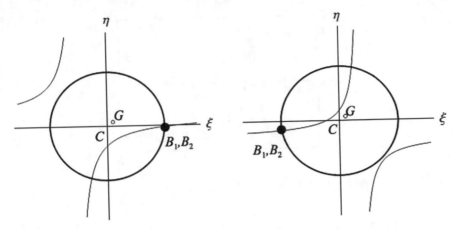

Fig. 48 Graphic solution of the system (5.45)

$$A_1 = (p_1^2 + p_2^2 + 2\chi p_1)(\rho_1^2 - \rho_2^2), \quad A_2 = \varepsilon\rho_2^2(p_2\cos\alpha - p_1\sin\alpha),$$
$$A_3 = \varepsilon\rho_1^2(p_1\cos\alpha + p_2\sin\alpha), \quad A_4 = 2\chi p_2\rho_1^2\rho_2^2.$$

System (5.45) can be reduced to a single fourth-degree equation, but it is more convenient to study it using the graphical method. To this end, let us express the first equation in the form of linear fractional functions of the form

$$\eta_0 = -\frac{A_2\xi_0 + A_4}{A_1\xi_0 + A_3} \quad \text{or} \quad \xi_0 = -\frac{A_3\eta_0 + A_1}{A_1\eta_0 + A_2},$$

whose graphs are symmetric hyperbolas with asymptotes

$$\xi = -\frac{A_3}{A_1}, \quad \eta = -\frac{A_2}{A_1}.$$

Thus, the intersection points of the symmetric hyperbola and the ellipse are the graphical solution of system (5.45).

Figure 48 shows the graphical solutions of system (5.45) calculated for the following values of dimensionless parameters: $\chi=0.02$, $\varepsilon=0.04$, $\rho_1=1.28$, $\rho_2=1.2$, $\delta_n=0.1$, $\alpha=0.6$. The left image corresponds to $\nu = 0.85$, when the dimensionless angular speed of the rotor is below the critical one and the right one corresponds to the supercritical frequency $\nu = 1.5$. In both cases, the graphs of the hyperbola and the ellipse intersect at two points, however, the actual position of the balancing balls, as shown in the figure, corresponds to the points further from the rotor rotation axis, connecting the bearing centers. The stability of these positions is a result of centrifugal forces acting on the balls. In the subcritical mode, the balls are located on the unbalanced side of the disk—that increases the overall imbalance of the rotor; in

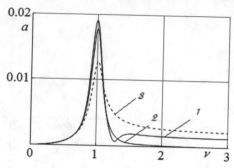

Fig. 49 Non-stationary passage of the critical region

the supercritical region they are located on the side opposite to the unbalance. Hence the total imbalance of the disk and balls becomes less than in the subcritical case.

Non-stationary motion modes. Let us study the non-stationary motion modes of a rotor equipped with an ABB with two balancing balls by numerically integrating the system of equations (5.30) and (5.31). Assuming that the rotor rotates at constant angular acceleration $\ddot{\theta} = \dot{\nu} = \text{const}$, consider the process of the system passing through the critical region. In Fig. 49, line 1 shows the amplitude of whirling motion of the point C versus the dimensionless angular speed $\nu = 0.16t$ for the dimensionless parameters: $\chi = 0.02$, $\varepsilon = 0.04$, $\alpha = 0.6$, $\delta_n = 0.18$, $\delta_b = 40$, $\rho_1 = 1.28$, $\rho_2 = 1.2$. For comparison, the same figure shows the amplitude-frequency responses for a rotor with an ideal circular ABB (line 2) and for a rotor without an ABB (dashed line 3). It can be seen from the figure that an elliptical ABB, unlike a circular one, does not balance the rotor in the supercritical region.

Figure 50 presents the graphs of changing the balancing balls position for the rotor with circular (a) and elliptical (b) ABB's races depending on the angular speed ν. The solid lines represent the associated coordinates ξ_1 and η_1 of the center of the first ball, and the dashed lines represent the associated coordinates of the center of the second ball ξ_2 and η_2. Figure also shows the steady-state position of the balancing balls in the race of circular and elliptical autobalancers.

Analysis of Figs. 49 and 50 gives that a change in the shape of the ABB's race from circular one to an elliptical one leads to the ABB losing functionality in the supercritical region, since in this case the balancing balls cannot take a position neutralizing the imbalance.

Transitional processes when the rotor spins at a constant angular speed are also interesting to consider. Figure 51 shows results of numerical integration of systems (5.30) and (5.31) in the case of subcritical ($\nu = 0.8$) and supercritical ($\nu = 1.5$) angular speeds. Figure (a) shows the change of the whirling motion amplitudes of the point C and figure (b) and (c) shows the change of balancing balls positions. The graphs demonstrate that in both cases when the balancing balls are at the same point the system stabilizes in unbalanced stationary modes.

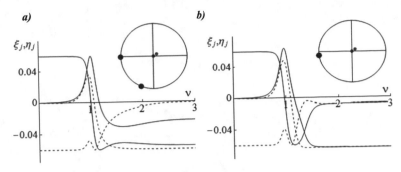

Fig. 50 Balancing balls movement: **a** circular ABB; **b** elliptic ABB

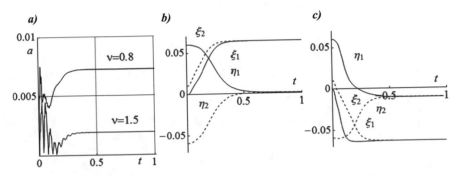

Fig. 51 Rotation of the rotor at a constant angular speed

Main conclusions.

1. An ABB, which is eccentrically mounted on a disk, cannot provide complete balancing of a statically unbalanced rotor with a variable unbalance, since a balanced steady-state mode is possible only for certain ratios between system parameters.

2. The best practicable result for an eccentric ABB is a supercritical, almost balanced steady-state mode in which the amplitude of whirling motion of the rotor does not depend on the angular speed and is exactly equal to the eccentricity value of the ABB.

3. The race of the autobalancer balancing balls should be shaped as close as possible to the ideal circle. An elliptical shape can lead to loss of functionality of the ABB.

Chapter 5
Elements of the Motion Control

P. E. Tovstik and N. V. Naumova ⓘ

In this chapter we state problems of control theory and give a brief survey of some methods of their solution. Problems in the control theory can be subdivided into two large classes. The first class is related with the choice of a control which is optimal in a sense, while the second one is related with the problem of the confinement of motion on a selected path or near it. The Pontryagin maximum principle is presented. The solution of some problems of control theory is given. The notions of controllability, stabilizability, and observability are introduced.[1]

1 Statements of Optimal Control Problems

We assume that the motion of a mechanical system is described by the system of first-order differential equations

$$\dot{x}_i = f_i(x_j, u_k, t), \quad i, j = \overline{1, n}, \quad k = \overline{1, m}, \tag{1.1}$$

where x_i are the phase coordinates including the generalized displacements and velocities, $u_k(t) \in U \subset R_m$ are the controls to be chosen from a closed subset U of the space R_m, and t is the time. We seek a solution of system (1.1) satisfying the following boundary conditions

$$x_i(t_0) = x_i^0, \quad x_i(t_1) = x_i^1, \quad i = \overline{1, n}.$$

---

[1] See the books by *L.S. Pontryagin, V.G. Boltyanskii, R.V. Gamkrelidze,* and *E.F. Mishchenko.* The Mathematical Theory of Optimal Processes. New York: Gordon and Breach. 1987; *G.A. Leonov* and *M.M. Shumafov.* Stabilization of Linear Systems. Cambridge Scientific Pub. 2012; *V.V. Aleksandrov, V.G. Boltyanskii, S.S. Lemak, N.A. Parusnikov,* and *V.V. Tikhomirov.* Optimization of Dynamics of Controlled Systems. Moscow: Izd-vo Mekh.-Mat. Fak. MGU. 2000. 304 p. [in Russian]; *V.E. Pasynkov, I.A. Pasynkova,* and *N.V. Naumova.* Extremal problems. St. Petersburg Univ. 2002. 78 p. [in Russian].

© Springer Nature Switzerland AG 2021
N. N. Polyakhov et al., *Rational and Applied Mechanics*,
Foundations of Engineering Mechanics,
https://doi.org/10.1007/978-3-030-64118-4_5

In the principal statement of the problem it is assumed that all the quantities

$$t_0, \quad t_1, \quad x_i^0, \quad x_i^1, \quad i = \overline{1, n} \tag{1.2}$$

are given. One also considers different variants of boundary conditions in which some of the quantities in (1.2) are not given and/or they are subject to constrains of the form

$$\varphi_p(x_i^0, t_0) = 0, \quad p = \overline{1, P}, \quad \psi_q(x_i^1, t_1) = 0, \quad q = \overline{1, Q}.$$

As an illustration of expediency of various statements of boundary conditions, we consider the motion control of a missile that should hit a moving target. The lift-off time of a missile (that is, the quantity t_0) may not be given; likewise, the missile velocity at the end of the path (that is, one of the quantities $x_i(t_1)$) may also be unknown. In addition, one also has to pose conditions of the form $\psi_q(x_i(t_1), t_1) = 0$, which depend on the motion law of the target.

System (1.1) may fail to have a solution satisfying the applied boundary conditions (for example, a plain is incapable of reaching Moscow from St. Petersburg in one minute). However, if solutions do exist, then as a rule there are infinitely many of them. Hence, one poses the problem of finding an optimal solution that delivers the maximum to the functional

$$J = \int_{t_0}^{t_1} f_0(x_i(t), u_k(t), t) \, dt, \tag{1.3}$$

where the functions $x_i(t)$ satisfy Eq. (1.1) and the boundary conditions, $u_k(t) \in U$ are the chosen controls. It is immaterial whether one searches for the maximum or the minimum of functional (1.3), because by changing the sign of functional (1.3) we get the minimum problem.

As an example of an extremal problem for functional (1.3) we mention the time-optimal problem (the problem of finding a solution with the smallest required time of motion). In this case one may put $f_0 \equiv -1$. As another example, we have the minimum-fuel problem to control the thrust of a moving object.

Remark 1 The above problem is called the *Lagrange problem*. There are various modifications of this problem, which have practically the same solution as the one described below. The *Mayer problem* differs from the Lagrange problem in having a term outside the integral in the functional to be minimized

$$J = g(x_i^0, t_0, x_i^1, t_1) + \int_{t_0}^{t_1} f_0(x_i(t), u_k(t), t) \, dt, \tag{1.4}$$

while in the *Bolz problem* one minimizes only the term outside the integral ($f_0 \equiv 0$ in (1.4)).

2 Solving the Optimal Control Problem by Methods of the Classical Variational Calculus. The Pontryagin Maximum Principle

We minimize the functional (1.3) under n nonholonomic constraints (1.1). We include these constraints in the functional (1.3) using the Lagrange multipliers $\lambda_i(t)$ (they depend on time because the constraints are nonholonomic). So, we have the functional

$$
J^* = \int_{t_0}^{t_1} \left(f_0 - \sum_{i=1}^{n} \lambda_i (\dot{x}_i - f_i) \right) dt =
$$
$$
= \int_{t_0}^{t_1} \left(H(x_i, \lambda_i, u_k, t) - \sum_{i=1}^{n} \lambda_i \dot{x}_i \right) dt \,, \tag{2.1}
$$

where we introduced the function

$$
H = f_0 + \sum_{i=1}^{n} \lambda_i f_i \,, \tag{2.2}
$$

which is commonly called the *Pontryagin-Hamilton function* due to the formal similarity of the below system of equations with the system of canonical equations.

Let us find the variation of functional (2.1)

$$
\delta J^* = \tilde{J}^* - J^* = \int_{\tilde{t}_0}^{\tilde{t}_1} \left(H(\tilde{x}_i, \lambda_i, \tilde{u}_k, t) - \sum_{i=1}^{n} \lambda_i \dot{\tilde{x}}_i \right) dt -
$$
$$
- \int_{t_0}^{t_1} \left(H(x_i, \lambda_i, u_k, t) - \sum_{i=1}^{n} \lambda_i \dot{x}_i \right) dt \,,
$$

where $\tilde{x}_i = x_i + \delta x_i$, $\tilde{t}_0 = t_0 + \delta t_0$, $\tilde{t}_1 = t_1 + \delta t_1$, $\tilde{u}_k = u_k + \delta u_k$, the tilde denotes the values to be compared, and δ denotes the variation. The inequality $\delta J^* \leqslant 0$ must be satisfied for a real motion under arbitrary variations.

Assuming that the variations δx_i, δt_0, δt_1 are small, let us find the variation δJ^* with the first-order accuracy with respect to δx_i, δt_0, δt_1. Integrating in parts and simplifying, this gives

$$
\delta J^* = \int_{t_0}^{t_1} \left(\sum_{i=1}^{n} \left(\dot{\lambda}_i + \frac{\partial H}{\partial x_i} \right) \delta x_i + \Delta H \right) dt +
$$
$$
+ \left[H \delta t - \sum_{i=1}^{n} \lambda_i \Delta x_i \right]_{t=t_0}^{t=t_1} \,, \tag{2.3}
$$

where we set

$$\Delta H = H(x_i, \lambda_i, \widetilde{u}_k, t) - H(x_i, \lambda_i, u_k, t), \qquad \Delta x_i = \delta x_i + \dot{x}_i \delta t, \quad i = \overline{1, n}.$$

Here Δx_i is the total variation.

From the condition $\delta J^* \leqslant 0$ and since δx_i is arbitrary we get the equations for $\dot{\lambda}_i$, which together with Eq. (1.1) form the system

$$\begin{cases} \dot{x}_i = \dfrac{\partial H}{\partial \lambda_i}, \\ \dot{\lambda}_i = -\dfrac{\partial H}{\partial x_i}, \end{cases} \quad i = \overline{1, n}, \tag{2.4}$$

which is similar to the system of canonical equations.

Changing ΔH by the expression

$$\Delta H = \sum_{k=1}^{m} \frac{\partial H}{\partial u_k} \delta u_k$$

with subsequent derivation of the equations

$$\partial H / \partial u_k = 0, \quad k = \overline{1, m}, \tag{2.5}$$

for the controls u_k is not always correct. This change is legitimate if the controls lie strictly inside the domain U. However, if the controls reach the boundary of U, then they are to be found from the condition

$$\min_{u_k \in U} H(x_i(t), \lambda_i(t), u_k, t),$$

where the functions $x_i(t)$, $\lambda_i(t)$ satisfy Eq. (2.4) and the boundary conditions.

The above algorithm is the core of the *Pontryagin maximum principle*. For historical reasons, here the maximum is written in lieu of the minimum, but this is immaterial. (We recall that for a minimum problem one has to change the sign in the functional (1.3).)

The corollaries consequent on the consideration of the terms outside the integral in (2.3) in minimizing the functional δJ^* will be discussed in the next subsection.

3 Boundary Conditions

.In accordance with the order of system (2.4), on each end-point of the time interval (t_0, t_1) there should be n conditions, but since the times t_0, t_1 may not be specified, the number of conditions on each end-point increases to $n + 1$.

The term outside the integral in (2.3) with $t = t_1$ gives (the case $t = t_0$ is dealt similarly)

$$\delta J_1^* = H\delta t_1 - \sum_{i=1}^{n} \lambda_i(t_1)\Delta x_i^1 \geqslant 0, \quad \Delta x_i^1 = \delta x_i^1 + \dot{x}_i \delta t_1 \quad (t = t_1). \qquad (3.1)$$

If the end-point $t = t_1$ is fixed (that is, if $x_i(t_1) = x_i^1$), then $\Delta x_i^1 = \delta t_1 = \delta J_1^* = 0$.

If the end-point $t = t_1$ is free, then the variations Δx_i^1 and δt_1 are arbitrary, the boundary conditions assuming the form

$$\lambda_i = 0, \quad i = \overline{1, n}, \quad H = 0 \quad (t = t_1).$$

In the intermediate case, if some of the quantities x_i^1, t_1 are not given, then the corresponding quantities λ_i, H vanish at $t = t_1$.

Let us consider a more involved case when the conditions

$$\psi_p(x_i, t) = 0, \quad p = \overline{1, P}, \quad P \leqslant n+1, \qquad (3.2)$$

should be satisfied at the end-point $t = t_1$, the conditions of fixing the separate coordinates $x_j(t_1) = x_j^1$ are written in the framework of the general scheme (3.2).

Introducing the variations of constraints (3.2) with the Lagrange multipliers Λ_p to expression (3.1), this gives

$$\delta J_1^* = \left(H - \sum_{p=1}^{P} \Lambda_p \frac{\partial \psi}{\partial t} \right) \delta t_1 - \sum_{i=1}^{n} \left(\lambda_i + \sum_{p=1}^{P} \Lambda_p \frac{\partial \psi}{\partial x_i} \right) \Delta x_i^1 \geqslant 0.$$

Now the variations δt_1 and Δx_i^1 become independent. Equating to zero the coefficients of these variations, we get the system of $(n + 1)$ equations

$$H - \sum_{p=1}^{P} \Lambda_p \frac{\partial \psi_p}{\partial t} = 0, \quad \lambda_i + \sum_{p=1}^{P} \Lambda_p \frac{\partial \psi_p}{\partial x_i} = 0, \quad i = \overline{1, n}, \qquad (3.3)$$

which, together with P Eq. (3.2), will be used to construct the boundary conditions with $t = t_1$. The total number of unknowns x_i, λ_i, Λ_p, t_1, H is $(2n + P + 2)$, while there are $(n + P + 1)$ Eqs. (3.2), (3.3). Hence, this gives us $(n + 1)$ boundary conditions.

4 Solving the Time-Optimal Problem Using the Pontryagin Maximum Principle

Let us consider the motion of a material point of mass $m = 1$ along the line Ox under the control force $u(t)$ with $|u(t)| \leqslant 1$. At the initial time the coordinate and the projection of the velocity of the point

$$x(0) = a, \quad \dot{x}(0) = v$$

are given. It is required to move the point at the origin with zero velocity in the smallest time $T: x(T) = \dot{x}(T) = 0$. For example, this problem is a simplification of the docking problem of a cargo ship with a space station.

Let us write the equation of motion $\ddot{x} = u$ as a system of first-order Eq. (1.1):

$$\dot{x}_1 = x_2, \quad \dot{x}_2 = u.$$

The Pontryagin-Hamilton function (2.2) reads as

$$H = -1 + \lambda_1 x_2 + \lambda_2 u. \tag{4.1}$$

We recall that in minimizing the functional one should replace the function f_0 by $(-f_0)$ in (1.3).

We write the system of Eq. (2.4)

$$\dot{x}_1 = x_2, \quad \dot{x}_2 = u,$$
$$\dot{\lambda}_1 = -\frac{\partial H}{\partial x_1} = 0, \quad \dot{\lambda}_2 = -\frac{\partial H}{\partial x_2} = -\lambda_1 \tag{4.2}$$

with the boundary conditions

$$x_1(0) = a, \quad x_2(0) = v, \quad x_1(T) = 0, \quad x_2(T) = 0. \tag{4.3}$$

Integrating the third and fourth equations in (4.2), we find that

$$\lambda_1 = c, \quad \lambda_2 = -ct + c_1, \tag{4.4}$$

where c, c_1 are arbitrary constants. From (4.1) it follows that the function $H(u)$ attains its minimum with $u(t) = -\text{sign}\,\lambda_2 = \pm 1$. Besides, from (4.4) it follows that the control switches at most once.

The next step in solving the problem is to integrate the first two equations in (4.2) with $u = 1$ and $u = -1$, construct all possible trajectories on the phase plane $x_1 x_2$, which are obtained after eliminating the time, and finally, satisfy the boundary condition (4.3).

The trajectories just mentioned look like parabolas (see Fig. 1)

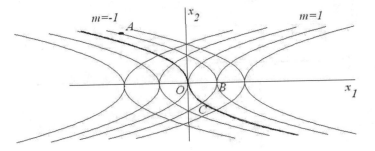

Fig. 1 Solution of the time-optimal problem

$$x_1 = \frac{x_2^2}{2} + D_1, \qquad u = 1,$$

$$x_1 = -\frac{x_2^2}{2} + D_2, \qquad u = -1,$$

where D_1, D_2 are arbitrary constants.

Assume that at the initial time $t = 0$ the point was located at position A in the phase space. The optimal trajectory for this point is $ABCO$, the point falls in the part ABC with $u = -1$, there is a switching at the point C, and the final part CO is traversed by the point with $u = 1$. The solid line in Fig. 1 shows the curve along which the point reaches the origin without switchings.

5 Control of the Horizontal Motion of a Cart with Pendula with the Use of the Pontryagin Maximum Principle

Statement of one problem in the control theory. Let us consider one of the principal problems in the control theory. Here we speak about finding an optimal control force that moves in a given time a mechanical system with finite number of degrees of freedom from one phase state with specific generalized coordinates and velocities to a new phase state with given generalized coordinates and velocities. As a model example, we shall seek a control force F that moves a cart of mass m that travels along the x-axis in the horizontal direction in a given time $\widetilde{T}$ at a distance S. A cart is equipped with s mathematical pendula with masses m_σ and lengths l_σ, $\sigma = \overline{1, s}$. For definiteness, we show in Fig. 2 a cart with two pendula. It is required to move this system from the initial state of rest in a new rest position (this problem is usually called the problem of oscillations suppression).

The system has $s + 1$ degrees of freedom. In order to describe its motion, we introduce the coordinate x, which characterizes the horizontal displacement of the cart, and the angles of rotation of pendula, φ_σ, $\sigma = \overline{1, s}$. In the case of small oscillations the kinetic and potential energies of the system read as

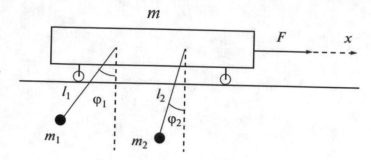

Fig. 2 Cart with two pendula

$$2T = m\dot{x}^2 + \sum_{\sigma=1}^{s} m_\sigma (\dot{x} - l_\sigma \dot{\varphi}_\sigma)^2 , \quad 2\Pi = g \sum_{\sigma=1}^{s} l_\sigma m_\sigma \varphi_\sigma^2$$

(here, g is the acceleration of gravity), and hence the Lagrange equations of the second kind assume the form

$$M\ddot{x} - \sum_{\sigma=1}^{s} m_\sigma l_\sigma \ddot{\varphi}_\sigma = F , \quad M = m + \sum_{\sigma=1}^{s} m_\sigma ,$$

$$\ddot{x} - l_\sigma \ddot{\varphi}_\sigma = g\varphi_\sigma , \quad \sigma = \overline{1, s} . \tag{5.1}$$

Clearly, here the first equation expresses the motion law of the center of mass of the entire system:

$$M\ddot{x}_c = F , \quad x_c = x - \frac{\sum_{\sigma=1}^{s} m_\sigma l_\sigma \varphi_\sigma}{M} .$$

To suppress the oscillation of the mechanical system under study (that is, in order to assure its state of rest at the initial time and to ensure that oscillations cease as the system stops), the control force should require the fulfillment of the following boundary-value conditions:

$$x(0) = \dot{x}(0) = \dot{x}(\widetilde{T}) = 0 , \quad x(\widetilde{T}) = S ,$$

$$\varphi_\sigma(0) = \varphi_\sigma(\widetilde{T}) = 0 , \quad \dot{\varphi}_\sigma(0) = \dot{\varphi}_\sigma(\widetilde{T}) = 0 , \quad \sigma = \overline{1, s} . \tag{5.2}$$

The mechanical system under study has zero frequency Ω_0 and s nonzero dimensional eigenfrequencies Ω_σ, $\sigma = \overline{1, s}$.

For further analysis it will be convenient to write system (5.1) in the principal coordinates (see Sect. 4 of Chap. 7 of Vol. I). It is worth pointing out that on substituting $\ddot{x}$ from the first equation of system (5.1) into the other ones, we shall obtain a system of s equations with respect to the unknowns φ_σ, $\sigma = \overline{1, s}$. Considering the so-obtained equations as an independent system of differential equations, we change in it to the dimensionless normal coordinates x_σ, $\sigma = \overline{1, s}$, which are linear combinations of the same angles of pendula deflection. Changing to the dimensionless time $\tau = \Omega_1 t$ and introducing the $(s + 1)$th dimensionless principal coordinate x_0, which is proportional to the displacement of the center of mass of the mechanical system, we finally obtain

$$\begin{cases} x_0'' = u \,, \\ x_\sigma'' + \omega_\sigma^2 x_\sigma = u \,, \quad \sigma = \overline{1, s} \,. \end{cases} \qquad (5.3)$$

Here, u is the control proportional to the force F, the prime indicates the derivatives in the dimensionless time τ, $\omega_\sigma = \Omega_\sigma / \Omega_1$, $\sigma = \overline{1, s}$. On the right of these equations there is the same dimensionless control u, this can be easily achieved by corresponding scale changes in the principal coordinates. The resulting system of differential Eq. (5.3), which satisfies requirements (5.2), will be solved under the following boundary conditions:

$$x_0(0) = x_0'(0) = 0 \,, \quad x_\sigma(0) = x_\sigma'(0) = 0 \,, \quad T = \Omega_1 \widetilde{T} \,,$$

$$x_0(T) = a \equiv \frac{S}{l_1} \,, \quad x_0'(T) = 0 \,, \quad x_\sigma(T) = x_\sigma'(T) = 0 \,,$$

$$\sigma = \overline{1, s} \,. \qquad (5.4)$$

System (5.3) contains $(s + 1)$ differential equations; from it one needs to find the unknown functions x_0, x_σ, $\sigma = \overline{1, s}$. But in the same system (5.3) the function $u(t)$ is indefinite. Hence, one more condition is required to solve the above problem (5.3)–(5.4). This condition should express the principle underlying the choice of the control $u(\tau)$ (the control force $F(t)$) from the entire set of all possible controls for which this problem has a solution. Usually, when solving similar problems, a control is taken from the condition that the functional be minimal[2]

$$J = \int_0^T u^2 d\tau \,. \qquad (5.5)$$

One of the most acceptable classical methods for solving the above optimal control problem (5.3)–(5.5) is the method depending on the Pontryagin maximum principle.

Solution of the problem with the use of the Pontryagin maximum principle. According to the theory of Sect. 2, Eq. (5.3) is rewritten as (1.1) when applying this principle. Next, one considers the Lagrange multipliers $\lambda_k(\tau)$, $k = \overline{1, 2s + 2}$, and writes down the Pontryagin-Hamilton function (2.2)

---

[2] See, for example, the book by F.L. Chernous'ko, L.D. Akulenko, and B.N. Sokolov. Control of Oscillations. Moscow: Nauka. 1980. 384 p. [in Russian].

$$H = -u^2 + \sum_{k=1}^{2s+2} \lambda_k f_k (q, u).$$

We recall that when searching for the minimum of functional (5.5) one replaces in it the integrand u^2 by $(-u^2)$.

The unknown functions $\lambda_k(\tau)$ obey the second group of equations in (2.4)

$$\lambda_k' = -\frac{\partial H}{\partial q_k}, \quad k = \overline{1, 2s+2},$$

while the sought-for control $u(\tau)$ is determined from condition (2.5):

$$\frac{\partial H}{\partial u} = 0. \tag{5.6}$$

In the case under consideration, we have

$$H = -u^2 + \lambda_1 q_2 + \lambda_2 u + \sum_{\sigma=1}^{s} \lambda_{2\sigma+1} q_{2\sigma+2} + \sum_{\sigma=1}^{s} \lambda_{2\sigma+2} (u - \omega_\sigma^2 q_{2\sigma+1}), \tag{5.7}$$

$$\lambda_1' = 0, \lambda_2' = -\lambda_1, \lambda_{2\sigma+1}' = \omega_\sigma^2 \lambda_{2\sigma+2}, \lambda_{2\sigma+2}' = -\lambda_{2\sigma+1},$$
$$\sigma = \overline{1, s}. \tag{5.8}$$

From expressions (5.6) and (5.7) it follows that

$$u(\tau) = \frac{1}{2} \sum_{\sigma=1}^{s+1} \lambda_{2\sigma} (\tau).$$

The functions $\lambda_{2\sigma}(\tau)$, $\sigma = \overline{1, s+1}$, satisfy, in accordance with system (5.8), the equations

$$\lambda_2'' = 0, \quad \lambda_{2\sigma+2}'' + \omega_\sigma^2 \lambda_{2\sigma+2} = 0, \quad \sigma = \overline{1, s}.$$

Hence, the functional (5.5) attains its minimum with

$$u(\tau) = C_1 + C_2 \tau + \sum_{\sigma=1}^{s} (C_{2\sigma+1} \cos \omega_\sigma \tau + C_{2\sigma+2} \sin \omega_\sigma \tau). \tag{5.9}$$

Here, C_k, $k = \overline{1, 2s+2}$, are arbitrary constants. Next, we substitute function (5.9) into the right-hand sides of Eq. (5.3). The general solution of this system of differential equations will contain arbitrary constants D_k, $k = \overline{1, 2s+2}$. Hence, the so-obtained general solution will depend on $4s+4$ arbitrary constants C_k, D_k,

$k = \overline{1, 2s + 2}$, which are determined from the given $4s + 4$ boundary conditions (5.4).

However, the above general approach to finding a solution satisfying the boundary-value conditions (5.4) is fairly cumbersome. Hence, we shall show how one may substantially simplify this issue of finding arbitrary constants in our setting.

Of special importance here is that we are solving the problem of suppression of oscillations in the mechanical system under consideration. Hence, ab initio $\tau_0 = 0$ the system is at rest; that is, all its principal generalized coordinates and velocities are zero. As a result, it is convenient to write the partial solution to system (5.3) with zero initial conditions in terms of the Duhamel integrals:

$$x_0(\tau) = \int_0^\tau u(\tau_1)(\tau - \tau_1)\,d\tau_1 ,$$

$$x_\sigma(\tau) = \frac{1}{\omega_\sigma}\int_0^\tau u(\tau_1)\sin\omega_\sigma(\tau - \tau_1)\,d\tau_1 , \quad \sigma = \overline{1, s} . \tag{5.10}$$

Now, substituting these expressions into the second group of boundary conditions (with $\tau = T$), taking into account (5.9), and calculating the integrals, we obtain a linear inhomogeneous algebraic system of equations with respect to only the unknowns C_σ, $\sigma = \overline{1, 2s + 2}$. Solving it, we finally obtain the required control.

It is worth pointing out that control (5.9) obtained via the Pontryagin maximum principle, first, depends only on the eigenfrequencies of the system, and second, is delivered by finding the sought-for control force as a series in eigenfrequencies of the system. However, for a long-time motion this will bring the mechanical system under study in resonance.

Numerical calculation. We give the results of calculations for $s = 2$ (see Fig. 2). The boundary-value problem being linear, its solution is proportional to a. Hence, without loss of generality we may put $a = 1$. As a result, the solution will depend only on the two dimensionless parameters

$$\frac{T}{T_2} \quad \text{and} \quad \frac{T_2}{T_1} ,$$

where T_1 and T_2 are the dimensionless periods of oscillations, which correspond to the first and second nonzero natural frequencies. We have $\omega_1 = 1$, and hence, $T_1 = 2\pi$ and $T_2 = 2\pi/\omega_2$.

Let us consider three cases of motion:

$$T = T_2 , \quad T = 8\,T_2 , \quad T = 16\,T_2 , \quad \text{and besides} \quad T_2 = 0.5\,T_1 . \tag{5.11}$$

We have $s = 2$, and hence (5.9) assumes the form

$$u(\tau) = C_1 + C_2\tau + C_3\cos\omega_1\tau + C_4\sin\omega_1\tau + C_5\cos\omega_2\tau + C_6\sin\omega_2\tau . \tag{5.12}$$

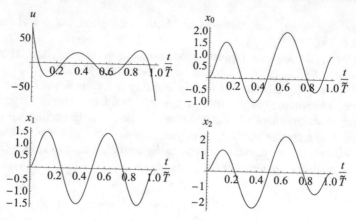

Fig. 3 Short-time motion of a mechanical system, $T = T_2$, $T_2 = 0.5\,T_1$

Substituting control (5.12) into the Duhamel integrals (5.10) (with $s = 2$), after satisfying the boundary conditions (5.4) with $\tau = T$ we obtain an algebraic system with respect to C_k, $k = \overline{1,6}$, from which one readily finds the numerical values of the sought-for arbitrary constants:

$$T = T_2 :$$
$$C_1 = 2174.6, \quad C_2 = -1384.4, \quad C_3 = -2102.12,$$
$$C_4 = 0, \quad C_5 = 0, \quad C_6 = 400.01;$$

$$T = 8\,T_2 :$$
$$C_1 = 0.0099, \quad C_2 = -0.0008, \quad C_3 = 0,$$
$$C_4 = -0.0016, \quad C_5 = 0, \quad C_6 = -0.0008;$$

$$T = 16\,T_2 :$$
$$C_1 = 0.0024, \quad C_2 = -0.0001, \quad C_3 = 0,$$
$$C_4 = -0.0002, \quad C_5 = 0, \quad C_6 = -0.0001.$$

The results of calculations with these values are depicted in Figs. 3, 4, 5.

As is seen from the graphs, all the curves are symmetric about the point of abscissa corresponding to the center of time interval of the motion. the middle time of the motion. In addition to this, the results of calculations have two characteristic peculiarities. First, the control has jumps at the beginning and end of the motion, which is clear, since $u(0)$ and $u(1)$ are both nonzero by formula (5.12). Hence, at the beginning and end of the process there appear large oscillations of the coordinates x_1 and x_2. Second, for a long-time motion a mechanical system features intense oscillations during the entire process of motion—the result expected, because the control

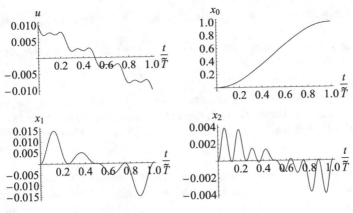

Fig. 4 Motion of a mechanical system with $T = 8\,T_2$, $T_2 = 0.5\,T_1$

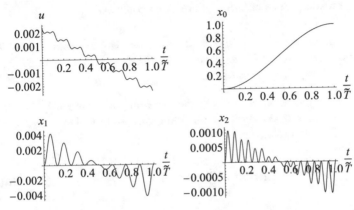

Fig. 5 Long-time motion of a mechanical system, $T = 16\,T_2$, $T_2 = 0.5\,T_1$

obtained by virtue of the Pontryagin maximum principle contains harmonics with natural frequencies of the system.

In the next chapter we shall introduce a new method of finding the control force with the help of the generalized Gauss principle. This method allows one to construct a control as a polynomial.

6 Linear Problems in the Control Theory. Controllability

In this and succeeding subsections we shall give without proofs the principal facts about the behavior of linear controllable (controlled) systems with constant coefficients. Here, we shall be concerned with the controllability.

We consider a controlled system in the n-dimensional phase space R_n

$$\dot{x} = A \cdot x + B \cdot u, \quad x = (x_1, \dots, x_n)^T \in R_n, \quad u = (u_1, \dots, u_m)^T \in R_m, \quad (6.1)$$

where A and B are constant matrices of sizes $(n \times n)$ and $(n \times m)$, respectively, and $u = u(t)$ is the control vector.

Definition 1 A system (A, B) is called *completely controllable* if, for any points x^0, $x^1 \in R_n$ and any time instant T, there exists a control $u(t)$ such that the solution to system (6.1) with the initial condition $x(0) = x^0$ assumes the value $x(T) = x^1$ at time T.

In other words, a control transforms a system from one point of the phase space into another one. Clearly, a control is not unique. Indeed, let $T_1 < T$. One may take any control with $0 \leqslant t \leqslant T_1$, and then satisfy the condition $x(T) = x^1$ at the expense of choosing the control on the interval $T_1 < t \leqslant T$.

The next result gives a criterion of complete controllability.

Theorem 1 *A necessary and sufficient condition that a system (A, B) be completely controllable is that the rank of the controllability matrix $U = (B, A \cdot B, A^2 \cdot B, \dots, A^{n-1} \cdot B)$ of size $n \times mn$ be n.*

Let us prove the necessity part of the theorem, simultaneously explaining the origin of the matrix U. We write the solution of system (6.1) satisfying the initial condition $x(0) = 0$ as follows:

$$x(T) = \int_0^T e^{A(T-t)} \cdot Bu(t)\, dt = \sum_{k=0}^{\infty} A^k \cdot Bf_k, \qquad f_k = \frac{1}{k!} \int_0^T (T-t) u(t)\, dt,$$

the series converging uniformly. Let $\mathrm{rank}(U) = n_1 < n$. We represent the space R_n as the sum of the subspace R_{n_1} spanned by the column vectors of the matrix U and the orthogonal complement R_{n-n_1} ($R_n = R_{n_1} + R_{n-n_1}$). We take $x_1 \neq 0$, $x_1 \in R_{n-n_1}$. By Cayley-Hamilton's theorem, the matrix satisfies the characteristic equation $|A - \lambda E| = 0$, which yields

$$A^n = \alpha_1 A^{n-1} + \alpha_2 A^{n-2} + \dots + \alpha_n E.$$

The subspace R_{n_1} agrees with the subspace spanned by the column vectors of all matrices $A^k B$, $k = 0, 1, 2, \dots$, and hence the equality $x(T) = x^1$ may not be true for any choice of f_k.

Example 1 Consider the equation $\ddot{x} = u$. We set $x_1 = x$, $x_2 = \dot{x}$. Hence,

$$A = \begin{pmatrix} 0 & 1 \\ 0 & 0 \end{pmatrix}, \qquad B = \begin{pmatrix} 0 \\ 1 \end{pmatrix}.$$

This system is completely controllable, because the rank of the controllability matrix

$$U = \begin{pmatrix} 0 & 1 \\ 1 & 0 \end{pmatrix}$$

is two.

Taking $B = (1, 0)^T$ one easily checks that the system (A, B) is not completely controllable.

7 Stabilizability and Observability

The purpose of the present section is to discuss the methods for confinement of the system on the chosen path. Let x be the deflection (departure) from this path. In the linear approximation, the problem is described by the same Eq. (6.1)

$$\dot{x} = A \cdot x + B \cdot u(x),$$
$$x = (x_1, \dots, x_n)^T \in R_n, \quad u = (u_1, \dots, u_m)^T \in R_m,$$
(7.1)

in which now the control $u = u(x)$ is a function of the deflections (departures) x. Let $u = K \cdot x$, where K is a constant $(m \times n)$ matrix.

Now Eq. (7.1) assumes the form

$$\dot{x} = (A + B \cdot K) \cdot x.$$
(7.2)

Definition 2 A system (A, B) is *completely stabilizable* if there exists matrix K such that the matrix $A + B \cdot K$ is Hurwitzian.

In other words, all the eigen numbers of the matrix $A + B \cdot K$ have negative real parts and $x(t) \to 0$ as $t \to \infty$ for any initial conditions.

The following result is a criterion for complete controllability.

Theorem 2 *A necessary and sufficient condition that a system (A, B) be completely stabilizable is that it be completely controllable.*

It can be shown that for a completely stabilizable system the matrix K can be chosen so that the roots of the characteristic equation $A + B \cdot K - \lambda E = 0$ be arbitrary. This enables one to control the rate with which the solution $x(t)$ approaches zero.

In order to realize the equality $u = K \cdot x$ one needs to know the vector x. Assume that we have observations $z = (z_1, \dots, z_p)^T$ over the motion of the system. The results of observations are related with x by the formula

$$z = H \cdot x,$$
(7.3)

where H is a constant $(p \times n)$ matrix.

If $p \geqslant n$ and rank$(H) = n$, then from the system of Eq. (7.3) one may find x, and then use the control $u = K \cdot x$ for stabilization. Otherwise, one has to solve the *problem of recovery* of x by the results of observations.

Assume that we first consider system (7.1) with $B = 0$, that is $\dot{x} = A \cdot x$, and that we know the values of $z(t)$ on some time interval $[t_0, t]$, $t_0 < t$. Differentiating (7.3) in time and taking into account the equation $\dot{x} = A \cdot x$, we obtain the system of equations with respect to $x(t)$

$$\frac{d^k z}{dt^k} = H \cdot A^k \cdot x, \quad k = 0, 1, \ldots, \tag{7.4}$$

in which the left-hand sides are considered to be known. System (7.4) has a unique solution if the rank of the matrix on the left is n. As in the controllability problem, we may confine ourselves to the consideration of the observability matrix V of size $(pn \times n)$

$$V = \begin{pmatrix} H \\ H \cdot A \\ \ldots \\ H \cdot A^{n-1} \end{pmatrix}.$$

A criterion of observability is given by the equality rank$(V) = n$.

In implementing this scenario we have the impediment that the differentiation of the results of observations leads to a loss in accuracy. So we propose to employ the *procedure of recovery* of $x(t)$, in which we construct an auxiliary function $\widehat{x}(t)$ such that $\widehat{x}(t) \to x(t)$ as $t \to \infty$. We shall assume that this function satisfies the auxiliary system of equations

$$\dot{\widehat{x}} = \widehat{A} \cdot \widehat{x} + R \cdot z, \tag{7.5}$$

where a constant matrix R of size $(m \times p)$ needs to be chosen. Let $y = \widehat{x} - x$. Then $\dot{y} = \widehat{A} \cdot y + (\widehat{A} - A + R \cdot H) \cdot x$. We set $\widehat{A} = A - R \cdot H$. Hence, $\dot{y} = (A - R \cdot H) \cdot y$. If rank$(V) = n$, then the system (A, H) is completely controllable (the rows of the matrix V become columns of the matrix U). As a result, one may find a matrix R in order that the matrix $A - R \cdot H$ be a Hurwitz matrix with any roots of the characteristic equation. Taking the roots with negative real parts, we find that $y(t) \to 0$ and $\widehat{x}(t) \to x(t)$ as $t \to \infty$.

Example 2 Let us consider the motion $X = \cos t$ satisfying the equation $\ddot{X} + X = 0$ and the initial conditions $X(0) = 1$, $\dot{X}(0) = 0$. It is required to recover this motion by the results of observation of $z = X + \dot{X}$. Reducing to first-order equations, we obtain the system $\dot{x} = A \cdot x$, where

$$A = \begin{pmatrix} 0 & 1 \\ -1 & 0 \end{pmatrix}, \quad H = (1, \ 1).$$

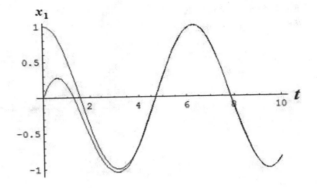

Fig. 6 Recovery of the solution by the results of observations

The system (A, H) is observable, because the rank of the observability matrix

$$V = \begin{pmatrix} 1 & 1 \\ -1 & 1 \end{pmatrix}$$

is two. We take, for example, $R = (r_1, r_2)^T = (1, 2)^T$. In this case, the matrix $A - R \cdot H$ is Hurwitzian. In order to recover the vector x we solve numerically the Cauchy problem

$$\dot{x} = A \cdot x, \quad \dot{y} = (A - R \cdot H) \cdot y, \quad x(0) = (1, 0)^T, \quad y(0) = (-1, 0)^T.$$

It should be noted that the initial conditions for the auxiliary vector y are arbitrary; they are set for definiteness.

Figure 6 gives the graphs of the functions $x_1(t)$ and $\widehat{x}_1(t)$ obtained by numerical integration. We see that $\widehat{x}_1(t) \to x_1(t)$ as $t \to \infty$.

One might be tempted to conclude that for stabilization one may use in (7.2) the vector $\widehat{x}$ thus constructed:

$$\dot{x} = A \cdot x + B \cdot K \cdot \widehat{x}. \tag{7.6}$$

However, this is not so, because now the vector of observations z was obtained for system (7.6), rather than for the system $\dot{x} = A \cdot x$ with $B = 0$. Hence, the procedure of recovery of the vector $\widehat{x}$ must be repeated again.

Considering jointly Eqs. (7.5) and (7.6), we arrive at the equation for the vector $y = \widehat{x} - x$

$$\dot{y} = (\widehat{A} - R \cdot H) \cdot y + (\widehat{A} - A - B \cdot K + R \cdot H) \cdot x.$$

Setting $\widehat{A} - A - B \cdot K + R \cdot H = 0$, we get the system of equations

$$\begin{aligned} \dot{y} &= (\widehat{A} - R \cdot H) \cdot y, \\ \dot{x} &= (A + B \cdot K) \cdot x + B \cdot K \cdot y, \end{aligned} \tag{7.7}$$

Fig. 7 Stabilization with
$a = 0.01$

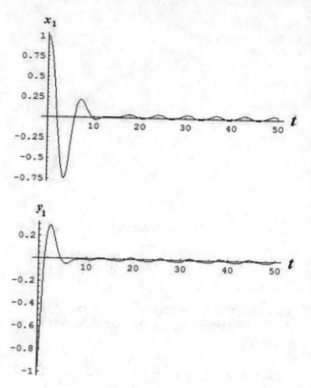

in which the matrices R and K are chosen from the condition that the matrices $A - R \cdot H$ and $A + B \cdot K$ be Hurwitzian. In this case, the recovery condition $y(t) \to 0$ as $t \to \infty$ and the stabilization condition $x(t) \to 0$ as $t \to \infty$ both hold.

The main purpose of stabilization is to offset the perturbations acting on the system. We first give an example. Assume that a freely moving point is subject to a perturbing force $\xi(t)$. The solution of the equation $\ddot{x} = \xi(t)$ with the initial conditions $x(0) = a$, $\dot{x}(0) = v$ is as follows:

$$x(t) = a + vt + \int_0^t \left(\int_0^{t_1} \xi(t_2)dt_2 \right) dt_1 .$$

In particular, for the constant perturbation $\xi(t) = \xi_0$ the solution is $x(t) = a + vt + \xi_0 t^2/2$. Hence, even for pretty small a, v, ξ_0 the solution becomes large with increasing time.

Let us consider the stabilization of a completely controllable system under the perturbations

$$\dot{x} = A \cdot x + B \cdot u + \xi(t) , \qquad z = H \cdot x + \eta(t) ,$$

Fig. 8 Stabilization with
$a = 0$

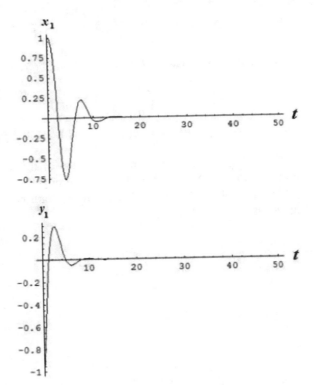

where $\xi(t)$ is a small perturbation and $\eta(t)$ is a small error in an observation. As before, we introduce the auxiliary vector $\widehat{x}$ and misadjustment vector $y = \widehat{x} - x$. Repeating the above calculations, we obtain instead of (7.7) the system of equations

$$\dot{y} = (\widehat{A} - R \cdot H) \cdot y + R \cdot H \cdot \eta(t) - \xi(t),$$
$$\dot{x} = (A + B \cdot K) \cdot x + B \cdot K \cdot y + \xi(t). \tag{7.8}$$

Of course, the solutions of Eq. (7.8) do not tend to zero, but irrespective of the initial conditions they become small in time (of the order of perturbations).

Example 3 Let

$$A = \begin{pmatrix} 0 & 1 \\ 0 & 0 \end{pmatrix}, \qquad B = \begin{pmatrix} 0 \\ 1 \end{pmatrix}, \qquad H = (1, 0).$$

This system is completely controllable and observable. We take

$$R = (r_1, r_2)^T = (1, 1)^T, \qquad K = (k_1, k_2) = (-1, -1).$$

Hence, the matrices

$$A - R \cdot H = \begin{pmatrix} -1 & 1 \\ -1 & 0 \end{pmatrix}, \qquad A + B \cdot K = \begin{pmatrix} 0 & 1 \\ -1 & -1 \end{pmatrix}$$

are Hurwitzian. We take the following perturbations and the initial conditions

$$\xi(t) = \begin{pmatrix} a \sin^2 t \\ a \cos^2 t \end{pmatrix}, \quad \eta(t) = \begin{pmatrix} a \cos t \\ a \sin t \end{pmatrix},$$

$$x(0) = 1, \dot{x}(0) = 0, \ y(0) = -1, \ \dot{y}(0) = 0.$$

Figure 7 gives the results of stabilization of the first three components of the vectors x and y. We see that in the course of time these functions become small (of the order of perturbations). If there are no perturbations ($a = 0$), then both these functions tend to zero (see Fig. 8).

Chapter 6
Generalized Chebyshev Problem. Nonholonomic Mechanics and Control Theory

V. V. Dodonov⊙, Sh. Kh. Soltakhanov⊙, P. E. Tovstik, M. P. Yushkov⊙, and S. A. Zegzhda

This chapter consists of two parts. In the first part we present two theories of motion of nonholonomic systems with high-order (program) constraints. In the first theory, we construct a consistent system of differential equations for the unknown generalized coordinates and the Lagrange multipliers. The second theory is based on the generalized Gauss principle. As an application of this theory, we solve the problem of motion of an artificial satellite with time-constant acceleration.

In the second part of the chapter, for one of the most principal problems of the control theory—the problem of optimal control force that transforms a given mechanical system in a given amount of time from one phase state into a different one—we employ the second theory of motion of nonholonomic systems with high-order constraints. It is shown that a high-order constraint will continuously take place for the solution of this problem with the Pontryagin maximum principle utilized for the minimization of the functional of the squared control force. Hence the generalized Gauss principle, which is peculiar to the theory of motion of nonholonomic systems with high-order constraints, can be conveniently used for solving this problem. This also allows one to construct a control force in the form of a polynomial of time. The application of this theory is illustrated on the model problem of oscillation suppression for a cart with pendulums. We pose and solve an extended boundary-value problem, in which one specifies the values and accelerations at the beginning and end of the motion of the system. Because of this, it proves possible to find a control force without jumps peculiar to solutions obtained via the Pontryagin maximum principle. At the end of the chapter, we show that the generalized Gauss principle can be used for the problem of oscillation suppression for a system with distributed parameters (for example, a flexible robotic arm).

© Springer Nature Switzerland AG 2021
N. N. Polyakhov et al., *Rational and Applied Mechanics*,
Foundations of Engineering Mechanics,
https://doi.org/10.1007/978-3-030-64118-4_6

I) Formulation of the Generalized Chebyshev Problem. Two Theories of Motion for Nonholonomic Systems with Linear High-Order Constraints

1 Formulation of the Generalized Chebyshev Problem

P.L. Chebyshev is[1] well known for his outstanding achievements in many different topics in mathematics and mechanics.[2] In particular, he created the theory of mechanism synthesis,[3] in which he posed the problem of machines whose separate parts should be capable of performing prescribed motions with given accuracy. Among such units one can mention, for example, multi-link mechanisms whose parts can stop in given positions.[4] We shall call such problems *Chebyshev problems*. Below in this section we shall give an extension of this problem to the case when under the same system of active generalized forces the resulting motion of the system should satisfy an additional system of high-order differential equations.

We thus pose the following problem. Assume that the motion of a mechanical system subject to generalized forces $Q = (Q_1, \ldots, Q_s)$ is described in the generalized coordinates $q = (q^1, \ldots, q^s)$ by the Lagrange equations of the second kind

$$\frac{d}{dt}\frac{\partial T}{\partial \dot{q}^\sigma} - \frac{\partial T}{\partial q^\sigma} = Q_\sigma, \quad \sigma = \overline{1, s},$$

$$T = \frac{M}{2}g_{\alpha\beta}\dot{q}^\alpha\dot{q}^\beta, \quad \alpha, \beta = \overline{0, s}, \quad q^0 = t, \quad \dot{q}^0 = 1, \tag{1.1}$$

where M is the mass of the entire system (see Chap. 6 of Vol. I).

---

[1] The present chapter is based on the book: *S.A. Zegzhda, M.P. Yushkov, Sh.Kh. Soltakhanov, E.A. Shatrov*. Nonholonomic mechanics and control theory. Moscow: Nauka. Fizmatlit. 2018. 236 p. [in Russian]. For a brief account, see the survey papers: *M.P. Yushkov*. Formulation and Solution of a Generalized Chebyshev Problem: First Part // Vest. St. Petersb. Univ. Math. 2019. Vol. 6 (64). Issue 4. Pp. 680–701; *M.P. Yushkov*. Formulation and Solution of a Generalized Chebyshev Problem: Second Part // Vest. St. Petersb. Univ. Mathematics. Mechanics. Astronomy. 2020. Vol. 7 (65). Issue 4. Pp. 515–537 [in Russian].

[2] See: *P.L. Chebyshev*. Collected works, edited by A.A. Markov and N.Ya. Sonin. Vol. I. St. Petersburg. 1899. 714 p.; Vol. II. 1907. 736 p. [in Russian].

[3] See: P.L. Chebyshev's heritage. Theory of Mechanisms (Edited by N.G. Bruevich and I.I. Artobolevskii). Moscow-Leningrad: Publishing House of the USSR Academy of Sciences. 1945. 192 p.

[4] Some mechanisms made of wood by P.L. Chebyshev himself or under his supervision and with his markings still visible can be found in the The Museum of History of St. Petersburg University, the Museum of History of Faculty of Mathematics and Mechanics, and at the Department of Theoretical and Applied Mechanics at St. Petersburg State University. See the paper: *G. Kuteeva, M. Yushkov, E. Rimushkina*. Pafnutii Lvovich Chebyshev as a mechanician (2015) 2015 International Conference on Mechanics—Seventh Polyakhov's Reading, art. no. 7106746 http://www.scopus.com/alert/results/record.url?AID=1979589&ATP=search&eid=2-s2. 0-84938238584&origin=SingleRecordEmailAlert.

We shall require that the motion of this mechanical system should simultaneously satisfy the system of differential equations ($n \geqslant 3$)

$$f_n^{\varkappa} \equiv a_{n\sigma}^{l+\varkappa}(t, q, \dot{q}, \dots, \overset{(n-1)}{q}) \overset{(n)}{q}{}^{\sigma} + a_{n0}^{l+\varkappa}(t, q, \dot{q}, \dots, \overset{(n-1)}{q}) = 0,$$

$$\sigma = \overline{1, s}, \quad \varkappa = \overline{1, k}, \quad k \leqslant s, \quad l = s - k, \tag{1.2}$$

where we set $\overset{(n)}{q}{}^{\sigma} = d^n q^{\sigma} / dt^n$. The above problem will be called the generalized Chebyshev problem. This problem is more involved than the above Chebyshev problem of mechanism synthesis, because now it is required to find not only the motion of some links of a mechanism, but also the motion of the entire system satisfying the additional system of high-order differential equations (1.2).

To solve this problem it is required to find the additional forces $R = (R_1, \dots, R_s)$ such that if we add them to the right-hand sides of equations (1.1); that is, we rewrite them in the form

$$\frac{d}{dt} \frac{\partial T}{\partial \dot{q}^{\sigma}} - \frac{\partial T}{\partial q^{\sigma}} = Q_{\sigma} + R_{\sigma}, \quad \sigma = \overline{1, s}, \tag{1.3}$$

then we would get a closed problem. So, by a *generalized Chebyshev problem* we shall mean the solution of the compatible system of equations (1.2), (1.3), in which $q = (q^1, \dots, q^s)$ and $R = (R_1, \dots, R_s)$ are unknown functions of time.

It should be noted that following the proposal of Academician S. S. Grigoryan, the generalized Chebyshev problem is sometimes called the *mixed dynamics problem*.[5] because it involves features of both direct and inverse dynamics problems. Indeed, on the one hand, we seek the motion with given forces $Q = (Q_1, \dots, Q_s)$, and on the other hand, under given motion characteristics (1.2) we seek the generalized forces $R = (R_1, \dots, R_s)$.

It should be pointed out that the generalized Chebyshev problem (1.2), (1.3) can be looked upon as a control problem in which the (mandatory!) motion program is specified as an additional system of differential equations of high-order $n \geqslant 3$. So, in control theory the generalized Chebyshev problems introduce a new class of problems with specifically defined motion program. In order to be able to deal with such problems, we propose in the present chapter to invoke the machinery of nonholonomic mechanics by considering the additional system of differential equations as high-order nonholonomic constraints.

In view of the above, we recall some facts from the classical theory of nonholonomic systems. If the motion of a system is subject to holonomic constraints

$$f_0^{\varkappa}(t, q) = 0, \quad \varkappa = \overline{1, k}, \tag{1.4}$$

or nonholonomic first-order constraints

---

[5] See: *S.A. Zegzhda, M.P. Yushkov.* A mixed problem of dynamics // Dokl. Phys. 2000. Vol. 45. Pp. 547–549.

$$f_1^\varkappa(t, q, \dot{q}) = 0, \quad \varkappa = \overline{1, k}, \tag{1.5}$$

or linear nonholonomic second-order constraints

$$f_2^\varkappa(t, q, \dot{q}, \ddot{q}) \equiv a_{2\sigma}^{l+\varkappa}(t, q, \dot{q}) \ddot{q}^\sigma + a_{2,0}^{l+\varkappa}(t, q, \dot{q}) = 0,$$
$$\sigma = \overline{1, s}, \quad \varkappa = \overline{1, k}, \quad l = s - k, \tag{1.6}$$

then, assuming that these constraints are ideal, their reaction forces have, respectively, the form (see Chap. 6 of Vol. I)

$$\mathbf{R} = \Lambda_\varkappa \nabla f_0^\varkappa, \quad \mathbf{R} = \Lambda_\varkappa \nabla' f_1^\varkappa, \quad \mathbf{R} = \Lambda_\varkappa \nabla'' f_2^\varkappa, \quad \varkappa = \overline{1, k}. \tag{1.7}$$

The generalized Hamilton operators ∇' and ∇'' from formulas (1.7) were introduced by N.N. Polyakhov.[6] From these formulas, in the particular case of holonomic constraints, we get the classical Hamilton operator (the nabla operator) (see Chap. 6 of Vol. I).

It is important that in the classical analytical mechanics with ideal constraints (1.4), (1.5), (1.6), the Lagrange multipliers $\Lambda_\varkappa$, $\varkappa = \overline{1, k}$, are found as known functions of $t, q, \dot{q}$ (see again Chap. 6 of Vol. I):

$$\Lambda_\varkappa = \Lambda_\varkappa(t, q, \dot{q}), \quad \varkappa = \overline{1, k}. \tag{1.8}$$

For holonomic constraints, functions (1.8) were obtained in the beginning of the twentieth century by A.M. Lyapunov and G.K. Suslov,[7] and for nonholonomic constraints, they were first proposed in the paper of N.N. Polyakhov, S.A. Zegzhda, M.P. Yushkov,[8] and later given in the first edition of this textbook[9]. Ten years later, this result was obtained by different methods in the USA, Italy, Poland, Sweden,

---

[6] See the papers: *N.N. Polyakhov*. Motion equations of mechanical systems with nonlinear non-holonomic constraints in a general case // Vestn. Leningr. Univ. Ser. 1. Mat. Mekh. Astron. 1972. № 1. Pp. 124–132; On differential principles in mechanics derived from motion equations of non-holonomic systems // Vestn. Leningr. Univ. 1974. Ser. 1. Mat. Mekh. Astron. № 3. Pp. 106–116 [in Russian].

[7] See the books: *A.M. Lyapunov*. Lectures on Theoretical Mechanics. Kiev: Naukova Dumka. 1982. 632 p. [in Russian] and *G.K. Suslov*. Fundamentals of Analytic Mechanics. Kiev: Imp. Univ. Svy-atogo Vladimira. 1900. Vol. 1. 287 p. [in Russian].

[8] *N.N. Polyakhov, S.A. Zegzhda, M.P. Yushkov* Dynamic equations as necessary conditions for the minimality of compulsion in Gauss, in Oscillations and Stability of Mechanical Systems // Applied Mechanics. Leningrad: Leningr. Gos. Univ. 1981. Vol. 5. pp. 9–16 [in Russian].

[9] *N.N. Polyakhov, S.A. Zegzhda, M.P. Yushkov* Theoretical Mechanics. Leningrad: Leningr. Gos. Univ. 1985. 536 p. [in Russian].

and the USSR.[10] Functions (1.8) can also be obtained under constraints in the form (1.6).[11]

In contrast, if the above generalized Chebyshev problem is considered as a non-holonomic problem with ideal linear constraints (1.2) of order $n \geqslant 3$, then the reactions of these constraints (which play the role of the sought-for control forces from the point of view of the above new class of control problems) should be sought as unknown functions of time.

These arguments withdraw the question raised by some well-known Soviet mechanicians when posing the generalized Chebyshev problem. The thing is that according to L.A. Pars and V.V. Rumyantsev,[12] the forces cannot depend on accelerations. This is why many people thought that specification of high-order constraints would also require specification of the reaction forces of these constraints, which depend not only on the accelerations, but also on derivatives of higher orders of the generalized coordinates. However, as was pointed out above, in generalized Chebyshev problems these reaction forces are unknown functions of time, which should be determined simultaneously with the generalized coordinates.

To solve the above generalized Chebyshev problem one can propose two theories of motion of nonholonomic systems with high-order constraints (see below[13]).

---

[10] See the papers: *J. Storch, S. Gates*. Motivating Kane's method for obtaining equations of motion for dynamic systems // J. of Guidance, Dynamics and Control. 1989. Vol. 12. No. 4. Pp. 593–595; *F.E. Udwadia, R.E. Kalaba*. A new perspective on constrained motion // Proceedings of the Royal Society. London. 1992. Vol. A439. No. 1906. Pp. 407–410; *M. Borri, C. Bottasso, P. Mantegazza*. Equivalence of Kane's and Maggi's equations // Meccanica. 1990. Vol. 25. No. 4. Pp. 272–274; *M. Borri, C. Bottasso, P. Mantegazza*, Acceleration projection method in multibody dynamics // Europ. J. Mech. A/Solids. 1992. Vol. 11. No. 3. Pp. 403–417; *W. Blajer*. A projetion method approach to constrained dynamic analysis // ASME. J. Appl. Mech. 1992. Vol. 59. No. 3. Pp. 643–649; *H. Essén*. Projecting Newton's equations onto non-ordinate tangent vectors of the configuration space; a new look at Lagrange's equations in ferms of quasicoordinates // 18th Int. Congr. Theor. and Appl. Mech. Haifa, Aug. 22-28, 1992. Haifa, 1992. P. 52; *H. Essén*. On the geometry of non-holonomic dynamics // ASME. J. Appl. Mech. 1994. No. 61. Pp. 689–694; *V.V. Velichenko*. Matrix equations of motion of nonholonomic systems // Dokl. Akad. Nauk SSSR. 1991. Vol. 321. № 3. Pp. 499–504 [in Russian]; *Yu.F. Golubev*. Basic principles of mechanics for systems with differential nonlinear constraints // In Proc. 2nd All-Russ. Meeting-Seminar by Heads of Theoretical Mechanics Departments. Moscow, Oct. 11–16, 1999. Pp. 14–15 [in Russian].

[11] Note that at present one can mention only one example of appearance of linear nonholonomic second-order constraint obtained by a contact method when a material point is located at the end of an inextensible string spiraling about a vertical circular cylinder: *F. Kitzka*. An example for the application of a nonholonomic constraint of 2nd order in particle mechanics // ZAMM. 1986. Vol. 66. No. 7. S. 312-314.

[12] See the book and paper: *L.A. Pars*. A Treatise on Analytical Dynamics. London: Heinemann. 1965; *V.V. Rumyantsev*. On the compatibility of the two basic principles of dynamics and on the Chetaev principle // In Problems of Analytical Mechanics. Theories of Stability and Control. Moscow: Nauka. 1975. Pp. 258–267 [in Russian]; *V.V. Rumyantsev*. On the compatibility of the differential principles of mechanics // In Aeromechanics and Gas Dynamics. Moscow: Nauka. 1976. Pp. 172–178 [in Russian].

[13] For a more detailed account of these theories and their applications in some problems of mechanics, see the books: *Sh.Kh. Soltakhanov, M.P. Yushkov, S.A. Zegzhda*. Mechanics of nonholonomic systems. A New Class of control problems. Berlin Heidelberg: Springer-Verlag. 2009. 329 p.;

2 The First Theory of Motion for Nonholonomic Systems with High-Order Constraints. Construction of a Compatible System of Differential Equations

The first theory of motion for nonholonomic systems with high-order constraints. Formation of control forces. Let us now consider the motion of a mechanical system described by the system of curvilinear coordinates $q = (q^1, \dots, q^s)$ with bases

$$\{e_1, \dots, e_s\}, \quad \{e^1, \dots, e^s\}$$

in the case when the motion of this system is subject to linear ideal high-order nonholonomic constraints (1.2).

As was already pointed out in the previous subsection, the solution of a generalized Chebyshev problem requires the knowledge of the additional generalized forces

$$R = (R_1, \dots, R_s), \tag{2.1}$$

which secure the execution of the program motion given by the system of differential equations (1.2). To be able to deal with this problem, one should extend the classical theory of motion for nonholonomic systems to the case of constraints of high-order $n \geqslant 3$. Assuming that constraints (1.2) are ideal,[14] their reaction, in analogy with the classical theory of motion for nonholonomic systems, could be written in the form

$$\mathbf{R} \equiv \mathbf{R}^K = \Lambda_\varkappa \boldsymbol{\nabla}^{(n)} f_n^\varkappa, \quad \varkappa = \overline{1, k}, \qquad \mathbf{R}_L = 0, \tag{2.2}$$

where we use the generalized Polyakhov operator

$$\boldsymbol{\nabla}^{(n)} f_n^\varkappa = \frac{\partial f_n^\varkappa}{\partial \overset{(n)}{q^\sigma}} \mathbf{e}^\sigma.$$

It is worth mentioning that in the classical nonholonomic mechanics the reaction forces (1.7) are automatically created by "materially implemented" constraints (1.4)–(1.6). In contrast, in the generalized Chebyshev problem under consideration, reactions (2.1) are control forces and are created by the control system. Let us now discuss the formation of control forces in the problem under consideration.

Assume that the control system creates some force with possible (virtual) elementary work

$$\delta A = \Lambda b_\sigma \delta q^\sigma, \quad \sigma = \overline{1, s}.$$

S.A. Zegzhda, M.P. Yushkov, Sh.Kh. Soltakhanov, E.A. Shatrov. Nonholonomic mechanics and control theory. Moscow: Nauka. Fizmatlit. 2018. 236 p. [in Russian].

[14] Here it is worth pointing out that the above problem has a nonunique solution, and hence a certain solution is singled out by introducing the requirement that the constraints be "ideal".

The quantity Λ involved in this expression will be called the *generalized control force*. If now the control system allows us the formation of k control forces $\Lambda_\varkappa$, $\varkappa = \overline{1, k}$, then we have

$$\delta A = \Lambda_\varkappa b^\varkappa_\sigma \delta q^\sigma, \quad \varkappa = \overline{1, k}, \quad \sigma = \overline{1, s}. \tag{2.3}$$

The coefficients $b^\varkappa_\sigma$ involved in formula (2.3) are usually implemented by the control system in the form of constant quantities or prescribed functions of the generalized coordinates.

Formulas (2.3) show that the sought-for additional generalized forces (2.1) are as follows:

$$R_\sigma = \Lambda_\varkappa b^\varkappa_\sigma. \tag{2.4}$$

We recall that in the framework of the theory under consideration, the Lagrange multipliers $\Lambda_\varkappa$, $\varkappa = \overline{1, k}$, involved in formula (2.4) are looked upon as unknown functions of time. Moreover, we shall see below that in the first theory of motion for nonholonomic systems with high-order constraints, the differential equation with respect to each of the functions $\Lambda_\varkappa$ will be of order $(n - 2)$.

If the generalized control forces (2.4) create a control vector $\mathbf{R}$ coinciding with the vector (2.2), then from the point of view of the nonholonomic mechanics the control thus obtained can be called an *ideal control*.

The first theory of motion for nonholonomic systems with high-order constraints. Construction of a compatible system of differential equations. Now the system of differential equations (1.3) in the tangent space to the manifold of all positions of the mechanical system, which are possible at a given time instant, is written as the single vector equation[15]

$$M\mathbf{W} = \mathbf{Y} + \Lambda_\varkappa \mathbf{b}^\varkappa, \quad \varkappa = \overline{1, k}, \tag{2.5}$$

in which we set

$$\mathbf{Y} = Q_\sigma \mathbf{e}^\sigma, \quad \mathbf{b}^\varkappa = b^\varkappa_\sigma \mathbf{e}^\sigma,$$

$$\mathbf{W} = \left(g_{\sigma\tau} \ddot{q}^\tau + \Gamma_{\sigma,\alpha\beta} \dot{q}^\alpha \dot{q}^\beta\right) \mathbf{e}^\sigma = \left(\ddot{q}^\sigma + \Gamma^\sigma_{\alpha\beta} \dot{q}^\alpha \dot{q}^\beta\right) \mathbf{e}_\sigma,$$

$$\Gamma^\sigma_{\alpha\beta} = g^{\sigma\tau} \Gamma_{\tau,\alpha\beta} = \frac{1}{2} g^{\sigma\tau} \left(\frac{\partial g_{\tau\beta}}{\partial q^\alpha} + \frac{\partial g_{\tau\alpha}}{\partial q^\beta} - \frac{\partial g_{\alpha\beta}}{\partial q^\tau}\right),$$

$$\sigma, \tau = \overline{1, s}, \quad \alpha, \beta = \overline{0, s}, \quad \varkappa = \overline{1, k}. \tag{2.6}$$

From (2.5) and (2.6) we can compose the system of differential equations solved for the generalized accelerations:

---

[15] Here it is worth recalling Chap. 6 of Vol. I.

$$\ddot{q}^{\sigma} = \mathfrak{F}_2^{\sigma}(t, q, \dot{q}, \Lambda),$$

$$\mathfrak{F}_2^{\sigma} = -\Gamma_{\alpha\beta}^{\sigma}\dot{q}^{\alpha}\dot{q}^{\beta} + (Q_{\tau} + \Lambda_{\varkappa}b_{\tau}^{\varkappa})g^{\sigma\tau}/M,$$

$$\sigma, \tau = \overline{1, s}, \qquad \alpha, \beta = \overline{0, s}, \qquad \varkappa = \overline{1, k}. \tag{2.7}$$

System (2.7) of s equations contains $(s + k)$ unknown functions q^{σ}, $\sigma = \overline{1, s}$, $\Lambda_{\varkappa}$, $\varkappa = \overline{1, k}$. Let us first show how this system can be augmented with differential equations with respect to the Lagrange multipliers $\Lambda_{\varkappa}$ for $n = 3$. In this case, the constraint equations (1.2) have the form

$$f_3^{\varkappa} \equiv a_{3\sigma}^{\varkappa}(t, q, \dot{q}, \ddot{q})\,\dddot{q}^{\sigma} + a_{3,0}^{\varkappa}(t, q, \dot{q}, \ddot{q}) = 0,$$

$$\sigma = \overline{1, s}, \qquad \varkappa = \overline{1, k}. \tag{2.8}$$

At this point, we need to derive one important formula. Since

$$e^{\rho} \cdot e_{\tau} = \text{const}, \quad \rho, \tau = \overline{1, s}, \tag{2.9}$$

we have

$$-\dot{e}^{\rho} \cdot e_{\tau} = e^{\rho} \cdot \dot{e}_{\tau}. \tag{2.10}$$

If for notational convenience we denote for the time being the vector $\dot{e}_{\tau}$ as $b = b^{\sigma} e_{\sigma}$, then the right-hand side of (2.10) will be equal to b^{ρ}. Hence b^{ρ} can be equated to the left-hand side of (2.10),

$$b^{\rho} = -\dot{e}^{\rho} \cdot e_{\tau} = -\dot{q}^{\alpha}\frac{\partial e^{\rho}}{\partial q^{\alpha}} \cdot e_{\tau}. \tag{2.11}$$

But at the same time, from (2.9) we get

$$-\frac{\partial e^{\rho}}{\partial q^{\alpha}} \cdot e_{\tau} = \frac{\partial e_{\tau}}{\partial q^{\alpha}} \cdot e^{\rho} \equiv \Gamma_{\tau\alpha}^{\rho},$$

and hence by (2.11) we can write

$$b^{\rho} = \Gamma_{\tau\alpha}^{\rho}\,\dot{q}^{\alpha}.$$

As a result, we get the following useful formula:

$$\dot{e}_{\tau} = \Gamma_{\tau\alpha}^{\rho}\,\dot{q}^{\alpha}\,e_{\rho}. \tag{2.12}$$

We now write down the acceleration vector of the system (2.6) in the form

$$\mathbf{W} = \left(\ddot{q}^{\tau} + \Gamma_{\alpha\beta}^{\tau}\dot{q}^{\alpha}\dot{q}^{\beta}\right)e_{\tau}$$

and differentiate it with respect to time. We get

$$\dot{\mathbf{W}} = \left(\dddot{q}^{\,\tau} + \frac{d}{dt} \left(\Gamma^\tau_{\alpha\beta} \dot{q}^\alpha \dot{q}^\beta \right) \right) \mathbf{e}_\tau + \left(\ddot{q}^{\,\tau} + \Gamma^\tau_{\alpha\beta} \dot{q}^\alpha \dot{q}^\beta \right) \dot{\mathbf{e}}_\tau . \qquad (2.13)$$

Consider the new vectors

$$\mathbf{a}^{\varkappa}_3 = a^{\varkappa}_{3\sigma} \mathbf{e}^\sigma , \qquad \sigma = \overline{1,s} , \qquad \varkappa = \overline{1,k} , \qquad (2.14)$$

which are completely defined by the constraint equations (2.8). Multiplying (2.13) by vectors (2.14), in which the vector $\dot{\mathbf{e}}_\tau$ is first replaced by its expression (2.12), we get

$$\dot{\mathbf{W}} \cdot \mathbf{a}^{\varkappa}_3 = \left(\dddot{q}^{\,\tau} + \frac{d}{dt} \left(\Gamma^\tau_{\alpha\beta} \dot{q}^\alpha \dot{q}^\beta \right) \right) a^{\varkappa}_{3\sigma} + \left(\ddot{q}^{\,\tau} + \Gamma^\tau_{\alpha\beta} \dot{q}^\alpha \dot{q}^\beta \right) \Gamma^\rho_{\tau\alpha} \dot{q}^\alpha ,$$

$$\varkappa = \overline{1,k} . \qquad (2.15)$$

Now we add $a^{\varkappa}_{3,0}(t, q, \dot{q}, \ddot{q})$ to both sides of (2.15) and take into account constraints (2.8). As a result, we get the following inner-product representation of the given nonholonomic constraints (2.8):

$$\dot{\mathbf{W}} \cdot \mathbf{a}^{\varkappa}_3 = \chi^{\varkappa}_3 (t, q, \dot{q}, \ddot{q}) ,$$

$$\chi^{\varkappa}_3 = -a^{\varkappa}_{3,0} + a^{\varkappa}_{3\sigma} \left(\frac{d}{dt} \left(\Gamma^\sigma_{\alpha\beta} \dot{q}^\alpha \dot{q}^\beta \right) + \left(\ddot{q}^{\,\tau} + \Gamma^\tau_{\alpha\beta} \dot{q}^\alpha \dot{q}^\beta \right) \Gamma^\sigma_{\tau\alpha} \dot{q}^\alpha \right) ,$$

$$\sigma, \tau = \overline{1,s} , \qquad \alpha, \beta = \overline{0,s} , \qquad \varkappa = \overline{1,k} . \qquad (2.16)$$

From this representation of the constraint equations (2.16) one can construct an additional system of differential equations with respect to the Lagrange multipliers. Differentiating the vector motion equation of system (2.5) with respect to time, we get

$$M \dot{\mathbf{W}} = \dot{\mathbf{Y}} + \dot{\Lambda}_\varkappa \mathbf{b}^\varkappa + \Lambda_\varkappa \dot{\mathbf{b}}^\varkappa , \qquad \varkappa = \overline{1,k} . \qquad (2.17)$$

We recall that if a vector $\mathbf{a} = a_\tau \mathbf{e}^\tau$ is given, then the covariant components of its derivative $\mathbf{b} = \dot{\mathbf{a}}$ are evaluated by the formula[16]

$$b_\rho = \dot{a}_\rho - \Gamma^\tau_{\rho\alpha} a_\tau \dot{q}^\alpha . \qquad (2.18)$$

Hence the derivatives of the vectors in (2.17) have the form

---

[16] See formula (A.52) in the book: *Sh.Kh. Soltakhanov, M.P. Yushkov, S.A. Zegzhda.* Mechanics of nonholonomic systems. A New Class of control problems. Berlin Heidelberg: Springer-Verlag. 2009. 329 p.

$$\dot{\mathbf{Y}} = (\dot{Q}_\tau - Q_\sigma \Gamma^\sigma_{\tau\alpha} \dot{q}^\alpha) \, \mathbf{e}^\tau , \qquad \dot{\mathbf{b}}^\varkappa = (\dot{b}^\varkappa_\tau - b^\varkappa_\sigma \Gamma^\sigma_{\tau\alpha} \dot{q}^\alpha) \, \mathbf{e}^\tau ,$$
$$\sigma, \tau = \overline{1, s} , \qquad \alpha, \beta = \overline{0, s} , \qquad \varkappa = \overline{1, k} .$$

If now we multiply equation (2.17) by the vectors $\mathbf{a}^\mu_3$ and take into account constraints (2.16), we get

$$\dot{\Lambda}_\varkappa h^{\varkappa\mu}_3 = B^\mu_3(t, q, \dot{q}, \ddot{q}, \Lambda) , \qquad B^\mu_3 = M\chi^\mu_3 - \dot{\mathbf{Y}} \cdot \mathbf{a}^\mu_3 - \Lambda_\varkappa \dot{\mathbf{b}}^\varkappa \cdot \mathbf{a}^\mu_3 ,$$
$$h^{\varkappa\mu}_3 = \mathbf{b}^\varkappa \cdot \mathbf{a}^\mu_3 = b^\varkappa_\sigma a^\mu_{3\tau} g^{\sigma\tau} , \qquad \sigma, \tau = \overline{1, s} , \qquad \varkappa, \mu = \overline{1, k} . \qquad (2.19)$$

Under the assumption that

$$\det \left[b^\varkappa_\sigma a^\mu_{3\tau} g^{\sigma\tau} \right] \neq 0 , \quad \sigma, \tau = \overline{1, s} , \quad \varkappa, \mu = \overline{1, k} ,$$

one can solve the linear inhomogeneous algebraic system (2.19) for $\dot{\Lambda}_\varkappa$,

$$\dot{\Lambda}_\varkappa = h^3_{\varkappa\mu}(t, q, \dot{q}, \ddot{q}) B^\mu_3(t, q, \dot{q}, \ddot{q}, \Lambda) , \qquad \varkappa, \mu = \overline{1, k} . \qquad (2.20)$$

Here $h^3_{\varkappa\mu}$ are the entries of the inverse of the matrix $(h^{\varkappa\mu}_3)$. It is important that using (2.7) one can exclude the derivatives $\ddot{q}^\sigma$ from the functions $h^3_{\varkappa\mu}$, B^μ_3 and write the right-hand sides of equations (2.20) in the form

$$\dot{\Lambda}_\varkappa = C^3_\varkappa(t, q, \dot{q}, \Lambda) , \qquad \varkappa = \overline{1, k} . \qquad (2.21)$$

Let us now consider the general case. With arbitrary n, there appear the functions $h^n_{\varkappa\mu}$, B^μ_n, from which one needs to exclude the derivatives $\ddot{q}^\sigma$, ..., $\overset{(n-1)}{q^\sigma}$. From (2.7) we get

$$\dddot{q}^\sigma = \frac{\partial F^\sigma_2}{\partial t} + \frac{\partial F^\sigma_2}{\partial q^\tau} \dot{q}^\tau + \frac{\partial F^\sigma_2}{\partial \dot{q}^\tau} \ddot{q}^\tau + \frac{\partial F^\sigma_2}{\partial \Lambda_\varkappa} \dot{\Lambda}_\varkappa , \quad \sigma, \tau = \overline{1, s} , \quad \varkappa = \overline{1, k} . \tag{2.22}$$

Using (2.7), one can exclude the derivatives $\ddot{q}^\tau$ from (2.22) and write them in the form

$$\dddot{q}^\sigma = \mathfrak{F}^\sigma_3(t, q, \dot{q}, \Lambda, \dot{\Lambda}) , \qquad \sigma = \overline{1, s} .$$

A similar analysis shows that

$$\overset{(n-1)}{q^\sigma} = \mathfrak{F}^\sigma_{n-1}(t, q, \dot{q}, \Lambda, \dot{\Lambda}, ..., \overset{(n-3)}{\Lambda}) , \qquad \sigma = \overline{1, s} .$$

So, in the general case we have

$$\overset{(n-2)}{\Lambda}_\varkappa = C^n_\varkappa(t, q, \dot{q}, \Lambda, \dot{\Lambda}, ..., \overset{(n-3)}{\Lambda}) , \qquad \varkappa = \overline{1, k} , \qquad n \geqslant 3 . \qquad (2.23)$$

System (2.21) is a particular case of these equations with $n = 3$.

Equations (2.7) and (2.23) form a closed system of equations for the functions $q^\sigma(t)$ and $\Lambda_\varkappa(t)$. For the initial data

$$\Lambda_\varkappa(t_0) = \Lambda_\varkappa^0, \quad \dot\Lambda_\varkappa(t_0) = \dot\Lambda_\varkappa^0, \quad \ldots, \quad \overset{(n-3)}{\Lambda_\varkappa}(t_0) = \overset{(n-3)}{\Lambda_\varkappa}{}^0,$$
$$q^\sigma(t_0) = q_0^\sigma, \quad \dot q^\sigma(t_0) = \dot q_0^\sigma, \quad \varkappa = \overline{1,k}, \quad \sigma = \overline{1,s}, \tag{2.24}$$

this system has a unique solution.

3 Motion of an Artificial Earth Satellite with Constant in Magnitude Acceleration. Dimensional Differential Equations of Motion[17]

The general motion theory of a satellite with fixed in magnitude acceleration (via the first theory of motion.) An interesting example of a real nonholonomic high-order constraint is the problem of motion of an artificial Earth satellite with constant in modulo acceleration in the Earth gravity field.[18] Let us examine in more detail the solution of this problem, because it provides a good illustration of the application of the first theory of motion for nonholonomic systems with high-order constraints (see the previous subsection).

Assume that a material point of mass m moves in the Earth gravity field (this point is an artificial Earth satellite). It is well known that in this case the point moves along an ellipse with the Earth gravitation center in one focus. The motion can be conveniently described in polar coordinates

$$q^1 = r, \quad q^2 = \varphi$$

with bases $\{e_r, e_\varphi\}$ and $\{e^r, e^\varphi\}$.

---

[17] The present section is based on the paper: *V.V. Dodonov, Sh.Kh. Soltakhanov, M.P. Yushkov.* The motion of an Earth satellite after imposition of a nonholonomic third-order constraint // AIP (American Institute of Physics) Conference Proceedings 1959, 030006 (2018); doi: 10.1063/1.5034586.

[18] It seems that the first example of a real nonholonomic constraint is given by the motion of an artificial Earth satellite with constant in modulo acceleration. This example was proposed in the papers by *Sh.Kh. Soltakhanov, M.P. Yushkov.* Application of the generalized Gauss principle for the compilation of motion equations of systems with third-order nonholonomic constraints // Vestn. Leningr. Univ. 1990. Ser. 1. Issue 3 (№ 15). Pp. 77–83; *Sh.Kh. Soltakhanov, M.P. Yushkov.* Motion equations of a certain nonholonomic system with a second-order constraint // Vestn. Leningr. Univ. 1991. Ser. 1. Issue 4 (№ 22). Pp. 26–29. Later, this example was investigated, in particular, in the paper *M.P. Juschkov, S.H. Soltachanov, S.A. Zegzhda.* Anwendung des generalisierten Gaußschen Prinzips auf die Untersuchung der Bewegung eines Satelliten mit konstanter Beschleunigung // Technische Mechanik. 2004. Bd. 24. Heft 3-4. S. 236-241.

The acceleration $\boldsymbol{w}$ of the satellite changes as it moves; its projections onto the axes of the above polar system of coordinates are given by the formulas

$$pr_{\mathbf{e}_r}\boldsymbol{w} = \ddot{r} - r\dot{\varphi}^2, \qquad pr_{\mathbf{e}_\varphi}\boldsymbol{w} = r\ddot{\varphi} + 2\dot{r}\dot{\varphi}.$$

It is clear that the squared acceleration w^2 is

$$w^2 = (\ddot{r} - r\dot{\varphi}^2)^2 + (r\ddot{\varphi} + 2\dot{r}\dot{\varphi})^2.$$

Assume that at some time instant $t = 0$ the acceleration of the satellite is w_0. We pose the following problem: find the motion of the satellite (in this setting, it is better to call it a spacecraft) for which the value of acceleration w_0 does not change (it is constant). Such requirement can be expressed by the nonlinear nonholonomic second-order constraint

$$f_2(q, \dot{q}, \ddot{q}) \equiv (\ddot{r} - r\dot{\varphi}^2)^2 + (r\ddot{\varphi} + 2\dot{r}\dot{\varphi})^2 - w_0^2 = 0, \qquad (3.1)$$

which is imposed on the further motion of the spacecraft.

So, we have a real space dynamics mechanical problem in which the motion of a material point is subject to the nonholonomic nonlinear second-order constraint (3.1). This constraint can be considered ideal by the problem formulation (here it is worth recalling the footnote preceding formula (2.2)). Moreover, the above problem can be looked upon as a control theory problem, in which the motion program requires that the nonlinear differential equation (3.1) be satisfied in the course of motion. We shall attack this problem using the machinery of nonholonomic mechanics with high-order constraints, which was given in the previous subsection. Hence the motion constraint (3.1) can be naturally called a *program constraint*, while the resulting reaction is the sought-for *control force* solving the above control problem.

To be able to apply the theory of motion for nonholonomic systems with high-order constraints (see the previous subsection), one should differentiate constraint (3.1) with respect to time and represent it in the form of a linear nonholonomic constraint of third order,

$$f_3 \equiv \dot{f}_2 = (\ddot{r} - r\dot{\varphi}^2)(\dddot{r} - \dot{r}\dot{\varphi}^2 - 2r\dot{\varphi}\ddot{\varphi}) +$$
$$+ (r\ddot{\varphi} + 2\dot{r}\dot{\varphi})(\dot{r}\ddot{\varphi} + r\dddot{\varphi} + 2\ddot{r}\dot{\varphi} + 2\dot{r}\ddot{\varphi}) = 0.$$

This linear constraint can be conveniently written in the standard form

$$f_3(q, \dot{q}, \ddot{q}, \dddot{q}) \equiv a_{3r} \dddot{r} + a_{3\varphi} \dddot{\varphi} + a_{3,0} = 0 \,,$$

$$a_{3r} = (\ddot{r} - r\dot{\varphi}^2) \,,$$

$$a_{3\varphi} = (r\ddot{\varphi} + 2\dot{r}\dot{\varphi})r \,,$$

$$a_{3,0} = (r\ddot{\varphi} + 2\dot{r}\dot{\varphi})(\dot{r}\ddot{\varphi} + 2\ddot{r}\dot{\varphi} + 2r\dot{\varphi}) - $$

$$- (\ddot{r} - r\dot{\varphi}^2)(\dot{r}\dot{\varphi}^2 + 2r\dot{\varphi}\ddot{\varphi}) \,. \tag{3.2}$$

The vector motion equation reads as

$$m\mathbf{w} = \mathbf{F} + \Lambda \nabla''' f_3 \equiv \mathbf{F} + \Lambda \left(\frac{\partial f_3}{\partial \dddot{r}} \mathbf{e}^r + \frac{\partial f_3}{\partial \dddot{\varphi}} \mathbf{e}^\varphi \right) . \tag{3.3}$$

Multiplying this equation by vectors of the principal basis $\mathbf{e}_r$, $\mathbf{e}_\varphi$, we get

$$\Lambda_*(\ddot{r} - r\dot{\varphi}^2) = \frac{\mu}{r^2} \,,$$

$$\Lambda_*(r\ddot{\varphi} + 2\dot{r}\dot{\varphi}) = 0 \,,$$

$$\Lambda_* = \frac{\Lambda}{m} - 1 \,. \tag{3.4}$$

Here μ is the Gauss constant for the Earth gravity field. From system (3.4), we get the differential equations of form (2.7):

$$\ddot{r} = r\dot{\varphi}^2 + \frac{\mu}{\Lambda_* r^2} \,, \qquad \ddot{\varphi} = -\frac{2\dot{r}\dot{\varphi}}{r} \,. \tag{3.5}$$

Next, we use Eq. (3.5) and apply the constraint equation (3.2) to find an additional differential equation with respect to Λ_*. To this aim, we first differentiate the first two equations of (3.4) with respect to time,

$$\dot{\Lambda}_*(\ddot{r} - r\dot{\varphi}^2) + \Lambda_*(\dddot{r} - \dot{r}\dot{\varphi}^2 - 2r\dot{\varphi}\ddot{\varphi}) + 2\frac{\mu}{r^3}\dot{r} = 0 \,, \tag{3.6}$$

$$\dot{\Lambda}_*(r\ddot{\varphi} + 2\dot{r}\dot{\varphi}) + \Lambda_*(\dot{r}\ddot{\varphi} + r\dddot{\varphi} + 2\ddot{r}\dot{\varphi} + 2\dot{r}\ddot{\varphi}) = 0 \,. \tag{3.7}$$

To single out the constraint equation (thereby getting rid of the third-order derivatives), we multiply equation (3.6) by a_{3r}, multiply equation (3.7) by $a_{3\varphi}$, and add the results. We have

$$\dot{\Lambda}_* = \frac{2\mu\dot{r} \left(r\dot{\varphi}^2 - \ddot{r} \right)}{r^3((\ddot{r} - r\dot{\varphi}^2)^2 + (r\ddot{\varphi} + 2\dot{r}\dot{\varphi})^2)} \,.$$

Using the constraint equation (3.2), we get the required equation of type (2.21):

$$\dot{\Lambda}_* = -\frac{2\mu}{w_0^2} \frac{\dot{r}}{r^5 \Lambda_*} \,. \tag{3.8}$$

Integration of the system of differential equations (3.5), (3.8) solves the above problem: with given value of gravity, find the motion of an artificial Earth satellite with constant in modulo acceleration. To solve this problem, one has to apply to it the additional force (control force) formed by the Lagrange multiplier Λ, which will be found as a function of time after the solution of the system of equations (3.5), (3.8).

Motions of Kosmos, Molniya and Tundra-type spacecrafts with constant accelerations (via the first theory of motion for nonholonomic systems with high-order constraints). To specify the initial data for the system of differential equations (3.5), (3.8), we shall use the known formulas (see Sect. 9 of Chap. 4 of Vol. I; we recall that e is the eccentricity of the orbit, p is the latus rectum (the focal parameter)):

$$r = \frac{p}{1 + e \cos \varphi}, \quad \dot\varphi = \frac{\sqrt{p\mu}}{r^2}, \quad \dot r = \frac{pe\dot\varphi \sin \varphi}{(1 + e \cos \varphi)^2},$$

$$\ddot\varphi = -\frac{2\dot r\dot\varphi}{r}, \quad \ddot r = r\dot\varphi^2 - \frac{\mu}{r^2}. \tag{3.9}$$

We first consider the motion of one Soviet Kosmos-type spacecraft with the following altitudes over the Earth surface: $H_\pi = 183\,\text{km}$ in the perigee and $H_\alpha = 244\,\text{km}$ in the apogee (here and what follows, the numerical data are taken from the Internet). We assume that the Earth radius is $R_E = 6371\,\text{km}$ and that the acceleration of gravity at the Earth surface is $g_0 = 9.82 \cdot 10^{-3}\,\text{km/s}^2$ (that is, we assume that the Earth has a spherical form with uniformly distributed mass). We have

$$r_\pi = 6554\,\text{km}, \quad r_\alpha = 6615\,\text{km}, \quad \mu = g_0 R_E^2 = 398590\,\text{km}^3/\text{s}^2,$$

$$e = \frac{r_\alpha - r_\pi}{r_\alpha + r_\pi} = 0.004632, \quad p = r_\pi(1 + e) = 6584.36\,\text{km}.$$

We assume that the acceleration of the satellite is fixed at its perigee. Hence the initial data for $t_0 = 0$ for numerical integration of the system of equations (3.5), (3.8) by formulas (3.9) assume the values (according to (2.24) we augment them by the condition $\Lambda(0) = 0$, which is equivalent to saying that $\Lambda_*(0) = -1$):

$$r(0) = 6554\,\text{km}, \quad \varphi(0) = 0, \quad \dot r(0) = 0,$$
$$\dot\varphi(0) = 0.00119263\,\text{s}^{-1}, \quad \ddot r(0) = 0.0000429824\,\text{km/s}^2,$$
$$\ddot\varphi(0) = 0, \quad \Lambda(0) = 0. \tag{3.10}$$

The results of numerical integration with the initial data (3.10) are given in Fig. 1. The left-hand side of the figure shows that in this problem the spacecraft moves practically along a circle depicted by the solid line (here and in what follows, the dash-dot circle shows the spherical Earth surface of radius $R_E = 6371\,\text{km}$). Note that initially the spacecraft moves along a nearly circular orbit (with given orbital parameters). In the early stage of space exploration, the majority of spacecrafts were

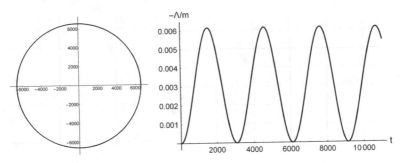

Fig. 1 Motion of a Kosmos-type spacecraft after the value of its acceleration is fixed at perigee via the first theory of motion

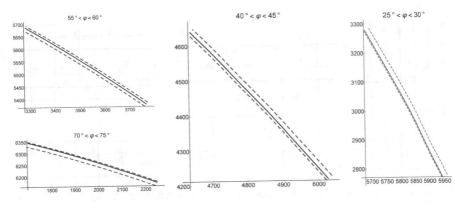

Fig. 2 Trajectory fragments of a Kosmos-type spacecraft with constant in modulo acceleration via the first theory of motion

launched in this way, which enabled one to use the smallness of the orbit eccentricity e in evaluating their motion.

The right panel shows the dependence on time of the Lagrange multiplier forming the control force and securing the execution of the program motion, as given by the differential equations (3.2). The solid line corresponds to the first revolution, the dashed line shows the second revolution, and the dotted line shows the beginning of the third revolution. Note that the required control force can be easily implemented by equipping the spacecraft with an additional engine.

Considering the trajectory of a Kosmos-type spacecraft in our problem in large scale, it can be noticed that once the constant in modulo acceleration is achieved, the spacecraft starts to rotate and touch alternately two concentric circles with centers at the Earth center (the dashed circular arcs in Fig. 2). Once the motion starts from the inner circle, the first such approach to the outer circle is manifested on the sequence of fragments of the trajectory (the solid line) in Fig. 2.

Spacecraft motion with constant acceleration between two concentric circles is more clearly visualized for spacecrafts on highly elliptical orbits. To illustrate this, we consider the motion of a Molniya-type spacecraft. The perigee of the orbit was above

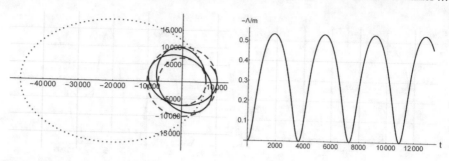

Fig. 3 Motion of a Molniya-type spacecraft after the value of its acceleration is fixed at perigee via the first theory of motion

Moscow, and the apogee, above Vladivostok. Due to the law of equal areas, such spacecrafts move rapidly above Moscow and travel slowly ("freezing") above Vladivostok. With the full constellation of such spacecrafts, there was always one spacecraft above Vladivostok providing good transmission of TV signal from Moscow.

So, for a Molniya-type spacecraft, we have

$$r_\pi = 6871\,\text{km}\,, \quad r_\alpha = 46371\,\text{km}\,,$$

$$e = \frac{r_\alpha - r_\pi}{r_\alpha + r_\pi} = 0.741895\,, \quad p = r_\pi(1 + e) = 11968.6\,\text{km}\,.$$

This gives us the initial conditions for numerical integration of the system of differential equations:

$$r(0) = 6871\,\text{km}\,, \quad \varphi(0) = 0\,, \quad \dot{r}(0) = 0\,,$$
$$\dot{\varphi}(0) = 0.001463\,\text{s}^{-1}\,, \quad \ddot{r}(0) = 0.00626368\,\text{km/s}^2\,,$$
$$\ddot{\varphi}(0) = 0\,, \quad \Lambda(0) = 0\,. \tag{3.11}$$

The results of integration of the motion equations with the initial data (3.11) are given in Fig. 3, where the initial elliptical orbit is shown by the dotted curve; the solid curve between two dashed concentric circles shows the spacecraft trajectory after its acceleration is fixed at perigee.

Let us compare the results obtained for the above two spacecraft. The orbit of a Kosmos-type spacecraft is nearly circular. Hence its acceleration changes a little. Therefore, a constant acceleration motion can be achieved by applying an insignificant generalized control force $\Lambda = \Lambda(t)$. In contrast, if a Molniya-type spacecraft fixes its acceleration at perigee, it starts to move along a nearly elliptical trajectory, which drastically differs from the original orbit. This requires much larger generalized control force, which is higher by two orders than the previous one (cf. the graphs on the right panels of Figs. 1 and 3).

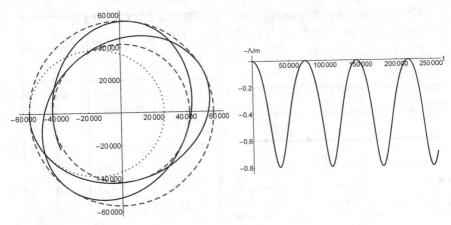

Fig. 4 Motion of a Tundra-type spacecraft after the value of its acceleration is fixed at apogee via the first theory of motion

Let us now consider a Tundra-type spacecraft with parameters[19]

$$r_\pi = 26371 \, \text{km}, \qquad r_\alpha = 56371 \, \text{km},$$

$$e = \frac{r_\alpha - r_\pi}{r_\alpha + r_\pi} = 0.362573, \quad p = r_\pi(1 + e) = 35932.4 \, \text{km}.$$

Tundra-type spacecrafts were aimed at providing stable communication service for Soviet north lands. If now, for a change, we would like to fix the acceleration of the spacecraft at apogee (rather than at perigee), then to integrate the system of differential equations (3.5), (3.8) the following initial conditions should be specified:

$$r(0) = 56371 \, \text{km}, \quad \varphi(0) = \pi, \quad \dot{r}(0) = 0,$$
$$\dot{\varphi}(0) = 0.00003766 \, \text{s}^{-1}, \quad \ddot{r}(0) = -0.00004548 \, \text{km/s}^2,$$
$$\ddot{\varphi}(0) = 0, \quad \Lambda(0) = 0. \tag{3.12}$$

The solution corresponding to the initial data (3.12) is shown in Fig. 4.

---

[19] See the paper: *V. V. Dodonov*. Motion of an artificial Earth satellite with fixed acceleration at apogee // Proc. Seminar "Computer Methods in Continuum Mechanics" 2019-2020. St. Petersburg: St. Petersburg Univ. Press. 2020. Pp. 51–61.

4 Motion of an Artificial Earth Satellite with Constant Modulo Acceleration. The Dimensionless Differential Motion Equations

The motion of a spacecraft in the Earth gravity field is described by the following vector equation:

$$\frac{d^2\boldsymbol{\rho}}{dt^2} = -\frac{\mu\boldsymbol{\rho}}{\rho^3}, \quad \mu = \gamma M, \quad \rho = |\boldsymbol{\rho}|. \tag{4.1}$$

Here $\boldsymbol{\rho}$ is the radius vector connecting the Earth center with the spacecraft, γ is the gravitational constant, M is the Earth mass. The constant μ can be written in the form (here and in what follows it is worth recalling Sect. 9 of Chap. 4 of Vol. I)

$$\mu = \frac{4\pi^2 a^3}{T^2},$$

where a is the semi-major axis of the spacecraft elliptical orbit, T is the time of complete revolution.

In the dimensionless variables

$$\mathbf{r} = x\mathbf{i} + y\mathbf{j} = \boldsymbol{\rho}/a, \quad \tau = 2\pi t/T$$

equation (4.1) assumes the form

$$\ddot{\mathbf{r}} = -\mathbf{r}/r^3, \quad r = |\mathbf{r}|. \tag{4.2}$$

Here and in what follows, a dot denotes the derivative with respect to the dimensionless time τ. The energy integral and the area integral of equation (4.2) have the form

$$v^2 = 2/r - 1, \quad v = |\dot{\mathbf{r}}|, \quad r^2\dot{\varphi} = \sqrt{1 - e^2}, \tag{4.3}$$

where e is the eccentricity of the elliptical orbit. Assume that the spacecraft is at the x-axis at the initial moment starting from which it should travel with constant acceleration. Without loss of generality one can assume that the initial data are as follows:

$$x(0) = x_0, \quad \dot{x}(0) = \dot{x}_0 = \sqrt{2x_0 - x_0^2 - 1 + e^2}/x_0,$$

$$y(0) = y_0 = 0, \quad \dot{y}(0) = \dot{y}_0 = \sqrt{1 - e^2}/x_0, \quad 1 - e \leqslant x_0 \leqslant 1 + e. \tag{4.4}$$

In this notation, the constraint equation reads as

$$\ddot{\mathbf{r}}^2 - 1/x_0^4 = 0 \,. \tag{4.5}$$

In particular, this equation will be satisfied in the case when the vector $\ddot{\mathbf{r}}$, which is collinear to the vector $\mathbf{r}$, has constant magnitude. Moreover, the derivative with respect to time of the vector $\ddot{\mathbf{r}}$ is orthogonal to the vector $\mathbf{r}$; that is, we have

$$\mathbf{e}_r \cdot \dddot{\mathbf{r}} = 0 \,, \quad \mathbf{e}_r = \mathbf{r}/r \,. \tag{4.6}$$

This equation is a linear nonholonomic program constraint of third order. So, the system of equations (1.2) in this problem reduces to one equation (4.6).

We shall assume that the spacecraft is equipped with the generalized control force Λ for which the vector of the control force is

$$\mathbf{R} = \Lambda \mathbf{e}_r \,. \tag{4.7}$$

From (4.6) it follows that with this force $\mathbf{R}$ the control is ideal.

The motion of the spacecraft starting from the moment when constraint (4.6) was imposed is described by the equation

$$\ddot{\mathbf{r}} = -\frac{\mathbf{r}}{r^3} + \Lambda \frac{\mathbf{r}}{r} \,. \tag{4.8}$$

The control force is absent at the moment the constraint was imposed; that is,

$$\Lambda(0) = 0 \,. \tag{4.9}$$

Differentiating expression (4.8) with respect to τ, we get

$$\dddot{\mathbf{r}} = -\frac{\dot{\mathbf{r}}}{r^3} + \frac{3\dot{r}}{r^4}\mathbf{r} + \dot{\Lambda}\frac{\mathbf{r}}{r} + \Lambda\frac{\dot{\mathbf{r}}}{r} - \Lambda\frac{\dot{r}\mathbf{r}}{r^2} \,.$$

Multiplying this equation by $\mathbf{r}$, taking into account the constraint equation (4.6), and since

$$r^2 = \mathbf{r}^2 \,, \quad \mathbf{r} \cdot \dot{\mathbf{r}} = r\dot{r} \,,$$

we get

$$\dot{\Lambda} = -\frac{2\dot{r}}{r^3} \,. \tag{4.10}$$

So, the system of equations (2.21) in this problem reduces to one equation (4.10). Setting

$$\dot{\Lambda} = \frac{d\Lambda}{dr}\dot{r}$$

in this equation, we get

$$\frac{d\Lambda}{dr} = -\frac{2}{\dot{r}^3}.$$

Integrating this equation and taking into account that in view of (4.4) and (4.9) we have $\Lambda = 0$ for $r = x_0$, we get that

$$\Lambda = \frac{1}{r^2} - \frac{1}{x_0^2}.$$

Substituting this expression into equation (4.8), we find that

$$\ddot{\mathbf{r}} = -\mathbf{r}/(rx_0^2). \tag{4.11}$$

Using equation (4.11) we can find for our problem the motion satisfying equation (4.5) without knowing the control force $\mathbf{R} = \Lambda\mathbf{r}/r$ responsible for this motion. However, to be able to implement this motion, one needs to know this force as a function of time. Hence, we shall consider equation (4.8) jointly with equation (4.10) without excluding the control force from equation (4.8).

Projecting the vector equation (4.8) onto the unit vectors of the polar coordinates $\mathbf{e}_r = \mathbf{r}/r$ and $\mathbf{e}_\varphi^0$, we get

$$\ddot{r} - r\dot{\varphi}^2 + \frac{1}{r^2} = \Lambda,$$
$$r\ddot{\varphi} + 2\dot{r}\dot{\varphi} = 0. \tag{4.12}$$

Augmenting these two equations by equation (4.10), we get a closed system of equations, from which one can find both the motion and the control force.

The system of equations (4.10), (4.12) was numerically integrated with the initial data

$$r(0) = x_0 = 1 - e, \quad \dot{r}(0) = 0, \quad \varphi(0) = 0,$$
$$\dot{\varphi}(0) = \sqrt{1 - e^2}/x_0^2, \quad \Lambda(0) = 0.$$

Calculations show that for any value of the eccentricity e (distinct from 0 and 1), the spacecraft moves along a curve lying between two concentric circles of radii r_1 and r_2 (these numbers are the positive roots of the equation)[20]

$$2r^3 - (4 - x_0)x_0 r^2 + x_0^2(1 - e^2) = 0.$$

Note that the motion between these circles is not periodic in the sense that the point will never return to its initial position after an integer number of rotations.

---

[20] See p. 140 of the book *Sh. Kh. Soltakhanov, M. P. Yushkov, S. A. Zegzhda.* Mechanics of nonholonomic systems. A New Class of control problems. Berlin Heidelberg: Springer-Verlag. 2009. 329 p.

Fig. 5 Spacecraft trajectory

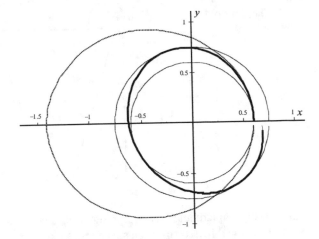

Fig. 6 Hodograph of the
control force

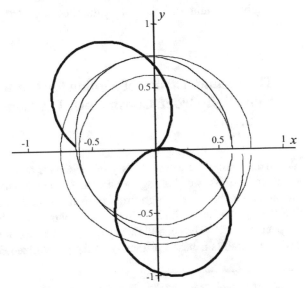

As an example, we show in Figs. 5, 6, 7 the calculation results in the time interval
$0 \leqslant t \leqslant T/2$ $(0 \leqslant \tau \leqslant \pi)$ for $e = 0.4$. The thin lines in Fig. 5 show the initial elliptical orbit and the concentric circles of radii $r_1 = 0.6$ and $r_2 = 0.754$, respectively,
between which the solution of equation (4.11) is located. This solution is shown by
the heavy line.

The hodograph of the control force $\mathbf{R} = \Lambda(\tau)\mathbf{r}/r$ providing this motion is shown
by the heavy line in Fig. 6. Here one should take into account that $\Lambda \leqslant 0$ during the
entire time of motion. The graph of the function $\Lambda(\tau)$ is depicted in Fig. 7. Note that
from Eqs. (4.1) and (4.8) it follows that Λ is measured in fractions of the gravity
force F, where $F = \mu m/a^2$. Here m is the spacecraft mass. Such a control force

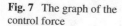

Fig. 7 The graph of the control force

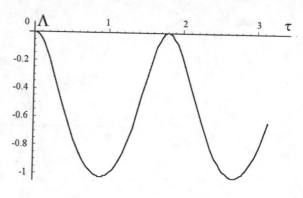

can be easily implemented by equipping the spacecraft with an additional jet engine whose thrust is directed from the spacecraft to the attracting center. This very force is given by formula (4.7) (note that the function $\Lambda(\tau)$ was found to be negative or zero).[21]

5 The Second Theory of Motion for Nonholonomic Systems with High-Order Constraints. The Generalized Gauss Principle

When discussing the second theory of motion for nonholonomic systems with high-order constraints, we shall again be concerned only with linear constraints that are linear with respect to the high-order derivatives. In the case of nonlinear constraints, the theory can be applied if these constraints are preliminary differentiated with respect to time.[22] The second theory of motion of nonholonomic systems with high-order constraints is based primarily on the generalized Gauss principle.[23]

---

[21] Another example reflecting the effect of a linear nonholonomic constraint of third order is considered in the book indicated in the previous footnote. This example considers a smooth transfer of a spacecraft from one circular orbit to a different one as a variant of motion providing an alternative to that along the Hohmann ellipse. In this transfer, its smoothness is characterized by parameters of a specially constructed generalized Sears equation.

[22] The effect of nonlinearity of constraints is discussed in detail in the book: *Sh. Kh. Soltakhanov, M.P. Yushkov, S.A. Zegzhda*. Mechanics of nonholonomic systems. A New Class of control problems. Berlin Heidelberg: Springer-Verlag. 2009. 329 p.

[23] The generalized Gauss principle was proposed already in 1983 in the papers: *N.N. Polyakhov, S.A. Zegzhda, M.P. Yushkov*, Generalization of the Gauss principle to the case of higher order nonholonomic systems // Dokl. Akad. Nauk SSSR. 1983. Vol. 269. № 6. Pp. 1328–1330 [in Russian]; *N.N. Polyakhov, S.A. Zegzhda, M.P. Yushkov*. Linear transformation of forces and the generalized Gauss principle // Vest. Leningr. Univ. 1984. № 1. Pp. 73–79 [in Russian]. It should also be noted that this principle was first formulated by M.A. Chuev in his little-known paper: *M.A. Chuev*. To the question of the analytical method of synthesis mechanism // In Proc. Vyssh. Uchebn. Zaved.

Discussion of the Gauss principle. We first recall the classical statement of the Gauss principle (see (4.13) and (4.14) in Chap. 9 of Vol. I)

$$\delta''Z = 0\,, \tag{5.1}$$

where the Gauss constraint (compulsion) reads as

$$Z = \frac{M}{2}\left(\mathbf{W} - \frac{\mathbf{Y}}{M}\right)^2. \tag{5.2}$$

The two primes in formula (5.1) stress that only the second-order derivatives of the generalized coordinates are varied. By varying these derivatives, we get another representation of the generalized Gauss principle:

$$\left(M\mathbf{W} - \mathbf{Y}\right)\cdot\delta''\mathbf{W} = 0\,. \tag{5.3}$$

Using the second Newton law

$$M\mathbf{W} = \mathbf{Y} + \mathbf{R} \tag{5.4}$$

and using (5.3), the Gauss principle can also be put in the form

$$\delta''(\mathbf{R})^2 = 0\,, \tag{5.5}$$

which can be interpreted as the minimality requirement of the reaction of ideal linear nonholonomic second-order constraints (1.6):

$$f_2^\varkappa(t, q, \dot q, \ddot q) \equiv a_{2\sigma}^{l+\varkappa}(t, q, \dot q)\,\ddot q^{\,\sigma} + a_{2,0}^{l+\varkappa}(t, q, \dot q) = 0\,, \quad \varkappa = \overline{1, k}\,. \tag{5.6}$$

The geometric interpretation of the Gauss principle is also worth mentioning. We recall that constraints (5.6) split the tangent space into two orthogonal subspaces K and L with bases $\{\varepsilon^{l+1}, \ldots, \varepsilon^s\}$ and $\{\varepsilon_1, \ldots, \varepsilon_l\}$ (see Chaps. 6 and 9 of Vol. I). The constraints in the space of generalized accelerations define the l-dimensional plane $\mathbb{T}(t, q, \dot q, \ddot q)$, for which t, q, and $\dot q$ are given parameters. This plane should contain the end-points of the acceleration vectors $\mathbf{W}$ of the mechanical system. The vectors ε_λ, $\lambda = \overline{1, l}$, also lie in this plane. We recall that the variation of the acceleration $\delta''\mathbf{W}$ is defined as the vector $\delta''\mathbf{W} = \delta''\ddot q^{\,\sigma}\mathbf{e}_\sigma$, which can be expanded in the basis of the space L and satisfies the conditions

$$\nabla'' f_2^\varkappa \cdot \delta''\mathbf{W} = 0\,, \quad \varkappa = \overline{1, k}\,. \tag{5.7}$$

_____

Mashinostr. 1974. № 8. Pp. 165–167 [in Russian]. Later, using a linear transformation of forces, this principle was rigorously formulated in the above two papers independently of Chuev.

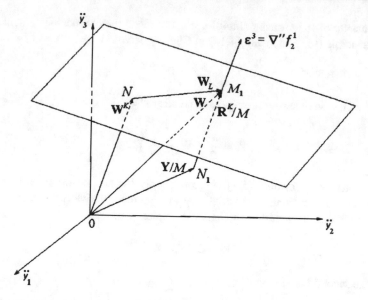

Fig. 8 Acceleration of the point with linear second-order constraint

From formulas (5.3) and (5.7) it easily follows that the reaction force of ideal non-holonomic second-order constraints is expressed by the formula

$$\mathbf{R}^K = \Lambda_{\varkappa} \varepsilon^{l+\varkappa},$$

that is, it lies in the space K.

The Gauss principle in the form (5.5) shows that in the case of ideal linear non-holonomic second-order constraints (5.6) their reaction $R/M = W - Y/M$ "forces" the mechanical system to move with the smallest magnitude of this reaction. This is why the Gauss principle is sometimes called the *principle of least constraint (compulsion)*.

Figure 8 illustrates this situation for the motion of a single material point under the ideal nonholonomic second-order constraint

$$f_2^1(t, y, \dot{y}, \ddot{y}) \equiv a_{2\sigma}^3(t, y, \dot{y}) \ddot{y}_\sigma + a_{2,0}^3(t, y, \dot{y}) = 0, \quad \sigma = \overline{1,3}, \tag{5.8}$$

where $y = (y_1, y_2, y_3)$ are the Cartesian coordinates of the point. In Fig. 8, in the space of accelerations of the point, there is the plane (5.8) defined by the vector W^K; on this plane the vector W of acceleration of the material point ends at the point M_1; the vector of the reciprocal basis $\varepsilon^3 = \nabla'' f_2^1$ is directed perpendicularly to the plane from the point M_1; the reaction $\mathbf{R}^K$ of the ideal constraint (5.8) should be directed along this vector. In Fig. 8, the constraint R^K/M itself is represented by the vector $\overrightarrow{N_1 M_1}$, which is perpendicular to the plane of the constraint and starts at the vector end of the active force Y/M acting on the material point. By construction, this vector has the smallest length; that is, by the Gauss principle in the form (5.5) the reaction securing the fulfillment of the ideal constraint (5.8) is indeed minimal.

The generalized Gauss principle. As was pointed out above, for a rigorous formulation of the generalized Gauss principle one should introduce a linear transformation of forces. However, the crux of this principle can be easily visualized by extending the above geometrical interpretation of the classical Gauss principle.

Let us consider the motion of a mechanical system subject to the linear third-order nonholonomic constraints

$$f_3^{\varkappa}(t,q,\dot{q},\ddot{q},\dddot{q}) \equiv a_{3\sigma}^{l+\varkappa}(t,q,\dot{q},\ddot{q})\,\dddot{q}^{\,\sigma} + a_{3,0}^{l+\varkappa}(t,q,\dot{q},\ddot{q}) = 0, \quad \varkappa = \overline{1,k}.$$
$$(5.9)$$

In the space of vectors $\dot{\mathbf{W}}$, these equations define an l-dimensional plane, which for satisfaction of constraints (5.9) should contain the end-points of these vectors. If now in Fig. 8 one replaces $\ddot{y}_1$, $\ddot{y}_2$, $\ddot{y}_3$ by $\dddot{y}_1$, $\dddot{y}_2$, $\dddot{y}_3$, and replace the vectors $\mathbf{W}$, $\mathbf{W}^K$, $\mathbf{W}_L$, $\mathbf{Y}$, $\mathbf{R}^K$ by the vectors $\dot{\mathbf{W}}$, $\dot{\mathbf{W}}^K$, $\dot{\mathbf{W}}_L$, $\dot{\mathbf{Y}}$, $\dot{\mathbf{R}}^K$, then as in the previous paragraph one can assert that constraints (5.9) minimize the quantity $\dot{\mathbf{R}}/M = \dot{\mathbf{R}}^K/M$, and hence, as in (5.1), we have

$$\delta'''Z_{(1)} = 0,\tag{5.10}$$

where we set

$$Z_{(1)} = \frac{M}{2}\left(\dot{\mathbf{W}} - \frac{\dot{\mathbf{Y}}}{M}\right)^2.\tag{5.11}$$

Equation (5.10) can be looked upon as the *the generalized Gauss principle*, which holds under constraints (5.9). The subscript "(1)" in (5.10) and (5.11) indicates the order of the generalized principle with respect to the classical Gauss principle, and the three primes in (5.10) show that only the third derivatives of the generalized coordinates are varied. As in (5.1) and (5.3) the generalized Gauss principle (5.10) can be written in the form

$$\left(M\dot{\mathbf{W}} - \dot{\mathbf{Y}}\right)\cdot\delta'''\dot{\mathbf{W}} = 0.\tag{5.12}$$

For linear nonholonomic constraints of order $(n+2)$, the above generalized first-order Gauss principle can be easily extended to the *the generalized Gauss principle of nth order*

$$\delta^{(n+2)}Z_{(n)} = 0,\tag{5.13}$$

where we set

$$Z_{(n)} = \frac{M}{2} \left(\overset{(n)}{W} - \frac{\overset{(n)}{Y}}{M} \right)^2.$$ (5.14)

The superscript (n) in (5.13), (5.14) indicates the order of the derivative of a vector with respect to time, and the superscript $(n+2)$ means that the partial differential is evaluated with fixed $t, q^\sigma, \dot{q}^\sigma, \ldots, \overset{(n+1)}{q^\sigma}$.

When using principle (5.13) one should indicate the initial conditions

$$\Lambda_\varkappa(t_0) = \Lambda_\varkappa^0, \quad \dot{\Lambda}_\varkappa(t_0) = \dot{\Lambda}_\varkappa^0, \quad \ldots, \quad \overset{(n-1)}{\Lambda_\varkappa}(t_0) = \overset{(n-1)}{\Lambda_\varkappa}{}^0,$$
$$q^\sigma(t_0) = q_0^\sigma, \quad \dot{q}^\sigma(t_0) = \dot{q}_0^\sigma, \quad \varkappa = \overline{1, k}, \quad \sigma = \overline{1, s}.$$

The vector $\overset{\rightarrow}{\mathfrak{R}} \equiv \overset{(n)}{R} = M \overset{(n)}{W} - \overset{(n)}{Y}$, whose magnitude is minimized in this paragraph, can be conventionally called the "reaction" of linear nonholonomic constraints of order $(n+2)$.

In the second theory of motion of nonholonomic systems with high-order constraints, the motion equations are composed using the generalized Gauss principle.

6 Motion of a Spacecraft with Constant Acceleration via the Second Theory of Motion for Nonholonomic Systems with High-Order Constraints. Dimensional Differential Equations[24]

Let us write down the generalized Gauss principle for an artificial Earth satellite under the linear nonholonomic constraint of third order (3.2):

$$\left(m\dot{\mathbf{w}} - \dot{\mathbf{F}} \right) \cdot \delta''' \dot{\mathbf{w}} = 0.$$ (6.1)

We rewrite the principle (6.1) as

$$\left(mU_\rho - P_\rho \right) \delta''' U^\rho = 0, \quad \rho = 1, 2, \quad U = \dot{\mathbf{R}}, \quad P = \dot{\mathbf{F}}$$ (6.2)

[24] See the papers: *Sh.Kh. Soltakhanov, M.P. Yushkov.* Application of the generalized Gauss principle for the compilation of motion equations of systems with third-order nonholonomic constraints // Vestn. Leningr. Univ. 1990. Ser. 1. Issue 3 (№ 15). Pp. 77–83 [in Russian]; *Sh.Kh. Soltakhanov, M.P. Yushkov.* Motion equations of a certain nonholonomic system with a second-order constraint // Vestn. Leningr. Univ. 1991. Ser. 1. Issue 4 (№ 22). Pp. 26–29 [in Russian]; *V.V. Dodonov, Sh.Kh. Soltakhanov, M.P. Yushkov.* The motion of an Earth satellite after imposition of a nonholonomic third-order constraint // AIP (American Institute of Physics) Conference Proceedings 1959, 030006 (2018).

and use the well-known formula (2.18) for the covariant components of the vectors
U and **P**

$$U_\rho = \dot{w}_\rho - \Gamma^\tau_{\rho\sigma} w_\tau \dot{q}^\sigma , \quad P_\rho = \dot{F}_\rho - \Gamma^\tau_{\rho\sigma} F_\tau \dot{q}^\sigma , \quad \rho, \sigma, \tau = 1, 2 , \qquad (6.3)$$

where $\Gamma^\tau_{\rho\sigma}$ are the Christoffel symbols of the second kind. The covariant components
of the acceleration and the force in formula (6.3) read as

$$w_1 = w_r = \ddot{r} - r\dot{\varphi}^2 , \quad w_2 = w_\varphi = r(r\ddot{\varphi} + 2\dot{r}\dot{\varphi}) ,$$

$$F_1 = -\frac{m\mu}{r^2} , \quad F_2 = 0 .$$

The variations $\delta''' U^1$ and $\delta''' U^2$ in our problem are as follows:

$$\delta''' U^1 = \delta''' \ddot{r} , \quad \delta''' U^2 = \delta''' \ddot{\varphi} ;$$

in view of constraint (3.2) they are related by the formula

$$\delta''' \ddot{\varphi} = -\frac{\ddot{r} - r\dot{\varphi}^2}{r(r\ddot{\varphi} + 2\dot{r}\dot{\varphi})} \delta''' \ddot{r} .$$

Hence principle (6.2) can be put in the form

$$\left(mU_1 - P_1 - \frac{\ddot{r} - r\dot{\varphi}^2}{r(r\ddot{\varphi} + 2\dot{r}\dot{\varphi})} (mU_2 - P_2) \right) \delta''' \ddot{r} = 0 . \qquad (6.4)$$

The symbols

$$\Gamma^2_{12} = \Gamma^2_{21} = \frac{1}{r} , \quad \Gamma^1_{22} = -r$$

are the only nonzero Christoffel symbols, and hence (6.3) can be written as

$$U_1 = \dddot{r} - 3\dot{r}\dot{\varphi}^2 - 3r\dot{\varphi}\ddot{\varphi} , \quad U_2 = 3r\ddot{r}\dot{\varphi} + 3r\dot{r}\ddot{\varphi} + r^2\dddot{\varphi} - r^2\dot{\varphi}^3 ,$$

$$P_1 = \frac{2\mu m\dot{r}}{r^3} , \quad P_2 = -\frac{\mu m\dot{\varphi}}{r} .$$

Now from the form (6.4) of the principle and since the variations $\delta''' \ddot{r}$ are arbitrary,
we get

$$\dddot{r} - 3r\dot{\varphi}\ddot{\varphi} - 3\dot{r}\dot{\varphi}^2 - \frac{2\mu\dot{r}}{r^3} -$$

$$-\frac{(\ddot{r} - r\dot{\varphi}^2)\left(3r\ddot{r}\dot{\varphi} + 3r\dot{r}\ddot{\varphi} + \frac{\mu\dot{\varphi}}{r} + r^2\dddot{\varphi} - r^2\dot{\varphi}^3 \right)}{r(2\dot{r}\dot{\varphi} + r\ddot{\varphi})} = 0 . \qquad (6.5)$$

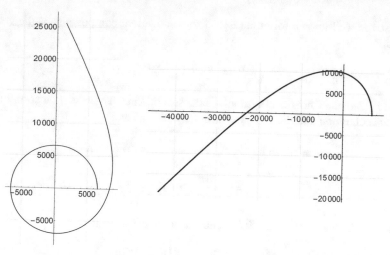

Fig. 9 Motion of Kosmos and Molniya-type spacecrafts after the value of its acceleration is fixed at perigee via the second theory of motion

Solving the system of equations (3.2) and (6.5) for $\ddot{r}$ and $\dddot{\varphi}$, we have

$$\ddot{r} = \frac{\dot{r}\left(2\mu + 3r^3\dot{\varphi}^2\right) - \frac{\mu(r\dot{\varphi}^2 - \ddot{r})\left(2\dot{r}(2r\dot{\varphi}^2 - \ddot{r}) + r^2\dot{\varphi}\ddot{\varphi}\right)}{\left(\ddot{r} - r\dot{\varphi}^2\right)^2 + 4\dot{r}^2\dot{\varphi}^2 + 4r\dot{r}\dot{\varphi}\ddot{\varphi} + r^2\ddot{\varphi}^2} + 3r^4\dot{\varphi}\ddot{\varphi}}{r^3},$$

$$\dddot{\varphi} = \frac{r^3\dot{\varphi}\left(r\dot{\varphi}^2 - 3\ddot{r}\right) - 3r^3\dot{r}\ddot{\varphi} - \frac{\mu(r\dot{\varphi}^2 - \ddot{r})\left(\dot{\varphi}(r(r\dot{\varphi}^2 - \ddot{r}) - 4\dot{r}^2) - 2r\dot{r}\ddot{\varphi}\right)}{\left(\ddot{r} - r\dot{\varphi}^2\right)^2 + 4\dot{r}^2\dot{\varphi}^2 + 4r\dot{r}\dot{\varphi}\ddot{\varphi} + r^2\ddot{\varphi}^2}}{r^4}. \tag{6.6}$$

Now the solutions of the above problems are obtained by integrating the system of differential equations (6.6) with the initial data (3.10) and (3.11).

Figures 9 show the trajectories of motion of Kosmos and Molniya-type spacecrafts obtained from the differential equations via the second theory for the case when the acceleration is fixed at perigee. These trajectories show that, once the spacecraft acceleration becomes fixed, the spacecraft first rotates about the Earth and moves asymptotically to the Earth along a straight line.

We now examine the motion of a Tundra-type spacecraft in the case its acceleration w_α is fixed at apogee.[25] In this case, the system of differential equations (6.6) should be integrated with the initial data (3.12). The resulting motion trajectory of a Tundra-type spacecraft is shown in Fig. 10. Note that this trajectory differs qualitatively from those shown in Fig. 9 (a curve knee appears).

Let us discuss this phenomenon. Figure 9 shows the trajectories of the spacecrafts after their accelerations are fixed at perigees of the trajectories. After these points, the

---

[25] See the paper: *V.V. Dodonov*, Motion of an artificial Earth satellite after the value of acceleration is fixed at apogee // Proc. Seminar "Computer Methods in Continuum Mechanics" 2019-2020, St. Petersburg: St. Petersburg Univ. Press. 2020. Pp. 51–61.

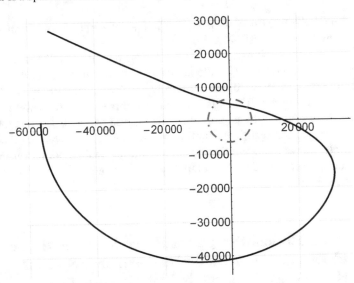

Fig. 10 Trajectory of a Tundra-type spacecraft after the value of its acceleration is fixed at apogee via the second theory of motion

spacecraft starts to move away from the attracting center. In turn, once the acceleration magnitude is fixed at apogee, the spacecraft starts to move toward the attracting center; the before mentioned knee of the trajectory is manifested once the zone of active attraction of this center is entered. Later, when moving away from the attracting center, the spacecraft starts to rapidly move along a straight line. Here another point is also pointing out. The above arguments pertain to a point source of attraction. If the same force is created by the Earth model in the form of a homogeneous ball of radius $R_3 = 6371$ km, then the by Fig. 10 the spacecraft will hit the Earth surface. We recall that in the Figure the Earth is shown by the dash-dot circle.

Discussion of the results. There is a principal difference of the trajectories of spacecrafts after their accelerations are fixed at perigee, as obtained by two theories of motion for nonholonomic systems with high-order constraints (the solutions of the generalized Chebyshev problems via two methods).

From the point of the mechanics of nonholonomic systems, the difference between the above solutions can be interpreted as follows. The first theory is based on transformations of the vector Newton equation in the case of ideal linear nonholonomic high-order constraints (in our example, transformation of equation (4.4)). To such motion equations, there corresponds the Mangeron-Deleanu principle,[26] which minimizes the absolute value of the constraint reaction. In turn, the second theory is based on the generalized Gauss principle, which minimizes the absolute value of the

---

[26] See, for example, the book: *Sh.Kh. Soltakhanov, M.P. Yushkov, S.A. Zegzhda*. Mechanics of non-holonomic systems. A New Class of control problems. Berlin Heidelberg: Springer-Verlag. 2009. 329 p.

corresponding derivative (the first derivative, in our setting) of the reaction vector of the imposed high-order constraints.

The following observations are worth pointing out. It is well known that the motion of a point with constant acceleration takes place either for a uniform motion in a circle or in the case of rectilinear uniformly accelerated motion. Elements of the first motion as the spacecraft travels between two concentric circles were obtained via the first theory, while the asymptotic convergence of the spacecraft motion to a uniformly accelerated rectilinear motion was obtained by the second theory. So, in our example the two different theories of motion for nonholonomic systems with high-order constraints successfully complement each other and in combination give the expected solutions.

7   Spacecraft Motion with Constant Accelerations via the Second Theory of Motion for Nonholonomic Systems with High-Order Constraints. The Dimensionless Differential Equations

Let us now use the generalized Gauss principle to compose the dimensionless differential equations of the above problem.[27] Differentiating the constraint equation (4.5) with respect to time, we get

$$\ddot{\boldsymbol{r}} \cdot \dddot{\boldsymbol{r}} = 0 . \tag{7.1}$$

Minimizing the function

$$Z_{(1)} = \frac{m}{2} \left| \dddot{\boldsymbol{r}} + \frac{\dot{\boldsymbol{r}}}{r^3} - \frac{3\dot{r}\boldsymbol{r}}{r^4} \right|^2$$

over $\dddot{\boldsymbol{r}}$ admissible in equations (7.1), we arrive at the equation

$$\dddot{\boldsymbol{r}} = -\frac{\dot{\boldsymbol{r}}}{r^3} + \frac{3\dot{r}}{r^4} \boldsymbol{r} + \Lambda^* \ddot{\boldsymbol{r}} , \tag{7.2}$$

where Λ^* is the sought-for Lagrange multiplier. From (4.5), (4.3) we have

$$\Lambda^* = \frac{x_0^4}{r^3} \dot{\boldsymbol{r}} \cdot \ddot{\boldsymbol{r}} - \frac{3x_0^4 \dot{r}}{r^4} \boldsymbol{r} \cdot \ddot{\boldsymbol{r}} .$$

Numerical integration of equation (7.2) with the above Λ^* was carried out in Cartesian coordinates. The initial data (4.4) were augmented with the initial data for

[27] See the book: *Sh.Kh. Soltakhanov, M.P. Yushkov, S.A. Zegzhda*. Mechanics of nonholonomic systems. A New Class of control problems. Berlin Heidelberg: Springer-Verlag. 2009. 329 p.

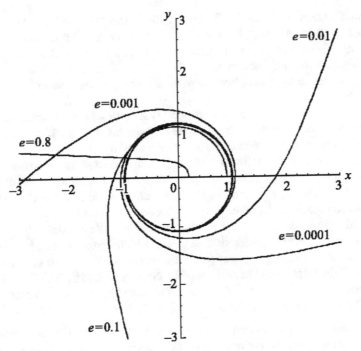

Fig. 11 Spacecraft motion with constant modulus of acceleration (the dimensionless differential equations)

accelerations;

$$\ddot{x}(0) = \ddot{x}_0 = -1/x_0^2, \qquad \ddot{y}(0) = \ddot{y}_0 = 0.$$

Calculations show that the trajectory goes off to infinity even for very small eccentricity (and independently of x_0). This trajectory asymptotically approaches the motion along a straight line with constant acceleration. Figure 11 shows that with increasing e the convergence to the rectilinear motion becomes faster. All the curves correspond to the case $x_0 = 1 - e$. An interesting feature of this solution is manifested in the reach of the rectilinear motion with constant acceleration after approximately three revolutions with $e \approx 4 \cdot 10^{-6}$, when the motion until the constraint was imposed satisfied this constraint with high accuracy. Let us try to understand why this happens.

The principle formulated by Gauss with the lack of active forces and constraints results in a motion with zero acceleration W, which is in conformity with the first Newton law. Note that from this principle the dynamics equations can be obtained.

The generalized Gauss principle, as applied to the case when the active forces and constraints are absent, results in a motion with nonzero acceleration W, but rather with zero nth order derivative of the vector W with respect to time, where n is the order of the principle. Therefore, for $n = 1$ the generalized Gauss principle with absence of active forces and constraints will lead to a rectilinear uniformly accelerated motion. The spacecraft tries to achieve this "natural" (for this principle) motion even for $e \approx 4 \cdot 10^{-6}$. It is clear that this satellite (spacecraft) motion problem with constant in modulo acceleration may have a solution for which it asymptotically approaches a rectilinear motion with this constant acceleration. We see that this solution is obtained by the application of the generalized first-order Gauss principle to this problem.

Let us compare once more these two theories of motion for nonholonomic systems with high-order constraints on an example of the motion of satellites (spacecrafts) when their accelerations are fixed (at some instant of time).

With the use of the first theory, we obtained in Sect. 4 the spacecraft motion between two concentric circles, which more and more approach each other with decreasing eccentricity e of the original elliptical orbit (this is clearly manifested in Sect. 3 on an example of a Kosmos-type spacecraft with eccentricity $e = 0.004632$). So, this theory establishes, for the involved problem under consideration, elements of motion with constant acceleration along a circle. In contrast, the analysis of this subsection shows that the application of the second theory (based on the generalized Gauss principle) results in an asymptotic convergence to a rectilinear uniformly accelerated motion. Thus, in the above problem these two theories complement each other and give the expected solutions.

The generalized Gauss principle proves especially successful in the transfer problem of a mechanical system over a fixed time interval from one phase state to another phase state. This approach was found to be instrumental in designing a new efficient method for the solution of the above problem, which is very important in control theory. In the next part of the present chapter, this method will be presented and developed.

II) Nonholonomic Mechanics and Control Theory

The second part of the present chapter establishes a link between two quite different topics in mechanics—the nonholonomic mechanics and control theory. It will be shown that based on the solution of the generalized Chebyshev problem one can construct a new method for finding an optimal control force that transfers a mechanical system over a given time interval from one phase state to a different one.

8 Formulation of One of the Most Important Problems of Control Theory

Let us consider one of the principal problems of control theory in which it is required to find an optimal control force that transforms in a given amount of time a mechanical

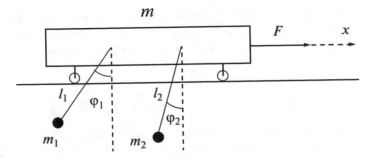

Fig. 12 Cart with pendula

system with the finite number of degrees of freedom from one phase state with specific generalized coordinates and velocities to a new phase state with given generalized coordinates and velocities. As a model example, we shall try to find a control force F that moves a cart of mass m horizontally along the x-axis in a given time $\widetilde{T}$ to a distance S (this problem was formulated above in Sect. 5 of Chap. 5). The cart accommodates the axes of s mathematical pendula of masses m_σ and lengths l_σ, $\sigma = \overline{1, s}$. In Fig. 12 (which is identical to Fig. 2 of Chap. 5) we show a cart with two pendula. It is required to transfer this system from the initial state of rest to a new state of rest (this is usually referred to as the problem of oscillation suppression).

The system has $s + 1$ degrees of freedom. As in Sect. 5 of Chap. 5, to describe its motion we introduce the coordinate x, which characterizes the horizontal displacement of the cart and the pendula rotation angles φ_σ, $\sigma = \overline{1, s}$. The kinetic and potential energies of the system in the case of small oscillations have the form (g is the acceleration of gravity)

$$2T = m\dot{x}^2 + \sum_{\sigma=1}^{s} m_\sigma(\dot{x} - l_\sigma\dot{\varphi}_\sigma)^2, \quad 2\Pi = g\sum_{\sigma=1}^{s} l_\sigma m_\sigma \varphi_\sigma^2.$$

Hence the Lagrange equations of the second kind read as

$$M\ddot{x} - \sum_{\sigma=1}^{s} m_\sigma l_\sigma \ddot{\varphi}_\sigma = F, \quad M = m + \sum_{\sigma=1}^{s} m_\sigma,$$

$$\ddot{x} - l_\sigma \ddot{\varphi}_\sigma = g\varphi_\sigma, \quad \sigma = \overline{1, s}. \tag{8.1}$$

It is clear that here the first equation expresses the motion law of the center of mass of the entire system:

$$M\ddot{x}_c = F, \quad x_c = x - \frac{\sum_{\sigma=1}^{s} m_\sigma l_\sigma \varphi_\sigma}{M}.$$

To suppress the oscillations of the above mechanical system (that is, the system should be at rest initially and the oscillations should vanish as the system stops), the control force should obey the following boundary-value conditions:

$$x(0) = \dot{x}(0) = \dot{x}(\widetilde{T}) = 0, \quad x(\widetilde{T}) = S,$$
$$\varphi_\sigma(0) = \varphi_\sigma(\widetilde{T}) = 0, \quad \dot{\varphi}_\sigma(0) = \dot{\varphi}_\sigma(\widetilde{T}) = 0, \quad \sigma = \overline{1,s}. \tag{8.2}$$

For further investigations, it is convenient to write system (8.1) in the principal coordinates. The mechanical system under consideration has zero frequency $\Omega_0 = 0$ and s nonzero natural frequencies Ω_σ, $\sigma = \overline{1,s}$. Using the natural modes of oscillations corresponding to these frequencies, we introduce the principal dimensionless coordinates x_σ, $\sigma = \overline{1,s}$, as linear combinations of the angles φ_σ, $\sigma = \overline{1,s}$. Changing to the dimensionless time $\tau = \Omega_1 t$ and introducing the $(s+1)$st dimensionless principal coordinate x_0, which is proportional to the displacement of the center of mass of the mechanical system, we get

$$\begin{cases} x_0'' = u, \\ x_\sigma'' + \omega_\sigma^2 x_\sigma = u, \quad \sigma = \overline{1,s}. \end{cases} \tag{8.3}$$

Here u is the control proportional to the force F; the primes mean the derivative with respect to the dimensionless time τ, $\omega_\sigma = \Omega_\sigma/\Omega_1$, $\sigma = \overline{1,s}$. The right-hand sides of the equations contain the same dimensionless control u; this can be easily achieved by scaling the principal coordinates. This system of differential equations (8.3), which satisfies requirements (8.2), will be solved with the following boundary conditions:

$$x_0(0) = x_0'(0) = 0, \quad x_\sigma(0) = x_\sigma'(0) = 0, \quad T = \Omega_1 \widetilde{T},$$
$$x_0(T) = a \equiv \frac{S}{l_1}, \quad x_0'(T) = 0, \quad x_\sigma(T) = x_\sigma'(T) = 0,$$
$$\sigma = \overline{1,s}. \tag{8.4}$$

System (8.3) has $(s+1)$ differential equations. From it one needs to find the unknown functions x_0, x_σ, $\sigma = \overline{1,s}$. But in system (8.3) the function $u(t)$ is also undefined. Hence, for the solution of problem (8.3)–(8.4), another condition is required. This condition should express the same principle that underlies the choice of the control $u(\tau)$ (the control force $F(t)$) from the entire set of possible controls for which the problem is solvable. Usually[28] in such problems the control is chosen to minimize the functional

---

[28] See, for example, the book: *F.L. Chernous'ko, L.D. Akulenko, and B.N. Sokolov*. Control of Oscillations. Moscow: Nauka. 1980. 384 p. [in Russian].

$$J = \int_0^T u^2 d\tau. \tag{8.5}$$

One of the most common classical methods of solution of the above optimal control problem (8.3)–(8.5) is the method based on the Pontryagin maximum principle, which was presented in Sect. 2 of Chap. 5.

9 Relation Between a Solution Obtained via the Pontryagin Maximum Principle and the Nonholonomic Problem[29]

According to Sect. 5 of Chap. 5, the representation of the dimensional differential equations (8.1) for the horizontal motion of a cart with s pendula in the form of dimensionless independent equations in the principal coordinates (8.3) (with given boundary conditions (8.4)) was proved very efficient. Because of this approach, in the problem of oscillation suppression of the mechanical system under consideration, with the use of the Pontryagin maximum principle with minimization of the functional (8.5), it proved possible to construct a dimensionless control in the following simple form (see (5.9) of Chap. 5):

$$u(\tau) = C_1 + C_2\tau + \sum_{\sigma=1}^{s} (C_{2\sigma+1} \cos \omega_\sigma \tau + C_{2\sigma+2} \sin \omega_\sigma \tau). \tag{9.1}$$

Using this formula, one can look at the solution obtained from the Pontryagin maximum principle from a new and interesting point of view, which is capable of linking two absolutely distinct branches of mechanics—the control theory and the nonholonomic mechanics.

To this end, we note, first of all, that control (9.1), as obtained from the Pontryagin maximum principle, can be looked upon as the solution of the differential equation

$$\frac{d^2}{d\tau^2}\left(\frac{d^2}{d\tau^2} + \omega_1^2\right)\left(\frac{d^2}{d\tau^2} + \omega_2^2\right) \cdots \left(\frac{d^2}{d\tau^2} + \omega_s^2\right) u = 0. \tag{9.2}$$

A particular case of solution (9.1) with $s = 2$ is the solution (5.12) of Chap. 5 corresponding to the particular case of equation (9.2) written in the form

$$\frac{d^2}{d\tau^2}\left(\frac{d^2}{d\tau^2} + \omega_1^2\right)\left(\frac{d^2}{d\tau^2} + \omega_2^2\right) u = 0. \tag{9.3}$$

Returning in (9.3) to the dimensional variables, we get

---

[29] A brief account of this and some subsequent sections can be found in the survey *M. Yushkov, S. Zegzhda, Sh. Soltakhanov, N. Naumova, T. Shugaylo*. A novel approach to suppression of oscillations // ZAMM. May 2018. Vol. 98. Issue 5. Pp. 781–788.

$$\frac{d^2}{dt^2}\left(\frac{d^2}{dt^2} + \Omega_1^2\right)\left(\frac{d^2}{dt^2} + \Omega_2^2\right) F = 0 . \tag{9.4}$$

Substituting in equation (9.4) the expression for F from the first equation of the original system (8.1) with $s = 2$, we get the eighth-order differential equation for the generalized coordinates x, φ_1 and φ_2:

$$a_{8,x}\frac{d^8 x}{dt^8} + a_{8,\varphi_1}\frac{d^8 \varphi_1}{dt^6} + a_{8,\varphi_2}\frac{d^8 \varphi_2}{dt^8} + a_{6,x}\frac{d^6 x}{dt^6} + a_{6,\varphi_1}\frac{d^6 \varphi_1}{dt^6} +$$
$$+ a_{6,\varphi_2}\frac{d^6 \varphi_2}{dt^6} + a_{4,x}\frac{d^4 x}{dt^4} + a_{4,\varphi_1}\frac{d^4 \varphi_1}{dt^4} + a_{4,\varphi_2}\frac{d^4 \varphi_2}{dt^4} = 0 . \tag{9.5}$$

The constant coefficients of this equation are related to the parameters of the mechanical system by the formulas

$$a_{8,x} = M + m_1 + m_2 , \quad a_{8,\varphi_1} = -m_1 l_1 , \quad a_{8,\varphi_2} = -m_2 l_2 ,$$
$$a_{6,x} = (\Omega_1^2 + \Omega_2^2)(M + m_1 + m_2) , \quad a_{6,\varphi_1} = -(\Omega_1^2 + \Omega_2^2) m_1 l_1 ,$$
$$a_{6,\varphi_2} = -(\Omega_1^2 + \Omega_2^2) m_2 l_2 , \quad a_{4,x} = \Omega_1^2 \Omega_2^2 (M + m_1 + m_2) ,$$
$$a_{4,\varphi_1} = -\Omega_1^2 \Omega_2^2 m_1 l_1 , \quad a_{4,\varphi_2} = -\Omega_1^2 \Omega_2^2 m_2 l_2 .$$

So, to the solution of the above problem with $s = 2$, as obtained from the Pontryagin maximum principle, there corresponds the solution of some nonholonomic problem with the eighth-order constraint (9.5).

Note that by specifying a different number s, from equation (9.2), which we write in the dimensional quantities, we get, instead of (9.5), the differential equation of order $2s + 4$ in the generalized coordinates, in which, as before, the constant coefficients will be evaluated in terms of the dimensional parameters of the mechanical system under consideration. This differential equation can again be looked upon as a nonholonomic constraint of order $2s + 4$ imposed on the motion of the mechanical system. In other words, if the mechanical system moves under the control found by the Pontryagin maximum principle, then in the course of this motion a nonholonomic constraint of order $2s + 4$ is continuously satisfied. Hence the presence of a high-order constraint, which follows from the minimization of the functional (8.5) by the Pontryagin maximum principle, enables one to consider the problem of the control force that suppresses the oscillations as some problem of nonholonomic mechanics with high-order constraints.

So, the solution of the boundary-value problem (8.3), (8.4) with minimization of the functional (8.5) by the Pontryagin maximum principle was found to be equivalent to the solution of the motion of a mechanical system under a nonholonomic constraint of order $2s + 4$. Hence one can pose a generalized Chebyshev problem, in which the motion program is specified in the form of a differential equation of order $2s + 4$. It seems reasonable to solve this generalized Chebyshev problem by invoking the theories of motion for nonholonomic systems with high-order constraints, which were developed in Part$I$) of the present chapter. According to these theories, given a constraint of order $2s + 4$, one can compose an equation of order $2s + 2$ with respect to the reaction of this constraint. So, if one considers the constraint of order $2s + 4$ as some motion program, which should be followed by the mechanical system, then the reaction of this constraint is a control force securing the fulfillment of this program. Therefore, in the general case the differential equation (9.2) of order $2s + 2$ with respect to the control can be interpreted as a differential equation with respect to the reaction of the constraint. However, if we continue using the theory of motion of nonholonomic systems with high-order constraints, then it is natural to invoke the variation principle of this theory in lieu of minimizing the functional (8.5) by the Pontryagin maximum principle. This is the generalized Gauss principle, which was presented in Sect. 5 of the present chapter. The next subsection is concerned with application of this principle to solving the above problem with $s = 2$.

10 Solution of the Problem via the Generalized Gauss Principle

Some general remarks. The system of equations (8.3) describes the controlled motion of a mechanical system that has zero natural frequency and s distinct nonzero natural frequencies. It is only necessary that the control force would actuate all natural modes of the oscillations. Of course, this is a fairly broad class of mechanical systems, which involves, as a classical example, the problem of a cart with pendula. System (8.3) is written in the dimensionless form, and hence is simple. Thanks to its simplicity, it was shown above by simple arguments that the minimality of the functional (8.5) via the Pontryagin maximum principle is achieved if the sought-for optima control has the form (9.1). It is plain that neither the generalized Gauss principle nor the Pontryagin maximum principle depends on the dimensional or the dimensionless form of the motion equations and on which of the coordinates (principal or standard) are used. Taking this into account, for simplicity of presentation we formulate the generalized Gauss principle in the context of the system of equations (8.3) taking the usual derivatives with respect to time t. We consider the problem of suppression of oscillations of a cart with two pendula, whose motion is subject to the eighth-order constraint (9.5).

Motion equation of a cart with two pendula ($s = 2$) in the principal coordinates. In view of the system of equations (8.1) with $s = 2$, the dimensional differential equations of motion of a cart with two pendula (see Fig. 12) have the form

$$M\ddot{x} - m_1 l_1 \ddot{\varphi}_1 - m_2 l_2 \ddot{\varphi}_2 = F, \quad M = m + m_1 + m_2,$$
$$\ddot{x} - l_1 \ddot{\varphi}_1 = g\varphi_1, \quad \ddot{x} - l_2 \ddot{\varphi}_2 = g\varphi_2. \tag{10.1}$$

We recall that here the first Lagrange equation of the second kind reflects the theorem of the motion of the center of mass,

$$M\ddot{x}_c = F, \quad x_c = x - \frac{\sum_{\sigma=1}^{2} m_\sigma l_\sigma \varphi_\sigma}{M}. \tag{10.2}$$

Hence the abscissa x_c of the center of mass can be taken as one of the principal dimensional coordinates of the mechanical system under consideration.

Expressing $\ddot{x}$ from the first equation of system (10.1) and substituting it into the second and third equations, we get

$$l_1\ddot{\varphi}_1 + g\varphi_1 = \frac{F}{M} + \frac{m_1 l_1 \ddot{\varphi}_1}{M} + \frac{m_2 l_2 \ddot{\varphi}_2}{M},$$

$$l_2\ddot{\varphi}_2 + g\varphi_2 = \frac{F}{M} + \frac{m_1 l_1 \ddot{\varphi}_1}{M} + \frac{m_2 l_2 \ddot{\varphi}_2}{M}. \tag{10.3}$$

Introducing the parameters

$$\alpha = \frac{l_2}{l_1}, \quad \beta = \frac{m_1}{M}, \quad \gamma = \frac{m_2}{M}, \quad k^2 = \frac{g}{l_1},$$

we write system (10.3) in the form

$$(1-\beta)\ddot{\varphi}_1 - \gamma\alpha\ddot{\varphi}_2 + k^2\varphi_1 = \frac{F}{Ml_1},$$

$$-\beta\ddot{\varphi}_1 + \alpha(1-\gamma)\ddot{\varphi}_2 + k^2\varphi_2 = \frac{F}{Ml_1}. \tag{10.4}$$

Now system (10.4) can be looked upon as a system of two differential equations for the unknowns φ_1, φ_2, which is better to write in the principal coordinates.

According to the theory of Chap. 7 of Vol. I, we put $F = 0$ in system (10.4) and try to find its partial solutions in the form

$$\varphi_\sigma = B_\sigma \sin(\Omega t + \delta), \quad \sigma = 1, 2,$$

where Ω is the sought-for dimensional natural frequency. From system (10.4) it follows that the constants B_1 and B_2 should satisfy the equations

$$(k^2 - (1-\beta)\Omega^2)B_1 + \gamma\alpha\Omega^2 B_2 = 0,$$

$$\beta\Omega^2 B_1 + (k^2 - \alpha(1-\gamma)\Omega^2)B_2 = 0. \tag{10.5}$$

Equating to zero the determinant of this system and setting $\Omega^2 = k^2\lambda^2$, we get the following equation for λ^2:

$$\alpha(1-\beta-\gamma)\lambda^4 - (1+\alpha-\beta-\alpha\gamma)\lambda^2 + 1 = 0. \tag{10.6}$$

Therefore, the sought-for natural frequencies (eigenfrequencies) are as follows:

$$\Omega_\nu^2 = k^2 \lambda_\nu^2, \quad \nu = 1, 2,$$

$$\lambda_{1,2}^2 = \frac{1 + \alpha - \beta - \alpha\gamma \mp \sqrt{(1 + \alpha - \beta - \alpha\gamma)^2 - 4\alpha(1 - \beta - \gamma)}}{2\alpha(1 - \beta - \gamma)}. \tag{10.7}$$

If λ_ν^2 assumes values (10.7), then Eq. (10.5) are linearly dependent; we discard the last equation. For each λ_ν^2, $\nu = 1, 2$, the arbitrary constants $B_{\nu\sigma}$, $\sigma = 1, 2$, which follow from the first equation, are proportional to the cofactor $\Delta_{\nu\sigma}$, $\nu, \sigma = 1, 2$, of the elements of the last row of the characteristic determinant of system (10.5). Hence they can be found by the formulas

$$\Delta_{\nu 1} = -\alpha\gamma\lambda_\nu^2, \quad \Delta_{\nu 2} = 1 - (1 - \beta)\lambda_\nu^2, \quad \nu = 1, 2. \tag{10.8}$$

According to the general theory of small oscillations, the presence of eigenvectors in the form (10.8) allows one to relate the coordinates φ_1, φ_2 to the principal coordinates ξ_1, ξ_2 as follows:

$$\varphi_1 = \sum_{\nu=1}^{2} \Delta_{\nu 1}\xi_\nu, \quad \varphi_2 = \sum_{\nu=1}^{2} \Delta_{\nu 2}\xi_\nu. \tag{10.9}$$

Substituting (10.8) into the first equation of system (10.4), we get

$$(1 - \beta)\sum_{\nu=1}^{2} \Delta_{\nu 1}\ddot{\xi}_\nu - \alpha\gamma \sum_{\nu=1}^{2} \Delta_{\nu 2}\ddot{\xi}_\nu + k^2 \sum_{\nu=1}^{2} \Delta_{\nu 1}\xi_\nu = \frac{F}{Ml_1}.$$

Hence using (10.7) and (10.8), we find that

$$-\alpha\gamma \sum_{\nu=1}^{2} (\ddot{\xi}_\nu + \Omega_\nu^2\xi_\nu) = \frac{F}{Ml_1}. \tag{10.10}$$

Let us now consider the second equation of system (10.4). In order to derive from it an equation similar in its structure to equation (10.10), one should take into account that the frequency equation (10.6) implies the following relations:

$$(1 - (1 - \beta)\lambda_\nu^2)(1 - \alpha(1 - \gamma)\lambda_\nu^2) = \alpha\beta\gamma\lambda_\nu^4, \quad \nu = 1, 2.$$

Hence we can write

$$\Delta_{\nu 2} = 1 - (1 - \beta)\lambda_\nu^2 = \frac{\alpha\beta\gamma\lambda_\nu^4}{1 - \alpha(1 - \gamma)\lambda_\nu^2}, \quad \nu = 1, 2. \tag{10.11}$$

Using (10.11) and substituting (10.9) into the second equation of (10.4), we get

$$\sum_{\nu=1}^{2} \frac{\alpha\beta\gamma\lambda_\nu^2}{1 - \alpha(1 - \gamma)\lambda_\nu^2} (\ddot{\xi}_\nu + \Omega_\nu^2\xi_\nu) = \frac{F}{Ml_1} . \tag{10.12}$$

Considering now Eqs. (10.10) and (10.12) as a system of two linear algebraic equations for the unknowns $y = \ddot{\xi}_1 + \Omega_1^2\xi_1$, $z = \ddot{\xi}_2 + \Omega_2^2\xi_2$, we get

$$\ddot{\xi}_1 + \Omega_1^2\xi_1 = \frac{F(1 + a_2)}{\alpha\gamma Ml_1(a_1 - a_2)} ,$$
$$\ddot{\xi}_2 + \Omega_2^2\xi_2 = \frac{F(1 + a_1)}{\alpha\gamma Ml_1(a_2 - a_1)} , \tag{10.13}$$

where

$$a_\nu = \frac{\beta\lambda_\nu^2}{1 - \alpha(1 - \gamma)\lambda_\nu^2} , \quad \nu = 1, 2 .$$

Now equation (10.2) and system (10.13) are sought-for equations in the principal dimensional coordinates x_c, ξ_1 and ξ_2. It is reasonable to change in these relations to the dimensionless variable $x_0 = x_c/l_1$, the dimensionless time $\tau = \Omega_1 t$, and put

$$x_1 = \frac{\alpha\gamma(a_1 - a_2)}{1 + a_2} \xi_1 , \quad x_2 = \frac{\alpha\gamma(a_2 - a_1)}{1 + a_1} \xi_2 .$$

As a result, we have

$$\begin{cases} x_0'' = u , \\ x_\sigma'' + \omega_\sigma^2 x_\sigma = u , \quad \sigma = 1, 2 . \end{cases} \tag{10.14}$$

Here the prime denotes the derivative with respect to the dimensionless time τ, and

$$u = \frac{F}{Mg\lambda_1^2} , \quad \omega_\sigma = \frac{\Omega_\sigma}{\Omega_1} , \quad \sigma = 1, 2 .$$

So, the system of differential equations (10.14) in the dimensionless principal coordinates is a particular case of the system of equations (8.3) with $s = 2$.

System (10.14) should be integrated with the boundary conditions

$$x_0(0) = x_0'(0) = 0 , \quad x_\sigma(0) = x_\sigma'(0) = 0 , \quad \sigma = 1, 2 , \quad T = \Omega_1\tilde{T} ,$$
$$x_0(T) = a \equiv \frac{S}{l_1} , \quad x_0'(T) = 0 , \quad x_\sigma(T) = x_\sigma'(T) = 0 . \tag{10.15}$$

In Chap. 5 the Pontryagin maximum principle was applied to find the dimensionless control u (see formula (9.1) above). Substituting control (9.1) into the Duhamel integrals (5.10) of Chap. 5 (we recall that $s = 2$), from the satisfaction of the boundary conditions (10.15) for $\tau = T$, we get six linear algebraic inhomogeneous equations for the coefficients C_σ, $\sigma = \overline{1, 6}$.

Determination of the control via the generalized Gauss principle. We have already seen in Chap. 6 of Vol. I that by using the concept of the tangent space to the manifold of all positions of a mechanical system which it can have at a given time instant, the Lagrange equations of the second kind can be written in the form of a single vector equation

$$MW = Y + R, \tag{10.16}$$

where

$$MW = \sum_{\sigma, \tau = 1}^{3} a_{\sigma\tau} \ddot{q}^\tau \, e^\sigma, \quad Y = - \sum_{\sigma, \tau = 1}^{3} c_{\sigma\tau} q^\tau \, e^\sigma, \quad R = \sum_{\sigma = 1}^{3} R_\sigma \, e^\sigma,$$

$$q^1 = \varphi_1, \quad q^2 = \varphi_2, \quad q^3 = x,$$

in the case of a cart with two pendula, and e^σ, $\sigma = \overline{1, 3}$, are vectors of the reciprocal basis introduced in the tangent space. Note that the sought-for control u satisfying the motion program given in the form of the differential equation (9.5) can be considered as the reaction of the linear nonholonomic eighth-order constraint. Hence in equation (10.16) the vector corresponding to the control u is denoted by the letter R, which is usually used in nonholonomic mechanics for denoting the reaction vector of the constraint. It is worth mentioning that in the context of the given problem of controlled motion it can be written in the form

$$R = u(t)\,b, \quad b = \sum_{\sigma = 1}^{3} b_\sigma \, e^\sigma,$$

where the coefficients (b_1, b_2, b_3) depend on the technical implementation of a specific unit creating the real control force.

The generalized Gauss principle with the presence of an eighth-order constraint asserts that

$$\delta^{(8)} \left(M \frac{d^6 W}{dt^6} - \frac{d^6 Y}{dt^6} \right)^2 = 0. \tag{10.17}$$

Here the symbol $\delta^{(8)}$ means that only the eighth derivatives of the generalized coordinates are varied. According to principle (10.17), the linear eighth-order constraint (9.5) is ideal (we can say that in this case the sought-for control can also be called ideal) if its "reaction" $\vec{\mathfrak{R}} \equiv R_{(6)}$ is minimal; that is, if

$$\left(\overrightarrow{\Re}\right)^2 \equiv \left(\mathbf{R}_{(6)}\right)^2 = \left(M\frac{d^6\mathbf{W}}{dt^6} - \frac{d^6\mathbf{Y}}{dt^6}\right)^2 \tag{10.18}$$

is minimized.

From all possible linear nonholonomic eighth-order constraints, we single out a subset such that the quantity $\left(\overrightarrow{\Re}\right)^2 \equiv \left(\mathbf{R}_{(6)}\right)^2$ is equal to its zero infimum on the elements of this set. From (10.18) it follows that to all these elements there correspond a single equation

$$\frac{d^6u}{dt^6} = 0$$

with general solution in the form

$$u(t) = \sum_{k=1}^{6} C_k t^{k-1}. \tag{10.19}$$

As distinct from the control given by formula (5.12) of Chap. 5, the control sought in the form of the polynomial (10.19) will not have oscillations corresponding to the natural frequencies of the system. The function thus found will be sufficiently smooth, which is a good advantage.[30]

Analysis of numerical calculations. We perform three calculations for the above mechanical system corresponding to three motion times (5.11) of Chap. 5. The values of the arbitrary constants in control (10.19) with each concrete value of the dimensionless motion time T can be found in the same way as it was done in the determination of the arbitrary constants for the solution obtained earlier with the use of the Pontryagin maximum principle. As a result, for the control (10.19) obtained via the generalized Gauss principle, we get three values:

---

[30] Note that using a different approach a control in a polynomial form was obtained by *G. V. Kostin, V. V. Saurin*, Integro-differential approach to solving problems of linear elasticity theory // Doklady Physics. 2005. Vol. 50. No. 10. Pp. 535–538; and *G. V. Kostin, V. V. Saurin*. Modeling and optimization of elastic system motions by the method of integro-differential relations // Doklady Mathematics. 2006. Vol. 73. No. 3. Pp. 469–472.

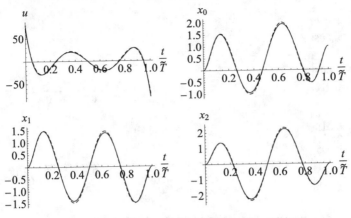

Fig. 13 Short-time motion of the mechanical system, $T = T_2$, $T_2 = 0.5\,T_1$

$$T = T_2\ :$$
$$C_1 = 78.876\,,\quad C_2 = -693.61\,,\quad C_3 = 1492.85\,,$$
$$C_4 = -1248.86\,,\quad C_5 = 445.03\,,\quad C_6 = -56.663\,;$$
$$T = 8\,T_2\ :$$
$$C_1 = 0.00002\,,\quad C_2 = 0.00254\,,\quad C_3 = 0.00004\,,$$
$$C_4 = -0.00005\,,\quad C_5 = 0\,,\quad C_6 = 0\,;$$
$$T = 16\,T_2\ :$$
$$C_1 = 0\,,\quad C_2 = 0.00034\,,\quad C_3 = 0\,,$$
$$C_4 = 0\,,\quad C_5 = 0\,,\quad C_6 = 0\,.$$

In Figs. 13, 14, and 15 we show graphically the results of calculations obtained via the above two different principles. As in Chap. 5 it is assumed that $T_2 = 0.5\,T_1$ and $\omega_1 = 1$. The solutions obtained via the Pontryagin maximum principle (see Figs. 3–4 of Chap. 5) are shown by dashed curves, and the solutions obtained via the generalized Gauss principle are shown by solid curves.

A comparison of these cases of motion shows that for a short-time motion (if the cart travel time is close to the period of the second form of oscillations, see Fig. 13) the solutions obtained via these two methods are practically equal (this can be looked upon as an indication of the good quality of the new calculation method proposed here), but they become differ substantially with increased motion time. This difference can be explained by the fact that the control obtained via the Pontryagin maximum principle contains, as was noted above, harmonics with natural frequencies of the system, which results in the resonance of the system. At the same time, the control obtained via the generalized Gauss principle is given by a polynomial, which is responsible for a relatively smooth motion of the system.

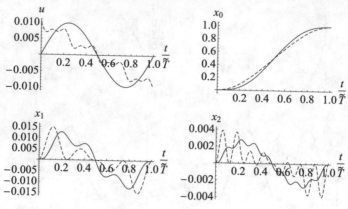

Fig. 14 Short-time motion of the mechanical system, $T = 8\,T_2$, $T_2 = 0.5\,T_1$

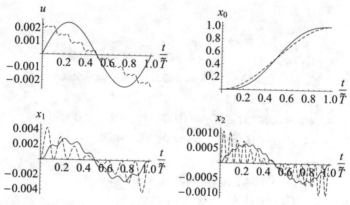

Fig. 15 Long-time motion of the mechanical system, $T = 16\,T_2$, $T_2 = 0.5\,T_1$

Another fact is also worth mentioning: the application of the Pontryagin maximum principle always produces jumps of the control force at the beginning and end of the motion. If, however, the generalized Gauss principle is used, then such jumps disappear in long-time motion. This leads to the question whether one can get rid of control jumps also for a short-time motion of the system. This question will be addressed in the next section.

11 The Extended (Generalized) Boundary-Value Problem

Our aim in this section is to get rid of jumps of the control force at the beginning and end of motion. These jumps appear in the application of the generalized Gauss principle when studying a short-time motion of a mechanical system (see Fig. 13).

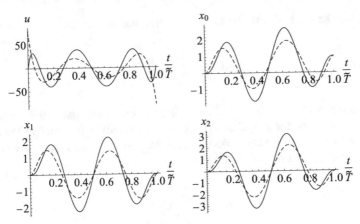

Fig. 16 Short-time motion without jumps of the control force control force, $T = T_2$, $T_2 = 0.5\,T_1$

To this aim, in addition to the boundary-value conditions (8.2), it suffices to require that the following conditions be also satisfied:

$$\dddot{x}(0) = 0, \quad \dddot{x}(\widetilde{T}) = 0.\qquad(11.1)$$

From Eq. (8.1) and the boundary conditions (8.2) it follows that if conditions (11.1) are satisfied at the beginning and end of the path, the accelerations of all points of the system will be zero. In the principal coordinates, the boundary conditions (8.4) are augmented with the boundary-value conditions

$$\dddot{x}_0(0) = 0, \quad \dddot{x}_0(T) = 0.\qquad(11.2)$$

Thus, Eq. (8.3) should be solved under the boundary conditions (8.4) and (11.2). The additional boundary conditions (11.2) distinguish the problem formulated in this subsection from the original classical boundary-value problem (8.3), (8.4). This is why problem (8.3), (8.4), and (11.2) will be called the *extended (generalized) boundary-value problem*.

To solve this problem, we shall use the generalized Gauss principle in which we increase its order by two in comparison with the principle employed in the previous subsection. So, we minimize

$$(\vec{\mathfrak{R}})^2 \equiv \left(\mathbf{R}_{(8)}\right)^2 = \left(M\frac{d^8\mathbf{W}}{dt^8} - \frac{d^8\mathbf{Y}}{dt^8}\right)^2,\qquad(11.3)$$

and in turn, the minimum of (11.3), which is zero, is achieved if the differential equation

$$\frac{d^8 u}{dt^8} = 0$$

is satisfied. The general solution of this equation has the form

$$u(t) = \sum_{k=1}^{8} C_k t^{k-1}.$$ (11.4)

The arbitrary constants C_k, $k = \overline{1, 8}$, will be found from the satisfaction of the boundary conditions for $\tau = T$ and if $\ddot{x}_0(0) = 0$. Moreover, it is easily seen $C_1 = 0$, because in view of the first equation in (8.3) and the additional boundary condition $\ddot{x}_0(0) = 0$ we see that $u(0) = 0$. Hence instead of (11.4) we get

$$u(t) = \sum_{k=2}^{8} C_k t^{k-1}.$$

So, to find the available unknown coefficients it suffices only to satisfy the boundary conditions with $\tau = T$.

Integrating and solving the linear algebraic inhomogeneous system of seventh-order equations, we get the following numerical values:

$$T = T_2 :$$
$$C_1 = 0, \quad C_2 = 669.77, \quad C_3 = -4266.12,$$
$$C_4 = 8837.38, \quad C_5 = -8337.44, \quad C_6 = 3958.27,$$
$$C_7 = -922.054, \quad C_8 = 83.8568.$$ (11.5)

Evaluating the particular solution of the system of differential equations (8.3) for $s = 2$ corresponding to the zero initial conditions in the form of Duhamel integrals with values (11.5), we get the graphs shown in Fig. 16. The graph of the dimensionless control shows that we indeed succeeded in getting rid of the jumps in the control force at the beginning and end of motion of the system.

It is also interesting to compare the results obtained via the generalized Gauss principle for the extended boundary-value problem and for the original classical boundary-value problem. The results of such calculations for various relations between the parameters are shown in Figs. 17, 18, 19, 20 by solid curves for the generalized problems. The dashed curves also correspond to the generalized Gauss principle, but for the usual unextended boundary-value problem. Note that if the solution of the usual boundary-value problem gives nearly zero values of the control at the beginning and end of motion, then there is no need to extend it (see Fig. 20).

It is worth pointing out that it is impossible to solve the extended (generalized) boundary-value problem of this section using the Pontryagin maximum principle, because the control obtained via this principle will contain an insufficient number of unknown arbitrary constants for satisfaction of all relevant boundary conditions. Nevertheless, as we have already seen, this boundary-value problem is amenable to the generalized Gauss principle with order increased by two.

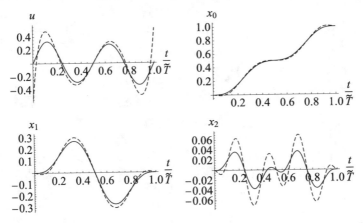

Fig. 17 Comparison of Gauss solutions corresponding to the extended and usual boundary-value problems with $T = 2T_2$, $T_2 = 0.25 T_1$

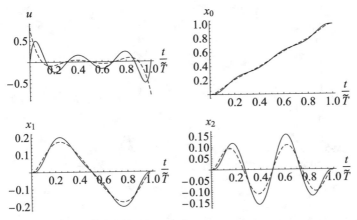

Fig. 18 Comparison of the Gauss solutions of the corresponding extended and usual boundary-value problems, for $T = 2T_2$, $T_2 = 0.5 T_1$

Thus, the principal qualitative difference of the new approach to the problem of damped oscillations is that it is capable of constructing a solution free from jumps in the control force both at the beginning and the end of motion of the mechanical system.

12 Singular Points of the Solution

As was pointed out several times, the application of the Pontryagin maximum principle to the study of the oscillation suppression problem always gives a control force

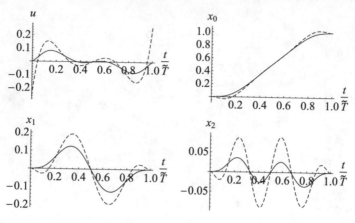

Fig. 19 Comparison of Gauss solutions corresponding to the extended and usual boundary-value problems, for $T = 3\,T_2$, $T_2 = 0.5\,T_1$

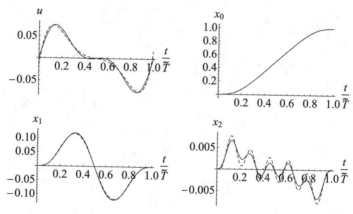

Fig. 20 Comparison of Gauss solutions corresponding to the extended and usual boundary-value problems, for $T = 6\,T_2$, $T_2 = 0.25\,T_1$

with jumps at the beginning and end of motion. However, the application (with the same aim of finding a control force) of the generalized Gauss principle produces such jumps only for short-time motion of the system. Such jumps decrease and then disappear as the motion time increases. These results are supported by the graphs in Figs. 13, 14 and 15. In Sect. 11, where the extended boundary-value problem was posed and solved, it proved possible to construct a control without jumps (Fig. 16, 17, 18, 19, and 20) via the generalized Gauss principle of increased order.

Thus, the generalized Gauss principle proves quite efficient. However, the formulation and solution of the extended boundary-value problem is not always useful. The thing is that according to calculations the results of motion of a mechanical system under a control obtained by solving a generalized boundary-value problem depend

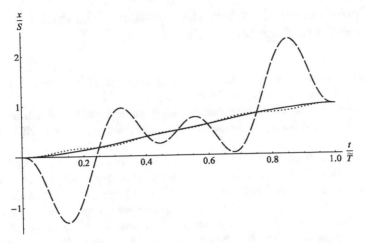

Fig. 21 Comparison of the Gauss solutions corresponding to various boundary-value problems, $T = 4.02\,\pi$, $\omega_2 = 1.734\,\omega_1$

substantially on the dimensionless parameter $K = T/T_1$. It turns out[31] that there exists a countable set of values of the parameter K such that intensive oscillations in the system develop as such values are approached. Such values of K were called *singular points* of solutions of extended boundary-value problems.

As an example, consider the problem of oscillation suppresion of a cart with two pendula with the following values of the dimensionless parameters:

$$l_2/l_1 = 1/5, \quad m_1/M = 2/3, \quad m_2/M = 1/12,$$
$$\omega_2 \equiv \Omega_2/\Omega_1 = 1.734, \quad T = 4.02\,\pi. \tag{12.1}$$

The motion of a cart, as measured in fractions of S and obtained via the generalized Gauss principle for ordinary boundary-value problem, is shown in Fig. 21 by a smooth solid line. At the same time, by solving the problem with the same parameters (12.1) for the extended boundary-value problem, we get the motion of the cart depicted in Fig. 21 by the dashed line. We see that even though the problem of oscillations suppression is solved, the system features intense large-size oscillations of the cart with pendula. Such unexpected behavior of the cart can be explained by the fact that to these values of parameters (12.1) there corresponds the value $K = 2.01$, which is close to the first singular point ($K = 2.01265$).

---

[31] See the paper: *S.A. Zegzhda, D.N. Gavrilov.* Oscillation suppression of a solid in its motion // Vestnik St. Petersburg University, Mathematics. 2012. Vyp. 3. Pp. 73–83.

13 Construction of an Analytic Solution of the Problem Free From Singular Points

We propose the following method of producing a control of a cart which is an analytic function of time for all values of the parameter K.

As before, we pose the extended boundary-value problem (which we call the *extended first-order boundary-value problem*. The resulting solution will be denoted by $u_1(\tau)$.

In parallel with this problem, we pose an even more involved extended problem (which will be called the *extended second-order boundary-value problem*), in which it is required in addition that at the beginning and end of motion of the cart the derivative of the acceleration with respect to time is also zero. The solution of this problem will be denoted by $u_2(\tau)$. For this new solution, there also exist singular values of the parameter K, but they are different from those corresponding to the previous solution. For any values of the parameter μ, the function

$$u(\tau) = u_1(\tau) + \mu(u_2(\tau) - u_1(\tau)) \tag{13.1}$$

is a solution to the above problem. In order to avoid evaluation of the singular values of the solutions u_1 and u_2 and to construct an analytic solution, which continuously depends on the parameter K, we should determine the parameter μ from the condition that the integral of the squared function $u(\tau)$ be minimal over the travel time T. This solution corresponds to the following value of the parameter μ

$$\mu = \frac{J_1 - J_2}{J_1 + J_3 - 2J_2},$$

where

$$J_1 = \int_0^T u_1^2(\tau)\,d\tau, \quad J_2 = \int_0^T u_1(\tau)u_2(\tau)\,d\tau, \quad J_3 = \int_0^T u_2^2(\tau)\,d\tau.$$

This approach will be used in calculations in the next section.

As a development of the method proposed above, one could also construct a solution $u_3(\tau)$ of the *extended boundary-value problem of third order*, in which there is an additional requirement that the fourth derivatives of the coordinates of the cart be zero at the beginning and end of the path. In analogy with formula (13.1), a control will be sought in the form

$$u(\tau) = u_2(\tau) + \mu(u_3(\tau) - u_2(\tau)), \tag{13.2}$$

where the new value of μ can again be found by minimizing the integral of the squared expression (13.2).

Remark. A control without singular points can be also obtained by applying instead of the generalized Gauss principle the generalized Hamilton-Ostrogradsky principle.[32] In this approach, the control will be constructed using the basis functions. In this case it is also polynomial in time, but the order of the polynomial will be equal to $4s + 3$.

14 Another Approach to the Problem of Oscillation Suppression of a Cart with Two Pendula

Motion equations with a different approach to the problem of oscillation suppression of a cart with two pendula. In the above problem of oscillation suppression of a cart with pendula, the sought-for control force is a horizontal force applied to the cart. In the above solution, it was necessary to find natural frequencies and natural modes of the oscillations to be able to change to the principal coordinates and write the system in the form (10.14).

We now propose a new simpler approach to the solution of the problem. In this approach, it is not anymore necessary to evaluate the natural frequencies and natural modes of the oscillations of this mechanical system. Note that these frequencies and modes depend substantially on the cart mass, whereas the problem of suppression of oscillations of the pendula depends only on how reasonably the cart travel law was chosen. Taking into account this important observation, we first start to search as a function of time not the force F, which is applied to the cart, but rather the acceleration $\ddot{x}$ of the cart under which, in a given amount of time $\widetilde{T}$, it will move in a given distance S assuming that at the beginning and end of motion both the velocities and accelerations of both the cart and the pendulum are zero.

With this formulation of the problem, instead of the system of differential equations (10.14), we naturally arrive at the following system of equations

$$\ddot{x} = U \,,$$
$$l_1 \ddot{\varphi}_1 + g\varphi_1 = U \,,$$
$$l_2 \ddot{\varphi}_2 + g\varphi_2 = U \,, \tag{14.1}$$

where U denotes the sought-for dimensional acceleration of the cart. It is interesting that the system thus obtained is a system of independent equations; that is, the displacement of the cart and the angles of rotation of the pendula are principal coordinates.

If we now introduce the dimensionless coordinates, the control, and the time by the formulas

---

[32] See the paper: *S.A. Zegzhda, P.E. Tovstik, M.P. Yushkov.* The Hamilton-Ostrogradsky generalized principle and its application for oscillations suppression // Dokl. Phys. 2012. Vol. 57. pp. 447–450.

$$\bar{x}_0 = \frac{x}{l_1}, \quad \bar{x}_1 = \varphi_1, \quad \bar{x}_2 = \alpha\varphi_2, \quad \bar{u} = \frac{\ddot{x}}{g}, \quad \bar{\tau} = \sqrt{\frac{g}{l_1}}\,t,$$

then the system of equations (14.1) assumes the form

$$\bar{x}_0'' = \bar{u},$$
$$\bar{x}_1'' + \bar{x}_1 = \bar{u},$$
$$\bar{x}_2'' + \frac{1}{\alpha}\bar{x}_2 = \bar{u}, \tag{14.2}$$

where the primes denote differentiation with respect to the dimensionless time $\bar{\tau}$, while $\bar{u}$ is the sought-for dimensionless acceleration of the cart in fractions of g, which is sought in the new approach.

System (14.2) has practically no differences from system (10.14). Hence one can invoke the above method based on the generalized Gauss principle to find the sought-for acceleration of the cart and the motion of the pendula. Consequently, having known the motion of the cart and of the pendula, one can easily determine the sought-for control force. The dimensional horizontal control force F_N, which is obtained by this new approach, will be evaluated by the formula

$$F_N = Mg\bar{u} - m_1 g\bar{x}_1'' - m_2 g\bar{x}_2'',$$

and the corresponding new dimensionless control u_N, as expressed in fractions of Mg, will assume the form

$$u_N \equiv \frac{F_N}{Mg} = \bar{u}\,(1 - \beta - \gamma) + \beta\bar{x}_1 + \frac{\gamma}{\alpha}\bar{x}_2.$$

Numerical evaluation of oscillation suppression. The results of numerical calculations for the motion of a cart with pendula, as obtained via the old and new approaches, were compared with the use of formula (13.1). The control force F and the angles of rotation of the pendula were calculated as functions of time. The force was expressed in fractions of Mg, where M is the mass of the entire system, the angles of rotation φ_1 and φ_2 were measured in degrees. As was mentioned above, these functions, as functions of the dimensionless time $\tau = t/\widetilde{T}$, depend substantially on the parameter K, which is equal to the ratio of the dimensionless time T to the period T_1 of the first mode of oscillations of the mechanical system. Of special interest is the case when the quantity K lies in the interval from one to two.[33] In view of this, for calculations we put $K=1.54$.

Since the sought-for functions are proportional to the displacement S of the cart in time $\widetilde{T}$, this displacement, as given in fractions of the length of the first, the longest, pendulum, was given so that the angles of rotation of the pendulum did not exceed 10

---

[33] See the reference in Sect. 12.

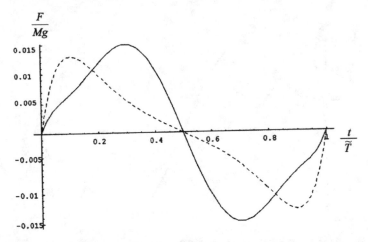

Fig. 22 Control force for the cart with one pendulum

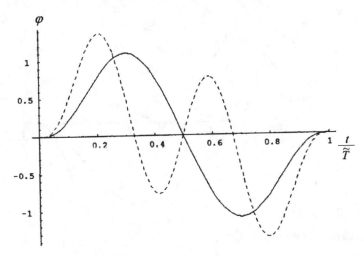

Fig. 23 Pendulum oscillations on the cart

degrees. In Figs. 22, 23, 24, 25, 26, and 27, the solid (dashed) lines correspond to the results obtained via the new (old) approach.

Figures 22 and 23 correspond to the case of a cart with one pendulum; the following parameters of the system were chosen:

$$\frac{m}{m_1} = \frac{1}{2}, \quad K = 1.54, \quad S = \frac{l}{5}.$$

Figures 24, 25, 26, and 27 correspond to the motion of a cart with two pendula. In this case, we have four independent parameters: K, the ratio of the mass of the cart m to the mass m_1 of the first pendulum and to the mass m_2 of the second pendulum, and

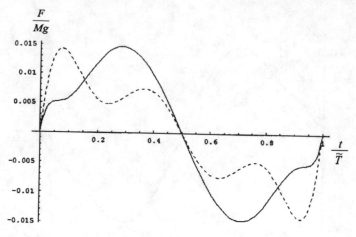

Fig. 24 Control force for the cart with two pendula (case $m_2/m_1 = 1/8$)

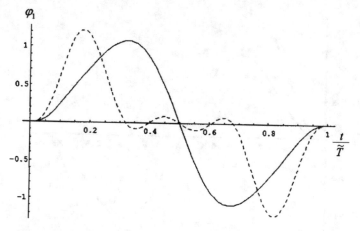

Fig. 25 Oscillations of the first pendulum

the ratio of the length l_2 of the second pendulum to that l_1 of the first one. Figures 24, 25, and 26 correspond to the parameters

$$\frac{l_2}{l_1} = \frac{1}{4}, \quad \frac{m_2}{m_1} = \frac{1}{8}, \quad \frac{m}{m_1} = \frac{1}{2}, \quad K = 1.54, \quad S = \frac{l_1}{5},$$

and for Fig. 27 we assume that

$$\frac{m_2}{m_1} = \frac{1}{64}.$$

Calculations show, as expected, that the greater the mass of the cart with respect to that of the pendula, the closer the results obtained via the first and second approaches.

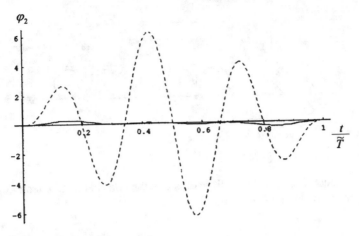

Fig. 26 Oscillations of the second pendulum

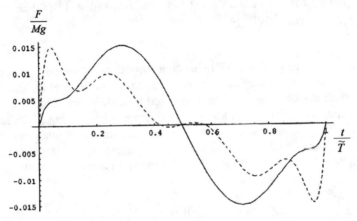

Fig. 27 Conrtol force for the cart with two pendula (case $m_2/m_1 = 1/64$)

Hence it was assumed that the mass of the cart is two times smaller than that of the first principal pendulum. The length of the second pendulum and its mass with respect to the first one were taken to be, respectively, four and eight times smaller. The displacement of the cart is $0.2 l_1$, where l_1 is the length of the main pendulum. It is seen that the new approach reduces the oscillations of the pendula both in amplitude and in frequency. A comparison of Figs. 24 and 22 shows that with the new approach an addition of the second pendulum has insignificant effect on the control force, whereas it changes substantially under the usual approach. The following result is also of special interest: assume that in the problem with two pendula the mass of the second one is 64 times (not 8 times) less than that of the first one. Comparing Figs. 27 and 24, we see that under the new approach this substantial decrease of the

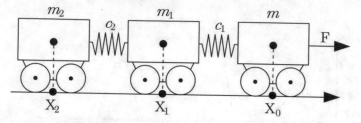

Fig. 28 A three-mass system with springs

mass of the second pendulum has little effect on the control force, whereas under the usual approach it changed substantially.

15 Suppression of Horizontal Oscillations of a Three-Mass System with Springs[34]

Formulation of the problem. The motion equations. Assume that we have a mechanical system consisting of three carts of masses m, m_1 and m_2 (see Fig. 28). The carts with masses m and m_1 are connected by a spring of stiffness c_1, and the carts with pendula of masses m_1 and m_2 are connected by a spring of stiffness c_2. It is required, in a given amount of time $\widetilde{T}$, to displace the mechanical from the state of rest through the distance S to the new state of rest by applying a control force F to the first cart. In this case, the conditions at the end-points of the path assume the form

$$X_0(0) = 0, \quad X_1(0) = 0, \quad X_2(0) = 0,$$
$$\dot{X}_0(0) = 0, \quad \dot{X}_1(0) = 0, \quad \dot{X}_2(0) = 0,$$
$$X_0(\widetilde{T}) = S, \quad X_1(\widetilde{T}) = S, \quad X_2(\widetilde{T}) = S,$$
$$\dot{X}_0(\widetilde{T}) = 0, \quad \dot{X}_1(\widetilde{T}) = 0, \quad \dot{X}_2(\widetilde{T}) = 0. \tag{15.1}$$

Here X_0, (X_1, and X_2 are, respectively, the displacements of the first, second, and third carts.

In the adopted generalized coordinates, the Lagrange equations of the second kind for our system have the form

---

[34] The present section is based on the paper: *K.M. Fazlyeva, T.S. Shugailo*. Control of the oscillation suppression for a three-mass system with horizontal motion // Proc. Seminar "Computer Methods in Continuum Mechanics" 2018-2019. St. Petersburg: St. Petersburg Univ. Press. 2019. Pp. 55–66. The notation of this paper are adopted in Sect. 15.

$$\begin{cases} m\ddot{X}_0 + c_1(X_0 - X_1) = F, \\ m_1\ddot{X}_1 + c_1(X_1 - X_0) + c_2(X_1 - X_2) = 0. \\ m_2\ddot{X}_2 + c_2(X_2 - X_1) = 0. \end{cases} \qquad (15.2)$$

Motion equations in principal coordinates. Using the theory of small oscillations, we rewrite the motion equation in normal coordinates, without singling out in advance the abscissa of the center of mass as one of the principal coordinates. For convenience, we rewrite system (15.2) in the matrix form

$$\mathbf{A}\ddot{\mathbf{X}} + \mathbf{C}\mathbf{X} = \mathbf{Y},$$

where

$$\mathbf{A} = \begin{pmatrix} m & 0 & 0 \\ 0 & m_1 & 0 \\ 0 & 0 & m_2 \end{pmatrix}, \quad \mathbf{C} = \begin{pmatrix} c_1 & -c_1 & 0 \\ -c_1 & c_1 + c_2 & -c_2 \\ 0 & -c_2 & c_2 \end{pmatrix},$$

$$\mathbf{X} = \begin{pmatrix} X_0 \\ X_1 \\ X_2 \end{pmatrix}, \quad \mathbf{Y} = \begin{pmatrix} F \\ 0 \\ 0 \end{pmatrix}.$$

We denote by Ω_i the natural frequencies of the system, and denote by $\mathbf{U}_i$ the natural modes of the oscillations. Note that Ω_i are roots of the characteristic equation

$$\det(\mathbf{C} - \Omega^2\mathbf{A}) = 0,$$

which gives the zero frequency $\Omega_0 = 0$ and two nonzero frequencies Ω_1 and Ω_2. In turn, the natural modes should be written as

$$\mathbf{U}_i = \begin{pmatrix} S \\ \dfrac{(c_1 - \Omega_i^2 m)S}{c_1} \\ \begin{vmatrix} c_1 - \Omega_i^2 m & -c_1 \\ -c_1 & c_1 + c_2 - \Omega_i^2 m_1 \end{vmatrix} \dfrac{S}{c_1 c_2} \end{pmatrix}.$$

In order to write the motion equations in principal dimensionless coordinates, we change the variables as

$$\mathbf{X} = \sum_{i=0}^{2} \mathbf{U}_i \frac{S^2 M}{\mathbf{U}_i^T \mathbf{A}\mathbf{U}_i} x_i. \qquad (15.3)$$

Now the motion equations (15.2) assume the form

$$\ddot{x}_i + \Omega_i^2 x_i = \frac{F}{SM}, \quad i = \overline{0, 2}, \quad M = m + m_1 + m_2.$$

We introduce the dimensionless time τ by

$$\tau = \Omega_1 t, \quad T = \Omega_1 \tilde{T}.$$

Now the dimensionless motion equations read as

$$\begin{cases} x_0'' = u, \\ x_1'' + x_1 = u, \\ x_2'' + \omega_2^2 x_2 = u. \end{cases} \tag{15.4}$$

Here $u = \dfrac{F}{SM\Omega_1^2}$ is the dimensionless control, $\omega_i^2 = \dfrac{\Omega_i^2}{\Omega_1^2}$, $i = \overline{0,2}$, $\Omega_0 = 0$. In the equations, the primes denote differentiation with respect to time τ.

The boundary conditions (15.1) in the dimensionless variables assume the form

$$\begin{aligned}
x_0(0) &= 0, & x_1(0) &= 0, & x_2(0) &= 0, \\
x_0'(0) &= 0, & x_1'(0) &= 0, & x_2'(0) &= 0, \\
x_0(T) &= 1, & x_1(T) &= 0, & x_2(T) &= 0, \\
x_0'(T) &= 0, & x_1'(T) &= 0, & x_2'(T) &= 0.
\end{aligned} \tag{15.5}$$

Solution of the problem via the Pontryagin maximum principle. In the system of equations (15.4), x_0, x_1, x_2, u are unknown functions of time, but there are only three equations. Hence, system (15.4) is underdetermined. So, the system should be augmented with another condition, which will express a criterion underlying the choice of the control force from all possible variants for which the system of equations (15.4) has a solution.

Criteria for selection of u may vary widely; most often they depend on practical preferences in the solution of the problem under consideration. For example, as in pervious sections, we require to minimize the functional

$$J = \int_0^T u^2(\tau)d\tau. \tag{15.6}$$

The functional (15.6) of the above boundary-value problem (15.4)–(15.6) will be minimized via the Pontryagin maximum principle, which is a classical principle in the control theory.

According to the general theory (see Sect. 2 of Chap. 5), Eq. (15.4) should be written as a system of first-order differential equations

$$z_k' = f_k, \quad k = \overline{1,6},$$

$$f_1 = z_2, \quad f_2 = u, \quad f_3 = z_4, \quad f_4 = u - z_3,$$

$$f_5 = z_6, \quad f_6 = u - {\omega_2}^2 z_5,$$

and then compose the Hamilton-Pontryagin function

$$H = u^2 + \sum_{k=1}^{6} \lambda_k f_k.$$

In our case, the system of equations for the Lagrange multipliers λ_k and the control u

$$\lambda_k' = -\frac{\partial H}{\partial z_k}, \quad k = \overline{1,6}, \quad \frac{\partial H}{\partial u} = 0$$

assumes the form

$$\begin{cases} \lambda_2'' = 0, \\ \lambda_4'' + \lambda_4 = 0, \\ \lambda_6'' + \omega_2^2 \lambda_6 = 0, \\ 2u + \lambda_2 + \lambda_4 + \lambda_6 = 0. \end{cases} \tag{15.7}$$

From system (15.7) it follows that the control reads as

$$u(\tau) = C_1 + C_2\tau + C_3 \sin\tau + C_4 \cos\tau + \\ + C_5 \sin\omega_2\tau + C_6 \cos\omega_2\tau. \tag{15.8}$$

Here $C_k, k = \overline{1,6}$, are the unknown arbitrary constants.

Application of the generalized Gauss principle. As in previous sections, using control (15.8) one can single out a nonholonomic high-order constraint continuously associated with the solution obtained via the Pontryagin maximum principle. Let us write down this constraint.

It is easily seen that control (15.8) is a solution of the differential equation

$$\frac{d^2}{d\tau^2}\left(\frac{d^2}{d\tau^2} + \omega_1^2\right)\left(\frac{d^2}{d\tau^2} + \omega_2^2\right)u = 0.$$

Returning in this equation from the dimensionless variables to the dimensional variables and substituting the expression F from the first equation of system (15.2), we get

$$\frac{m}{\Omega_1^2}\frac{d^8X_0}{dt^8} + \left(m + \frac{c_1}{\Omega_1^2} + m\frac{\Omega_2^2}{\Omega_1^2}\right)\frac{d^6X_0}{dt^6} - \frac{c_1}{\Omega_1^2}\frac{d^6X_1}{dt^6} +$$

$$+ \left(c_1 + m\Omega_2^2 + c_1\frac{\Omega_2^2}{\Omega_1^2}\right)\frac{d^4X_0}{dt^4} - \left(c_1 + c_1\frac{\Omega_2^2}{\Omega_1^2}\right)\frac{d^4X_1}{dt^4} +$$

$$+ c_1\Omega_2^2\frac{d^2X_0}{dt^2} - c_1\Omega_2^2\frac{d^2X_1}{dt^2} = 0. \tag{15.9}$$

The above differential equation (15.9) can be looked upon as a nonholonomic eighth-order constraint, which is continuously satisfied in the course of motion subject to the control obtained by minimization of the functional (15.6).

Thus, for the control found via the Pontryagin maximum principle, the nonholonomic high-order constraint is continuously satisfied, and so in this case one can try to solve the above control problem via the second theory of motion for nonholonomic systems with high-order constraints, which is based on the generalized Gauss principle. According to this principle, under the nonholonomic eighth-order constraint (15.9), the magnitude of the sixth derivative

$$\left(R_{(6)}\right)^2 = \left(M\frac{d^6W}{dt^6} - \frac{d^6Y}{dt^6}\right)^2$$

of the constraint reaction should be minimal. So, among all possible nonholonomic high-order constraints we choose a family, for which the minimum of $(R_{(6)})^2$ is zero. Since the control force (or the reaction of nonholonomic eighth-order constraint) can be written as

$$R = u(t)b, \quad b = \sum_{\sigma=1}^{3} b_\sigma e^\sigma$$

(see Sect. 10 of the present chapter), from the generalized Gauss principle we can write

$$\frac{d^6u}{dt^6} = 0.$$

Therefore, the control force should be sought as a fifth-degree polynomial

$$u = C_1 + C_2\tau + C_3\tau^2 + C_4\tau^3 + C_5\tau^4 + C_6\tau^5. \tag{15.10}$$

The solutions $x_0(\tau)$, $x_1(\tau)$ and $x_2(\tau)$ of system (15.4) with inhomogenuity (15.8) or (15.10) can be conveniently found using Duhamel integrals,

$$x_0(\tau) = \int_0^\tau u(\tau_1)(\tau - \tau_1)d\tau_1,$$

$$x_1(\tau) = \frac{1}{\omega_1}\int_0^\tau u(\tau_1)\sin(\omega_1(\tau - \tau_1))d\tau_1,$$

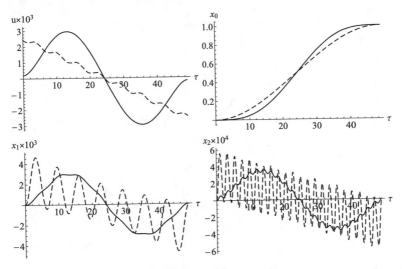

Fig. 29 Motion in the dimensionless coordinates with $T = 48.5$

$$x_2(\tau) = \frac{1}{\omega_2} \int_0^\tau u(\tau_1) \sin(\omega_2(\tau - \tau_1)) d\tau_1 \,.$$

The convenience of this approach is because the Duhamel integrals automatically satisfy the zero initial conditions. Hence, in order to find the unknown constants C_i, it suffices to substitute x_0, x_1 and x_2 into the boundary conditions on the right-hand end-point,

$$x_0(T) = 1 \,, \quad x_1(T) = 0 \,, \quad x_2(T) = 0 \,,$$

$$x_0'(T) = 0 \,, \quad x_1'(T) = 0 \,, \quad x_2'(T) = 0 \,.$$

Numerical calculation. Let us consider some case of motion control and compare the solutions obtained via the Pontryagin maximum principle and the generalized Gauss principle. As an example, we construct solutions for the system with parameters $m = 100\,\text{kg}$, $m_1 = 300\,\text{kg}$, $m_2 = 30\,\text{kg}$, $c_1 = 15\,\text{N/m}$, $c_2 = 45\,\text{N/m}$.

A solution will be sought in two scales for the time-distance pair. For the first calculation, we take large motion time $\widetilde{T} = 110\,s$ and large distance $S = 600\,m$. For the second calculation, we take small motion time $\widetilde{T} = 17\,s$ and small distance $S = 3\,m$. In each of the cases, we get the dimensionless motion time $T = 48.5$ and $T = 7.5$, respectively.

We start our calculations in the dimensionless quantities, and then, changing the variables by (15.3), we get the motion in the original curvilinear coordinates. The dimensionless results of calculation are depicted in Figs. 29 and 30, the dimensional calculations are shown in Figs. 31 and 32. The solutions obtained via the generalized Gauss principle (the Pontryagin maximum principle) are shown by solid (dashed) curves.

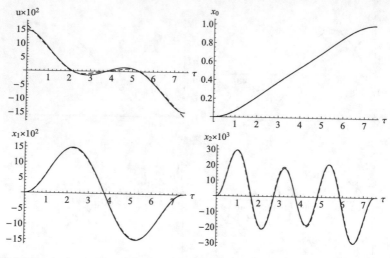

Fig. 30 Motion in the dimensionless coordinates with $T = 7.5$

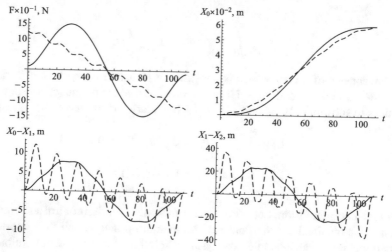

Fig. 31 Motion of the mechanical system with $\widetilde{T} = 110\,s$ and $S = 600\,m$

It can be seen from the graphs that, for a long-time motion, the mechanical system features strong oscillations in the solution obtained via the Pontryagin maximum principle—however, this is quite logical, because the form of the control obtained by this method contains harmonic terms that act on the system with its natural frequencies. Note also that the control obtained by the classical method is much more involved, which hinders its practical implementation. However, the solution obtained via the generalized Gauss principle has polynomial form, and hence looks more reliable from this side. From the point of view of nonholonomic mechanics, the method

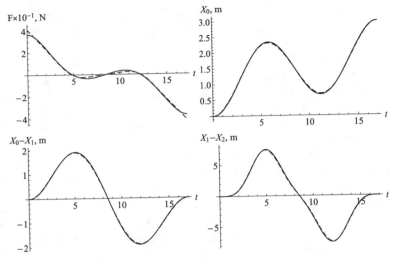

Fig. 32 Motion of the mechanical system with $\widetilde{T} = 17\,s$ and $S = 3\,m$

has the advantage that, for a long-time motion, the control obtained by this method starts with a smooth increase of the control force, whereas the control force obtained via the Pontryagin maximum principle has abrupt jumps at the beginning and end of motion. However, despite the very different approaches to the solution of the problem, these two methods give solutions that more and more approach each other with decreasing the time span in which the system should reach the required phase state.

16 Suppression of Oscillations of a Cantilever[35]

Introduction. In this section, on an example of oscillations of a cantilever, we shall show that the generalized Gauss principle can be successfully applied not only in studying the suppression of oscillations of mechanical systems with finite number of degrees of freedom, but also for systems with distributed parameters.

In the present Part II of this chapter, for the studying the suppression of oscillations of a mechanical systems with finite number of degrees of freedom, we have frequently used the method based on the Pontryagin maximum principle. A relation of this approach to the generalized Chebyshev problem was found. Analysis of this relation shows that the generalized Gauss principle can be successfully used for determining the acceleration of an elastic system in order to travel through a given distance in

---

[35] See the paper and the book: *Sh. Kh. Soltakhanov*. Suppression of oscillations of the cantilever // Vestnik St. Petersburg University. 2009. Ser. 1. Issue 4. Pp. 105–112 [in Russian]; *Sh. Kh. Soltakhanov*. Determination of control forces with high-order constraints. Moscow: Nauka. 2014. 240 p. [in Russian].

a given amount of time from the initial state of rest to be again at rest at the end of motion. The sought-for acceleration, for which the oscillations are suppressed at the end of motion, can be represented in powers of the time t. The number of terms of this series is $(2s + 2)$, where s is the number of nonzero frequencies of the elastic system. These frequencies were assumed to be different. The frequencies are arranged in an increasing order, and to the frequency ω_σ, $\sigma = \overline{1, s}$, there correspond the terms of the series $a_{2\sigma}t^{2\sigma} + a_{2\sigma+1}t^{2\sigma+1}$, where $a_{2\sigma}$ and $a_{2\sigma+1}$ are the sought-for constants. The displacement of the elastic system as an absolutely rigid body is taken into account by the terms $a_0 + a_1 t$.

An elastic body, for example, a cantilever, has an infinite number of frequencies. The natural question here is as follows: is it reasonable, when considering the motion of the cantilever base, to suppress the oscillations at the end of the displacement for the *entire* spectrum of frequencies. The thing is that the contribution of the higher modes of oscillations of the cantilever into its total energy is small at the time the motion stops. A constructive solution of this problem was given by G.V. Kostin and V.V. Saurin.[36] In these papers, it was proposed to determine the acceleration, which is also given as a power series in t, by minimizing the total energy of the cantilever at the moment its base stops its motion. The deflection curve, which is involved in the expression of this energy, was determined by the method of integro-differential relations. In the present subsection, this energy is evaluated by applying the Lagrange equations of the second kind to the solution of the problem. The acceleration jumps of the cantilever as an absolutely rigid body at the beginning and end of motion are removed, because the power series expansion in time t of the acceleration starts with t^2. The energy of oscillations is considered not only at the end of the path, but also during the travel of the cantilever. This extended energy approach provides a new viewpoint on this problem.

Application of the Lagrange equations of the second kind to the problem of oscillations suppression of the cantilever. Following G.V. Kostin and V.V. Saurin (2006), we assume for simplicity that the cantilever is homogeneous and has a constant cross section. The efficiency of application of the Lagrange equations to the above problem stems from the fact that if the deflection of the cantilever is expanded as a series in eigenfunctions $X_\sigma(x)$, $\sigma = \overline{1, \infty}$,

$$y_r(x, t) = \sum_{\sigma=1}^{\infty} q_\sigma(t) X_\sigma(x), \quad 0 \leqslant x \leqslant l$$

[36] See the papers: *G.V. Kostin, V.V. Saurin.* Integro-differential approach to solving problems of linear elasticity theory // Dokl. Phys. 2005. Vol. 50. pp. 535–538; *G.V. Kostin, V.V. Saurin.* Modeling and optimization of elastic system motions by the method of integro-differential relations // Dokl. Math. 2006. Vol. 73. pp. 469–472.

with travel time close to the period of the first mode of oscillations or exceeding this value, then the kinetic and potential energies of the cantilever are given by the rapidly converging series

$$K_r = \sum_{\sigma=1}^{\infty} \frac{M_\sigma \dot{q}_\sigma^2}{2}, \qquad \Pi = \sum_{\sigma=1}^{\infty} \frac{M_\sigma \omega_\sigma^2 q_\sigma^2}{2},$$

$$M_\sigma = \frac{m}{l} \int_0^l X_\sigma^2(x)\,dx, \quad \omega_\sigma^2 = \frac{EJ}{ml^3}\lambda_\sigma^4, \quad \sigma = \overline{1,\infty}. \tag{16.1}$$

Here l is the cantilever length, m is its mass, E is the Young modulus, J is the moment of inertia of the cross section, λ_σ are roots of the equation

$$\cos \lambda \cosh \lambda = -1.$$

The natural modes and the reduced masses M_σ, $\sigma = \overline{1,\infty}$, are as follows[37]:

$$X_\sigma(x) = \sin \frac{\lambda_\sigma x}{l} - \sinh \frac{\lambda_\sigma x}{l} + A_\sigma \left(\cosh \frac{\lambda_\sigma x}{l} - \cos \frac{\lambda_\sigma x}{l} \right),$$

$$A_\sigma = \frac{\sinh \lambda_\sigma + \sin \lambda_\sigma}{\cosh \lambda_\sigma + \cos \lambda_\sigma}, \qquad M_\sigma = m A_\sigma^2. \tag{16.2}$$

Assume that the function $\xi(t)$ is given by the displacement of the base of the cantilever in the direction orthogonal to the rod axis. Then the absolute displacement $y_a(x, t)$ of the section x of the cantilever assumes the form

$$y_a(x, t) = \xi(t) + y_r(x, t),$$

where $y_r(x, t)$ is the displacement of the sections of the cantilever with respect to the undeformed state.

Finding the kinetic energy of the system

$$K = \frac{m}{2l} \int_0^l \left(\frac{\partial y_a}{\partial t} \right)^2 dx$$

---

[37] See, for example: *I.M. Babakov.* Theory of oscillations. Moscow: Nauka. 1965. 559 p. [in Russian]; *S.P. Timoshenko.* Vibration Problems in Engineering. New York: McGraw-Hill. 1955.

and substituting it in the Lagrange equation

$$\frac{d}{dt}\frac{\partial K}{\partial \dot{q}_\sigma} - \frac{\partial K}{\partial q_\sigma} = -\frac{\partial \Pi}{\partial q_\sigma}, \qquad \sigma = \overline{1, \infty},$$

we arrive at the equations

$$\ddot{q}_\sigma + \omega_\sigma^2 q_\sigma = -\frac{a_\sigma}{A_\sigma^2}\ddot{\xi}, \qquad \sigma = \overline{1, \infty}.$$

Here

$$a_\sigma = \frac{1}{l}\int_0^l X_\sigma(x)\,dx, \qquad \sigma = \overline{1, \infty}.$$

Changing to the dimensionless variables by the formulas

$$x_0 = \frac{\xi}{l}, \quad x_\sigma = -\frac{A_\sigma^2}{a_\sigma}\frac{q_\sigma}{l}, \quad \sigma = \overline{1, \infty}, \qquad \tau = \omega_1 t, \tag{16.3}$$

and denoting for simplicity the derivative with respect to the dimensionless time τ by dot, we get the equations

$$\ddot{x}_0 = u, \quad \ddot{x}_\sigma + \overline{\omega}_\sigma^2 x_\sigma = u, \quad \overline{\omega}_\sigma = \frac{\omega_\sigma}{\omega_1} = \left(\frac{\lambda_\sigma}{\lambda_1}\right)^2, \quad \sigma = \overline{1, s}. \tag{16.4}$$

Here u is the sought-for dimensionless acceleration of the cantilever base and s is the number of considered natural modes of the oscillations.

According to Rayleigh,[38] in impacting spheres the elastic oscillations are very little excited, because at the beginning and end of the impact the acceleration of the center of mass of each of the spheres, as well as its derivative with respect to time are zero. In view of this, the displacement x_0 will be subject to the following boundary-value conditions:

$$x_0(0) = \dot{x}_0(0) = \ddot{x}_0(0) = u(0) = \ddot{x}_0(0) = \dot{u}(0) = 0,$$
$$x_0(T) = a, \quad \dot{x}_0(T) = \ddot{x}_0(T) = u(T) = \ddot{x}_0(T) = \dot{u}(T) = 0. \tag{16.5}$$

Here T is the dimensionless motion time and a is the dimensionless displacement magnitude.

---

[38] See, for example: *S.A. Zegzhda. Impacts of Elastic Bodies.* St. Petersburg: St. Petersburg Univ. Press. 1997. 316 p. [in Russian].

From (16.1)–(16.3) it follows that the energy of oscillations of the cantilever reads as

$$K_r + \Pi = \frac{ml^2\omega_1^2}{2} \sum_{\sigma=1}^{\infty} \frac{a_\sigma^2}{A_\sigma^2} (\dot{x}_\sigma^2 + \overline{\omega}_\sigma^2 x_\sigma^2). \tag{16.6}$$

Assume that the rod is an absolutely rigid body. Then the acceleration u^*, which is sought in the form

$$u^* = C_1\tau^2 + C_2\tau^3 + C_3\tau^4 + C_4\tau^5$$

will be uniquely determined from the boundary conditions (16.5). We denote by x_0^* the displacement corresponding to this acceleration.

The function $u^*(\tau)$ has the following property

$$u^*(\tau) = -u^*(T - \tau).$$

It follows that $u^*(T/2) = 0$. Hence, since $u^*(\tau) > 0$ for $0 < \tau < T/2$, we find the maximal velocity of the base is

$$v_{max} = l\omega_1 v_m, \quad v_m = \dot{x}_0^*(T/2).$$

Taking

$$K_* = \frac{mv_{max}^2}{2}$$

for the measure of energy, and using (16.6), we get

$$En(\tau) = \frac{K_r + \Pi}{K_*} = \frac{1}{v_m^2} \sum_{\sigma=1}^{\infty} \frac{a_\sigma^2}{A_\sigma^2} (\dot{x}_\sigma^2 + \overline{\omega}_\sigma^2 x_\sigma^2). \tag{16.7}$$

Consider the maximal acceleration of the cantilever as an absolutely rigid body:

$$\ddot{\xi}_{max}^* = l\omega_1^2 u_m, \quad u_m = u^*(\tau_*).$$

Here τ_* is the time instant at which the function $u^*(\tau)$ is maximal.

The sought-for acceleration $\ddot{\xi}$ of the flexible cantilever, as calculated in fractions of this acceleration, is as follows:

$$\bar{u} = \frac{\ddot{\xi}}{\ddot{\xi}^{*}_{max}} = \frac{u}{u_m}.$$

The function $u(\tau)$, which is the solution of the above problem, depends directly on both a and T. The new dimensionless quantity $\bar{u}$, which is considered as a function of the argument τ/T, is independent of a; the only parameter of this function is the ratio of the displacement time to the period of the first mode of oscillations. Since the sought-for solution is independent of a, we took $a = 1$ for calculations.

Suppression of oscillations of the cantilever as a boundary-value problem and minimization of $En(T)$. At first, we consider the problem of oscillation suppression of the cantilever at time instant T as a boundary-value problem; that is, we augment conditions (16.5) with the conditions

$$x_\sigma(0) = \dot{x}_\sigma(0) = x_\sigma(T) = \dot{x}_\sigma(T) = 0, \quad \sigma = \overline{1, s}. \tag{16.8}$$

The boundary-value problem (16.4), (16.5), (16.8) can be solved if the sought-for function $u(\tau)$ is represented as a sum of any linearly independent functions of number $(2s + 6)$.

Above, we established a link between the boundary-value conditions imposed on the motion described by Eq. (16.4) and the generalized Chebyshev problem. Consequently, if a system of linearly independent functions was chosen when solving the boundary-value problem under consideration, then to each such a system of functions there correspond a constraint of order $(2s + 8)$.

It was also shown that in the case of minimization of the functional of the squared dimensionless control via the Pontryagin maximum principle, from the set of equations appearing with this high-order constraints one can single out a subset such that to all elements of this set there corresponds one unique differential equation with respect to the control, whose structure depends only on the spectrum of natural frequencies of the system and is independent of the choice of generalized coordinates.

The following question is natural: is it possible to select a new subset of the set of constraint equations to which there also corresponds a unique differential equation (in our case, of order $(2s + 6)$) with constant coefficients.

According to the generalized Gauss principle of order $(2s + 6)$,

$$\left(\mathbf{R}_{(2s+6)}\right)^2 = \left(M \overset{(2s+6)}{W} - \overset{(2s+6)}{Y}\right)^2 = (\overset{(2s+6)}{u} b)^2$$

and so the least "compulsion" with constraints of order $(2s + 8)$ will occur in the above problem only in the case when the sought-for function $u(\tau)$ satisfies the equation

$$\overset{(2s+6)}{u} = 0 .$$

Hence, since $u(0) = \dot{u}(0) = 0$, it follows that in accordance with the generalized Gauss principle the acceleration should be sought in the form

$$u(\tau) = \sum_{k=1}^{2s+4} C_k \tau^{k+1} , \qquad (16.9)$$

where C_k are the sought-for constants, which can be determined by the algorithm proposed in pervious subsections.

We recall that the functions $u(\tau)$, $x_0(\tau)$, $x_\sigma(\tau)$, $\sigma = \overline{1, s}$, which give a solution to this problem, have the following properties:

$$U(\tau) = -u(T - \tau) ,$$
$$x_0(\tau) = a - x_0(T - \tau) ,$$
$$x_\sigma(\tau) = -x_\sigma(T - \tau) ,$$
$$\sigma = \overline{1, s} .$$

Let us now consider the method of minimization of the quantity $En(T)$ given by formula (16.7). As in the paper by G.V. Kostin and V.V. Saurin (2006), we shall be concerned only with the case when the function $u(\tau)$ satisfying conditions (16.5) has either two or four free parameters. So, let

$$u(\tau) = \sum_{k=1}^{4} C_k \tau^{k+1} + \alpha \tau^6 + \beta \tau^7 + \gamma \tau^8 + \delta \tau^9 . \qquad (16.10)$$

For simplicity of the description of the methods, we consider only two parameters; that is, we set $\gamma = \delta = 0$. Using the equation $\ddot{x}_0 = u$ and conditions (16.5) to find the constants C_k, $k = \overline{1, 4}$, as linear functions of the parameters α and β, we get

$$u = u(\tau, \alpha, \beta) ,$$

$$En(T, \alpha, \beta) = \frac{1}{v_m^2} \sum_{k=1}^{N} \frac{a_\sigma^2}{A_\sigma^2} (\dot{x}_k^2(T, \alpha, \beta) + \overline{\omega}_k^2 x_k^2(T, \alpha, \beta)) . \qquad (16.11)$$

Here

$$x_k(T, \alpha, \beta) = \frac{1}{\overline{\omega}_k} \int_0^T u(\tau, \alpha, \beta) \sin \overline{\omega}_k(T - \tau) \, d\tau \,,$$

$$\dot{x}_k(T, \alpha, \beta) = \int_0^T u(\tau, \alpha, \beta) \cos \overline{\omega}_k(T - \tau) \, d\tau \,,$$

$$k = \overline{1, N} \,.$$

The number N was chosen from the condition that the error of calculation of the total energy of the cantilever be majorized by $0.01\,\%$. Calculations show that for $T/T_1 \geqslant 0.6$ this is achieved with $N = 8$. The required parameters α and β were defined from the system of linear equations

$$a_{11}\,\alpha + a_{12}\,\beta = -f_1(0, 0)\,,$$
$$a_{21}\,\alpha + a_{22}\,\beta = -f_2(0, 0)\,,$$

where

$$f_1(\alpha, \beta) = \frac{\partial En(T, \alpha, \beta)}{\partial \alpha}\,, \quad f_2(\alpha, \beta) = \frac{\partial En(T, \alpha, \beta)}{\partial \beta}\,,$$

$$a_{11} = \frac{\partial f_1}{\partial \alpha}\,, \quad a_{12} = a_{21} = \frac{\partial f_1}{\partial \beta}\,, \quad a_{22} = \frac{\partial f_2}{\partial \beta}\,.$$

The construction of a solution with four free parameters is similar.

Analysis of the calculation results. From (16.9) and (16.10) it follows that the solution obtained by minimization of two (four) free parameters should be compared with that corresponding to suppression of one mode ($s = 1$) (respectively, $s = 2$).

The total energy of oscillations of the cantilever left after suppression of s forms is as follows:

$$En^{(s)}(T) = \frac{1}{v_m^2} \sum_{k=s+1}^N \frac{a_k^2}{A_k^2} (\dot{x}_k^{(s)}(T) + \overline{\omega}_k^2 (x_k^{(s)}(T))^2)\,,$$

where

$$x_k^{(s)}(T) = \frac{1}{\overline{\omega}_k} \int_0^T u^{(s)}(\tau) \sin \overline{\omega}_k(T - \tau) \, d\tau \,,$$

$$\dot{x}_k^{(s)}(T) = \int_0^T u^{(s)}(\tau) \cos \overline{\omega}_k(T - \tau) \, d\tau \,,$$

$$k = \overline{1, N} \,.$$

Table 1 Values of $En^{(s)}(T)$

s \ T/T_1	0.6	0.8	1	2
1	0.3954	0.008575	$8.205 \cdot 10^{-7}$	$1.331 \cdot 10^{-7}$
2	0.01593	$0.1146 \cdot 10^{-3}$	$3.059 \cdot 10^{-7}$	$2.059 \cdot 10^{-10}$

Table 2 Values of $En^{(s)}(T)/En^*(T)$

s \ T/T_1	0.6	0.8	1	2
1	0.3245	0.005475	$5.167 \cdot 10^{-7}$	$3.771 \cdot 10^{-6}$
2	0.01308	$0.7320 \cdot 10^{-4}$	$1.926 \cdot 10^{-7}$	$5.831 \cdot 10^{-9}$

Table 3 The difference in percents between the boundary-value problem and the minimization problem

s \ T/T_1	0.6	0.8	1
1	29.9	0.437	$0.576 \cdot 10^{-4}$
2	4.66	0.555	0.0633

Here $u^{(s)}(\tau)$ is the solution of the boundary-value problem with suppression of s forms. In evaluation of the energy, the number N was taken to be the same as in the minimization method.

The quantities $En^{(s)}(T)$ rapidly decay with increasing ratio T/T_1. Their values are given in Table 1.

The degree of suppression is characterized by the quantity $En^{(s)}(T)/En^*(T)$, where $En^*(T)$ is the energy of oscillations at the end of the path if the cantilever is considered as an absolutely rigid body. These values are given in Table 2.

The quantities $En(T, \alpha, \beta)$ and $En(T, \alpha, \beta, \gamma, \delta)$ are, respectively, smaller than $En^{(1)}(T)$ and $En^{(2)}(T)$. These differences in percents

$$\frac{En^{(1)}(T) - En(T, \alpha, \beta)}{En(T, \alpha, \beta)} \, 100\% , \quad \frac{En^{(2)}(T) - En(T, \alpha, \beta, \gamma, \delta)}{En(T, \alpha, \beta), \gamma, \delta} \, 100\%$$

are given in Table 3.

Fig. 33 Results of calculations with $T/T_1 = 0.8$

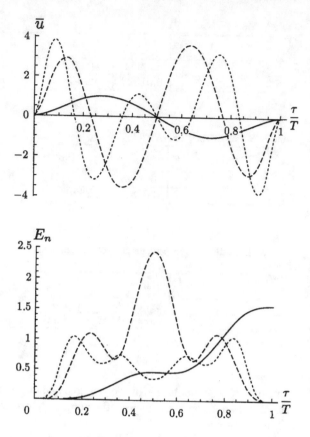

Calculations show that the quantities from Table 3 characterize quite well the dependence on the parameter T/T_1 of the difference between the accelerations obtained via these two methods.

The results of calculations given in Table 3 show that the ratio $T/T_1 = 0.8$ can be looked upon as a special one in the sense that if it decreases, then the difference between these two methods of suppression of oscillations of the cantilever abruptly increases, and vice versa, the difference rapidly decreases if this ratio is increased. The problem of oscillation suppression of the cantilever for $T/T_1 < 0.8$ requires a special attention. The thing is that with small ratio T/T_1 the suppression of the first mode is effected for quite large values of the modulus of the function $\bar{u}(\tau/T)$, and, as a corollary, with very large energy of oscillations in the process of displacement. Hence the case with $T/T_1 < 0.8$ is not considered in what follows.

The results of calculations for three typical values of T/T_1 are given in Figs. 33, 34, and 35, in which the solid curves correspond to a rod as an absolutely rigid body, the curves with long dashes show the suppression of the first mode, and the curves with short dashes, the suppression of two modes. The typical value $T/T_1 = 0.8$ was already discussed above.

Fig. 34 Results of
calculations with
$T/T_1 = 1.12$

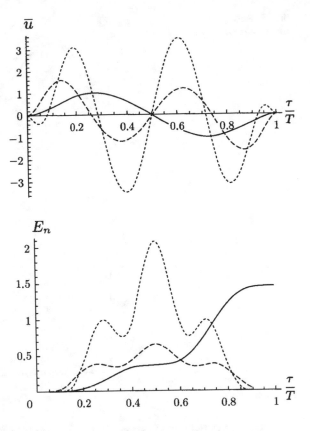

For $T/T_1 = \lambda_1^2/\pi = 1.12$, calculations were performed by G.V. Kostin and V.V. Saurin (2006). Note that in this paper the acceleration of the cantilever as an absolutely rigid body was given by a first degree polynomial, and the minimization of the total energy at time T with two and four parameters was effected, respectively, with polynomials of degree three and five.

The third value $T/T_1 = 2$ was chosen because, as is seen from Tables 1 and 2, the oscillations of the cantilever at time T are so small that there is no need for suppression.

The following conclusions can be made from the graphs of the functions $\bar{u}(\tau/T)$ and $En(\tau/T)$:

for $0.8 \leqslant T/T_1 < 1$ it is expedient to suppress the first two modes;

for $1 \leqslant T/T_1 \leqslant 2$ is suffices to suppress the first mode;

for $T/T_1 > 2$, there is no need to suppress oscillations.

Fig. 35 Results of calculations with $T/T_1 = 2$

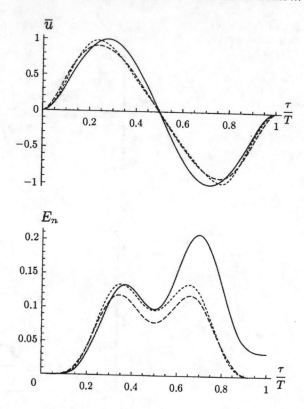

It should be noted that the above theory can be also applied directly to calculation of suppresion of oscillations of a flexible robotic arm as it stops after its basis was moved through a given distance within a given time interval.

Chapter 7
Mechanics with Random Forces

P. E. Tovstik and T. M. Tovstik ⓘ

The present chapter gives a brief account on the methods of determination of probabilistic characteristics of motion of mechanical systems subject to random forces. In the introductory sections we give the required definitions of random variables and processes. The auxiliary material presented here is not a treatise on the probability theory. In the definition of the probabilistic characteristics, we shall be mostly concerned with the correlation level when the expectations and correlation function are determined under the condition that these characteristics are given for the exterior forces. For the stationary processes, the Fourier transform is used and the spectral densities are defined. For statistically linear systems it proves possible to find an exact solution. However, nonlinear systems can be treated only by approximate methods—these are the methods of statistical linearization and statistical modeling. The method of the solution of the Fokker–Planck–Kolmogorov equation, mentioned at the end of this chapter, gives an exact solution; however, the area of its practical application is very narrow.[1]

1 Elements of Probability Theory

The auxiliary material given in this section is not aimed at providing a background in probability.

The basic objects of probability theory are a *random event* **A**, which could or could not take place. To an event **A** there corresponds the *probability*

---

[1] Brief list of references: *V.S. Pugachev.* Random functions theory and its applying to automatic control problems. Moscow: Fizmatgiz. 1962, 884 p. [in Russian]; *V.A. Svetlitsky.* The random oscillations of mechanical systems. Moscow: Mashinostroen i.e. 1976, 216 p. [in Russian]; *M.F. Dimentberg.* Nonlinear stochastic problems of mechanical oscillations. Moscow: Nauka. 1980, 368 p. [in Russian]; *S.M. Ermakov, G.A. Mikhailov.* Statistical modelling. Moscow: Nauka. 1982, 206 p. [in Russian].

© Springer Nature Switzerland AG 2021
N. N. Polyakhov et al., *Rational and Applied Mechanics*,
Foundations of Engineering Mechanics,
https://doi.org/10.1007/978-3-030-64118-4_7

$$p = P(A), \quad 0 \leqslant p \leqslant 1.$$

For $p = 0$ we have an *impossible event* and for $p = 1$, a *certain event*.

If the result of an event is a number x, then we talk about a *random variable X*. The basic characteristic of a random variable X is its *distribution function $F(x)$*,

$$F(x) = P(X < x). \tag{1.1}$$

We assume that $F(-\infty) = 0$, $F(\infty) = 1$. We have $P(x_1 \leqslant X < x_2) = F(x_2) - F(x_1)$, and therefore $F(x)$ is an (unstrictly) increasing function.

If a variable X assumes only a discrete number of values, then one talks about a *discrete random variable*. For example, the results produced by rolling a dice is discrete random variable. In mechanics we are primarily interested in random variables that may take any value in the range $-\infty < x < \infty$ or in some more narrow range. Such variables are called *continuous variables*. For such random variables (with the exception of some degenerate cases) one considers the *probability density function $f(x)$*,

$$f(x) = \frac{dF}{dx}, \text{ where } f(x) \geqslant 0,$$

$$F(x) = \int_{-\infty}^{x} f(x)\,dx, \quad P(x_1 \leqslant X < x_2) = \int_{x_1}^{x_2} f(x)\,dx. \tag{1.2}$$

In addition to the functions $F(x)$ and $f(x)$, which contain comprehensive information about a random variable, one also considers the initial and central moments, and in particular, the *expectation m_x* and the *variance D_x*,

$$m_x = E(X) = \int_{-\infty}^{\infty} x f(x)\,dx,$$

$$D_x = E(|X - m_x|^2) = \int_{-\infty}^{\infty} |x - m_x|^2 f(x)\,dx. \tag{1.3}$$

In what follows, the expectation of a random variable Y will be denoted as $E(Y)$. The expectation have the same dimension as a random variable, while the dimension of the variance is the squared dimension of a variable. For example in monetary transactions it has the dimension of the square roubles. In order to return to the original dimension one introduces the *standard deviation $\sigma_x = \sqrt{D_x}$*.

There is an analog between the above probabilistic and mechanical concepts. Let us consider a bar of weight $M = 1$ and linear density $f(x)$. Then the expectation is equal to the coordinate of its gravity center ($x_c = m_x$), the variance is equal to the moment of inertia relative to the gravity center, and the standard deviation agrees with the radius of inertia. For example, for a random variable uniformly distributed on the interval $[a, b]$

$$f(x) = \begin{cases} 1/(b-a), & x \in [a, b], \\ 0, & x \notin [a, b], \end{cases} \qquad m_x = \frac{b-a}{2}, \qquad D_x = \frac{(b-a)^2}{12}.$$

The probability density function of a *normal (Gaussian) random variable* with expectation $m_x = a$ and standard deviation $\sigma_x = \sigma$ is given by

$$f(x) = \frac{1}{\sqrt{2\pi}\sigma} e^{-\frac{(x-a)^2}{2\sigma^2}}. \qquad (1.4)$$

The *expectation* m_x is the mean value of a random variable X, the *standard deviation* σ_x is the measure of deviation from the expectation. There is a remarkable *Chebyshev's theorem*, which holds for any random variable

$$p = P(|X - m_x| > N\sigma_x) \leqslant \frac{1}{N^2}$$

and states that the large deviations from the expectation occur fairly rarely. For example, $p \leqslant 1/9$ for $N = 3$. For a random variable with a specific distribution this estimation can be refined. For a normal distribution one has $p \leqslant 0.003$.

2 Multivariate Random Variables

Given several random variables $X_1, \dots, X_n$ related with each other, one talks about an n-dimensional random variable or a n-dimensional random vector $X = (X_1, \dots, X_n)^T$ (the sign ^{T} means the transposition). For each of the components X_k of the vector X, the one-dimensional distribution functions $F_k(x_k)$, the probability density functions $f_k(x_k)$, the expectations m_{x_k}, the variances D_{x_k}, and the standard deviations σ_{x_k} are introduced using formulas similar to (1.1)–(1.3) as in Sect. 1.

One also considers new concepts and probability characteristics. For simplicity, we first consider the two-dimensional random variable X, Y. We will observe a two-dimensional distribution function $F_2(x, y)$ and the corresponding two-dimensional probability density function $f_2(x, y)$,

$$F_2(x, y) = P(X < x, Y < y), \quad f_2(x, y) = \frac{\partial^2 F_2}{\partial x \partial y},$$

where the function $F_2(x, y)$ is defined as the probability of occurrence of two events $X < x$ and $Y < y$ together. If these events are *independent*, $F_2(x, y) = F_x(x)F_y(y)$ and $f_2(x, y) = f_x(x)f_y(y)$, where $F_x(x)$, $F_y(y)$ and $f_x(x)$, $f_y(y)$ are the one-dimensional distribution functions and the probability density function.

The probability that a point (x, y) falls in some region G in the XY-plane is equal to

$$P((x, y) \in G) = \int_G f(x, y) \, dx \, dy.$$

The *correlation* K_{xy} of random variables X and Y is defined as

$$K_{xy} = E(X^0 \overline{Y^0}), \quad X^0 = X - m_x, \quad m_x = E(X), \quad (x \to y). \quad (2.1)$$

Here X^0 denotes the centered random variable (the variable from which its expectation is subtracted), the bar over Y^0 denotes the complex conjunction. Note that

$$E(X^0) = 0, \quad K_{yx} = \overline{K_{xy}}, \quad K_{xx} = D_x. \quad (2.2)$$

It was the last relation in (2.2) which resulted in the introduction of the complex-conjugated variable $\overline{Y^0}$ in the definition of the correlation K_{xy} in (2.1).

If $K_{xy} = 0$, then variables X and Y are called *uncorrelated*. Independent random variables are uncorrelated, but the converse is false.

The correlation satisfies the inequality $K_{xy}^2 \leqslant D_x D_y$. The dimensionless *coefficient of correlation* is defined as

$$r_{xy} = \frac{K_{xy}}{\sqrt{D_x D_y}}, \quad |r_{xy}| \leqslant 1. \quad (2.3)$$

The above can be extended to a random vector $X = (X_1, \dots, X_n)^T$, for which the n-dimensional distribution function and the probability density function are as follows:

$$F_n(x_1, \dots, x_n) = P(X_1 < x_1, \dots, X_n < x_n), \quad f_n(x_1, \dots, x_n) = \frac{\partial^n F_n}{\partial x_1 \dots \partial x_n}.$$

The matrix

$$K_X = E\left(X^0 \cdot \overline{X^{0T}}\right) = \begin{pmatrix} D_{x_1} & K_{x_1 x_2} & \dots & K_{x_1 x_n} \\ K_{x_2 x_1} & D x_2 & \dots & K_{x_2 x_n} \\ \dots & \dots & \dots & \dots \\ K_{x_n x_1} & K_{x_n x_2} & \dots & D_{x_n} \end{pmatrix} \quad (2.4)$$

is called the correlation matrix; it has variances of separate processes on the principal diagonal. In view of (2.2) the correlation matrix is seen to satisfy the relation $K_X^T = \overline{K_X}$.

The matrix K_X is positive. It means that all of its minor determinants, which are formed by deletion of the columns and lines with the same numbers, are nonnegative. To prove this fact it suffices to consider the random variable $Z = \alpha \cdot X$, $\alpha = (\alpha_1, \dots, \alpha_n)$, which is equal to the random linear combination of the elements of the vector X. We have

$$D_z = E\left(|Z|^2\right) = \alpha \cdot E\left(X^0 \cdot \overline{X^{0T}}\right) \cdot \overline{\alpha^T} = \alpha \cdot K_X \cdot \overline{\alpha^T} \geqslant 0,$$

and hence, since the variance is nonnegative, we have the required result.

3 Random Processes

A *random process* or a *random function* $X(t)$ is any random one-parameter class of random variables, where the parameter t is time. Having fixed the event we get the time function $x(t)$ known as the *realization*.

With fixed time t, we get the random variable for which one may introduce the above probability characteristics: the distribution function $F_1(x, t) = P(X(t) < x)$, the expectation $m_x(t) = E(X(t))$ and the variance

$$D_x(t) = E\left(|X^0(t)|^2\right), \quad X^0(t) = X(t) - m_x(t),$$

where $X^0(t)$ is a centered random process.

With fixed several time instants $t_1, t_2, \ldots, t_n$ one obtains several random variables or a random vector

$$X(t) = (X(t_1), \ldots, X(t_n))^T,$$

for which one may introduce the n-dimensional distribution function

$$F_n(x_1, t_1; \ldots; x_n, t_n) = P\left(X(t_1) < x_1, \ldots, X(t_n) < x_n\right). \tag{3.1}$$

With increasing of n every next distribution function contains (in general) more information about the process than the previous one. It may be supposed that the class of all finite distributions contains exhaustive information about the process. However, the use of distribution functions with $n \geqslant 2$ is fairly difficult, because as a rule they are unknown and have at least 4 arguments.

As a result, one deals with the *correlation matrix* $K_X(t_1, \ldots, t_n)$ *of a process* $X(t)$, which is constructed in the same way as matrix (2.4)

$$K_X(t_1, \ldots, t_n) = E\left(X^0(t) \cdot \overline{X^{0T}(t)}\right) =$$

$$= \begin{pmatrix} D_x(t_1) & k_x(t_1, t_2) & \ldots & k_x(t_1, t_n) \\ k_x(t_2, t_1) & D_x(t_2) & \ldots & k_x(t_2, t_n) \\ \ldots & \ldots & \ldots & \ldots \\ k_x(t_n, t_1) & k_x(t_n, t_2) & \ldots & D_x(t_n) \end{pmatrix}, \tag{3.2}$$

where $k_x(u, v) = E\left(X^0(u)\overline{X^0(v)}\right)$ is the *correlation function of a process* $X(t)$.

Note that $D_x(u) = k_x(u, u)$, $k_x(v, u) = \overline{k_x(u, v)}$. It is worth noting that not any function of two variables may serve as the correlation function of a random process, because matrix (3.2) is positive. In particular, the inequality $|k_x(u, v)| \leqslant \sqrt{D_x(u)D_x(v)}$ must hold.

The correlation function $k_x(v, u)$ describes the measure of probability relation between random variables $X(u)$ and $X(v)$. As a rule, this relation decreases with increasing difference $|u - v|$. Figure 1 shows possible graphs of correlation functions with fixed v.

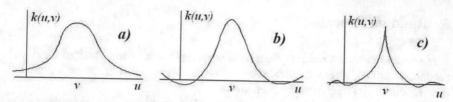

Fig. 1 Possible graphs of correlation functions

4 Calculus of Random Variables and Random Processes

There are several definitions of the limit in probability. We shall use the mean-square limit.

Definition 1 Given an infinite sequence $X_1, X_2, \dots, X_n, \dots$ of random variables, a random variable X is called its *mean-square limit* if

$$\lim_{n \to \infty} E\left(|X - X_n|\right) = 0.$$

In formulas like $X = \underset{n \to \infty}{\text{l.i.m.}} X_n$ "l.i.m." is an abbreviation for the phrase "limit in mean".

Definition 2 A random function $X(t)$ is called *mean-square continuous* with $t = \tau$ if $\underset{t \to \tau}{\text{l.i.m.}} X(t) = X(\tau)$.

Using Definition 1 one may write down a test for continuity of a random variable: A necessary and sufficient condition that a random function $X(t)$ in be mean-square continuous with $t = \tau$ is that its expectation $m_x(t)$ and the correlation function $k_x(t_1, t_2)$ be continuous at this point.

In the proof of this result (which is not given here) one has to rewrite the notion of limit from the "language of sequences" to the "language of ε–δ".

Definition 3 A random function $Y(t)$ is called the *mean-square derivative* of a function $X(t)$ if

$$Y(t) = \underset{\Delta t \to 0}{\text{l.i.m.}} \frac{X(t + \Delta t) - X(t)}{\Delta t}.$$

The criterion for the existence of the derivative of a function $X(t)$ is the existence of the continuous derivatives dm_x/dt and $\partial^2 k_x/(\partial t_1 \partial t_2)$. Besides,

$$m_y(t) = \frac{dm_x}{dt}, \qquad k_y(t_1, t_2) = \frac{\partial^2 k_x}{\partial t_1 \partial t_2}. \tag{4.1}$$

The integrals

$$J = \int_a^b X(t)dt, \quad Z(t) = \int_a^t X(\tau)\,d\tau, \quad Y(t) = \int_a^t g(t, \tau)X(\tau)\,d\tau, \tag{4.2}$$

involving a random function $X(t)$ are defined as the mean-square limits of the Riemannian sums (here $g(t, \tau)$ is a continuous nonrandom function). For example,

$$J = \text{l.i.m.} \sum_{k=1}^{n} X(t_k) \Delta t_k , \qquad t_0 = a , \quad t_n = b , \quad \Delta t_k = t_k - t_{k-1}$$

as $\max_k \Delta t_k \to 0$.

Unlike the continuity and differentiability of processes, we shall assume that the observed integrals always exist, because the integration increases the smoothness of functions $Y(t)$ and $Z(t)$ versus the function $X(t)$.

Let us find the expectation $m_y(t)$ and the correlation function $k_y(t_1, t_2)$ of a random process $Y(t)$. Since the expectation is additive, we have

$$m_y(t) = E(Y(t)) = \int_a^t g(t, \tau) m_x(\tau) \, d\tau ,$$

$$k_y(t_1, t_2) = E\left(Y^0(t_1)\overline{Y^0(t_2)} \right) , \quad Y^0(t) = \int_a^t g(t, \tau) X^0(\tau) \, d\tau . \tag{4.3}$$

Rearranging

$$k_y(t_1, t_2) = E\left(Y^0(t_1) Y^0(t_2) \right) =$$

$$= E\left(\int_a^{t_1} g(t_1, \tau) X^0(\tau) \, d\tau \int_a^{t_2} g(t_2, \tau) X^0(\tau) \, d\tau \right) =$$

$$= \int_a^{t_1} \int_a^{t_2} g(t_1, \tau_1)\overline{g(t_2, \tau_2)} E\left(X^0(\tau_1)\overline{X^0(\tau_2)} \right) d\tau_1 d\tau_2 ,$$

we finally have

$$k_y(t_1, t_2) = \int_a^{t_1} \int_a^{t_2} g(t_1, \tau_1) \, \overline{g(t_2, \tau_2)} \, k_x(\tau_1, \tau_2) \, d\tau_1 \, d\tau_2 . \tag{4.4}$$

Using (4.4) we find the variance of the random process $Y(t)$

$$D_y(t) = k_y(t, t) = \int_a^t \int_a^t g(t, \tau_1) \, \overline{g(t, \tau_2)} \, k_x(\tau_1, \tau_2) \, d\tau_1 \, d\tau_2 , \tag{4.5}$$

which results, in particular, in the fact that it is necessary to know the correlation function of the initial process for the definition of the integral variance.

Formula (4.2) represents the generalization of the Duhamel. integral for the process $Y(t)$, which describes the response of a linear system with one degree of freedom to the force $X(t)$.

The results represented in formulas (4.1)–(4.4) show that several levels of analysis are possible in solving linear problems in mechanics under the action of random forces. The first one is that we are concerned only with the definition of the expectations of processes. In this setting we actually deal with the nonrandom functions. The next level, which we shall focus mainly below, is the *correlation level*, where we define the expectations and the correlation functions. This level of analysis is closed under the operations of differentiation and integration. Besides, the variance is determined automatically, which enables one to assess the possible departure from the expectation. At the same time, as follows from formulas (4.1), (4.5), the expectation and variance do not create the closed level of analysis.

5 Mechanical System with One Degree of Freedom Subject to a Random Force

Consider the Cauchy problem

$$m\ddot{x} + n\dot{x} + cx = F(t), \quad x(0) = a, \quad \dot{x}(0) = v,$$

which describes small forced oscillations of a weight on a spring with the consideration of the resistance. The solution of this problem reads as

$$x(t) = \left(a\cos\omega t + \frac{v+ha}{\omega}\sin\omega t\right)e^{-ht} + \int_0^t g(t-\tau)F(\tau)\,d\tau,$$
$$h = \frac{n}{2m}, \quad \omega = \sqrt{k^2 - h^2}, \quad k^2 = \frac{c}{m},$$
$$g(t-\tau) = \frac{1}{m\omega}e^{-h(t-\tau)}\sin\omega(t-\tau).$$

Let $F(t)$ be a random force with given expectation $m_F(t)$ and correlation function $k_F(t_1, t_2)$. It is required to find the expectation $m_x(t)$ and the correlation function $k_x(t_1, t_2)$ of the solution. For the sake of simplicity, the initial data a and v will be assumed to be nonrandom.

The expectation and the centered perturbation are as follows:

$$m_x(t) = \left(a\cos\omega t + \frac{v+ha}{\omega}\sin\omega t\right)e^{-ht} + \int_0^t g(t-\tau)m_F(\tau)\,d\tau,$$
$$F^0(t) = F(t) - m_F(t). \tag{5.1}$$

Now we find the correlation function with the help of formula (4.4)

$$k_x(t_1, t_2) = \int_0^{t_1}\int_0^{t_2} g(t_1 - \tau_1)g(t_2 - \tau_2)k_F(\tau_1, \tau_2)\,d\tau_1\,d\tau_2 \tag{5.2}$$

(the sign of the complex conjugation is omitted, because we consider the real-valued force $F(t)$). Assuming that $t_1 = t_2 = t$, we find the variance $D_x(t) = k_x(t, t)$.

For the sake of better transparency of the analysis, we consider the perturbation subject to a *white noise*.

If $k_F(t_1, t_2) = H(t_1)\delta(t_1 - t_2)$, where $\delta(t)$ is the Dirac delta function, then a process $F(t)$ is called a *white noise*. Here, a bounded function $H(t_1)$ is the *intensity of a white noise*.

The basic properties of the delta function are as follows: $\delta(t) = 0$ with $t \neq 0$; $\delta(0) = \infty$; $\int_{-\varepsilon}^{\varepsilon} \delta(t)dt = 1$; $\int_{-\infty}^{\infty} f(t)\delta(t - c)dt = f(c)$.

A real process $F(t)$ can be approximately assumed to be a white noise if its correlation function $k_F(t_1, t_2)$ tends to zero quickly enough with increasing the difference $|t_1 - t_2|$; to be more exact, if it is possible to approximately assume that $k_F(t_1, t_2) \approx 0$ with $|t_1 - t_2| > \Delta t$, where Δt is significantly smaller then the characteristic time of a problem. Besides, for the intensity we have

$$H(t_1) = \int_{t_1-\Delta t}^{t_1+\Delta t} k_F(t_1, t_2)dt_2 .$$

In this example the approximation of the process $F(t)$ by a white noise is meaningful if $\Delta t \ll T, T = 2\pi/\omega$.

Let us find the variance of the process $X(t)$:

$$D_x(t) = \int_0^t H(\tau)g^2(t - \tau)\, d\tau . \tag{5.3}$$

Under the assumption that $H(\tau) = H_0 = \text{const}$ and that there are no resistance forces ($h = 0$), we find by (5.3) that

$$D_x(t) = \frac{H_0}{mc}\left(\frac{t}{2} - \frac{\sin 2\omega t}{4\omega}\right) . \tag{5.4}$$

Therefore, with the absence of the resistance the variance of the forced oscillations increases linearly with time, oscillating at the same time with frequency 2ω. If $a = v = m_F = 0$, then the oscillations of the function $X(t)$ will increase proportionally to the root-mean-square deviation $\sigma_x(t) = \sqrt{D_x(t)} \sim \sqrt{t}$. It is interesting to compare this result with the rate of resonance increase of the amplitude under the action of the harmonic excitation, which is proportional to t.

With the presence of resistance ($h > 0$) formula (5.3) gives more cumbersome appearance of the variance $D_x(t)$ in comparison with (5.4), which is not given there. However, as $t \to \infty$ the variance tends to a finite values, which is equal to

$$D_x(\infty) = \frac{H_0\omega^2}{4hc^2} .$$

In the course of time the amplitudes of oscillation first increase and later stabilize.

Figures 2 and 3 give the results of the numerical modeling of the random process $X(t)$ with $a = v = m_F = 0$, $m = c = 1$, $h = 0.1$ under the action of "white noise" with intensity $H = 1$ on time intervals $0 \leqslant t \leqslant 10$ and $0 \leqslant t \leqslant 100$, respectively.

Fig. 2 Modeling of the
process $X(t)$ on the interval
$0 \leqslant t \leqslant 10$

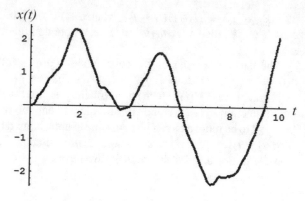

Fig. 3 Modeling of the
process $X(t)$ on the interval

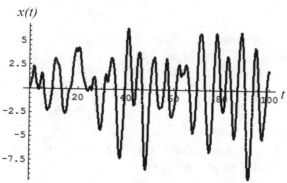

For the adopted values of the parameters the period of free oscillations is close to 2π, which can be proved by counting the number of maxima of the curve on Fig. 3.

6 Correlation Analysis of a Linear Mechanical System with Several Degrees of Freedom

Consider a system with n degrees of freedom of the form

$$A \cdot \ddot{q} + B \cdot \dot{q} + C \cdot q = Q(t), \quad q(0) = \dot{q}(0) = 0, \tag{6.1}$$

where $q(t) = (q_1, \ldots, q_n)^T$ is the sought-for vector of generalized coordinates, $Q(t) = (Q_1, \ldots, Q_n)^T$ are the given vector of generalized forces, A, B, C are the given square matrices of size n with $\det A \neq 0$. The subsequent analysis does not assume the matrices to be symmetric or positive definite.

The solution of the Cauchy problem (6.1) reads as

$$q(t) = \int_0^t G(t, \tau) \cdot Q(\tau) \, d\tau, \tag{6.2}$$

where $G(t, \tau)$ is the square matrix impulse response function. The element $G_{ij}(t, \tau)$ of this matrix is the reaction of the generalized ith coordinate at time t per unit impulse, which is exerted instead of the generalized jth force at time τ.

With $t < \tau$ the matrix $G(t, \tau) \equiv 0$, and for $t \geqslant \tau$ the elements of this matrix satisfy the homogeneous systems of equations

$$A \cdot \frac{\partial^2 G_j}{\partial t^2} + B \cdot \frac{\partial G_j}{\partial t} + C \cdot G_j = 0,$$

$$G_j = \left(G_{1j}, G_{2j}, \ldots, G_{nj} \right)^T, \quad j = \overline{1, n}, \tag{6.3}$$

and the initial conditions with $t = \tau$

$$G_j(\tau, \tau) = 0, \quad A \cdot \frac{\partial G_j}{\partial t} \bigg|_{t=\tau} = \delta_j, \quad \delta_j = \left(\delta_{1j}, \delta_{2j}, \ldots, \delta_{nj} \right)^T,$$

where δ_{ij} is the Kronecker delta.

If the matrices A, B, C are constant, then the system (6.3) can be solved analytically. In this case the matrix impulse response function depends on the difference $G(t, \tau) = \hat{G}(t - \tau)$. If the matrices A, B, C are time-dependent, then system (6.3) should be solved numerically to construct the matrix $G(t, \tau)$.

Let $Q(t)$ and $q(t)$ be vector-functions with random components, where the probability characteristics of perturbation $Q(t)$ are defined and the probability characteristics of the solution $q(t)$ are unknown. In the correlation analysis the vector of the expectation

$$m_Q(t) = E(Q(t)) = \left(m_{Q_1}(t), m_{Q_2}(t), \ldots, m_{Q_n}(t) \right)^T,$$

which consists of expectations of the several components, and the correlation matrix

$$K_Q(t_1, t_2) = E\left(Q^0(t_1) \cdot \overline{Q^{0T}(t_2)} \right) =$$

$$= \begin{pmatrix} K_{Q_1}(t_1, t_2) & K_{Q_1 Q_2}(t_1, t_2) & \ldots & K_{Q_1 Q_n}(t_1, t_2) \\ K_{Q_2 Q_1}(t_1, t_2) & K_{Q_2}(t_1, t_2) & \ldots & K_{Q_2 Q_n}(t_1, t_2) \\ \ldots & \ldots & \ldots & \ldots \\ K_{Q_n Q_1}(t_1, t_2) & K_{Q_n Q_2}(t_n, t_2) & \ldots & K_{Q_n}(t_1, t_2) \end{pmatrix} \tag{6.4}$$

are the sought-for characteristics. On the principal diagonal of (6.4) there are correlation functions of several components of the vector $Q(t)$, and the remaining elements being *relative correlation functions*, which describe the probabilistic relation between the components. They are determined as

$$K_{Q_1 Q_2}(t_1, t_2) = E\left(Q_1^0(t_1)\overline{Q_2^0(t_2)}\right), \quad Q_j^0(t) = Q_j(t) - m_{Q_j}(t).$$

The appearance of the probabilistic relation between the components of the vector $Q(t)$ is caused, in particular, by the following circumstance. The generalized forces Q_j are expressed via the original forces X_i, specified in the Cartesian reference system by the relations

$$Q_j = \sum_{i=1}^{3N} X_i \frac{\partial x_i}{\partial q_j}, \tag{6.5}$$

where the transfer equations from the orthogonal coordinates to generalized ones are equal to $x_i = x_i(q_j, t)$. It follows from (6.5), that the generalized forces Q_j are interdependent, even if the forces X_i are independent.

The correlation matrix $K_Q(t_1, t_2)$ is positive and obeys the condition $K_Q(t_2, t_1) = \overline{K_Q(t_1, t_2)}$.

The substitution of the random vector $Q(t)$ in (6.2) repeats the analysis of the previous section but with a slight difference—the scalar functions are substituted by the vector and matrix ones. So, we get

$$
\begin{aligned}
m_q(t) &= \int_0^t G(t, \tau) \cdot m_Q(\tau) \, d\tau, \\
K_q(t_1, t_2) &= \int_0^{t_1} \int_0^{t_2} G(t_1, \tau_1) \cdot K_Q(\tau_1, \tau_2) \cdot \overline{G^T(t_2, \tau_2)} \, d\tau_1 \, d\tau_2.
\end{aligned}
\tag{6.6}
$$

Formulas (6.6) extend formulas (5.1) and (5.2) in the case when a system has several degrees of freedom.

7 Stationary Random Processes

Stationary random processes constitute an important subclass of random processes. Probabilistic characteristics of stationary random processes are independent of the origin of time and are often encountered in applications. For example, if the noise in a moving train or in a flying plane is looked upon as a random process, then the noise can be described as a stationary process in the long time intervals. The noise by acceleration, deacceleration, and other changes of the motion mode will be nonstationary.

There are two definitions of a stationary process.

Definition 4 A random process $X(t)$ is called *stationary in the narrow sense* if all of its n-dimensional distribution functions (3.1) are independent of time t_1 and depend only on the differences $t_2 - t_1, \ldots, t_n - t_1$.

Definition 5 A random process $X(t)$ is called *stationary in the wide sense* if it has finite expectation and correlation function and, besides, the expectation is constant $(m_x(t) = m_x^0)$ and the correlation function depends only on the difference of arguments $(K_x(t_1, t_2) = k_x(\tau), \tau = t_2 - t_1)$.

None of these definition is not more generalized than the other one. For example, a process may have an infinite expectation in a the narrow sense and the distribution functions of the process in the wide sense may disobey the condition in Definition 4.

A process can be estimated via the sum, as calculated from the set of realizations $X^{(n)}(t)$ of the process

$$\widehat{m}_x(t) = \frac{1}{N} \sum_{n=1}^{N} X^{(n)}(t) .$$

For a specific class of stationary *ergodic processes*, the ensemble averaging can be replaced by the time averaging

$$\widehat{m}_x(t) = \frac{1}{T} \int_0^T X(t) \, dt .$$

The strict definition is as follows.

Definition 6 A process $X(t)$ is called *ergodic* relative to the expectation if

$$m_x = \underset{T \to \infty}{\text{l.i.m.}} \frac{1}{T} \int_0^T X(t) \, dt . \tag{7.1}$$

The ergodicity criteria relative to the expectation is the equation $\lim_{\tau \to \infty} k_x(\tau)$ $\to 0$.

The ergodicity can be introduced relative to the correlation function

$$k_x(\tau) = \underset{T \to \infty}{\text{l.i.m.}} \frac{1}{T} \int_0^T X^0(t) \int_0^T X(t)\overline{X(t+\tau)} \, dt . \tag{7.2}$$

Below it will be supposed that a stationary processes under consideration arc ergodic. The ergodicity allows one to perform a probabilistic analysis even with only one realization at the disposal. For example, one can carry out a probabilistic analysis of the Solar activity (these researches were made earlier and are continued now).

8 The Spectral Density

The use of Fourier integers to represent stationary processes is very convenient for their description and making of linear operations with them.

Recall that for a nonrandom absolutely integrable function $X(t)$

$$\int_{-\infty}^{\infty} |X(t)| \, dt < \infty \tag{8.1}$$

its *Fourier transform* is introduced as

$$A(\omega) = \frac{1}{\sqrt{2\pi}} \int_{-\infty}^{\infty} X(t) e^{i\omega t}\, dt .\tag{8.2}$$

In view of Dirichlet's theorem, the inverse transform

$$X(t) = \frac{1}{\sqrt{2\pi}} \int_{-\infty}^{\infty} A(\omega) e^{-i\omega t}\, d\omega \tag{8.3}$$

is well defined if on every finite interval the function $X(t)$ has a finite number of points of discontinuity of first kind and does not have other singularities. In addition, the integral (8.3) converges as a Cauchy principal value; at points of discontinuity of $X(t)$ formula (8.3) gives the half-sums of the left and right limits.

Clearly condition (8.1) does not hold for a stationary process. This calls for the introduction of the auxiliary function

$$X_T(t) = \begin{cases} X(t), & |t| \leqslant T, \\ 0, & |t| > T, \end{cases}$$

for which formulas (8.2) and (8.3) can be used.

Using the ergodicity and assuming that $m_x = 0$, we write the variance as

$$D_x = \lim_{T \to \infty} \frac{1}{2T} \int_{-T}^{T} X(t)\, \overline{X(t)}\, dt .$$

Transforming this integral with the help of (8.2) and (8.3) we find that

$$D_x = \lim_{T \to \infty} \int_{-\infty}^{\infty} \frac{1}{2T} |A_T(\omega)|^2\, d\omega , \qquad A_T(\omega) = \frac{1}{\sqrt{2\pi}} \int_{-\infty}^{\infty} X_T(t)\, e^{i\omega t}\, dt .$$

We set

$$s_x(\omega) = \lim_{T \to \infty} \frac{1}{2T} |A_T(\omega)|^2 .\tag{8.4}$$

Now, for the variance, we have

$$D_x = \int_{-\infty}^{\infty} s_x(\omega)\, d\omega .\tag{8.5}$$

A similar analysis gives the expression for the correlation function

$$k_x(\tau) = \int_{-\infty}^{\infty} s_x(\omega)\, e^{-i\omega\tau}\, d\omega .\tag{8.6}$$

The function $s_x(\omega)$ is called the *spectral density* or the *power spectral density of the process* $X(t)$. Let us consider, for example, the work A of the resisting forces $F(t) = -nv(t)$, where $v(t)$ is a point velocity. Then the work during time T is equal to $A = -n \int_0^T v^2(t)dt$, and the average power is as follows

$$N = \frac{A}{T} = -\frac{k}{T} \int_0^T v^2(t)\, dt \ .$$

We see that the power is calculated by the same formula as the variance. Expanding the process $v(t)$ in the Fourier integral and singling out the part with the frequencies in the interval of $[\omega, \omega + d\omega]$, the power of the highlighted part will be $s_v(\omega)d\omega$, which explains the origin of the name "power spectral density".

Formula (8.4) is unfit for calculation of the spectral density. However, if the function of correlation is known, then it can be calculated by formula (8.6)

$$s_x(\omega) = \frac{1}{2\pi} \int_{-\infty}^{\infty} k_x(\tau)\, e^{i\omega\tau} d\tau \ . \tag{8.7}$$

According to (8.4), the spectral density is nonnegative $s_x(\omega) \geqslant 0$.

For a real process the correlation function and the spectral density are even functions, which are related as follows:

$$k_x(\tau) = 2 \int_0^{\infty} s_x(\omega) \cos(\omega\tau)\, d\omega \ , \quad s_x(\omega) = \frac{1}{\pi} \int_0^{\infty} k_x(\tau) \cos(\omega\tau)\, d\tau \ . \tag{8.8}$$

Below $D_x = k_x(0)$ and besides, $|k_x(\tau)| \leqslant k_x(0)$ for a real process by (2.3).

Example 1 Consider the process $X(t)$ with the spectral density of the kind

$$s_x(\omega) = \begin{cases} S_0, & |\omega| \leqslant \Omega, \\ 0, & |\omega| > \Omega. \end{cases}$$

According to formula (8.8) the correlation function is proportional to the Dirichlet function

$$k_x(\tau) = 2S_0 \frac{\sin(\Omega\tau)}{\tau}, \quad k_x(0) = D_x = 2S_0\Omega \ .$$

With increasing Ω the maximum of the function $k_x(\tau)$, which is equal to $2S_0\Omega$, increases, the function $k_x(\tau)$ converging to the origin. The graphs of the functions $s_x(\omega)$ and $k_x(\tau)$ are shown in Fig. 4. In the limit

$$\lim_{\Omega \to \infty} k_x(\tau) = S_0 \delta(\tau) \ ,$$

where $\delta(\tau)$ is the delta function, the process $X(t)$ becomes a white noise, and S_0 is its intensity. The name "white noise" is given by analogy with the white light, which incorporates all the frequencies of the visible range. The white noise has infinite power. The power of a white light is finite, because the relation $s_x(\omega) = S_0$ is violated in the ultraviolet and infrared regions.

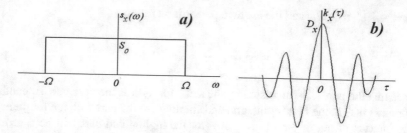

Fig. 4 The spectral density and correlation function of the process $X(t)$

9 The Spectral Decomposition of a Stationary Process

Here we give a sequence of formal operations on random processes that give correct final results.

In spite of the fact that condition (8.1) is not satisfied, we suppose that

$$X(t) = \frac{1}{\sqrt{2\pi}} \int_{-\infty}^{\infty} V_x(\omega) e^{-i\omega t} \, d\omega \,, \qquad V_x(\omega) = \frac{1}{\sqrt{2\pi}} \int_{-\infty}^{\infty} X(t) e^{i\omega t} \, dt \,. \quad (9.1)$$

Let us find the correlation function of the random function $V_x(\omega, \omega_1)$

$$K_{V_x}(\omega, \omega_1) = E(V_x(\omega)\overline{V_x(\omega_1)}) =$$

$$= E\left(\frac{1}{2\pi} \iint_{-\infty}^{\infty} X(t)\, \overline{X(t_1)}\, e^{i\omega t} e^{-i\omega_1 t_1} \, dt \, dt_1 \right) .$$

Assuming that

$$t_1 - t = \tau, \quad E\left(X(t)\overline{X(t_1)}\right) = k_x(\tau), \quad \frac{1}{2\pi} \int_{-\infty}^{\infty} k_x(\tau)\, e^{-i\omega_1 \tau} \, d\tau = s_x(\omega_1),$$

we get

$$K_{V_x}(\omega, \omega_1) = s_x(\omega_1) \int_{-\infty}^{\infty} e^{i(\omega_1 - \omega)t} \, dt = 2\pi s_x(\omega_1)\, \delta(\omega_1 - \omega). \quad (9.2)$$

Substituting the divergent integral in (9.2) by the delta function is justified by the equality

$$V_x(\omega) = \int_{-\infty}^{\infty} \left(\frac{1}{2\pi} \int_{-\infty}^{\infty} e^{i(\omega_1 - \omega)t} dt \right) V_x(\omega_1) \, d\omega_1 \,,$$

which is obtained on excluding $X(t)$ from (9.1).

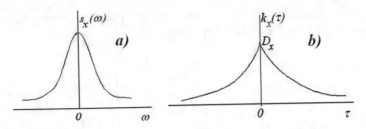

Fig. 5 The spectral density and correlation function of the nondifferentiable process $X(t)$

Decomposition (9.1) allows one to differentiate and integrate stationary random processes. Let us consider the differentiation. Assume that $Y(t) = dX/dt$ and that we know the spectral density $s_x(\omega)$. Our aim is to find $s_y(\omega)$.

We write $Y(t)$ as

$$Y(t) = \frac{1}{\sqrt{2\pi}} \int_{-\infty}^{\infty} V_y(\omega) e^{-i\omega t} d\omega.$$

Differentiating the first equality in (9.1) with respect to t, we obtain $V_y(\omega) = i\omega V_x(\omega)$. Calculating the correlation function $V_y(\omega)$ in two ways we get the equality

$$K_{V_y}(\omega, \omega_1) = 2\pi s_y(\omega_1)\delta(\omega_1 - \omega) = \omega\omega_1 K_{V_x}(\omega, \omega_1) = 2\pi\omega\omega_1 s_x(\omega_1) \delta(\omega_1 - \omega),$$

which secures the sought-for formula

$$s_y(\omega) = \omega^2 s_x(\omega). \tag{9.3}$$

Note that we have left aside the question about existence of the derivative. A criteria for its existence is the following condition

$$D_y = \int_{-\infty}^{\infty} s_y(\omega) d\omega < \infty. \tag{9.4}$$

Assume for example that $s_x(\omega) = D/(\omega^2 + \alpha^2)$. We have $s_y(\omega) = D\omega^2/(\omega^2 + \alpha^2)$ and so condition (9.4) is not fulfilled. Therefore, the random function $X(t)$ does not have the derivative. The graphs of the functions $s_x(\omega)$ and $k_x(\tau)$ are given in Fig. 5. If $s_x(\omega) = D/((\omega^2 - \omega_0)^2 + \alpha^2)$, then the function $X(t)$ has the first derivative and does not have the second one.

Formula (9.3) can be also derived without using the spectral decomposition. From formula (4.1) we have for a stationary process

$$k_y(\tau) = -\frac{d^2 k_x}{d\tau^2}, \quad Y = \frac{dX}{dt}.$$

Hence,

$$s_y(\omega) = \frac{1}{2\pi} \int_{-\infty}^{\infty} k_y(\tau) e^{i\omega\tau} d\tau = -\frac{1}{2\pi} \int_{-\infty}^{\infty} \frac{d^2 k_x}{d\tau^2} e^{i\omega\tau} d\tau.$$

Integrating by parts twice and taking into the consideration the ergodicity of $k_x(\tau) \to 0$ with $\tau \to \pm\infty$ we get

$$s_y(\omega) = \omega^2 \frac{1}{2\pi} \int_{-\infty}^{\infty} k_x(\tau) e^{i\omega\tau} d\tau = \omega^2 s_x(\omega).$$

10 Oscillations of a Mechanical System with One Degree of Freedom Under Stationary Random Perturbation

Consider the equation of second order with constant coefficients

$$m\frac{d^2 X}{dt^2} + n\frac{dX}{dt} + cX = F(t), \quad m > 0 \tag{10.1}$$

where we assume that the function $F(t)$ is a stationary random process with the given expectation m_F and the spectral density $s_F(\omega)$. It is required to find the conditions for the existence of a stationary solution $X(t)$ of Eq. (10.1) and find m_x, $s_x(\omega)$. Taking into account that the solution is a sum of the general solution of the homogeneous equation and the specific solution of the inhomogeneous equation, it is necessary to require that the solution of the general solution would tend to zero (the condition that $n > 0$, $c > 0$ is a necessary and sufficient for this). The resulting motion will become a stationary process once the free oscillations are decayed.

The expectation $m_x = m_F/c$ of the solution is found by calculating the expectation of both parts of Eq. (10.1) and considering that the expectations of the derivatives are all zero.

To calculate $s_x(\omega)$ we set

$$X(t) = \frac{1}{\sqrt{2\pi}} \int_{-\infty}^{\infty} V_x(\omega) e^{-i\omega t} d\omega, \quad F(t) = \frac{1}{\sqrt{2\pi}} \int_{-\infty}^{\infty} V_F(\omega) e^{-i\omega t} d\omega.$$

Using Eq. (10.1), this gives

$$(c - m\omega^2 - in\omega)V_x = V_F, \quad V_x = S(\omega)V_F(\omega), \quad S(\omega) = \frac{1}{c - m\omega^2 - in\omega},$$

where the function $S(\omega)$ is called the *transfer function*, because with a perturbation of $F(t) = e^{-i\omega t}$ the solution of Eq. (10.1) is $X(t) = S(\omega)e^{-i\omega t}$.

As in the previous paragraph, expressing the correlation function $K_{V_x}(\omega, \omega_1)$ in terms of $K_{V_F}(\omega, \omega_1)$ we finally obtain

$$s_x(\omega) = s_F(\omega)|S(\omega)|^2 = \frac{s_F(\omega)}{(c - m\omega^2)^2 + n^2\omega^2}. \tag{10.2}$$

If the spectral density $s_x(\omega)$ is known, it is possible to find the variance D_x by formula (8.5), and hence, estimate the oscillation amplitude.

We note that the process $X(t)$ is unstationary without resistance ($n = 0$); formula (8.5) gives that $D_x = \infty$.

Example 2 The rolling ship motion is described by the equation

$$J\frac{d^2\varphi}{dt^2} + n\frac{d\varphi}{dt} + c\varphi = M(t),$$

where φ is the angle of roll, J is the ship moment of inertia about the roll axis with the inertia torque of the added water masses, n is the coefficient of resistance, $-c\varphi$ is the restoring torque, $M(t)$ is the moment of wave action. The function $M(t)$ is assumed to be a stationary random process with spectral density $s_M(\omega)$. According to the recommendation of the International Congress at Delft (1964) the function $s_M(\omega)$ can be written as

$$s_M(\omega) = D_M\Phi(z), \quad z = \frac{\omega}{\omega_a}, \quad \Phi(z) = 0.88z^{-5}e^{-0.44z^{-4}},$$

where the variance D_M and the frequency ω_a characterize the waves height and their average frequency; the dimensionless function $\Phi(z)$ is shown in Fig. 6.

According to formula (10.2) the variance of the rolling motion D_φ is calculated by

$$D_\varphi = 2\int_0^\infty s_M(\omega)|S(\omega)|^2 \, d\omega = 2\int_0^\infty \frac{s_M(\omega)\,d\omega}{(c - J\omega^2)^2 + n^2\omega^2}.$$

Fig. 6 Typical spectral
density of rough sea

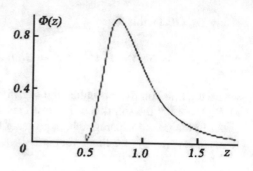

The maximum of the function $s_M(\omega)$ is attained with $\omega = \omega_*$ (see Fig. 6). With small values of n the function $|S(\omega)|^2$ has a maximum near the resonance frequency $\omega_0 = \sqrt{c/J}$. The calculations show that the maximum variance of the rolling motion D_φ is attained with $\omega_* \approx \omega_0$.

Therefore *the definition of resonance is extended to oscillations with random stationary perturbation.*

11 Oscillations of a Mechanical System with Several Degrees of Freedom Under a Stationary Random Perturbation

As the initial one we consider the system (6.1) of differential equations with constant coefficients

$$A \cdot \ddot{q} + B \cdot \dot{q} + C \cdot q = Q(t), \qquad (11.1)$$

where $q(t) = (q_1, \dots, q_n)^T$ is the unknown vector of generalized coordinates, $Q(t) = (Q_1, \dots, Q_n)^T$ is a given vector of generalized forces, A, B, C are the given square positive definite matrices of order n. Unlike (6.1), Q is a stationary random vector of generalized forces with the given vector of expectations m_Q and the spectral matrix $S_Q(\omega)$. No initial data are imposed. Our aim is to find the expectation m_q and the spectral matrix $S_q(\omega)$ of the particular stationary solution $q(t)$.

As in (8.7), the spectral matrices are related with the correlation matrices by the equations

$$S_Q(\omega) = \frac{1}{2\pi} \int_{-\infty}^{\infty} K_Q(\tau) e^{i\omega\tau} d\tau \,,$$
$$S_q(\omega) = \frac{1}{2\pi} \int_{-\infty}^{\infty} K_q(\tau) e^{i\omega\tau} d\tau \,.$$

Along the principal of the spectral functions there are the spectral functions of elements of random vectors; the remaining elements are the mutual spectral functions $s_{Q_k Q_m}(\omega)$. The spectral matrices are Hermitian; that is,

$$S_Q^T = \overline{S_Q},$$
$$z^T \cdot S_Q \cdot \overline{z} \geqslant 0, \quad z = (z_1, \ldots, z_n)^T.$$

In other words the transposed matrix is equal to the conjugated one, and the quadric form, as constructed from this matrix, is nonnegative.

In analogy with the one degree of freedom, it is easily verified that

$$m_q = C^{-1} \cdot m_Q,$$
$$S_q(\omega) = S(\omega) \cdot S_Q(\omega) \cdot \overline{S^T(\omega)},$$
(11.2)

where $S(\omega)$ is the matrix of the transposition functions,

$$S(\omega) = ((i\omega)^2 A + i\omega \, B + C)^{-1}.$$

It is worth noting that the matrix impulse response function $G(t_1, t_2)$ (see (6.2)) for a system with constant coefficients depends only on the difference between the arguments $\tau = t_2 - t_1$ and is related with the matrix of the transposition functions by the formula

$$S(\omega) = \int_0^\infty G(\tau) \, e^{-i\omega\tau} d\tau.$$

Example 3 Let us consider oscillations of a two-wheeler (for instance, a motorbike) moving with a constant velocity v along a rough road. Small oscillations and translations in the vertical plane will be modeled by a system with two degrees of freedom. The motion equations can be written as

$$m_1\ddot{x}_1 + m_{12}\ddot{x}_2 + n\dot{\Delta}_1 + c\Delta_1 = 0, \; x_1 = \Delta_1 - \xi_1,$$
$$m_{12}\ddot{x}_1 + m_2\ddot{x}_2 + n\dot{\Delta}_2 + c\Delta_2 = 0, \; x_2 = \Delta_2 - \xi_2,$$
(11.3)

where x_1, x_2 are the vertical displacements of the frame above the wheels, Δ_1, Δ_2 are the deformations of the springs, ξ_1, ξ_2 are the sizes of the irregularities under the wheels (Fig. 7), m_1, m_{12}, m_2 are the mass characteristics ($m_{12}^2 < m_1 m_2$), n is

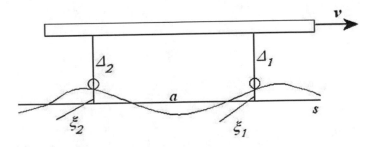

Fig. 7 Car motion on a road of random roughness

the coefficient of resistance, c is the spring stiffness. The road irregularities are described by a random function $\xi(s)$ with the given correlation function $k_\xi(\lambda) = E\{\xi(s)\xi(s+\lambda)\}$. If the velocity is v, then $\xi_1(t) = \xi(vt)$, $\xi_2(t) = \xi(vt - a)$, where a is the space between the wheels.

Rewriting system (11.3) in the form (11.1) with $n = 2$, this gives

$$A = \begin{pmatrix} m_1 & m_{12} \\ m_{12} & m_2 \end{pmatrix}, \qquad B = \begin{pmatrix} \bar{n} & 0 \\ 0 & \bar{n} \end{pmatrix}, \qquad C = \begin{pmatrix} c & 0 \\ 0 & c \end{pmatrix},$$

$$q = \begin{pmatrix} x_1 \\ x_2 \end{pmatrix}, \qquad Q = \begin{pmatrix} Q_1 \\ Q_2 \end{pmatrix} = \begin{pmatrix} \bar{n}\dot{\xi}_1 + c\xi_1 \\ \bar{n}\dot{\xi}_2 + c\xi_2 \end{pmatrix}.$$

In order to employ formula (11.2) one needs to calculate the matrix of spectral densities for the perturbation $S_Q(\omega)$. We first find

$$s_{\xi_1}(\omega) = \frac{1}{2\pi} \int_{-\infty}^{\infty} k_{\xi_1}(\tau) e^{i\omega\tau} d\tau = \frac{1}{2\pi} \int_{-\infty}^{\infty} k_\xi(v\tau) e^{i\omega\tau} d\tau =$$

$$= \frac{1}{2\pi v} \int_{-\infty}^{\infty} k_\xi(\lambda) e^{i\lambda(\omega/v)} d\lambda = \frac{1}{v} s_r(\omega/v).$$

Here,

$$s_r(p) = \frac{1}{2\pi} \int_{-\infty}^{\infty} k_\xi(\lambda) e^{i\lambda p} d\lambda$$

is the road spectral characteristic. The both wheels travel along the same road surface, and therefore $s_{\xi_2}(\omega) = s_{\xi_1}(\omega)$. Similar calculations show that $s_{\xi_1\xi_2}(\omega) = s_{\xi_1}(\omega)e^{i\omega a/v}$, $s_{\xi_2\xi_1}(\omega) = \overline{s_{\xi_1\xi_2}(\omega)}$.

Using the spectral decomposition from Sect. 9, we finally get

$$S_Q(\omega) = \frac{1}{v} s_r(\omega/v)(n^2\omega^2 + c^2) \cdot \begin{pmatrix} 1 & e^{i\omega a/v} \\ e^{-i\omega a/v} & 1 \end{pmatrix}. \qquad (11.4)$$

Formula (11.4) allows one judge about the dependence of perturbation on the velocity of motion v and to find the spectral matrix of the solution with the help of formula (11.2).

12 Nonlinear and Statistically Nonlinear Problems

So far we were concerned only with linear problems of mechanical oscillations under the action of random forces. For such problems, the solutions were constructed on the correlation level of description; that is, probabilistic characteristics of the solution were found using the given expectations and correlation functions of perturbation. Here, no information about the distribution functions as a given perturbation as well as desired solution was used. In nonlinear and statistically nonlinear problems the results depend on the information (or the assumptions) about the distribution functions.

A sufficiently general form of a nonlinear system with n degrees of freedom with a random perturbation is as follows:

$$m_k \ddot{x}_k = F_k(x_j, \dot{x}_j, \xi_p(t), t), \qquad k, j = \overline{1, n}, \quad p = \overline{1, P}, \qquad (12.1)$$

here F_k are random nonlinear functions of their arguments, $\xi_p(t)$ are random functions of time with given probabilistic characteristics. It is required to find the probabilistic characteristics of the unknown functions $x_k(t)$.

Statistically, nonlinear problems are linear ones with random parametric perturbation—for example, the Hill equation

$$\ddot{x} + \xi_1(t) \, x = 0 \qquad (12.2)$$

or the more general equation

$$\ddot{x} + (a + b\,\xi_1(t))\,\dot{x} + (c + d\,\xi_2(t))\,\dot{x} = \xi_3(t) \qquad (12.3)$$

with random functions $\xi_j(t)$. The name "statistic nonlinearity" is connected with two products of random functions involved in Eqs. (12.2) and (12.3). The correlation level of the solution is inappropriate for the solution of problems (12.1)–(12.3).

A discussion about the solution methods of these problems is beyond the scope of this book.[2] Here we shall be only focused on two approximate solution methods the method of statistic linearization and the method of the numeric modeling. In Sect. 13 the method of construction of an exact solution for Markov random processes will be presented.

The method of the statistic linearization. Let us consider this method in the context of the equation

$$\ddot{x} + n\dot{x} + f(x) = \xi(t), \qquad (12.4)$$

where $f(x)$ is a nonlinear function. Assume that the random process $\xi(t)$ is stationary and that its expectation m_ξ and the spectral density $s_\xi(\omega)$ are known.

---

[2] Solutions to these problems can be found in the three books given in the previous remark.

The idea of the method of the statistic linearization is the following. Let us write the unknown function in the form $x(t) = m_x + x^0(t)$, where m_x is its expectation. We assume that $f(x) = k_0 m_x + k_1 x^0$ approximately, where the coefficients k_0 and k_1 (the gain coefficients of the useful signal and the fluctuations; see the first monograph in the previous footnote) have to be determined. For the stationary solution of the linearized problem we find

$$m_x = \frac{m_\xi}{k_0}, \quad s_x(\omega, k_1) = \frac{s_\xi(\omega)}{(\omega^2 - k_1^2)^2 + n^2 \omega^2}.$$

The next stage of the solution is to evaluate the coefficients k_0 are k_1 using the minimum condition of the variance of the deviation $f(x)$ of $k_0 m_x + k_1 x^0$

$$\min_{k_0, k_1} E\{(f(x) - k_0 m_x - k_1 x^0)^2\}.$$

We get

$$k_0 = \frac{E\{f(x)\}}{m_x}, \quad k_1 = \frac{E\{f(x)(x - m_x)\}}{D_x(k_0, k_1)}, \quad D_x = \sigma_x^2 = \int_{-\infty}^{\infty} s_x(\omega)\, d\omega.$$

To complete the calculations requires accepting the hypothesis about the density $p(x, m_x, \sigma_x)$ of the single measured distribution of a random variable x with still unknown expectation m_x and the root-mean-square deviation σ_x. Most often one uses here the normal law of distribution (11.4), but sometimes if one knows the exact solution, other distribution laws give a more accurate result. Now the problem is reduced to the system of four algebraic equations

$$k_0 m_x = m_f, \quad k_0 m_x = \int_{-\infty}^{\infty} f(x) p(x, m_x, \sigma_x) dx,$$

$$k_1 \sigma_x^2 = \int_{-\infty}^{\infty} f(x)(x - m_x) p(x, m_x, \sigma_x) dx,$$

$$\sigma_x^2 = \int_{-\infty}^{\infty} s_x(\omega, k_1) d\omega \tag{12.5}$$

in the unknowns k_0, k_1, m_x, σ_x.

Let us consider a particular case, when there is an exact solution of this problem. Assume that process $\xi(t)$ is a white noise with $m_\xi = 0$, $s_\xi(\omega) = S_0$, and that the nonlinear function is $f(x) = x - x^3$. Then $k_0 = m_x = 0$, and so system (12.5) keeps only the third and the fourth equations, the integrals can be calculated explicitly. Using these equations, we find that

$$\sigma_x^2 = \frac{\pi S_0}{k_1^2 n}, \qquad k_1 \sigma_x^2 = \sigma_x^2 - 3\sigma_x^4,$$

whence for k_1 we get the cubic equation

$$k_1^2(1 - k_1) = a, \qquad a = \frac{3\pi S_0}{n}. \qquad (12.6)$$

Only the roots of the equation from the interval $I = 0 < k_1 < 1$ are meaningful. The left-hand side of this equation assumes a maximum value $a_* = 4/27$ with $k_1 = k_1^* = 2/3$. For $a < a_*$ Eq. (12.6) has two roots $k_1^{(1)}$, $k_1^{(2)}$ on the interval I, and also $1 > k_1^{(1)} > k_1^* > k_1^{(2)} > 0$. To these roots there correspond the standard deviations $\sigma_x^{(j)} = \sqrt{a/3}/k_1^{(j)}$, $j = 1, 2$, $\sigma_x^{(1)} < \sigma_x^{(2)}$. One of the other type of motion is realized depending on the initial data.

If $a > a_*$, then there are no stationary solutions. It is clear, because for a sufficiently large level of perturbation (in comparison with the resistance forces) the solution falls in the region of $|x| > 1$, while the solutions of the homogeneous equation (12.4) unboundedly increase with $|x| > 1$.

Of course, the so-constructed solution is only approximate. It correctly predicts the presence of perturbation limits for which only the stationary solution exists. There are certain doubts that there exist two stationary solutions. This question can be solved, in particular, by numerical modeling, with which we shall be concerned below.

Modeling of motion of mechanical systems with random perturbation. Assume that we need to model the solution of a system subject to random perturbations (12.1) under given initial data $\xi_p(t)$ and given probabilistic characteristics. The first step here is to construct realizations $\xi_p^{(j)}(t)$ of the random processes $\xi_p(t)$. Next, the system (12.1) is integrated numerically and we find the realization $x_k^{(j)}(t)$. This procedure is repeated until we get the necessary number of realizations $j = \overline{1, J}$, which are further analyzed by the machinery of mathematical statistics. In this way we find estimates of the probabilistic characteristics of the solution. The estimates $\widehat{m}_x(t)$ of the expectation and $\widehat{k}_x(t_1, t_2)$ of the correlation function of the solution can be found from the formulas

$$\widehat{m}_x(t) = \frac{1}{J} \sum_{j=1}^{J} x_k^{(j)}(t),$$

$$\widehat{k}_x(t_1, t_2) = \frac{1}{J} \sum_{j=1}^{J} (x_k^{(j)}(t_1) - \widehat{m}_x(t_1))(x_k^{(j)}(t_2) - \widehat{m}_x(t_2)).$$

If one is expected to get a stationary ergodic solution, then it would be sufficient to construct only one realization, discarding the initial time interval on which the initial data have some effect and make statistical analysis with the help of formulas (7.1), (7.2).

The methods of random processes modeling are very diverse and depend on their probabilistic characteristics (see the previous footnote). Let us now consider the problem of modeling a stationary centered random process $\xi(t)$ with the prescribed spectral density $s_\xi(\omega)$.

Let $[\Omega_0, \Omega_1]$ be a frequency interval whose interior contains the frequencies under consideration. The interval $[\Omega_0, \Omega_1]$ is chosen in such a way the effect of the terms in (12.7) with lower and higher frequencies can be neglected. The process $\xi(t)$ is modeled as follows:

$$\xi(t) = \sum_{n=1}^{N} \sqrt{2s_\xi(\omega_n)\Delta\omega} \left(\eta_n \cos \omega_n t + \zeta_n \sin \omega_n t\right),$$
$$\Delta\omega = \frac{\Omega_1 - \Omega_0}{N}, \quad \omega_n = \Omega_0 + \left(n - \tfrac{1}{2}\right)\Delta\omega. \tag{12.7}$$

Here ω_n are the uniformly distributed frequencies on the interval $[\Omega_0, \Omega_1]$, η_n and ζ_n are the independent random variables with zero average and unit variance. There are random number generator that produce independent uniformly distributed numbers from the interval $(0,1)$. Let α_n and β_n be such numbers. Then the numbers

$$\eta_n = \sqrt{-2\log\alpha_n}\,\cos 2\pi\beta_n, \quad \zeta_n = \sqrt{-2\log\alpha_n}\,\sin 2\pi\beta_n$$

are independent and normally distributed, have zero average and unit variance.

13 Markov Processes. The Fokker–Planck–Kolmogorov (FPK) Equation

Let us firstly consider the values $X^{(1)}, X^{(2)}, \ldots, X^{(k-1)}, X^{(k)}$ of an n-dimensional vector process $X = \{x_1, x_2, \ldots, x_n\}$ at time $t_1 < t_2 < \ldots < t_{k-1} < t_k$. In the general case the value $X^{(k)}$ of the process X depends on all of the values $X^{(1)}, X^{(2)}, \ldots, X^{(k-1)}$.

Definition 7 A process $X(t)$ is called a *Markov process* if the value $X^{(k)}$ of the process X depends on $X^{(k-1)}$ and is independent of the previous values.

The probabilistic characteristics of a Markov process are completely determined by the density $p(Y, \tau, X, t)$ of the transition probability from the value of the process $Y(\tau)$ to $X(t)$.

Let us consider the Cauchy problem for the system of stochastic differential equations of first order

$$\frac{dx_i}{dt} = a_i(x_j, t) + \sum_{k=1}^{n} \sigma_{ik}(x_j, t)\, \xi_k(t)\,, \quad x_i(\tau) = y_i\,, \quad i, j = \overline{1, n}\,, \tag{13.1}$$

where $\xi_k(t)$ are the independent standard processes (white noise with the correlation function $k_{x_{i_k}}(t_1, t_2) = \delta(t_2 - t_1)$; $\delta(t)$ is delta function. Problem (13.1) describes the vector Markov process with probability density of the transition $p(y_1, \ldots, y_n, \tau, x_1, \ldots, x_n, t)$, satisfying the FPK partial differential equation (see the third monograph in the last footnote)

$$\frac{\partial p}{\partial t} = -\sum_{i=1}^{n} \frac{\partial}{\partial x_i}(a_i\, p) + \frac{1}{2} \sum_{i,j=1}^{n} \frac{\partial^2}{\partial x_i \partial x_j}(\sigma_{ij}\, p)\,, \quad x_i(\tau) = y_i\,.$$

For a stationary process the density $p_0(x_1, \ldots, x_n)$ is the target value, which satisfies the equation

$$-\sum_{i=1}^{n} \frac{\partial}{\partial x_i}(a_i\, p_0) + \frac{1}{2} \sum_{i,j=1}^{n} \frac{\partial^2}{\partial x_i \partial x_j}(\sigma_{ij}\, p_0) = 0\,. \tag{13.2}$$

Let us apply the above arguments to Eq. (12.4), which we first rewrite as the first-order system of equations

$$\begin{aligned} \frac{dx_1}{dt} &= x_2\,, \\ \frac{dx_2}{dt} &= -n\, x_2 - f(x_1) + 2\,\pi S_0\, \xi(t)\,. \end{aligned} \tag{13.3}$$

The FPK equation (13.2) reads as

$$-x_2 \frac{\partial p}{\partial x_1} + \frac{\partial}{\partial x_2}((nx_2 + f(x_1))p) + \pi S_0 \frac{\partial^2 p}{\partial x_2^2} = 0\,. \tag{13.4}$$

Integrating (13.4) we find

$$p(x_1, x_2) = C \exp\left\{ -\frac{n}{\pi S_0}\left(\frac{x_2^2}{2} + \int_0^{x_1} f(x)\, dx \right) \right\}\,, \tag{13.5}$$

where the constant C is found from the normalization condition

$$\iint_{-\infty}^{\infty} p(x_1, x_2) \, dx_1 \, dx_2 = 1 \,. \tag{13.6}$$

In particular, the variance of steady-state oscillations is evaluated by the formula

$$\sigma_x^2 = \iint_{-\infty}^{\infty} x_1^2 \, p(x_1, x_2) \, dx_1 \, dx_2 \,.$$

Substituting (13.5) and integrating in x_2 we get

$$\sigma_x^2 = \pi C \sqrt{\frac{2 S_0}{n}} \iint_{-\infty}^{\infty} x_1^2 \exp\left\{ -\frac{n}{\pi S_0} \int_0^{x_1} f(x_1) \, dx_1 \right\} dx_1 \,. \tag{13.7}$$

The further analysis depend on the form of the function $f(x)$. If, for example, $f(x) = x + ax^3, a > 0$, then we can proceed with formulas (13.6), (13.7). If $a < 0$, then the integrals in these formulas are divergent, and hence the FPK equation is unapplicable for the construction of a steady-state solution.

In system (13.1) the processes $\xi_k(t)$ are white noises, which is a serious constraint. However, in a number of cases this constraint can be circumvented by increasing the order of the system of equations.

Let us illustrate this on Eq. (12.4), in which $\xi(t)$ is a stationary random process with the fractionally rational spectral density $S_\xi(\omega)$. Consider

$$S_\xi(\omega) = \frac{S_0}{(\omega^2 - \omega_0^2)^2 + \nu^2 \omega^2} \,.$$

For small ν the function $S_\xi(\omega)$ has a maximum near $\omega = \omega_0$. The function $x_i(t)$ satisfies the equation

$$\frac{d^2 \xi}{dt^2} + \nu \frac{d\xi}{dt} + \omega_0^2 \xi = 2 \pi S_0 \, \xi_0(t) \,, \tag{13.8}$$

where $\xi_0(t)$ is the *standard white noise*. Augmenting system (13.3) with Eq. (13.8), we get the fourth-order system of the form (13.1)

$$\frac{dx_1}{dt} = x_2 \,, \qquad\qquad \frac{dx_3}{dt} = x_4 \,,$$
$$\frac{dx_2}{dt} = -n x_2 - f(x_1) + x_3 \,, \quad \frac{dx_4}{dt} = -\nu x_4 - \omega_0^2 x_3 + 2 \pi S_0 \, \xi_0(t) \,.$$

The corresponding stationary FPK equation reads as

$$-x_2\frac{\partial p}{\partial x_1} + \frac{\partial(nx_2 + f(x_1) - x_3)p}{\partial x_2} - x_4\frac{\partial p}{\partial x_3} +$$
$$+\frac{\partial(\nu x_4 + \omega_0^2 x_3)p}{\partial x_4} + \pi S_0\frac{\partial^2 p}{\partial x_4^2} = 0. \tag{13.9}$$

In contrast to Eqs. (13.4), (13.9) is not solved explicitly. This is why the described method is rarely used in applications.

Chapter 8
Physical Impact Theory

S. A. Zegzhda

This chapter concerns the classical impact theory, although it starts with the Hertz theory of the impact of elastic spheres. We do it in order to clearly outline the scope of the problems where the basic preconditions of the classical impact theory are valid. We discuss in detail the concept of the restitution coefficient introduced by Newton. We give a new deduction of the algebraic system of first and second kind Lagrange equations, corresponding to the classical impact theory of mechanical systems with ideal constraints. It is essential that this system of equations takes a particularly simple form in a number of tasks due to the use of quasivelocities. As an example we consider the impact on a straight chain of rods and on a chain situated on a circular arc. In these problems the Lagrange equations written in terms of quasivelocities have the same form that the finite-difference equations. The use of well-developed methods of solving finite-difference equations allowed us to build an analytical solution of the two given problems. We also consider other important examples of application of the considered methods of classical impact theory.[1]

1 Central Impact of Two Bodies

Hertz theory of balls collision. The collision of bodies is a complex process where the shape of the bodies and the impact velocity are the main factors. The most simple and perfect body shape is a ball. Therefore, the Hertz theory of balls collision is actually the most simple and perfect of all the impact theories. Consider now this theory.

Suppose that before the collision the balls move translationally with the velocities directed along the straight line joining their centers. Assume that this line coincides with the fixed axis z (Fig. 1), and the coordinate z_1 of the mass center of the first

---

[1] For more details on various aspects of the theory of shock, see the book *S. A. Zegzhda. Collision of Elastic Bodies. St. Petersburg: Izd. St. Petersb. Un-ta. 1997. 316 p. [in Russian]*.

© Springer Nature Switzerland AG 2021
N. N. Polyakhov et al., *Rational and Applied Mechanics*,
Foundations of Engineering Mechanics,
https://doi.org/10.1007/978-3-030-64118-4_8

Fig. 1 Balls collision

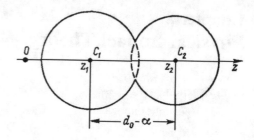

ball is always less than the coordinate z_2 of the mass center of the second ball. If the velocity $v_1 = \dot{z}_1$ of the first ball with the radius R_1 is greater than the velocity $v_2 = \dot{z}_2$ of the second ball with the radius R_2, then at some instant of time $t_0 = 0$ the balls will collide touching one another at one point. We denote $P(t)$ the contact force at the collision. The force $P(t)$ is called the *impact force*. The motion equations of the mass centers of the balls have in our notations the form

$$m_1 \ddot{z}_1 = -P(t), \quad m_2 \ddot{z}_2 = P(t).$$

It follows that

$$\ddot{z}_1 - \ddot{z}_2 = -\left(\frac{1}{m_1} + \frac{1}{m_2}\right) P(t) = -\frac{1}{m} P(t), \tag{1.1}$$

where $m = m_1 m_2 / (m_1 + m_2)$ is the reduced mass.

The distance $d = z_2 - z_1$ between the mass centers of the balls, equal to $d_0 = R_1 + R_2$ at the time of the contact at one point, becomes $d_0 - \alpha$ during the collision, i.e., $\alpha = d_0 + z_1 - z_2$.

Given the above, Eq. (1.1) becomes

$$m \ddot{\alpha} = -P(t). \tag{1.2}$$

The α value equal to the relative displacement of the mass centers of the balls is called the *local plastic compression*. The use of the term "compression" is not caused by the assumption of the plastic deformation, but by the fact that the relative displacement of the balls during the force interaction is mainly due to their deformation (compression) in the contact zone.

The problem of a static compression of balls in the framework of the classical theory of elasticity was solved by Hertz in 1881. His solution is very fine but complicated. The problem is reduced to an integral equation whose solution can be built using the potential theory. It follows from the obtained solution that the deformations decrease rapidly with the distance from the point of contact. The relationship between the contact force P and the displacement α established by Hertz has the form

$$P = K \alpha^{3/2}, \quad K = \left(\frac{1 - \nu_1^2}{E_1} + \frac{1 - \nu_2^2}{E_2}\right)^{-1} \frac{4}{3} \sqrt{\frac{R_1 R_2}{R_1 + R_2}}. \tag{1.3}$$

Here ν_1 and ν_2 are Poisson's moduli, and E_1 and E_2 Young's moduli of the first and the second ball, respectively.

Strictly speaking, the formula (1.3) is approximative, since it is obtained assuming that the balls are deformed in the contact area as flat elastic half-spaces. However, since the radius of the contact area is much smaller than the radii of the balls, this assumption greatly simplifies the solution of the problem and leads to the results which are accurate enough and in good agreement with the experimental data. Moreover, as shown by the latest theoretical and experimental studies, the formula (1.3) can also be used for a variable force $P(t)$, i.e., in the dynamics (for example, for the collision of balls), provided that the run time of a transverse wave across the radius of the contact area is much smaller than the rise time of the force. At the impact velocities when plastic deformations of the balls do not occur, this condition is always satisfied. Thus, we can assume that at the elastic balls collision the values P and α in (1.2) are related by (1.3), i.e.,

$$m\ddot{\alpha} = -K\alpha^{3/2}. \tag{1.4}$$

Note that the same equation results from the solution of the problem of a mass m falling down with the velocity $v_0 = \dot{\alpha}(0) = \dot{z}_1(0) - \dot{z}_2(0) = v_1 - v_2$ on a fixed massless nonlinear spring with the restoring force proportional to the power $3/2$ of the displacement.

The nonlinear Eq. (1.4) is an equation of the form $\ddot{x} = -f(x)$, which admits the existence of the energy integral

$$\frac{1}{2}m(\dot{\alpha}^2 - v_0^2) = -K \int_0^\alpha \alpha^{3/2} d\alpha = -\frac{2}{5}K\alpha^{5/2}. \tag{1.5}$$

At the time when the relative displacement velocity $\dot{\alpha}$ zeroises, the value α reaches its maximum α_{max}, hence

$$\alpha_{max} = \left(\frac{5mv_0^2}{4K}\right)^{2/5}, \tag{1.6}$$

$$P_{max} = K\alpha_{max}^{3/2} = K\left(\frac{5mv_0^2}{4K}\right)^{3/5}. \tag{1.7}$$

At the first stage of the collision, when the value of α increases with time, Eq. (1.5) can be written as

$$\frac{d\alpha}{dt} = \sqrt{v_0^2 - \frac{4K}{5m}\alpha^{5/2}} = v_0\sqrt{1 - \left(\frac{\alpha}{\alpha_{max}}\right)^{5/2}}. \tag{1.8}$$

Its integration gives

Fig. 2 The graph of the contact force

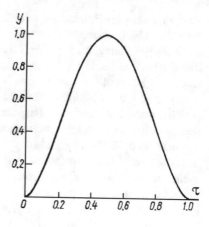

$$t = \frac{\alpha_{max}}{v_0} \int_0^{\eta} \frac{d\eta}{\sqrt{1 - \eta^{5/2}}}, \quad \eta = \frac{\alpha}{\alpha_{max}}.$$

The indefinite integral of the function $(1 - \eta^{5/2})^{-1/2}$ cannot be expressed in terms of known functions. Therefore, the dependence of t on η can only be found by numerical integration. The definite integral at $\eta = 1$ is expressed in terms of $\Gamma(x)$ function. It results in

$$I = \int_0^1 \frac{d\eta}{\sqrt{1 - \eta^{5/2}}} = \frac{2\sqrt{\pi}\,\Gamma(\frac{2}{5})}{5\Gamma(\frac{9}{10})} = 1.4716.$$

The time of rise of the contact force $P(t)$ is equal to the time of its decrease, so the time of collision t_* is expressed in terms of the calculated integral I by the formula

$$t_* = \frac{2I\alpha_{max}}{v_0} = \frac{2.9432\,\alpha_{max}}{v_0}. \tag{1.9}$$

In order to determine the dependence of the impact force on time in the dimensionless variables

$$\tau = \frac{t}{t_*}, \quad y = \frac{P}{P_{max}} = \left(\frac{\alpha}{\alpha_{max}}\right)^{3/2}, \quad \eta = y^{2/3}$$

we use Eq. (1.8). We obtain new variables

$$\frac{dy}{d\tau} = 3Iy^{1/3}\sqrt{1 - y^{5/3}}, \quad 0 \leqslant \tau \leqslant \frac{1}{2}.$$

The function $y(\tau)$ obtained by numeric integration of this equation is shown on Fig. 2. A characteristic feature of this function is that its derivative at $\tau = 0$ and $\tau = 1$ is equal to zero.

Note that the formulas (1.6), (1.7), and (1.9) can be applied to the collision of bodies with spherical roundings with the radii R_1 and R_2, respectively, only in the vicinity of their initial contact point.

Assuming that both colliding balls are identical and $\nu_1 = \nu_2 = 0.3$, we obtain the main parameters of the collision P_{max}, t_*, α_{max}:

$$P_{max} = 1.37ER^2\left(\frac{v_0}{a}\right)^{6/5}, \quad t_* = \frac{5.63R}{a}\left(\frac{a}{v_0}\right)^{1/5},$$

$$\alpha_{max} = 1.91R\left(\frac{v_0}{a}\right)^{4/5}. \tag{1.10}$$

Here $a = \sqrt{E/\rho}$ (ρ is the density) is the velocity of longitudinal wave propagation in a thin rod made of the same material as the balls. For the steel $E = 2.05 \cdot 10^{11}\,\text{N/m}^2$, $a = 5.13 \cdot 10^3\,\text{m/s}$. At $R = 1\,\text{cm}$ ($m = 32.7\,\text{g}$) and $v_0 = 5.13\,\text{cm/s}$ we obtain $P_{max} = 28.2\,\text{N}$, $t_* = 110 \cdot 10^{-6}\,\text{s} = 110\,\mu\text{s}$, $\alpha_{max} = 1.91 \cdot 10^{-6}\,\text{m} = 1.91 \cdot 10^{-3}\,\text{mm}$.

Estimate now for comparison how these values will change in case of the velocities 10 and 100 times higher. We have in the first case $v_0 = 0.513\,\text{m/s}$, $P_{max} = 446\,\text{N}$, $t_* = 69.5\,\mu\text{s}$, $\alpha_{max} = 1.21 \cdot 10^{-2}\,\text{mm}$, in the second case $v_0 = 5.13\,\text{m/s}$, $P_{max} = 7.1\,\text{kN}$, $t_* = 44\,\mu\text{s}$, $\alpha_{max} = 0.076\,\text{mm}$.

At high velocities the Hertz theory is used only for an approximate estimation of basic parameters of the collision, as even the hardened balls made of high quality steel at the impact velocities of about 5–8 m/s are subject to local plastic deformations.

As seen from (1.10), the time of collision and the maximum relative displacement of the balls is proportional to their radius, while the maximum value of the impact force is proportional to the square of the radius. The given values of P_{max}, t_*, α_{max} are obtained at $R = 1\,\text{cm}$, so one can easily found the appropriate values for spheres of an arbitrary radius.

Considering the formulas (1.10) and the values thus calculated we note that the displacements of the balls during the collision measured in the fractions of their radii (more precisely, of the value α_{max}/R) are of the order of 10^{-3}–10^{-4}. Consequently, during a typical collision the balls, with the positions almost unchanged, get the new velocities at the end of the collision. In order to find these velocities we integrate the equations of balls motion in the range from $t = 0$ to $t = t_*$. We obtain

$$m_1 v_1' - m_1 v_1 = -\widehat{P}, \quad m_2 v_2' - m_1 v_2 = -\widehat{P}, \tag{1.11}$$

where $v_k' = \dot{z}(t_*)$ are the balls velocities after the collision, $\widehat{P} = \int_0^{t_*} P(t)dt$ the *momentum of the impact force*.

The integration of the same Eq. (1.2) in the same limits gives

$$m(\dot{\alpha}(t_*) - \dot{\alpha}(0)) = -\widehat{P}. \tag{1.12}$$

Due to the symmetry of the curve $P(t)$ with respect to the straight line $t = t_*/2$ (Fig. 2)

$$\dot{\alpha}(t_*) = -\dot{\alpha}(0) = -v_0,$$

hence,

$$\widehat{P} = 2mv_0.$$

Substituting the value $\widehat{P}$ into (1.11) we obtain

$$v_1' = v_1 - \frac{2m_2(v_1 - v_2)}{m_1 + m_2}, \qquad v_2' = v_2 + \frac{2m_1(v_1 - v_2)}{m_1 + m_2}.$$

In particular, if $m_1 = m_2$, then

$$v_1' = v_2, \quad v_2' = v_1.$$

In this case the balls exchange the velocities.

We will relate now the momentum $\widehat{P}$ to the product $P_{max}t_*$. According to the formulas (1.9), (1.6), and (1.7) we have

$$P_{max}t_* = \frac{2I}{v_0}P_{max}\alpha_{max} = 2.5\,Imv_0 = 3.68\,mv_0. \tag{1.13}$$

It follows that

$$\widehat{P} = 2\,mv_0 = \frac{P_{max}t_*}{1.25I} = 0.544\,P_{max}t_*.$$

These expressions allow to use the value α_{max} found from the formula (1.6) in order to define the collision parameters P_{max} and t_* from the simple relations (1.13).

Impact momentum. Kelvin theorem. As we have shown in the example of the balls collision, the most characteristic feature of the collision is that the system acquires the new velocities almost with no change of position. This is achieved due to the forces with the finite momentum. Consider these forces in details.

Suppose that the force $\mathbf{F}(t)$ applied to the body with the mass m is acting for such a short period of time t_* that its displacement in this time can be neglected. Suppose that the momentum of this force

$$\widehat{\mathbf{P}} = \int_0^{t_*} \mathbf{F}(t)dt$$

is finite. Since this force $\mathbf{F}$ can result mainly from an impact of another body, then the momentum $\widehat{\mathbf{P}}$ is usually called the *impact momentum*.

Integrating the motion equation of the mass center of the considered body, we obtain

$$m\mathbf{v}' - m\mathbf{v} = \widehat{\mathbf{P}}. \tag{1.14}$$

Here $\mathbf{v}$, $\mathbf{v}'$ is the mass center velocity before and after the action of the impact momentum, respectively.

The scalar product of (1.14) by the vector $(\mathbf{v}' + \mathbf{v})/2$ gives

$$\frac{m(\mathbf{v}')^2}{2} - \frac{m\mathbf{v}^2}{2} = \widehat{\mathbf{P}} \cdot \frac{\mathbf{v}' + \mathbf{v}}{2}.$$

On the other hand, according to the work energy principle we find

$$\frac{m(\mathbf{v}')^2}{2} - \frac{m\mathbf{v}^2}{2} = \int_0^b \mathbf{F} \cdot d\mathbf{r} \equiv A. \tag{1.15}$$

Comparing these two equations, we see that the work of the force $\mathbf{F}$ in the time t_* is related to the momentum $\widehat{\mathbf{P}}$ by the following relation:

$$A = \widehat{\mathbf{P}} \cdot \frac{\mathbf{v}' + \mathbf{v}}{2}. \tag{1.16}$$

This equality is an analytical expression of the *Kelvin's theorem*. It is valid for any force $\mathbf{F}(t)$.

A following observation is to be done. On the one hand, Eq. (1.16) results from the principle of linear momentum (1.14), on the other hand it results from the work energy principle (1.15). The expression (1.14) is applicable to any mechanical system if $\mathbf{v}$ and $\mathbf{v}'$ mean the velocities of the mass center of the system. The relation (1.15) is valid in the case when the considered body with the mass m is absolutely solid and moves translationally.

Consider now the possibility to apply Kelvin's theorem to the phenomenon of collision. Suppose that the impact force $\mathbf{F}(t)$ is applied to the body with the mass m along the line which passes through its mass center. This is called a *central impact*. In this case, if the body moved translationally before the impact, it will move translationally after it as well.

Note that in the immediate neighborhood of the point of application of the impact momentum some plastic deformation may occur. However the area of significant deformation will be assumed small compared to the volume of the body. Only in this case we can neglect the presence of relative velocities related to the deformation of the body, when we calculate its kinetic energy at the time $t = t_*$, i.e., we assume that the kinetic energy of the body after the application of the momentum is equal to $m(\mathbf{v}')^2/2$.

The work energy principle, written under the form (1.15), is a consequence of the principle of motion of the mass center, so in the right side of (1.15), when the work is calculated, the force $\mathbf{F}$ is considered to be applied to the mass center. But in the final Eq. (1.16) we can consider the momentum applied directly to the point of impact, since its velocity before and after the impact within the accepted model is $\mathbf{v}$ and $\mathbf{v}'$, respectively.

It is essential that the work of the momentum $\widehat{\mathbf{P}}$ is calculated without calculating the displacements of the body during the collision. Hence we can use an approach where the impact is instantaneous, and the displacements are null. Obviously this approach allows only to study the issues related to the change in the velocity field under the action of the impact momentum. These questions relate to the *classical impact theory*.

Direct central impact of two bodies. Coefficient of restitution. Consider the solution of the problem of collision of two massive bodies in the framework of classical theory. Assume that the point of the initial contact of the bodies A is on the straight line joining their mass centers. In this case the impact is central. If, moreover, before the impact the bodies moved translationally and their velocities were lying on the straight line joining their centers, the impact is called *direct central*.

We have already considered above such a collision of two balls and we have found the dependence of the impact force on the time. The problem is formulated much easier in the framework of the classical impact theory. It does not account for the change of the impact force in the time, and one needs only to determine the velocity of the bodies after the collision and the momentum of the impact force. When solving this problem we use the same notations as in the problem of the collision of balls.

Consider now Fig. 1 as a scheme, where z_1 and z_2 are the coordinates of the mass centers of the first and second bodies, respectively. The initial Eqs. (1.1) and (1.2) have in this case the same form as in the case of balls. In order to determine the unknown variables v_1', v_2' and $\widehat{P}$ we will use Eqs. (1.11) and (1.12). The third equation of the given system contains the velocity $\dot{\alpha}(t_*)$ of the mass centers flying away after the impact. In the Hertz theory this velocity is i$-v_0$. In general case it is unknown.

The ratio of the velocity $-\dot{\alpha}(t_*)$ to the velocity $\dot{\alpha}(0) = v_0$ as an important parameter of the two bodies collision was first introduced by Newton. Assuming that this relation defined by the formula

$$k = \frac{-\dot{\alpha}(t_*)}{\dot{\alpha}(0)} = \frac{\dot{z}_2(t_*) - \dot{z}_1(t_*)}{\dot{z}_1(0) - \dot{z}_2(0)} = \frac{v_2' - v_1'}{v_1 - v_2} \tag{1.17}$$

is given, we obtain

$$\begin{aligned}
\widehat{P} &= (1+k)mv_0 = \frac{(1+k)m_1m_2v_0}{m_1 + m_2}, \\
v_1' &= v_1 - \frac{(1+k)mv_0}{m_1} = \frac{(m_1 - km_2)v_1 + (1+k)m_2v_2}{m_1 + m_2}, \\
v_2' &= v_2 + \frac{(1+k)mv_0}{m_2} = \frac{(1+k)m_1v_1 + (m_2 - km_1)v_2}{m_1 + m_2}.
\end{aligned} \tag{1.18}$$

The value k is called the *coefficient of restitution*. Its physical meaning is simple and clear and it represents the ratio of the velocity of the bodies flying away after the collision to the velocity of their approach before the collision. If $k = 1$, then the collision is called *absolutely elastic*, for $0 < k < 1$ *inelastic* or *not fully elastic*, and finally, at $k = 0$ *absolutely inelastic*.

The names of the first and the last cases ($k = 1$ and $k = 0$) exactly reflect the nature of the observed phenomenon, while the term "inelastic collision" requires additional explanation. The fact is that the coefficient of restitution cannot be the unity, not only because of the irreversible deformations during the collision, but also because of the elastic vibrations excited in the colliding bodies. In particular, the aftershock vibration energy strongly affects the coefficient of restitution, if one of the bodies is a thin rod. For example, such vibration energy is obviously significant in case of the impact of a massive ball against a long rod, if the direction of motion of the ball is perpendicular to the rod axis (transverse impact). But if the dimensions of the bodies along all the three axes are of the same order, then the energy of elastic vibrations in solids after the collision can be neglected compared to the kinetic energy of the system before it. In the simplest case of a collision between two identical spheres the relationship between these energies, as shown by Rayleigh, is

$$\frac{T}{T_0} = 0.02 \, \frac{v_0}{a} .$$

Here T is a maximum kinetic energy of vibration of balls, $T_0 = m v_0^2 / 2$, m the mass of a ball, v_0 the impact velocity, a the velocity of longitudinal waves propagation in the material of the balls (for the steel $a = 5.13 \cdot 10^3$ m/s). Local deformations are elastic only at the impact velocities with the ratio v_0 / a of the order of 10^{-2}–10^{-3}. In this case the relation T / T_0 is of the order of 10^{-4}–10^{-5}.

Consequently, for balls and bodies with the same order of magnitude of the three sizes, the value k is different from the unity mainly due to the fact that the impact is followed by plastic deformations, therefore for $0 < k < 1$ it is natural to call it *inelastic*. It should also be pointed out that the value of k is very sensitive to the quality of the surface finishing of the colliding bodies. The microroughnesses in the contact zone lead to local plastic deformations which are very difficult to measure and evaluate. The overall shape of the colliding bodies even in the vicinity of the point of impact is not changed as resulted from rough measurements. However the value of k is different from the unity due to irreversible local deformations.

The first experiments aiming to determine the coefficient of restitution conducted by Newton for balls made of different materials gave the following results: for wool, cork, and iron $k = 5/9$, for ivory $k = 8/9$, for glass $k = 15/16$. These k values correspond to the impact velocity of approximately 2.8 m/s. Further experiments have shown that the value of k depends not only on the material but also on other factors: the impact velocity, the shapes of the bodies, the quality of their surface finishing in the vicinity of the point of impact.

Two phases, in other words two periods, are discerned during a collision between two bodies. In the first phase the mass centers of the bodies approach, in the second phase they move away from each other. The first phase ends at the time $t = t_{**}$, when the relative displacement of α reaches its maximum value $\alpha = \alpha_{max}$, while the second phase is ended at the time of the end of the collision.

Integrating Eq. (1.2) from $t = 0$ to $t = t_{**}$, and then from $t = t_{**}$ to $t = t_*$ and taking into account that $\dot{\alpha}(t_{**}) = 0$, we obtain

Fig. 3 The graph of the
impact force

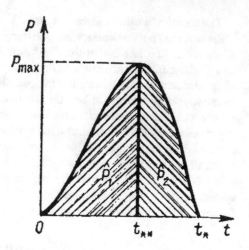

$$m\dot{\alpha}(0) = \int_0^{t_{**}} P(t)dt = \widehat{P}_1 , \quad -m\dot{\alpha}(t_*) = \int_{t_{**}}^{t_*} P(t)dt = \widehat{P}_2 . \qquad (1.19)$$

By definition, the coefficient of restitution k is equal to the relation $-\dot{\alpha}(t_*)/\dot{\alpha}(0)$. It follows from (1.19) that the value k can be represented as

$$k = \widehat{P}_2/\widehat{P}_1 . \qquad (1.20)$$

Hence,

$$\widehat{P} = (1+k)\widehat{P}_1 = (1+k)mv_0 .$$

We can see from this formula that the momentum P at a perfectly elastic collision ($k = 1$) is twice higher than at a perfectly inelastic one ($k = 0$).

The time dependence of the impact force is, generally speaking, determined not only by local, but also by general deformations of the colliding bodies. In the case of significant effect of common deformations the instant of time when the impact force reaches its maximum $P = P_{max}$ differs from the instant $t = t_{**}$ when $\alpha = \alpha_{max}$. In this case the approaching of the mass centers of the colliding bodies continues for a while even when the impact force decreases. Note that in this case the value of the coefficient of restitution can be found using the known function $P(t)$ from the formulas (1.11) and (1.17). Thus, the formula (1.20) is convenient only if the overall deformation of colliding bodies can be neglected, i.e., when $P(t_{**}) = P_{max}$ (Fig. 3).

So far we analyzed the case of the collision of two free solids. Consider now the case of the collision of a body having the mass m with an obstacle. If we consider this obstacle as a body rigid everywhere except for a small impact area, then this case can be considered as a special case of a collision of two bodies at $m_1 = m$ and $m_2 = \infty$. Assuming that the obstacle is fixed ($v_2 = v_2' = 0$), and introducing for simplicity the notation $v = v_1 \equiv v_0$ (the velocity of fall), $v' = v_1'$ (the velocity of rebound), according to (1.17) and (1.18) we have

$$v' = -kv, \quad \widehat{P} = (1+k)mv.$$

In particular, when a body with the mass m falls from the height h on a fixed horizontal plane, the impact velocity is equal to $\sqrt{2gh}$. With the rebound velocity $v' = -kv$ it rises to a height $h' = (v')^2/(2g)$. It follows that the coefficient of restitution can be found from the formula

$$k = \sqrt{h'/h},$$

which is commonly used for the experimental determination of the k value.

Kinetic energy change during the collision of bodies. Carnot theorem. Applying the Kelvin theorem (1.15), (1.16) to the first and then to the second body, we obtain

$$\Delta T_1 = \frac{m_1(v_1')^2}{2} - \frac{m_1 v_1^2}{2} = -\widehat{P}\frac{v_1'+v_1}{2},$$

$$\Delta T_2 = \frac{m_2(v_2')^2}{2} - \frac{m_2 v_2^2}{2} = \widehat{P}\frac{v_2'+v_2}{2},$$

thus obtaining

$$\Delta T = \Delta T_1 + \Delta T_2 = \frac{\widehat{P}(v_2' - v_1' - v_1 + v_2)}{2}. \tag{1.21}$$

Taking into account the relations

$$\widehat{P} = (1+k)mv_0, \quad v_2' - v_1' = k(v_1 - v_2) = kv_0, \tag{1.22}$$

we obtain

$$\Delta T = \frac{(1+k)mv_0(k-1)}{2} = -\frac{(1-k^2)mv_0}{2}.$$

Now we will establish the physical meaning of the energy $mv_0^2/2$. At the instant $t = t_{**}$, when $\dot\alpha(t_{**}) = 0$, the mass centers of the first and second bodies have the same velocity equal to the velocity v_c of the mass center of the whole system. Integrating the equations of motion of bodies in the range from $t = 0$ to $t = t_{**}$ we obtain

$$m_1(v_c - v_1) = -\widehat{P}_1, \quad m_2(v_c - v_2) = \widehat{P}_1,$$

where, according to the formula (1.19),

$$\widehat{P}_1 = m\dot\alpha(0) = m(v_1 - v_2) = mv_0.$$

Using these expressions we find out that the kinetic energy of the system in case of its relative motion with respect to a system moving translationally along with the mass center of the system of both bodies at a velocity v_c before the collision is

$$T_r = \frac{m_1(v_1 - v_c)^2}{2} + \frac{m_2(v_2 - v_c)^2}{2} = \frac{\widehat{P}_1^2}{2}\left(\frac{1}{m_1} + \frac{1}{m_2}\right) =$$

$$= \frac{\widehat{P}_1^2}{2m} = \frac{mv_0^2}{2} = \frac{m(v_1 - v_2)^2}{2}.$$

Similarly, it can be shown that

$$T_r' = \frac{m_1(v_1' - v_c)^2}{2} + \frac{m_2(v_2' - v_c)^2}{2} = \frac{m(v_2' - v_1')^2}{2}.$$

It follows that the coefficient of restitution can be given also under the form

$$k = \sqrt{T_r'/T_r}.$$

In order to found ΔT not from T_r, but from the kinetic energy of the lost velocities, i.e., from the value

$$T_* = \frac{m_1(v_1 - v_1')^2}{2} + \frac{m_2(v_2 - v_2')^2}{2},$$

we use the relations (1.11). They give immediately

$$T_* = \widehat{P}(v_2' - v_2 - v_1' + v_1)/2.$$

Based on this formula and on the formulas (1.21), (1.22) we found that

$$\frac{\Delta T}{T_*} = \frac{v_2' - v_1' - v_1 + v_2}{v_2' - v_1' + v_1 - v_2} = \frac{k - 1}{k + 1}.$$

The relation

$$|\Delta T| = \frac{1 - k}{1 + k}T_*,$$

relating the loss of the kinetic energy at the impact with the kinetic energy of the lost velocities is the analytic form of the *Carnot theorem*.

Oblique impact. Consider now the case when the impact is central, but the velocities of the translational motion of the bodies before the impact are not on the same line z (Fig. 4). This is called an *oblique central* impact.

Assume for simplicity that the velocities v_1 and v_2 lie in the same plane yz. They can be written under the form

$$\mathbf{v}_1 = v_{1\tau}\boldsymbol{\tau} + v_{1n}\mathbf{n}, \quad \mathbf{v}_2 = v_{2\tau}\boldsymbol{\tau} + v_{2n}\mathbf{n},$$

where $\boldsymbol{\tau}$ and $\mathbf{n}$ are the unit vectors of the axes y and z, respectively.

The force at the oblique impact has in general not only a normal component, but also a tangential one. Let the impact force applied to the second body have the form:

Fig. 4 Oblique central impact

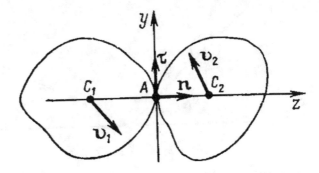

$$\mathbf{P}(t) = P_\tau(t)\boldsymbol{\tau} + P_n(t)\mathbf{n}\,.$$

Then the equations of motion of the mass centers of the first and second bodies can be written as

$$m_1\ddot{y}_1 = -P_\tau(t)\,, \quad m_2\ddot{y}_2 = P_\tau(t)\,,$$
$$m_1\ddot{z}_1 = -P_n(t)\,, \quad m_2\ddot{z}_2 = P_n(t)\,.$$

Here (y_k, z_k) are the coordinates of the points C_k, $k = 1, 2$. Integrating these equations within the range from $t = 0$ to $t = t_*$, we obtain

$$m_1(v'_{1\tau} - v_{1\tau}) = -\widehat{P}_\tau\,, \quad m_2(v'_{2\tau} - v_{2\tau}) = \widehat{P}_\tau\,, \tag{1.23}$$
$$m_1(v'_{1n} - v_{1n}) = -\widehat{P}_n\,, \quad m_2(v'_{2n} - v_{2n}) = \widehat{P}_n\,,$$

where

$$\widehat{P}_\tau = \int_0^{t_*} P_\tau(t)dt\,, \quad \widehat{P}_n = \int_0^{t_*} P_n(t)dt\,.$$

We introduce the coefficient of restitution k by the formula

$$k = \frac{v'_{2n} - v'_{1n}}{v_{1n} - v_{2n}}\,, \tag{1.24}$$

and consider it as a given value. In order to close the system of five Eqs. (1.23), (1.24) with respect to six unknown values $v'_{1\tau}, v'_{2\tau}, v'_{1n}, v'_{2n}, \widehat{P}_\tau, \widehat{P}_n$ we have to complete it with one more relation with respect to the unknown values $v'_{1\tau}, v'_{2\tau}, \widehat{P}_\tau$.

There exist several theories of oblique impact based on various assumptions about the dependence of the mentioned unknown values on other parameters of the impact. These theories closely related to the vibrotransportation problem are of semi-empirical nature. Consider the simplest model of an oblique impact. This model is based on the assumption that the force P_τ caused by the friction of the bodies against each other in the neighborhood of the A point can be neglected compared to the force P_n. In this case the translational motion of the bodies along the normal

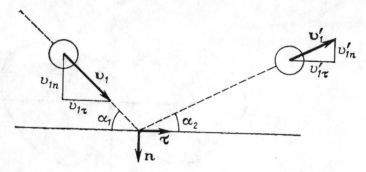

Fig. 5 Definition of the coefficient of restitution

n can be considered independently on their motion along the tangent τ. Solving the system (1.23), (1.24) at $\widehat{P}_\tau = 0$ we obtain

$$v'_{1\tau} = v_{1\tau}, \quad v'_{2\tau} = v_{2\tau},$$
$$\widehat{P}_n = \frac{(1+k)m_1 m_2 (v_{1n} - v_{2n})}{m_1 + m_2},$$
$$v'_{1n} = \frac{(m_1 - km_2)v_{1n} + (1+k)m_2 v_{2n}}{m_1 + m_2},$$
$$v'_{2n} = \frac{(1+k)m_1 v_{1n} + (m_2 - km_1)v_{2n}}{m_1 + m_2}.$$

In particular, in case of an oblique impact of a body with the mass $m_1 = m$ against a fixed flat barrier ($v_{2n} = 0$, $m_2 = \infty$) we have

$$v'_{1\tau} = v_{1\tau}, \quad \widehat{P}_n = (1+k)m v_{1n}, \quad v_{1n} = -k v_{1n}.$$

Consider the incident angle α_1 and the reflection angle α_2 (Fig. 5) defined, respectively, by the formulas

$$\tan \alpha_1 = \left| \frac{v_{1n}}{v_{1\tau}} \right|, \quad \tan \alpha_2 = \left| \frac{v'_{1n}}{v'_{1\tau}} \right|.$$

Since $v'_{1\tau} = v_{1\tau}$, then

$$\frac{\tan \alpha_2}{\tan \alpha_1} = \left| \frac{v'_{1n}}{v_{1n}} \right| = k.$$

Thus the coefficient of restitution can be considered also as $\tan \alpha_2 / \tan \alpha_1$.

2 Application of General Theorems of Dynamics to the Study of Collisions of Solids

Principle of linear momentum and principle of angular momentum in case of impact. We have already considered the interaction of two solids in case of a central impact. In order to determine the field of velocities of the system after the impact it was sufficient to use one theorem concerning the motion of the mass center of each body. But in order to analyze the influence of the impact in more complex mechanical systems we must take into account both basic theorems of dynamics: the principle of linear momentum and the principle of angular momentum simultaneously.

The principle of linear momentum can be formulated as a theorem of change of the momentum of the system, if we write it under the form

$$\frac{dK}{dt} = \mathbf{F}^{(e)}, \quad \mathbf{K} = \sum_{\nu} m_{\nu}\mathbf{v}_{\nu}, \quad \mathbf{F}^{(e)} = \sum_{\nu} \mathbf{F}_{\nu}^{(e)}, \tag{2.1}$$

and also as the principle of motion of the mass center of the system

$$M\frac{d\mathbf{v}_c}{dt} = \mathbf{F}^{(e)}, \quad M = \sum_{\nu} m_{\nu}, \quad M\mathbf{v}_c = \sum_{\nu} m_{\nu}\mathbf{v}_{\nu}, \tag{2.2}$$

where $\mathbf{F}_{\nu}^{(e)}$ is the resultant of all the external forces applied to the mass m_{ν}.

While formulating the principle of angular momentum we introduced a point O with respect to which are calculated the angular momentum of the system $\mathbf{l} = \sum_{\nu} \mathbf{r}_{\nu} \times m_{\nu}\mathbf{v}_{\nu}$ and the moment of all the external forces $\mathbf{L} = \sum_{\nu} \mathbf{r}_{\nu} \times \mathbf{F}_{\nu}^{(e)}$. Provided that the point O is fixed, we have (the principle of angular momentum)

$$\frac{d\mathbf{l}}{dt} = \mathbf{L}. \tag{2.3}$$

Integrating Eqs. (2.1)–(2.3) in the range from $t = 0$ to $t = t_*$, we obtain the

$$M\Delta\mathbf{v}_c = M\mathbf{v}_c\,\big|_0^{t_*} = \mathbf{K}\,\big|_0^{t_*} = \widehat{\mathbf{F}}^{(e)}, \tag{2.4}$$

and the

$$\Delta\mathbf{l} = \mathbf{l}\,\big|_0^{t_*} = \widehat{\mathbf{L}}. \tag{2.5}$$

The following notations have been introduced:

$$\widehat{\mathbf{F}}^{(e)} = \int_0^{t_*} \mathbf{F}^{(e)}dt = \sum_{\nu} \int_0^{t_*} \mathbf{F}_{\nu}^{(e)}dt = \sum_{\nu} \widehat{\mathbf{F}}_{\nu}^{(e)}, \quad \widehat{\mathbf{L}} = \int_0^{t_*} \mathbf{L}dt. \tag{2.6}$$

In this integral form the principle of linear momentum and the principle of angular momentum can be written for any mechanical system. If for the considered process

Fig. 6 Kinetic momentum
vector at impact

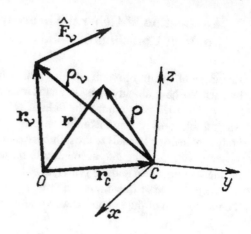

during the time $\Delta t = t_*$ the system typically acquires new velocities almost without changing the position, then the formulas (2.4), (2.5) can be regarded as the basic relations of the classical theory of impact. Recall that the main starting point of this theory is the assumption of immutability of the system position at a time $\Delta t = t_*$. The field of velocities is changed in the same position due to the fact that some forces $\mathbf{F}_\nu^{(e)}$ at the time $\Delta t = t_*$ have a finite momentum $\widehat{\mathbf{F}}_\nu^{(e)}$. These forces are called the *impact forces*. The forces whose momentum can be neglected during the time $\Delta t = t_*$ are called *finite forces*.

Note that the vector $\widehat{\mathbf{L}}$ used in the principle of angular momentum (2.5) can be represented in the classical impact theory as follows:

$$\widehat{\mathbf{L}} = \sum_\nu \mathbf{r}_\nu \times \widehat{\mathbf{F}}_\nu^{(e)} , \qquad (2.7)$$

where the summation is performed only over the points of application of the impact momentums $\widehat{\mathbf{F}}_\nu^{(e)}$. Such a simple representation of the vector $\widehat{\mathbf{L}}$ appears to be possible due to the fact that the radius-vector $\mathbf{r}_\nu$ of the point of application of an external force $\widehat{\mathbf{F}}_\nu^{(e)}$ is considered to be constant. Hence,

$$\int_0^{t_*} (\mathbf{r}_\nu \times \mathbf{F}_\nu^{(e)})dt = \mathbf{r}_\nu \times \int_0^{t_*} \mathbf{F}_\nu^{(e)}dt = \mathbf{r}_\nu \times \widehat{\mathbf{F}}_\nu^{(e)} .$$

Application of impact momentums to a free rigid body. The principle of linear momentum (2.4) and the principle of angular momentum (2.5) allow to solve various problems related to the change of the field of velocities of a solid subject to the action of impact momentums. Let the body be free for simplicity. As shown in "Kinematics", the field of velocities of a rigid body is defined by two vectors: the vector of velocity of the mass center $\mathbf{v}_c$ and the vector of instantaneous angular velocity $\boldsymbol{\omega}$.

Express now the vector of the angular momentum $\mathbf{l}$ about an arbitrary point O in terms of the vectors $\mathbf{v}_c$ and $\boldsymbol{\omega}$. By definition (see Fig. 6, see also Sect. 1 of the Chap.

8 of Vol. I), we have

$$\mathbf{l} = \int_m \mathbf{r} \times \mathbf{v} dm \,.$$

Taking into account that

$$\mathbf{r} = \mathbf{r}_c + \boldsymbol{\rho} \,, \quad \mathbf{v} = \mathbf{v}_c + \boldsymbol{\omega} \times \boldsymbol{\rho} \,,$$

$$\int_m \boldsymbol{\rho} dm = 0 \,, \quad M \mathbf{r}_c = \int_m \mathbf{r} dm \,, \quad M = \int_m dm \,,$$

we obtain

$$\mathbf{l} = \int_m \mathbf{r} \times \mathbf{v}_c dm + \int_m (\mathbf{r}_c + \boldsymbol{\rho}) \times (\boldsymbol{\omega} \times \boldsymbol{\rho}) dm = \mathbf{r}_c \times M \mathbf{v}_c + \int_m \boldsymbol{\rho} \times (\boldsymbol{\omega} \times \boldsymbol{\rho}) dm \,,$$

and finally

$$\mathbf{l} = \mathbf{r}_c \times M \mathbf{v}_c + \{J_{ik}\} \boldsymbol{\omega} \,, \tag{2.8}$$

where

$$\{J_{ik}\} \boldsymbol{\omega} = \int_m \boldsymbol{\rho} \times (\boldsymbol{\omega} \times \boldsymbol{\rho}) dm \,.$$

In the rigid body dynamics the tensor of inertia $\{J_{ik}\}$ about a point C is defined usually in a moving coordinate system $Cxyz$, rigidly connected to the rigid body. It is essential that the values of J_{ik} are unchanged in this case. Since the tensor $\{J_{ik}\}$ is defined in a moving system, the vectors $\boldsymbol{\omega}$, $\mathbf{l}$ should also be defined in the same system, although it is not always convenient.

The classical collision theory assumes that the colliding body position is unchanged during the impact. This allows to consider the relation (2.8) as written in a unique absolute coordinate system before the impact and after it. It follows that the change of the vector $\mathbf{l}$ during the impact can be represented under the form

$$\Delta \mathbf{l} = \mathbf{r}_c \times M \Delta \mathbf{v}_c + \{J_{ik}\} \Delta \boldsymbol{\omega} \,. \tag{2.9}$$

Substituting the relations (2.6), (2.7), and (2.9) in (2.4), (2.5), we obtain

$$M \Delta \mathbf{v}_c = \sum_\nu \widehat{\mathbf{F}}_\nu \,, \quad \mathbf{r}_c \times M \Delta \mathbf{v}_c + \{J_{ik}\} \Delta \boldsymbol{\omega} = \sum_\nu \mathbf{r}_\nu \times \widehat{\mathbf{F}}_\nu \,. \tag{2.10}$$

The system being analyzed consists of a single rigid body. The internal forces in such a system are the internal stresses which are not considered here. The impact momentums applied to the body are certainly the external forces, but since they are the only forces concerned, the index (e) at $\mathbf{F}_\nu$ can be omitted, as in case of the system (2.10). In particular, if the point O coincides with the point C (Fig. 6), then $\mathbf{r}_c = 0$, $\boldsymbol{\rho}_\nu = \mathbf{r}_\nu$. Therefore, the system (2.10) in this case takes the form (Fig. 7)

Fig. 7 Pplane motion of a system

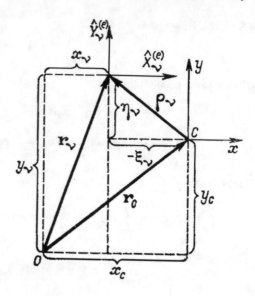

$$M\Delta\mathbf{v}_c = \sum_\nu \widehat{\mathbf{F}}_\nu, \quad \{J_{ik}\}\Delta\boldsymbol{\omega} = \sum_\nu \boldsymbol{\rho}_\nu \times \widehat{\mathbf{F}}_\nu. \tag{2.11}$$

If the axes x, y, z with respect to which the inertia moments J_{ik} are calculated are the main central axes of inertia, then the value $\{J_{ik}\}\Delta\boldsymbol{\omega}$ used in the systems (2.10) and (2.11) is found from the formula

$$\{J_{ik}\}\Delta\boldsymbol{\omega} = A\Delta\omega_x\mathbf{i} + B\Delta\omega_y\mathbf{j} + C\Delta\omega_z\mathbf{k},$$

where

$$A = J_{11} = J_{xx}, \quad B = J_{22} = J_{yy}, \quad C = J_{33} = J_{zz}.$$

In the simplest case of a flat motion the systems (2.10) and (2.11) take the form

$$M\Delta\dot{x}_c = \sum_\nu \widehat{X}_\nu, \quad M\Delta\dot{y}_c = \sum_\nu \widehat{Y}_\nu,$$

$$M(x_c\Delta\dot{y}_c - y_c\Delta\dot{x}_c) + J_c\Delta\omega = \sum_\nu (x_\nu\widehat{Y}_\nu - y_\nu\widehat{X}_\nu), \tag{2.12}$$

$$M\Delta\dot{x}_c = \sum_\nu \widehat{X}_\nu, \quad M\Delta\dot{y}_c = \sum_\nu \widehat{Y}_\nu,$$

$$J_c\Delta\omega = \sum_\nu (\xi_\nu\widehat{Y}_\nu - \eta_\nu\widehat{X}_\nu), \tag{2.13}$$

where $J_c = J_{zz} = J_{33}$, $\Delta\omega = \Delta\omega_z$, $\omega_x = \omega_y = 0$. The remaining notations are explained in Fig. 1.

Fig. 8 Cube hitting a stop

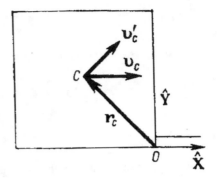

Note that the transition from the systems (2.10) and (2.11) to the systems (2.12) and (2.13) is possible under the following assumptions: all the impact momentums lie in the xy plane; the axis z perpendicular to the plane of motion is the dynamic symmetry axis, that is, for any choice of the axes x and y we have

$$J_{xz} = J_{yz} = 0.$$

In particular, if only one impact momentum $\widehat{\mathbf{F}}_1 = \widehat{\mathbf{F}} = \widehat{X}\mathbf{i} + \widehat{Y}\mathbf{j}$ is applied to the rigid body, and its application point coincides with the point O, then $\mathbf{r}_1 = 0$, so the system (2.12) can be written under the form

$$M\Delta\dot{x}_c = \widehat{X}, \quad M\Delta\dot{y}_c = \widehat{Y},$$
$$M(x_c\Delta\dot{y}_c - y_c\Delta\dot{x}_c) + J_c\Delta\omega = 0. \tag{2.14}$$

Consider the examples of the use of the system (2.14).

Example 1 A uniform cube with the mass M and the side a moves translationally with the velocity v along a horizontal plane. At some instant it hits against a stop O (Fig. 8). The translation transforms then into the rotation around the point O. Found the momentum of the impact force applied to the cube by the stop.

In this problem the field of velocities before the impact is given, as well as the nature of the field of velocities after it. Unknown are the angular velocity after the impact $\omega' = \Delta\omega$ and the impact momentum $\widehat{\mathbf{F}} = \widehat{X}\mathbf{i} + \widehat{Y}\mathbf{j}$ applied to the point O. The three equations of the system (2.14) allow us to find the unknown values ω', $\widehat{X}$ and $\widehat{Y}$.

The velocity of the mass center after the impact is

$$v'_c = \frac{a\sqrt{2}}{2}|\omega'|.$$

The cube rotates after the impact clockwise, hence the value ω' is negative and v'_c can be represented under the form

$$v'_c = -\frac{a\sqrt{2}}{2}\omega' \, .$$

One can see directly from Fig. 8 that

$$\Delta \dot{x}_c = v'_c \cos 45° - v = -\frac{a}{2}\omega' - v \, ,$$
$$\Delta \dot{y}_c = v'_c \sin 45° = -\frac{a}{2}\omega' \, ,$$
$$x_c = -a/2, \quad y_c = a/2 \, .$$

Substituting these values in the last equation of the system (2.14), we obtain

$$M\left(\frac{a^2}{4}\omega' + \frac{a^2}{4}\omega' + \frac{a}{2}v\right) + J_c\omega' = 0 \, .$$

Taking into account that $J_c = Ma^2/6$ we obtain

$$a\omega' = -\frac{3}{4}v \, .$$

It follows that

$$\Delta \dot{x}_c = -\frac{5}{8}v, \quad \Delta \dot{y}_c = \frac{3}{8}v \, ,$$

and, hence,

$$\widehat{X} = -\frac{5}{8}Mv, \quad \widehat{Y} = \frac{3}{8}Mv \, .$$

Example 2 A uniform rod being in a plane-parallel motion strikes by its end O against a horizontal plane whose normal is in the plane of motion (Fig. 9). Assuming that after the impact the end O of the rod does not lose contact with the plane, and the impact momentum applied to the rod is directed along the normal to the plane, determine the velocity of the mass center of the rod and its angular velocity after the impact, as well as the impact momentum. The rod length is l, its mass M, and the motion before the impact is considered to be given.

The unknown values in this problem are the velocities $\dot{x}'_c$, $\dot{y}'_c$, ω' and the momentum $\widehat{Y}$. In order to determine the four unknown values we use three equations of the system (2.14). In our case they can be written as

$$\dot{x}'_c = \dot{x}_c, \quad \widehat{X} = 0, \quad \widehat{Y} = M(\dot{y}'_c - \dot{y}_c) \, ,$$
$$M\tfrac{l}{2}\cos\alpha \dot{y}'_c + J_c\omega' = M\tfrac{l}{2}\cos\alpha \dot{y}_c + J_c\omega \, .$$

The fourth equation will be the condition that the velocity of the end O of the rod in the y direction after the impact is zero, i.e.,

Fig. 9 Rod hitting a fixed
plane

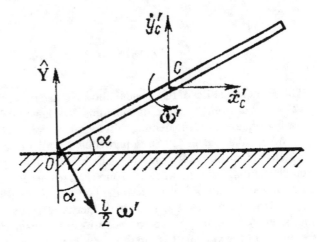

$$\dot{y}_c' - \frac{l}{2}\omega' \cos \alpha = 0 \, ,$$

thus obtaining

$$\dot{y}_c' = \tfrac{l}{2} \cos \alpha \left(\omega + \tfrac{6 \cos \alpha \dot{y}_c}{l} \right) \Big/ (1 + 3 \cos^2 \alpha) \, ,$$

$$\omega' = \left(\omega + \tfrac{6 \cos \alpha \dot{y}_c}{l} \right) \Big/ (1 + 3 \cos^2 \alpha) \, ,$$

$$\widehat{Y} = M \left(\tfrac{l}{2} \omega \cos \alpha - \dot{y}_c \right) \Big/ (1 + 3 \cos^2 \alpha) \, .$$

**Application of the impact momentums to a body rotating around a fixed
point.** The O point with respect to which the momentum **l** is calculated, was up to
now considered as an arbitrary point of the space (see Fig. 6). Assume now that it is
a fixed point of the body. Then the vector **l** can be represented as

$$\mathbf{l} = \{J_{ik}\}_O \boldsymbol{\omega} \, , \tag{2.15}$$

where J_{ik} are the inertia moments with respect to the system $Oxyz$.

Denote $\widehat{\mathbf{R}}$ the impact momentum of the reaction forces applied to the body in the
point O. Adding this momentum to the external impact momentums $\widehat{\mathbf{F}}_\nu$, the body can
be considered as free, and Eq. (2.10) can be used. Taking into account the expression
(2.15) and the fact that in this case $\Delta \mathbf{v}_c = \Delta \boldsymbol{\omega} \times \mathbf{r}_c$, we represent the system (2.10)
under the form

$$M(\Delta\omega \times \mathbf{r}_c) = \widehat{\mathbf{R}} + \sum_\nu \widehat{\mathbf{F}}_\nu ,$$

$$\{J_{ik}\}_0 \Delta\omega = \sum_\nu \mathbf{r}_\nu \times \widehat{\mathbf{F}}_\nu . \tag{2.16}$$

The second equation of this system allows to determine the variation of the angular velocity $\Delta\omega$ from the given momentums $\widehat{\mathbf{F}}_\nu$. Having calculated $\Delta\omega$, we find from the first equation the momentum of the reaction forces $\widehat{\mathbf{R}}$.

We use the system (2.16), in particular, in order to clarify the conditions under which the momentum $\widehat{\mathbf{R}}$ is missing. The solution of this problem is in a general case cumbersome. Therefore we restrict ourselves to the particular case when one external impact momentum $\widehat{\mathbf{F}}$ is applied to the body at the point with the coordinates (x, y, z), and the straight line OC is the main inertia axis. Let the straight line OC be the z axis, and the axes x, y will also be considered as main axes. In this case the second equation of the system (2.16) in the scalar notation can be written as

$$\Delta\omega_x = (y\widehat{F}_z - z\widehat{F}_y)/J_{xx} ,$$
$$\Delta\omega_y = (z\widehat{F}_x - x\widehat{F}_z)/J_{yy} , \tag{2.17}$$
$$\Delta\omega_z = (x\widehat{F}_y - y\widehat{F}_x)/J_{zz} .$$

Since $\mathbf{r}_c = z_c\mathbf{k}$, then

$$\Delta\omega \times \mathbf{r}_c = \Delta\omega_y z_c \mathbf{i} - \Delta\omega_x z_c \mathbf{j} ,$$

so it follows from the first equation of the system (2.16) that

$$\widehat{R}_x = -\widehat{F}_x + M\Delta\omega_y z_c ,$$
$$\widehat{R}_y = -\widehat{F}_y - M\Delta\omega_x z_c , \tag{2.18}$$
$$\widehat{R}_z = -\widehat{F}_z .$$

As seen from the last equation, the vector $\widehat{\mathbf{R}}$ can be zero only when the external momentum $\widehat{\mathbf{F}}$ is perpendicular to the z axis.

Substituting the expressions (2.17) into the relations (2.18) and assuming $\widehat{R}_z = \widehat{F}_z = 0$, we obtain

$$\widehat{R}_x = -\widehat{F}_x + Mz_c\frac{z\widehat{F}_x}{J_{yy}} , \quad \widehat{R}_y = -\widehat{F}_y + Mz_c\frac{z\widehat{F}_y}{J_{xx}} .$$

It follows that $\widehat{\mathbf{R}} = 0$ for any $\widehat{F}_x$, if

$$z = z_1 = J_{yy}/(Mz_c) , \quad \widehat{F}_y = \widehat{F}_z = 0 , \quad J_{xx} \neq J_{yy} ,$$

and for any $\widehat{F}_y$, if

$$z = z_2 = J_{xx}/(Mz_c) , \quad \widehat{F}_x = \widehat{F}_z = 0 , \quad J_{xx} \neq J_{yy} .$$

Fig. 10 Hitting a
parallelepiped

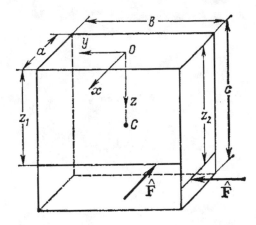

If $J_{xx} = J_{yy}$, then $\widehat{\mathbf{R}} = 0$ for any $\widehat{F}_x$ and $\widehat{F}_y$ simultaneously, i.e., for any momentum $\widehat{\mathbf{F}}$ lying in the plane $z = J_{xx}/(Mz_c)$.

Example 3 Consider a uniform rectangular parallelepiped, suspended over the center of one of the faces. In the notations shown on Fig. 10 we have

$$z_c = \frac{c}{2}, \quad J_{xx} = \frac{M}{12}(b^2 + 4c^2), \quad J_{yy} = \frac{M}{12}(a^2 + 4c^2),$$

$$z_1 = \frac{J_{yy}}{Mz_c} = \frac{a^2 + 4c^2}{6c}, \quad z_2 = \frac{b^2 + 4c^2}{6c}. \tag{2.19}$$

The lines defined by the formulas (2.19) are traced on the faces parallel to the OC axis. The momentum $\widehat{\mathbf{F}}$ applied perpendicularly to the respective face and passing through these lines is not transferred to the suspension point O.

If the parallelepiped is a cube ($a = b = c$), then any momentum perpendicular to the axis OC is not transferred to the suspension point if it is not lying in the plane $z = 5a/6$.

If the parallelepiped is a thin rod ($a \to 0$, $b \to 0$) then we obtain

$$z_1 = z_2 = 2l/3, \quad l \equiv c.$$

Hence in the case of the impact on a thin rod suspended by its end, the impact momentum applied at the distance of 1/3 length from the other end is not transmitted to the suspension point.

Application of the impact momentums to a body rotating around a fixed axis. Center of percussion. If a body is fixed at two points O and O_1, then under the action of the impact momentums $\mathbf{F}_\nu$, two reactive momentums $\widehat{\mathbf{R}}$ and $\widehat{\mathbf{R}}_1$ appear applied at the same points.

Consider the point O as fixed, while the momentum $\widehat{\mathbf{R}}_1$ will be considered as external. We can use in this case the system (2.16). If the momentum $\widehat{\mathbf{R}}_1$ is added to

Fig. 11 Hitting a rotating body

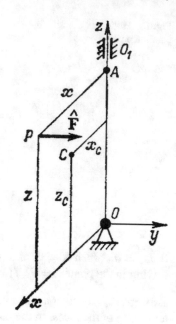

the right sides of this system it takes the form

$$M(\Delta\boldsymbol{\omega} \times \mathbf{r}_c) = \widehat{\mathbf{R}} + \widehat{\mathbf{R}}_1 + \sum_{\nu} \widehat{\mathbf{F}}_{\nu},$$

$$\{J_{ik}\}_O \Delta\boldsymbol{\omega} = \overrightarrow{OO}_1 \times \widehat{\mathbf{R}}_1 + \sum_{\nu} \mathbf{r}_{\nu} \times \widehat{\mathbf{F}}_{\nu}. \tag{2.20}$$

Note that in this case the vectors $\boldsymbol{\omega}$ and $\Delta\boldsymbol{\omega}$ lie at the straight line OO_1. Let it be the z axis (Fig. 11). In order to simplify the subsequent calculations we direct the x axis so that the mass center of the body lies in the xz plane. We obtain in this case

$$\Delta\boldsymbol{\omega} \times \mathbf{r}_c = \begin{vmatrix} \mathbf{i} & \mathbf{j} & \mathbf{k} \\ 0 & 0 & \Delta\omega \\ x_c & 0 & z_c \end{vmatrix} = x_c \Delta\omega\,\mathbf{j}. \tag{2.21}$$

Six scalar equations of the system (2.20) contain seven scalar values $\Delta\omega$, $\widehat{R}_x$, $\widehat{R}_y$, $\widehat{R}_z$, $\widehat{R}_{1x}$, $\widehat{R}_{1y}$, $\widehat{R}_{1z}$. The seventh missing equation should reflect the nature of fixation of the body at the points O and O_1. The simplest case is with a spherical hinge in the point O, and a cylindrical hinge in the point O_1. In this case of fixation $\widehat{R}_{1z} = 0$. The six remaining unknown values are found from the system (2.20) which in the case of one external momentum $\widehat{\mathbf{F}}$ and due to the relation (2.21) takes the form

2 Application of General Theorems of Dynamics to the Study ...

377

$$Mx_c\Delta\omega\mathbf{j} = \widehat{\mathbf{R}} + \widehat{\mathbf{R}}_1 + \widehat{\mathbf{F}},$$
$$\{J_{ik}\}_0\Delta\omega = \overrightarrow{OO_1} \times \widehat{\mathbf{R}}_1 + \mathbf{r} \times \widehat{\mathbf{F}}.$$

(2.22)

Since in this case, according to (1.18) of Chap. 8 of Vol. I, we have

$$\{J_{ik}\}_0\Delta\omega = -J_{xz}\Delta\omega\mathbf{i} - J_{yz}\Delta\omega\mathbf{j} + J_{zz}\Delta\omega\mathbf{k},$$

$$\overrightarrow{OO_1} \times \widehat{\mathbf{R}}_1 = \begin{vmatrix} \mathbf{i} & \mathbf{j} & \mathbf{k} \\ 0 & 0 & l \\ \widehat{R}_{1x} & \widehat{R}_{1y} & 0 \end{vmatrix} = -l\widehat{R}_{1y}\mathbf{i} + l\widehat{R}_{1x}\mathbf{j},$$

$$\mathbf{r} \times \widehat{\mathbf{F}} = \begin{vmatrix} \mathbf{i} & \mathbf{j} & \mathbf{k} \\ x & y & z \\ \widehat{F}_x & \widehat{F}_y & \widehat{F}_z \end{vmatrix},$$

then the system (2.22) is equivalent to the following six scalar equations:

$$\widehat{R}_x + \widehat{R}_{1x} + \widehat{F}_x = 0,$$
$$\widehat{R}_y + \widehat{R}_{1y} + \widehat{F}_y = Mx_c\Delta\omega,$$
$$\widehat{R}_z + \widehat{F}_z = 0,$$
$$-J_{xz}\Delta\omega = -l\widehat{R}_{1y} + y\widehat{F}_z - z\widehat{F}_y,$$
$$-J_{yz}\Delta\omega = l\widehat{R}_{1x} + z\widehat{F}_x - x\widehat{F}_z,$$
$$J_{zz}\Delta\omega = x\widehat{F}_y - y\widehat{F}_z.$$

The last equation of this system allows to determine the instantaneous variation of the angular velocity $\Delta\omega$, thus allowing to find from the remaining equations the momentums $\widehat{\mathbf{R}}$ and $\widehat{\mathbf{R}}_1$. We will now establish at what conditions they are missing. Assuming $\widehat{\mathbf{R}} = \widehat{\mathbf{R}}_1 = 0$, we have

$$\widehat{F}_x = \widehat{F}_z = 0, \quad \widehat{F}_y = Mx_c\Delta\omega,$$
$$-J_{xz}\Delta = -z\widehat{F}_y, \quad -J_{yz}\Delta\omega = 0, \quad J_{zz}\Delta\omega = x\widehat{F}_y.$$

As seen from this system, the momentums in the hinges are null if the external momentum $\widehat{\mathbf{F}}$ is perpendicular to the xz plane of the mass center, and its line of force crosses this plane at the point P with the coordinates

$$x = \frac{J_{zz}\Delta\omega}{\widehat{F}_y} = \frac{J_{zz}}{Mx_c}, \quad z = \frac{J_{xz}}{Mx_c}.$$

In addition, the relation $J_{yz} = 0$ must be fulfilled. The point P satisfying these conditions is called the *center of percussion*.

Eccentric impact. We have considered above the collisions of two bodies moving translationally before the impact and with the point of contact at the impact lying on the line connecting their mass centers. If the bodies rotate before the impact and the point of contact is not on the line connecting their mass centers, then the analysis of the impact even in the framework of the classical theory of impact becomes very

378

Fig. 12 Eccentric impact of
a ball on a rod

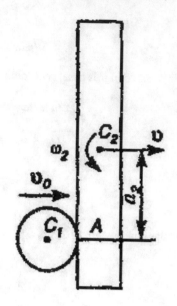

complicated. The problem is greatly simplified if the bodies in the vicinity of the
initial point of contact are limited by smooth surfaces and their velocities in the
vicinity of this point during the impact are directed along the same normal. In this
case we can assume that at sufficiently low velocity of impact the deformations at
the point of contact propagate in accordance with the theory of Hertz. A typical
example of such impact is a transversal impact of a ball on an arbitrary point of a free
rod (Fig. 12). The applicability of the Hertz impact theory to these cases essentially
depends on the extent to which we can neglect the general deformations of bodies
compared to their local deformations. For example, in case of a transverse impact on
a free rod its elastic vibrations can be obviously neglected, provided that the length
of the rod is comparable with the dimensions of its cross section.

Consider now the example of the collision of a ball with a rod as an example of
the definition of the reduced mass m in the formulas (1.6), (1.7) of the Hertz theory.

Let the center of the ball and the mass center of the rod be in the plane of symmetry
of the rod. Moreover, let the rod be at rest before the impact, and the ball has the
velocity v_0 perpendicular to the axis of the rod. Denote as above $\widehat{P}_1$ the momentum
of the impact force corresponding to the first phase. At the end of this phase the ball
and the point A of the rod will have the same velocity v_A. Denote ω_2 and v_2 the
angular velocity of the rod and the velocity of its mass center at this time. Using the
principle of linear momentum and the principle of angular momentum we obtain

$$m_1 v_A - m_1 v_0 = -\widehat{P}_1, \quad m_2 v_2 = \widehat{P}_1, \quad J_2 \omega_2 = \widehat{P}_1 a_2. \tag{2.23}$$

Here J_2 is the inertia moment of the rod with respect to its mass center, and a_2 the
distance between the mass center and the line of impact on the rod.

2 Application of General Theorems of Dynamics to the Study ...

379

Introducing the radius of inertia of the rod ρ_2 by the formula

$$\rho_2^2 = J_2/m_2 \, ,$$

and taking into account that the velocity is related to the velocities v_2 and ω_2 by the formula

$$v_A = v_2 + \omega_2 a_2 \, ,$$

we obtain

$$v_A = \frac{\widehat{P}_1}{m_2} + \frac{\widehat{P}_1 a_2^2}{J_2} = \frac{\widehat{P}_1}{m_2}\left(1 + \frac{a_2^2}{\rho_2^2}\right), \qquad (2.24)$$

or

$$\lambda_2 m_2 v_A = \widehat{P}_1 \, , \qquad \lambda_2 = \rho_2^2/(a_2^2 + \rho_2^2) \, .$$

Substituting the found value of the velocity v_A in the last equation of the system (2.23), we obtain

$$\widehat{P}_1 = \frac{\lambda_2 m_1 m_2}{m_1 + \lambda_2 m_2}\, v_0 \, . \qquad (2.25)$$

Comparing this expression to (1.19) we see that the reduced mass m in this case is

$$m = \frac{m_1 m_2^*}{m_1 + m_2^*} \, , \qquad m_2^* = \lambda_2 m_2 \, .$$

The mass m_2^* is called the *reduced mass of the rod*. The impact on the rod is, as follows from (2.24), equivalent to the impact on the mass m_2^*.

Similarly, we can determine the reduced mass for the impacts on bodies with more complex shapes than the rod. For example, for the plane motion of bodies shown in Fig. 13 the reduced mass of the system is

$$m = \frac{\lambda_1 \lambda_2 m_1 m_2}{\lambda_1 m_1 + \lambda_2 m_2} \, , \qquad \lambda_j = \frac{\rho_j^2}{a_j^2 + \rho_j^2} \, , \qquad j = 1, 2 \, .$$

Note that the velocity $v_0 = \dot{\alpha}(0)$ in (2.25) is just the impact velocity of the bodies in the point of contact. For example, if the ball was at rest before the impact, and the rod was rotating around its mass center, then $v_0 = a_2 \omega$.

Having defined the reduced mass m and the velocity v_0 and having then calculated the value K by the formula (1.3), where the values R_1 and R_2 are now to be considered as the radii of curvature of bodies in the vicinity of the point of contact, we can found all the initial values for to calculate the main parameters of the impact α_{max}, P_{max}, t_* by the formulas (1.6), (1.7), and (1.9).

If the impact is accompanied by local plastic deformations, then we can introduce the coefficient of restitution by the formula (1.20) and represent the full momentum of the impact force under the form

Fig. 13 Impact of two
complex shape bodies

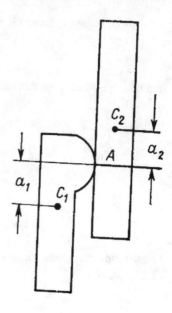

$$\widehat{P} = (1+k)mv_0 \,.$$

Defining $\widehat{P}$ by the formulas (2.14), we find the motion of each body after the impact.

3 Impact Theory of Mechanical Systems with Ideal Constraints

Generalized velocities and generalized impact momentums. Consider first a free mechanical system consisting of n material points. Let the equations of motion of the points of the system have the form

$$m_\nu \ddot{\mathbf{r}}_\nu = \mathbf{F}_\nu \,, \quad \nu = \overline{1, n} \,. \tag{3.1}$$

With the notations

$$\mathbf{r}_\nu = x_{3\nu-2}\mathbf{i} + x_{3\nu-1}\mathbf{j} + x_{3\nu}\mathbf{k} \,,$$
$$\mathbf{F}_\nu = X_{3\nu-2}\mathbf{i} + X_{3\nu-1}\mathbf{j} + X_{3\nu}\mathbf{k} \,,$$
$$m_\mu = m_{3\nu-2} = m_{3\nu-1} = m_{3\nu} \,,$$

the system (3.1) can be written under the form

$$m_\mu \ddot{x}_\mu = X_\mu \,, \quad \mu = \overline{1, 3n} \,. \tag{3.2}$$

Assume that the forces X_μ act for such a short time τ that the change in the position of the system during this time can be neglected. Integrating Eq. (3.2) in the range from 0 to τ and introducing the notations $\widehat{X}_\mu = \int_0^\tau X_\mu dt$ we obtain

$$m_\mu \dot{x}_\mu \big|_0^\tau = \widehat{X}_\mu, \quad \mu = \overline{1, 3n} . \tag{3.3}$$

Expressing the momentums $m_\mu \dot{x}_\mu$ in terms of the kinetic energy of the system

$$T = \sum_{\mu=1}^{3n} \frac{m_\mu \dot{x}_\mu^2}{2} ,$$

we can rewrite Eq. (3.3) under the form

$$\frac{\partial T}{\partial \dot{x}_\mu} \bigg|_0^\tau = \widehat{X}_\mu, \quad \mu = \overline{1, 3n} . \tag{3.4}$$

Assume that the velocities $\dot{x}_\mu$ can be unambiguously defined by a new set of velocities $\widehat{v}^\sigma$ by the expressions

$$\dot{x}_\mu = a_{\mu\sigma} \widehat{v}^\sigma + a_{\mu 0} , \quad \mu, \sigma = \overline{1, 3n} . \tag{3.5}$$

This linear transformation will be considered nondegenerate, therefore we can write

$$\widehat{v}^\sigma = b^{\sigma\mu} x_\mu + b^{\sigma 0} , \quad \mu, \sigma = \overline{1, 3n} . \tag{3.6}$$

In the classical impact theory there is no need to consider the velocities as the derivatives of the corresponding coordinates. It is sufficient to consider the change of the velocity field in a fixed position. Therefore, the transition to the new velocities is not needed to be associated with the transition to the new coordinates. The latter is very important as it can greatly simplify the solution of many specific problems. Naturally, if the transition to the new coordinates is known and is given under the form

$$x_\mu = x_\mu(t, q) , \quad q^\sigma = q^\sigma(t, x) , \quad \mu, \sigma = \overline{1, 3n} ,$$

then, introducing the new velocities by the formulas

$$\widehat{v}^\sigma = \dot{q}^\sigma = \frac{\partial q^\sigma}{\partial x_\mu} \dot{x}_\mu + \frac{\partial q^\sigma}{\partial t} , \tag{3.7}$$

we obtain

$$\dot{x}_\mu = \frac{\partial x_\mu}{\partial q^\sigma} \dot{q}^\sigma + \frac{\partial x_\mu}{\partial t} = \frac{\partial x_\mu}{\partial q^\sigma} \widehat{v}^\sigma + \frac{\partial x_\mu}{\partial t} . \tag{3.8}$$

Hence in this case

$$a_{\mu\sigma} = \frac{\partial x_\mu}{\partial q^\sigma}, \quad a_{\mu 0} = \frac{\partial x_\mu}{\partial t}, \quad b^{\sigma\mu} = \frac{\partial q^\sigma}{\partial x_\mu}, \quad b^{\sigma 0} = \frac{\partial q^\sigma}{\partial t}.$$

Formulas (3.5) and (3.6) can be considered as a generalization of the transformations (3.7), (3.8) to the case where the position of the system is fixed and it is not necessary to know how it changes over the time. In other words, we are talking about a fixed tangent space and linear transformations in it.

We call the velocities $\widehat{v}^\sigma$ the *generalized velocities*. The velocities of the points of the system can be expressed in terms of these velocities by (3.5). Note that

$$a_{\mu\sigma} = \frac{\partial \dot{x}_\mu}{\partial \widehat{v}^\sigma}, \quad b^{\sigma\mu} = \frac{\partial \widehat{v}^\sigma}{\partial \dot{x}_\mu}.$$

Multiplying Eq. (3.4) by $\partial \dot{x}_\mu / \partial \widehat{v}^\sigma$ and summing over μ, we obtain

$$\left.\frac{\partial T}{\partial \widehat{v}^\sigma}\right|_0^\tau = \widehat{V}_\sigma, \quad \sigma = \overline{1, s}, \quad s \equiv 3n. \tag{3.9}$$

Here

$$\widehat{V}_\sigma = \widehat{X}_\mu \frac{\partial \dot{x}_\mu}{\partial \widehat{v}^\sigma}. \tag{3.10}$$

We call the value $\widehat{V}_\sigma$ the *generalized impact momentum*, corresponding to the generalized velocities $\widehat{v}^\sigma$.

Equation (3.9) where the kinetic energy T is considered as a function of the generalized velocities defined as

$$T = \frac{M}{2}(g_{\rho\sigma}\widehat{v}^\rho\widehat{v}^\sigma + 2g_{\sigma 0}\widehat{v}^\sigma + g_{00}) \tag{3.11}$$

represent a system of linear algebraic equations

$$Mg_{\sigma\rho}\Delta\widehat{v}^\rho = \widehat{V}_\sigma, \quad \rho, \sigma = \overline{1, s} \tag{3.12}$$

with respect to the increments $\Delta\widehat{v}^\rho$ of the generalized velocities during the time of the collision. The determinant of this system is not null, since the expression $(1/2)g_{\rho\sigma}\widehat{v}^\rho\widehat{v}^\sigma$ is a positively definite quadratic form.

The system of differential equations of motion (3.2) when the motion is set by the generalized coordinates q^σ becomes a system of Lagrange equations of second kind. In the impact theory the problem is reduced not to the differential but to the algebraic equations. It is essential that for these linear algebraic equations we could find such a notation which, as seen from the comparison of the expressions (3.4) and (3.9), does not depend on the choice of the generalized velocities.

The differential equations of motion expressed in terms of the Lagrange operator of the kinetic energy are the Lagrange equations of second kind. Similarly to the above, Eq. (3.9) will also be called the *algebraic Lagrange equations of second kind in the impact theory.* So far these equations were derived only for a free mechanical system. The elegance and the generality of the Lagrange equations of second kind consist in the fact that they keep their form in presence of holonomic ideal constraints as well. Significantly, the above generalization also applies to the mechanical systems of general type consisting not only of material points, but also of material bodies. In this general case the generalized coordinates q^σ are the parameters that unambiguously determine the position of the mechanical system. Similarly, Eq. (3.9), as will be shown below, apply to the mechanical systems of general type. The generalized velocities $\widehat{v}^\sigma$ play here the role of the parameters which allow to define the velocities of all the points of the system in its given fixed position. Note that these parameters should allow to represent the kinetic energy of the system under the form (3.11).

Algebraic systems of Lagrange equations of first and second kind in the impact theory. In Sect. 3 of the Chap. 6 of Vol. 1 the first and second kind Lagrange equations for holonomic systems with ideal constraints were obtained as a corollary of the main assumption that the motion in which one of the generalized coordinates is a given function of time can be ensured by only one additional generalized force corresponding to this coordinate. The impact theory contains a similar major statement expressed by the following theorem:

In order to keep unchanged the generalized velocity $\widehat{v}^p$ when the impact momentums are applied to the system (or to let this velocity have a given increment), it is sufficient to add to the given generalized impact momentums $\widehat{V}_1, \widehat{V}_2, \ldots, \widehat{V}_s$ only one generalized impact momentum $\widehat{\Lambda}_\rho^*$ corresponding to the velocity $\widehat{v}^p$.

Assume for simplicity the values $\rho = s$ and the momentums $\widehat{V}_\lambda$, $\lambda = \overline{1, s-1}$ unchanged, the value $\Delta \widehat{v}^s$ being a given value. Then the first $s-1$ equations of the system (3.12) can be represented as

$$Mg_{\lambda\mu}\Delta\widehat{v}^\mu = \widehat{V}_\lambda - Mg_{\lambda s}\Delta\widehat{v}^s, \quad \lambda, \mu = \overline{1, s-1}.$$

The determinant of this system with respect to the increments $\Delta\widehat{v}^\mu$ is not null, since the expression $(1/2)g_{\lambda\mu}\widehat{v}^\lambda\widehat{v}^\mu$ is a positively definite quadratic form (the sum $(1/2)g_{\rho\sigma}\widehat{v}^\rho\widehat{v}^\sigma$ at $\widehat{v}^\lambda \neq 0$ and $\widehat{v}^s = 0$ can be only positive). It follows that the increments $\Delta\widehat{v}^\mu$ can be found unambiguously. Substituting the calculated values $\Delta\widehat{v}^\mu$ into the last equation of the system (3.12)

$$Mg_{s\mu}\Delta\widehat{v}^\mu = \widehat{V}_s + \widehat{\Lambda}_s^* - Mg_{ss}\Delta\widehat{v}^s,$$

we found the value $\widehat{\Lambda}_s^*$, which must be added to the generalized impact momentum $\widehat{V}_s$, in order to ensure a given increment to the velocity $\widehat{v}^s$. Thus the theorem is proved.

A direct consequence of this theorem is a more general statement that the specified increments of the generalized velocities $\widehat{v}^{l+1}, \widehat{v}^{l+2}, \ldots, \widehat{v}^s$, where $l = s - k$ and $k \leqslant s$, can be provided by the additional generalized impact momentums $\widehat{\Lambda}_\varkappa = \widehat{\Lambda}_{l+\varkappa}^*$, $\varkappa = \overline{1, k}$, corresponding to the given velocities. The system of equations which allows to solve this problem can be written under a compact form

$$\frac{\partial T}{\partial \widehat{v}^\lambda}\bigg|_0^\tau = \widehat{V}_\lambda , \quad \lambda = \overline{1,l}, \quad l = s - k ,$$ (3.13)

$$\frac{\partial T}{\partial \widehat{v}^{l+\varkappa}}\bigg|_0^\tau = \widehat{V}_{l+\varkappa} + \widehat{\Lambda}_\varkappa , \quad \varkappa = \overline{1,k} .$$ (3.14)

In order to consider the impact on non-free systems we use now Eqs. (3.13), (3.14). Consider first the systems with holonomic constraints.

Let the system consist of n points and the constraint equations have the form

$$f^\varkappa(t, x) = 0 , \quad \varkappa = \overline{1,k} .$$

In this case, when the external impact momentums $\widehat{X}_\mu$ are applied, the impact momentums $\widehat{R}_\mu$ of the reaction forces R_μ appear, so instead of (3.4), (3.9) we have the following equations:

$$m_\mu \dot{x}_\mu\big|_0^\tau = \frac{\partial T}{\partial \dot{x}_\mu}\bigg|_0^\tau = \widehat{X}_\mu + \widehat{R}_\mu ,$$ (3.15)

$$\frac{\partial T}{\partial \widehat{v}^\sigma}\bigg|_0^\tau = \widehat{V}_\sigma + \widehat{\Lambda}_\sigma^* ,$$ (3.16)

where $\widehat{\Lambda}_\sigma^*$ are the generalized impact momentums of the reaction forces calculated using the formula

$$\widehat{\Lambda}_\sigma^* = \widehat{R}_\mu \frac{\partial \dot{x}_\mu}{\partial \widehat{v}^\sigma} .$$ (3.17)

The formulas of the back transition from the momentums $\widehat{V}_\sigma$ and $\widehat{\Lambda}_\sigma^*$ to the momentums $\widehat{X}_\mu$ and $\widehat{R}_\mu$, due to the equivalence of the velocities $\dot{x}_\mu$ and $\widehat{v}^\sigma$, have the same structure that the formulas (3.10), (3.17), i.e.,

$$\widehat{X}_\mu = \widehat{V}_\sigma \frac{\partial \widehat{v}^\sigma}{\partial \dot{x}_\mu} , \quad \widehat{R}_\mu = \widehat{\Lambda}_\sigma^* \frac{\partial \widehat{v}^\sigma}{\partial \dot{x}_\mu} .$$ (3.18)

Assume now that the constraints are absent and the system is free. This assumption is introduced in order to make Eqs. (3.13), (3.14) applicable to the case when the generalized velocities $\widehat{v}^{l+\varkappa}$ are given under the form

$$\widehat{v}^{l+\varkappa} = \dot{q}^{l+\varkappa} = \frac{\partial f^\varkappa}{\partial x_\mu}\dot{x}_\mu + \frac{\partial f^\varkappa}{\partial t} , \quad l = s - k , \quad \varkappa = \overline{1,k} ,$$

and the remaining velocities $\widehat{v}^\lambda$ are chosen arbitrarily, however, under the condition that the set of the velocities $\widehat{v}^\sigma$ defines unambiguously the velocities of all the points of the system.

When the constraints given by the equations $f^\varkappa = 0$ are imposed to the system, the generalized velocities $\widehat{v}^{l+\varkappa}$ are null before as well as after the application of the external impact momentums $\widehat{X}_\mu$. The additional generalized impact momentums needed to keep the null values of the velocities $\widehat{v}^{l+\varkappa}$ during the impact are found in this case from Eqs. (3.13), (3.14). Comparing the system (3.16) to Eqs. (3.13), (3.14), we see that $\widehat{\Lambda}^*_\lambda = 0$, $\lambda = \overline{1,l}$. This is why the momentums of the reaction forces sufficient to satisfy the equations of the constraints in accordance with the formulas (3.18) are as follows:

$$\widehat{R}_\mu = \widehat{\Lambda}^*_{l+\varkappa} \frac{\partial \widehat{v}^{l+\varkappa}}{\partial \dot{x}_\mu} = \widehat{\Lambda}_\varkappa \frac{\partial f^\varkappa}{\partial x_\mu}. \tag{3.19}$$

To these impact momentums correspond the reaction forces given in the form

$$R_\mu = \Lambda_\varkappa \frac{\partial f^\varkappa}{\partial x_\mu}.$$

Recall that the constraints whose reactions can be represented under such form are ideal.

Equation (3.13) which do not contain the momentums of the reactions allow to find the field of velocities of the system after the impact from the given external impact momentums $\widehat{X}_\mu$. These equations contain only independent velocities $\widehat{v}^\lambda$ and have the same structure that Eq. (3.9) for a free system. Therefore they are called the *algebraic Lagrange equations of second kind.* But, strictly speaking, for a non-free system they represent only the first part of the Lagrange equations of second kind. The second part is Eq. (3.14), which allow to find the generalized impact reactions $\widehat{\Lambda}_\varkappa$. The required momentums $\widehat{R}_\mu$ are obtained in this case from the formulas (3.19). Note that for the use of Eq. (3.14) we need to assume that the constraints are absent and to express the kinetic energy in terms of all the velocities $\widehat{v}^\sigma$, i.e., represent it in the form (3.11). As follows from (3.12), Eq. (3.14) have the form :

$$M g_{l+\varkappa,\,\sigma} \Delta \widehat{v}^\sigma = \widehat{V}_{l+\varkappa} + \widehat{\Lambda}_\varkappa.$$

Here $\Delta \widehat{v}^{l+\varkappa} = 0$, while $\Delta \widehat{v}^\lambda$ must be considered as values found from Eq. (3.13), hence,

$$\widehat{\Lambda}_\varkappa = M g_{l+\varkappa,\,\lambda} \Delta \widehat{v}^\lambda - \widehat{V}_{l+\varkappa}, \quad \varkappa = \overline{1,k}. \tag{3.20}$$

In order to calculate the values $\widehat{V}_{\lambda+\varkappa}$ in these formulas, which are defined according to (3.10) as

$$\widehat{V}_{l+\varkappa} = \widehat{X}_\mu \frac{\partial \dot{x}_\mu}{\partial \widehat{v}^{l+\varkappa}},$$

we must assume the absence of the constraints and express the velocities $\dot{x}_\mu$ in terms of all the generalized velocities $\widehat{v}^\sigma$.

Taking into account (3.10) and

$$\frac{\partial T}{\partial \widehat{v}^{\sigma}} = \frac{\partial T}{\partial \dot{x}_{\mu}} \frac{\partial \dot{x}_{\mu}}{\partial \widehat{v}^{\sigma}}, \quad \widehat{R}_{\mu} = \frac{\partial T}{\partial \dot{x}_{\mu}}\Big|_0^{\tau} - \widehat{X}_{\mu},$$

we represent Eqs. (3.13), (3.14) as follows:

$$\widehat{R}_{\mu} = \frac{\partial \dot{x}_{\mu}}{\partial \widehat{v}^{\lambda}} = 0, \quad \lambda = \overline{1, l}, \quad l = s - k,$$

$$\widehat{R}_{\mu} \frac{\partial \dot{x}_{\mu}}{\partial \widehat{v}^{l+\varkappa}} = \widehat{\Lambda}_{\varkappa}, \quad \varkappa = \overline{1, k}.$$

The ensemble of these equations is considered as a system of algebraic equations for $\widehat{R}_{\mu}$. Its solution, according to the formulas (3.18), is

$$\widehat{R}_{\mu} = \widehat{\Lambda}_{\varkappa} \frac{\partial \widehat{v}^{l+\varkappa}}{\partial \dot{x}_{\mu}}.$$

These formulas can be rewritten in more detail:

$$m_{\mu}(\dot{x}_{\mu 1} - \dot{x}_{\mu 0}) = \widehat{X}_{\mu} + \widehat{\Lambda}_{\varkappa} \frac{\partial f^{\varkappa}}{\partial x_{\mu}}. \tag{3.21}$$

Here $\dot{x}_{\mu 0}$ and $\dot{x}_{\mu 1}$ are the velocities of the points of the system before and after the impact, respectively.

Equation (3.21) complemented by the constraint equations

$$\frac{\partial f^{\varkappa}}{\partial x_{\mu}}(\dot{x}_{\mu 1} - \dot{x}_{\mu 0}) = 0, \quad \varkappa = \overline{1, k}$$

are called the *algebraic Lagrange equations of first kind*. In these equations the unknown values are the velocities $\dot{x}_{\mu 1}$ after the impact and the factors $\widehat{\Lambda}_{\varkappa}$.

As seen from the deduction of the system (3.21), the algebraic Lagrange equations of first and second kind can be considered as two following reciprocal systems of algebraic equations:

$\widehat{\Lambda} \frac{\partial \widehat{v}^{\sigma}}{\partial \dot{x}_{\mu}} = \widehat{R}_{\mu}$ the algebraic Lagrange equations of first kind,

$\widehat{R}_{\mu} \frac{\partial \dot{x}_{\mu}}{\partial \widehat{v}^{\sigma}} = \widehat{\Lambda}_{\sigma}^{*}$ the algebraic Lagrange equations of second kind,

where in accordance with the assumption of ideal constraints $\widehat{\Lambda}_{\lambda}^{*} = 0, \lambda = \overline{1, l}$.

Basic equation of the impact theory. As we have seen, the idealness of the constraints is revealed in the fact that the generalized impact momentums $\widehat{\Lambda}_{\lambda}^{*}$ of the constraint forces corresponding to the free velocities are null.

The existence of independent generalized velocities $\widehat{v}^\lambda$, which allow, using the linear equations

$$\dot{x}_\mu = a_{\mu\lambda}\widehat{v}^\lambda + a_{\mu0} , \tag{3.22}$$

to find unambiguously the velocities at all the points of the system, means that in the differential forms

$$\delta\dot{x}_\mu = a_{\mu\lambda}\delta\widehat{v}^\lambda = \frac{\partial\dot{x}_\mu}{\partial\widehat{v}^\lambda}\delta\widehat{v}^\lambda \tag{3.23}$$

all the differentials $\delta\widehat{v}^\lambda$ are independent. Hence the condition of ideality of the constraints, expressed in the fact that the values $\widehat{\Lambda}^*_\lambda$, corresponding to $\widehat{v}^\lambda$, are null, can be written under the form of one equation

$$\widehat{\Lambda}^*_\lambda \delta\widehat{v}^\lambda = 0 . \tag{3.24}$$

Taking into account that, according to the definition,

$$\widehat{\Lambda}^*_\lambda = \widehat{R}_\mu\frac{\partial\dot{x}_\mu}{\partial\widehat{v}^\lambda} ,$$

we get the equation

$$\widehat{R}_\mu\frac{\partial\dot{x}_\mu}{\partial\widehat{v}^\lambda}\delta\widehat{v}^\lambda = 0 ,$$

which according to the formulas (3.23), (3.15) can also be rewritten as

$$\sum_\mu \widehat{R}_\mu\delta\dot{x}_\mu = \sum_\mu (m_\mu(\dot{x}_{\mu1} - \dot{x}_{\mu0}) - \widehat{X}_\mu)\,\delta\dot{x}_\mu = 0 . \tag{3.25}$$

This equation with $\delta\dot{x}_\mu$ as variations (differentials) of the velocities allowed by the constraints can be taken as a basic initial relation in the impact theory of mechanical systems with ideal constraints. It is called also the *basic equation of the impact theory*. Indeed, it leads immediately to Eq. (3.24) which in turn may be expressed under the form of the algebraic Lagrange equations of both first and second kinds.

Principle of least forcing in the impact theory. We show now that a direct consequence of the basic equation (3.25) is the following statement:

The expression

$$G = \frac{1}{2}\sum_\mu m_\mu\left(\dot{x}_\mu - \dot{x}_{\mu0} - \frac{\widehat{X}_\mu}{m_\mu}\right)^2 ,$$

defined on the set of values $\dot{x}_\mu$ admitted by the constraints has a minimum at the values $\dot{x}_\mu = \dot{x}_{\mu1}$, which are really taken by the system under the action of the external impact momentums $\widehat{X}_\mu$.

This statement similar to the principle of Gauss can be named the *principle of least forcing in the impact theory*. It is proven as follows.

We represent an arbitrary value of the velocity $\dot{x}_\mu$ admitted by the constraints under the form

$$\dot{x}_\mu = \dot{x}_{\mu 1} + \delta \dot{x}_\mu .$$

Write the difference

$$\delta G = \tfrac{1}{2} \sum_\mu m_\mu \left\{ \left(\dot{x}_{\mu 1} - \delta \dot{x}_\mu - \dot{x}_{\mu 0} - \frac{\widehat{X}_\mu}{m_\mu} \right)^2 - \left(\dot{x}_{\mu 1} - \dot{x}_{\mu 0} - \frac{\widehat{X}_\mu}{m_\mu} \right)^2 \right\} =$$
$$= \sum_\mu (m_\mu (\dot{x}_{\mu 1} - \dot{x}_{\mu 0}) - \widehat{X}_\mu) \delta \dot{x}_\mu + \tfrac{1}{2} \sum_\mu m_\mu (\delta \dot{x}_\mu)^2 .$$

This formula shows that, in accordance with the basic equation (3.25)

$$\delta G = \frac{1}{2} \sum_\mu m_\mu (\delta \dot{x}_\mu)^2 > 0 ,$$

if $\dot{x}_\mu \neq \dot{x}_{\mu 1}$. That means that the function G takes its minimum at $\dot{x}_\mu = \dot{x}_{\mu 1}$.

The above expression for δG shows also that the function G takes its minimum at $\dot{x}_\mu = \dot{x}_{\mu 1}$ only if the basic equation (3.25) is satisfied. That means that this principle can be taken as a basis for the impact theory of mechanical systems with ideal constraints. However, the algebraic Lagrange equations of second and first kind, the basic equation of impact and, finally, the last principle were obtained earlier as a consequence of the main theorem proved in this section. Therefore, it can be stated that this theorem is the basis of the principle of least forcing and reveals its physical content.

Impact theory of mechanical systems of general type. The most typical example of a holonomic constraint between material points is the constraint expressed by the condition that the distance between two points is unchanged. As was shown in Sect. 3 of the Chap. 6 of Vol. 1, such constraints are ideal if the forces of reaction allowing to maintain unchanged the distance between the points are directed along the line connecting these points. A perfectly rigid body can be considered as a mechanical system of an infinite number of points with the unchanged distances between them. When the impact forces are applied to the solid, the forces of reaction appear allowing the body to keep its shape and not to be destroyed. They are the internal stresses. They can be defined only if the solid is considered as a continuous medium. Since the nature of these forces corresponds to the assumption of ideal links between the particles of the body, then a perfectly rigid body, as noted in Chap. 6 of Vol. 1, can be considered as a typical mechanical system of general type with six degrees of freedom.

The velocity field of the rigid body is determined by two vectors: the velocity of the pole $\mathbf{v}_0$ and the angular velocity $\boldsymbol{\omega}$ independent on the choice of the pole. The velocity of an arbitrary point of the body is expressed through $\mathbf{v}_0$ and $\boldsymbol{\omega}$ by the formula

$$\mathbf{v} = \mathbf{v}_0 + \boldsymbol{\omega} \times \mathbf{r} . \tag{3.26}$$

Fig. 14 Choice of
generalized velocities

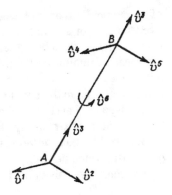

The generalized velocities $\widehat{v}^\lambda$ in this case are the three components of the vector $\mathbf{v}_0$
and the three components of the vector $\boldsymbol{\omega}$. The Euler formula (3.26) can be regarded
as a generalization of (3.22) for a rigid body.

Consider another possible method of choice of generalized velocities of a rigid
body. Let the body pivotally connected in the points A and B with other bodies. A
convenient choice for generalized velocities is the three components of the velocity
of the body at the point A (Fig. 14), with $\widehat{v}^3$ directed along the vector $\overrightarrow{AB}$. Since the
distance $|\overrightarrow{AB}|$ is not changed, the projection of the velocity of the point B to the
vector $\overrightarrow{AB}$ is $\widehat{v}^3$. Hence the point B has only two independent velocities $\widehat{v}^4$ and $\widehat{v}^5$.
The missing sixth velocity is the angular velocity of rotation of the body around the
axis $\overrightarrow{AB}$.

In the full system of the algebraic Lagrange equations of second kind (3.13), (3.14)
Eq. (3.13) can be considered as independent of the system (3.14). That expresses the
elegancy and the advantage of the Lagrange equations of second kind compared to
the equations of first kind. For a perfect rigid body, when the number of constraints
is infinite, this advantage is of particular importance. The problem can be reduced
to the solution of the system (3.13) which has the same structure as (3.9) for the free
system. Hence the study of the impact on a perfect rigid body is similar to the study
of the impact on a free system. The same assertion concerns as well any mechanical
system of general type with l degrees of freedom.

Consider now the calculation of generalized impact momentums $\widehat{V}_\lambda$ for mechanical systems of general type. According to the definition (3.10) we have

$$\widehat{V}_\lambda = \widehat{X}_\mu \frac{\partial \dot{x}_\mu}{\partial \widehat{v}^\lambda} . \tag{3.27}$$

In this case the summation over μ means the summation over the points of application
of the impact momentums $\widehat{X}_\mu$. The velocity $\dot{x}_\mu$ must be regarded as the velocity in
the direction of the impact momentum $\widehat{X}_\mu$ of the point of the system at which this
momentum is applied. Recall that the momentum can be applied at one point in three
different directions. Note that in this case the velocities of the points of application

of the momentums are not necessarily associated with one coordinate system $Oxyz$, as in the case of a system of n points. At each point where the impact momentum is applied, it is advisable to select a local coordinate system by directing the axes in the way that minimizes the overall number of the impact momentums $\widehat{X}_\mu$. Having then expressed in terms of $\widehat{v}^\lambda$ the velocities $\dot{x}_\mu$ directed along $\widehat{X}_\mu$, we will found the factors $\partial \dot{x}_\mu / \partial \widehat{v}^\lambda$ in the sum (3.27).

Consider now the mechanical systems of general type with additional constraints. Let s be the number of degrees of freedom of the system without additional constraints, and its field of velocities in the given fixed position be determined by the generalized velocities $\widehat{u}^\rho$, $\rho = \overline{1, s}$.

The constraints given by the equations

$$\varphi^\varkappa = b_\rho^{l+\varkappa} \widehat{u}^\rho + b_0^{l+\varkappa} = 0, \quad l = s - k, \quad \varkappa = \overline{1, k}, \tag{3.28}$$

will be considered as ideal, and their equations linearly independent. That means that the rank of the matrix formed by the coefficients $b_\rho^{l+\varkappa}$ is k. In this case, using the formulas

$$\widehat{v}^\lambda = b_\rho^\lambda \widehat{u}^\rho + b_0^\lambda, \quad \lambda = \overline{1, l},$$
$$\widehat{v}^{l+\varkappa} = b_\rho^{l+\varkappa} \widehat{u}^\rho + b_0^{l+\varkappa}, \quad \varkappa = \overline{1, k}, \tag{3.29}$$

and selecting the appropriate values b_ρ^λ, we can pass to the new generalized velocities $\widehat{v}^\sigma$.

The algebraic systems of Lagrange equations of second kind expressed in terms of new velocities $\widehat{v}^\sigma$ coincides with the systems (3.13), (3.14). In order to write Eq. (3.13) in an explicit form we have first to express the kinetic energy of the system through the independent velocities $\widehat{v}^\lambda$, and to calculate the values $\widehat{V}_\lambda$ using the formulas (3.27).

Assume that the expression for the kinetic energy in terms of the initial velocities $\widehat{u}^\rho$ is known:

$$T = \frac{M}{2} \left(\widetilde{g}_{\rho\sigma} \widehat{u}^\rho \widehat{u}^\sigma + 2 \widetilde{g}_{\rho 0} \widehat{u}^\rho + \widetilde{g}_{00} \right). \tag{3.30}$$

Solving the system (3.29) with respect to $\widehat{u}^\rho$ we obtain

$$\widehat{u}^\rho = a_\sigma^\rho \widehat{v}^\sigma + a_0^\rho. \tag{3.31}$$

In the presence of links given by Eq. (3.28) we obtain

$$\widehat{u}^\rho = a_\lambda^\rho \widehat{v}^\lambda + a_0^\rho.$$

Substituting these expressions in (3.30) we obtain the desired expression for the kinetic energy.

The values $\widehat{V}_\lambda$, as follows from the expressions (3.27), (3.31), can be represented under the form

$$\widehat{V}_\lambda = \widehat{X}_\mu \frac{\partial \dot{x}_\mu}{\partial \widehat{u}^\rho} \frac{\partial \widehat{u}^\rho}{\partial \widehat{v}^\lambda} = \widehat{X}_\mu \frac{\partial \dot{x}_\mu}{\partial \widehat{u}^\rho} a^\rho_\lambda.$$

If we compose and solve the system (3.13), we can found, using the formulas (3.20), the generalized impact momentums $\widehat{\Lambda}_\varkappa$ of the constraint reaction forces. To calculate the factors $g_{l+\varkappa,\lambda}$ entering in this formula we need, using the expressions (3.31) and (3.30), to represent the kinetic energy under the form (3.11). At the same time the values $\widehat{V}_{l+\varkappa}$ should be calculated using the formulas

$$\widehat{V}_{l+\varkappa} = \widehat{X}_\mu \frac{\partial \dot{x}_\mu}{\partial \widehat{u}^\rho} a^\rho_{l+\varkappa}.$$

Demonstrate now how we can express in terms of $\widehat{\Lambda}_\varkappa$ the reaction forces of the given constraint applied to the points of the system at which this constraint acts.

For a mechanical system consisting of a finite number of points, in accordance with (3.18) we have

$$\widehat{R}_\mu = \widehat{\Lambda}_\varkappa \frac{\partial \widehat{v}^{l+\varkappa}}{\partial \dot{x}_\mu} = \widehat{\Lambda}_\varkappa \frac{\partial \varphi^\varkappa}{\partial \dot{x}_\mu}. \tag{3.32}$$

In this formula, the constraint equations in the notation used in (3.6) should be considered as given as

$$\varphi^\varkappa = b^{l+\varkappa,\mu} \dot{x}_\mu + b^{l+\varkappa,0}. \tag{3.33}$$

The right side of the expression (3.32) is a sum over $\varkappa$ from 1 to k. That means that all the constraints act simultaneously at this point. However it is first necessary to select one of them and to establish to which reactions it leads. Hence we assume below that in the expression for the constraint reaction momentums there is no summation over $\varkappa$. In other words, instead of (3.32) we consider the following expression (no summation over $\varkappa$!)

$$\widehat{R}_\mu = \widehat{\Lambda}_\varkappa \frac{\partial \varphi^\varkappa}{\partial \dot{x}_\mu}. \tag{3.34}$$

Here $\widehat{R}_\mu$ and $\dot{x}_\mu$ are the orthogonal projections of the momentum of the constraint reaction and of the velocity of the point of its application, respectively, on the same coordinate axis and on the same direction, and $\widehat{R}_\mu$ appears only in the case when $\dot{x}_\mu$ enters into the equation of the constraint. Formula (3.34) can be simply and naturally generalized based on the physical meaning, under the following condition. The generalized velocities $\widehat{u}^\rho$, entering into the constraint equation, are equal to the orthogonal components of the velocities of the points of the system covered by this constraint. Thus, in accordance with the formula (3.34) we obtain (no summation over $\varkappa$!)

$$\widehat{R}^u_\rho = \widehat{\Lambda}_\varkappa \frac{\partial \varphi^\varkappa}{\partial \widehat{u}^\rho}. \tag{3.35}$$

Here $\widehat{R}^u_\rho$ is the orthogonal projection of the constraint reaction momentum on the direction of the velocity $\widehat{u}^p$ of the point of its application.

Passing from the generalized velocities $\widehat{u}^p$ to the generalized velocities $\widehat{v}^\sigma$ using the formulas (3.29), the function $\varphi^\varkappa$ as a function of the new generalized velocities $\widehat{v}^\sigma$ has the following simple form:

$$\varphi^\varkappa = \widehat{v}^{l+\varkappa}. \tag{3.36}$$

Due to the equivalence of the generalized velocities $\widehat{u}^p$ and $\widehat{v}^\sigma$, the structure of the main formula (3.35) expressed in new velocities is the same, and, hence, in consideration of (3.36) we obtain

$$\widehat{R}^v_{l+\varkappa} = \widehat{\Lambda}_\varkappa \frac{\partial \varphi^\varkappa}{\partial \widehat{v}^{l+\varkappa}} = \widehat{\Lambda}_\varkappa. \tag{3.37}$$

This implies that if the value $\widehat{v}^{l+\varkappa}$ is equal to the velocity in some direction of a point to which is applied a constraint expressed by the equation $\varphi^\varkappa = 0$, then $\widehat{\Lambda}_\varkappa$ is the impact momentum applied to this point by the constraint. It should be noted that both generalized velocities $\widehat{v}^\sigma$ and $\widehat{u}^p$ can have different physical meaning. In particular, they can be absolute or relative velocities of some points of the system. The basic requirement to be considered when choosing $\widehat{u}^p$ and $\widehat{v}^\sigma$ is the necessity to unambiguously define with their use the velocity field of the whole system and thus represent its kinetic energy under the form (3.11). Taking this remark into account, we can investigate in the following way the important special case of the constraint when the distance between two points is unchanged.

Consider the velocity $\widehat{v}^{l+\varkappa}$ of the point B with respect to the point A in the direction of the vector $\overrightarrow{AB}$. The condition of stability of the distance between the points B and A can be written under the form

$$\varphi^\varkappa = \widehat{v}^{l+\varkappa} = 0.$$

As follows from the formula (3.37), in this case the value $\widehat{\Lambda}_\varkappa$ is the constraint impact momentum applied to the point B in the direction of the vector $\overrightarrow{AB}$. Based on third Newton law the impact momentum applied to the point A has the opposite direction.

The above approach can be used, in particular, when considering the impact on the system of rigid bodies linked by spherical or cylindrical hinges and weightless inextensible rods with hinges on their ends (Fig. 15). The dimensions of the hinges and the friction in them will be neglected, that is, assume that the constraint in a hinge is expressed in the fact that two different bodies have a common point, and their velocities at this point are the same. Obviously, such constraint is ideal.

The ensemble of generalized velocities $\widehat{u}^p$ of the considered system when releasing it from all constraints in hinges should be taken equal to the ensemble of the velocities shown on Fig. 14 for each body. Recall that the points A and B on Fig. 14 correspond to the points with the hinges. In the absence of another hinge the second point can

Fig. 15 Impact on a system
of rigid bodies

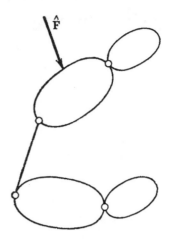

be chosen arbitrarily. It is suitable to situate it in the body in such a manner to obtain the most simple expression of the kinetic energy.

With this choice of the generalized velocities of a formally free system of rigid bodies the values $\widehat{u}^p$ in the constraint Eq. (3.28) are equal to the components of the velocities of the points to which the constraint reaction momentums are applied. That means that the impact momentums in the hinges can be defined according to (3.35). At the same time the impact momentums in thin inextensible rods must be found by the formula (3.37).

As an example, consider the impact on the chain of rods. Assume for simplicity that the motion is plane.

Impact on a straight chain of rods. Suppose that an impact momentum S directed perpendicular to the line of the chain of pivotally suspended initially motionless rods is applied to the lowest point of the chain (Fig. 16). Let the chain consist of n uniform thin rods interconnected by hinges. Let $2a$ be the length of the rod and M its mass. It is required to determine the velocity field of the system after the impact and the impact momentums transmitted up the chain through the hinges.

The use of the algebraic Lagrange equations of second kind (3.13) in the considered problem at a suitable choice of the generalized velocities $\widehat{v}^\lambda$ allows to build up a compact analytic solution for any number of rods. Knowing the dependence of the velocity field on n and S, we can easily found the momentums in the hinges. Therefore, in this case there is no need to refer to Eq. (3.14) to determine the reactions. The independent generalized velocities $\widehat{v}^\lambda$ in such tasks should be chosen so that the kinetic energy of each body of the chain has the same structure and depends on a minimum number of generalized velocities. These requirements are satisfied by the generalized velocities $\widehat{v}^\lambda$ equal to the velocities of the hinges v_λ. The values $\lambda = 0$ and $\lambda = n$ correspond to the velocities of the ends of the chain. For the upper end hinge $v_0 = 0$. If the upper end is free, then v_0 is the independent generalized velocity.

The kinetic energy of the kth rod is

Fig. 16 Impact on a straight
chain of rods

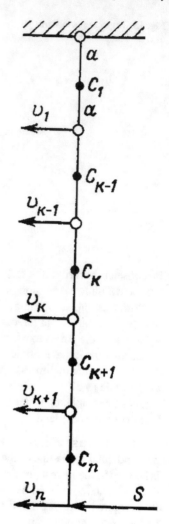

$$T_k = \frac{M}{2}\left(\frac{v_{k-1} + v_k}{2}\right)^2 + \frac{J\omega_k^2}{2},$$

where

$$J = \frac{Ma^2}{3}, \quad \omega_k = \frac{v_k - v_{k-1}}{2a}.$$

It follows that

$$T = \sum_{k=1}^{n} \frac{M(v_{k-1} + v_k)^2}{8} + \sum_{k=1}^{n} \frac{M(v_k - v_{k-1})^2}{24}. \tag{3.38}$$

The velocity of the point of application of the momentum S is equal to the velocity v_n, hence according to the formula (3.27) in this case $\widehat{V}_\lambda = 0$, $\lambda = \overline{1, n-1}$, and $\widehat{V}_n = S$. Substituting the expression (3.38), we obtain

$$v_0 = 0, \tag{3.39}$$

$$v_{\lambda-1} + 4v_\lambda + v_{\lambda+1} = 0, \quad \lambda = \overline{1, n-1}, \tag{3.40}$$

$$v_{n-1} + 2v_n = 6S/M. \tag{3.41}$$

Note that the considered chain consists of homogeneous rods, that is, of homogeneous elements. The number of these elements and their size is finite. The investigation of the behavior of many mechanical systems consisting of homogeneous finite elements, at an appropriate choice of the variables $\widehat{v}_\lambda$ characterizing the state of the system, is reduced to the equations of the type (3.40). It is therefore particularly important to master the method of solving the system of Eqs. (3.39)–(3.41).

Equation (3.40) belong to the *finite-difference equations*, or to the *linear difference equations*. They are also found in the numerical solution of ordinary differential equations. The arbitrary values in these equations are expressed in terms of finite differences.

A particular solution of Eq. (3.40) is found in the form

$$v_\lambda = z^\lambda.$$

Substituting it to Eq. (3.40), we see that the values z must satisfy the equation

$$z^2 + 4z + 1 = 0.$$

It follows $z_1 = -2 - \sqrt{3}$, $z_2 = -2 + \sqrt{3} = 1/z_1$.

A general solution of Eq. (3.40) depending on the arbitrary constants C_1 and C_2 is

$$v_\lambda = C_1 z_1^\lambda + C_2 z_2^\lambda. \tag{3.42}$$

The values C_1 and C_2 are found from Eqs. (3.39), (3.41) playing the role of the boundary conditions. It follows from the first Eq. (3.39) that

$$C_1 + C_2 = 0.$$

Assuming $2 + \sqrt{3} = e^x$, $2 - \sqrt{3} = e^{-x}$, $\cosh x = 2$, $\sinh x = \sqrt{3}$, $C_1 = -C_2 = C/2$, we obtain

$$v_\lambda = (-1)^\lambda \frac{C}{2}(e^{\lambda x} - e^{-\lambda x}) = C(-1)^\lambda \sinh \lambda x.$$

Substituting this expression to Eq. (3.41) and taking into account that $\sinh(n-1)x = \sinh nx \cosh x - \cosh nx \sinh x = 2\sinh nx - \sqrt{3}\cosh nx$, we obtain

$$C = \frac{2\sqrt{3}(-1)^n S}{M \cosh nx}.$$

Finally

$$v_\lambda = \frac{2\sqrt{3}(-1)^{n+\lambda} S \sinh \lambda x}{M \cosh nx}. \tag{3.43}$$

If the second end of the chain of rods is free, then the equation for v_0 is

$$2v_0 = v_1 = 0.$$

Substituting the expression (3.42), we obtain

$$C_1 = C_2 = C/2.$$

After the calculations similar to the above we get

$$v_\lambda = \frac{2\sqrt{3}(-1)^{n+\lambda} S \cosh \lambda x}{M \sinh nx}.$$

It follows from the solution (3.43) that, in particular, the momentum $S = S_n$ applied to the system of n rods is associated with the velocity v_n by a relation

$$v_n = \frac{2\sqrt{3}}{M} n \tanh nx. \tag{3.44}$$

It follows that the momentum S_λ transferred by the hinge having the velocity v_λ can be written as

$$S_\lambda = \frac{M v_\lambda}{2\sqrt{3}\tanh \lambda x} = \frac{(-1)^{n+\lambda}\cosh \lambda x}{\cosh nx} S_n.$$

Since the second end of the chain is free, then

$$S_\lambda = \frac{(-1)^{n+\lambda}\sinh \lambda x}{\sinh nx} S_n, \quad v_n = \frac{2\sqrt{3}}{M} S_n \coth nx. \tag{3.45}$$

The dependencies (3.44), (3.45) allow to develop the following approach to the problem of the impact to an arbitrary rod of a straight chain.

Consider a pivotally suspended chain of $n+m+1$ rods. The impact momentum S is applied to the $(n+1)$th rod at a distance b from its end to which m other rods are pivotally suspended (Fig. 17). The velocities of the ends of this rod are v_n and v_m. The indexes n and m indicate the number of rods attached to this end. The momentums

Fig. 17 Rod number $(n + 1)$

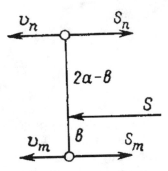

S_n and S_m, as follows from the formula (3.44), (3.45), are related to the velocities v_n and v_m by the formulas

$$S_n = \frac{M v_n}{2\sqrt{3}} \coth nx, \quad S_m = \frac{M v_m}{2\sqrt{3}} \tanh mx. \qquad (3.46)$$

Apply to the selected rod the Lagrange equations of second kind (3.13), considering S_n and S_m as the given external momentums. Since they are directed oppositely to the velocities of the points of their application, the corresponding generalized momentums are $(-S_n)$ and $(-S_m)$.

The velocity v of the point of application of the momentum S is

$$v = v_m + \frac{(v_n - v_m)b}{2a}. \qquad (3.47)$$

It follows from (3.27), (3.47) that the Lagrange equations (3.13) in this case have the form

$$\frac{\partial T}{\partial v_n}\bigg|_0^T = -S_n + S\frac{\partial v}{\partial v_n} = -S_n + \frac{Sb}{2a},$$

$$\frac{\partial T}{\partial v_m}\bigg|_0^T = -S_m + S\frac{\partial v}{\partial v_m} = -S_m + \frac{S(2a - b)}{2a}. \qquad (3.48)$$

The kinetic energy of the rod T, as seen from the expression (3.38), is

$$T = \frac{M(v_m + v_n)^2}{8} + \frac{M(v_m - v_n)^2}{24}. \qquad (3.49)$$

The system (3.48) with respect to the unknown v_n and v_m and taking into account the relations (3.46) and (3.49) can be written under the form

$$(2 + \sqrt{3}\coth nx) v_n + v_m = \frac{3bS}{Ma},$$

$$v_n + (2 + \sqrt{3}\tanh mx) v_m = \frac{3(2a - b)S}{Ma},$$

thus giving

$$v_n = \frac{3S}{M a \Delta} \left(\sqrt{3}\, b\, (\sqrt{3} + \tanh mx) - 2a \right),$$

$$v_m = \frac{3S}{M a \Delta} \left(\sqrt{3}(2a - b)(\sqrt{3} + \coth nx) - 2a \right),$$

where $\Delta = 3 + 2\sqrt{3}\,(\tanh mx + \coth nx) + 3 \tanh mx \coth nx$.

For the chains free at both ends the value $\coth nx$ in these formulas should be replaced by $\tanh nx$.

Expressing the velocities v_n and v_m in terms of the external momentum S and the parameters of the chain, we find using the formulas (3.46) the momentums S_n and S_m. In turn, the momentums S_n and S_m allow to define all the needed values.

Impact on a chain of rods with hinges located on an arc of a circle. Suppose that at the instant of the impact on the uniform chain of rods their ends are located on a circle (Fig. 18). Express the kinetic energy of the jth rod in terms of the velocities u_j, v_j, w_j of its ends which are directed as shown on Fig. 18. The velocity of the mass center of the rod along itself is v_j, and perpendicularly to the rod $(u_j - w_j)/2$. The angular velocity is $(u_j + w_j)/(2a)$. Hence

$$T_j = \frac{M v_j^2}{2} + \frac{M(u_j - w_j)^2}{8} + \frac{M a^2}{6} \left(\frac{u_j + w_j}{2a} \right)^2 =$$
$$= \frac{M}{6} \left(u_j^2 - u_j w_j + w_j^2 + 3 v_j^2 \right). \tag{3.50}$$

If the chain consists of n rods, then the overall number s of the velocities u_j, v_j, w_j is $3n$. The chain with free ends has $n - 1$ hinge. Each hinge has two constraints $(k = 2n - 2)$. Hence the number of degrees of freedom $l = s - k$ is $n + 2$.

Express now the velocities v_{j-1} and v_{j+1} of the two rods pivotally connected with the jth rod in terms of the velocities u_j, v_j and w_j. One can see directly from Fig. 18 that

$$v_{j-1} = v_j \cos \alpha - u_j \sin \alpha,$$
$$v_{j+1} = v_j \cos \alpha - w_j \sin \alpha. \tag{3.51}$$

Imagine the massless rods attached to both ends of the chain and introduce the velocities v_0 and v_{n+1} as for all other rods. Based on Eq. (3.51) the velocities u_j and w_j of all the rods can be found using the formulas

$$u_j = \frac{v_j \cos \alpha - v_{j-1}}{\sin \alpha}, \quad w_j = \frac{v_j \cos \alpha - v_{j+1}}{\sin \alpha}. \tag{3.52}$$

It follows that in this case the velocities $v_0, v_1, \ldots, v_{n+1}$ can be taken as independent generalized velocities $\hat{v}^\lambda$.

Fig. 18 Impact on a chain of
rods located on a circular arc

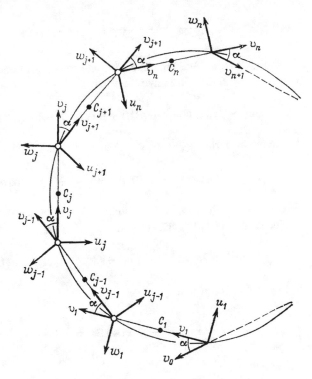

The kinetic energy of the system with the account of (3.50) and (3.52) can be
written as

$$
T = \sum_{j=1}^{n} \frac{M}{6} \left[\left(\frac{v_j \cos \alpha - v_{j-1}}{\sin \alpha} \right)^2 - \frac{(v_j \cos \alpha - v_{j-1})(v_j \cos \alpha - v_{j+1})}{\sin^2 \alpha} + \right.
$$
$$
\left. + \left(\frac{v_j \cos \alpha - v_{j+1}}{\sin \alpha} \right)^2 + 3v_j^2 \right].
$$

$$(3.53)$$

Hence,

$$
\frac{\partial T}{\partial v_0} = \frac{M}{6 \sin^2 \alpha} \left(2v_0 - v_1 \cos \alpha - v_2 \right), \tag{3.54}
$$

$$
\frac{\partial T}{\partial v_1} = \frac{M}{6 \sin^2 \alpha} \left(-v_0 \cos \alpha + 4v_1 (1 + \sin^2 \alpha) - 2v_2 \cos \alpha - v_3 \right), \tag{3.55}
$$

$$
\frac{\partial T}{\partial v_\lambda} = \frac{M}{6 \sin^2 \alpha} \left(-v_{\lambda-2} - 2v_{\lambda-1} \cos \alpha + v_\lambda (6 + 4 \sin^2 \alpha) - \right.
$$
$$
\left. - 2v_{\lambda+1} \cos \alpha - v_{\lambda+2} \right), \quad \lambda = \overline{2, n-1}, \tag{3.56}
$$

$$
\frac{\partial T}{\partial v_n} = \frac{M}{6 \sin^2 \alpha} \left(-v_{n-2} - 2v_{n-1} \cos \alpha + 4v_n (1 + \sin^2 \alpha) - v_{n+1} \cos \alpha \right), \tag{3.57}
$$

$$\frac{\partial T}{\partial v_{n+1}} = \frac{M}{6 \sin^2 \alpha} \left(-v_{n-1} - v_n \cos \alpha + 2v_{n+1} \right).$$ (3.58)

It follows from (3.56) that the homogeneous algebraic Lagrange equations similar to Eq. (3.40) in this case have the form

$$-v_{\lambda-2} - 2v_{\lambda-1} \cos \alpha + v_\lambda (6 + 4 \sin^2 \alpha) - 2v_{\lambda+1} \cos \alpha - v_{\lambda+2} = 0.$$ (3.59)

Assuming as earlier $v_\lambda = z^\lambda$, we obtain

$$z^2 + 2z \cos \alpha - (6 + 4 \sin^2 \alpha) + 2z^{-1} \cos \alpha + z^{-2} = 0.$$ (3.60)

Note that if z is the root of this equation, then $1/z$ is also its root.

Equation (3.60) after the substitution $z = e^x$ takes the form

$$e^{2x} + 2e^x \cos \alpha - (6 + 4 \sin^2 \alpha) + 2e^{-x} \cos \alpha + e^{-2x} = 0.$$

Taking into account that

$$\cosh x = \frac{e^x + e^{-x}}{2}, \quad \cosh 2x = \frac{e^{2x} + e^{-2x}}{2} = 2 \cosh^2 x - 1,$$

we obtain

$$\cosh^2 x + \cos \alpha \cosh x - (2 + \sin^2 \alpha) = 0.$$

Hence,

$$\cosh x_1 = (\sqrt{9 + 3 \sin^2 \alpha} - \cos \alpha)/2 > 1,$$
$$\cosh x_2^* = -(\sqrt{9 + 3 \sin^2 \alpha} + \cos \alpha)/2 < 0.$$

Assuming $x_2^* = x_2 + \pi i$, we obtain

$$-\cosh x_2^* = \cosh x_2 = (\sqrt{9 + 3 \sin^2 \alpha} + \cos \alpha)/2.$$

The general solution of Eq. (3.59) is

$$v_\lambda = C_1 e^{x_1 \lambda} + C_2 e^{-x_1 \lambda} + C_3 e^{x_2^* \lambda} + C_4 e^{-x_2^* \lambda}.$$

Taking into account that $e^{\pm \pi i \lambda} = (-1)^\lambda$, we obtain

$$v_\lambda = C_1 e^{x_1 \lambda} + C_2 e^{-x_1 \lambda} + C_3 (-1)^\lambda e^{x_2 \lambda} + C_4 (-1)^\lambda e^{-x_2 \lambda}.$$ (3.61)

Fig. 19 First rod in the chain

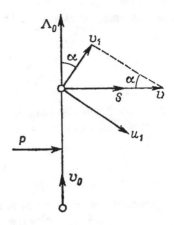

The arbitrary constants C_1, C_2, C_3, C_4 are found from the condition of fixation of the ends of the chain or from the Lagrange equations for the velocities v_0, v_1, v_n, v_{n+1}. These equations are easy to obtain, given the expressions (3.54)–(3.58).

We show now how the general solution (3.61) can be applied to the next task. Let an external impact momentum P be applied to any chain rod perpendicular to its axis. If the line of impact is a symmetry axis of the system, then the velocity of the rod itself along its axis is zero. In the absence of complete symmetry this velocity can be approximately taken equal to zero. Therefore, it is advisable first to solve the following problem.

The end of the first rod of the chain is moved so that the velocity is $v_0 = 0$. This condition is regarded as an ideal constraint whose reaction is directed along v_0. The active impact momentum is the momentum S, applied to the end of the first rod in the direction of its possible velocity v (Fig. 19). The other end of the chain may be free or hinged. The solution of this auxiliary problem is in a general case cumbersome. Therefore we confine ourselves to a particular case, when the number of rods is sufficiently large. In other words, we believe that $v_\lambda \to 0$ at $\lambda \to \infty$. As follows from (3.61), this is possible only when $C_1 = C_3 = 0$.

Then we found from the limit condition $v_0 = 0$ that

$$C_2 + C_4 = 0, \quad C_2 = -C_4 = C.$$

Hence,

$$v_\lambda = C(e^{-x_1\lambda} - (-1)^\lambda e^{-x_2\lambda}). \tag{3.62}$$

The constant C is found from the Lagrange equation with respect to the velocity v_1. The velocity of the point of application of the momentum S is $v_1/\sin\alpha$ (Fig. 19). Hence it follows from (3.27) that the corresponding generalized momentum is $S/\sin\alpha$. Taking it into account along with the relation (3.55) we get

$$4v_1(1 + \sin^2\alpha) - 2v_2\cos\alpha - v_3 = 6\sin\alpha S/M.$$

Substituting the expression (3.62), we obtain

$$C = 6 \sin \alpha S/(\eta M),$$
$$\eta = 4 (1 + \sin^2 \alpha)(e^{-x_1} + e^{-x_2}) -$$
$$- 2 \cos \alpha (e^{-2x_1} - e^{-2x_2}) - e^{-3x_1} - e^{-3x_2}.$$

It follows from the obtained solution (3.62), in particular, that the momentum S is related to the velocity $v = v_1/\sin \alpha$ of the point of its application by the formula:

$$M_* v = S, \quad M_* = \frac{\eta M}{6(e^{-x_1} + e^{-x_2})}. \tag{3.63}$$

Return now to the problem of the impact of the chain by an external momentum P. Assume for simplicity that the momentum P is applied to the middle of the rod (see Fig. 19). If we imagine this rod separated from the chain and apply to it the principle of linear momentum, we obtain

$$Mv = P - 2S.$$

Since S and v are related by (3.63), we obtain

$$(M + 2M_*) v = P, \quad S = \frac{M_* P}{2M_* + M}.$$

As follows immediately from this expression, the mass $2M_* + M$ is the reduced mass of a long enough chain of rods. It can be applied, in particular, when using the theory of Hertz in the case of an impact of a ball on a chain of rods.

Expressing S in terms of the given external momentum P, we can use (3.62) for to find the velocity field.

Consider now how we can found out the momentums in the hinges from the known velocity field. First we calculate the momentum Λ_0 of the constraint reaction $v_0 = 0$. In the Lagrange equation

$$\frac{\partial T}{\partial v_0}\bigg|_0^\tau = V_0 + \Lambda_0, \tag{3.64}$$

containing this momentum, the generalized momentum V_0 corresponding to the external momentum S should be found out using the expression (3.10), with a formal assumption of the absence of the constraint $v_0 = 0$.

Let v be the projection of the velocity of the point of application of the momentum S on the direction of its action. As follows from Fig. 19,

$$v = v_1 \sin \alpha + u_1 \cos \alpha.$$

Taking into account that according to (3.52) $u_1 = (v_1 \cos \alpha - v_0)/\sin \alpha$, we obtain

$$v = \frac{v_1}{\sin \alpha} - \frac{v_0 \cos \alpha}{\sin \alpha},$$

hence,

$$V_0 = S \frac{\partial v}{\partial v_0} = -S \cot \alpha. \qquad (3.65)$$

From Eq. (3.64) according to the formulas (3.54) and (3.65) we have

$$\Lambda_0 = S \cot \alpha - \frac{M}{6 \sin^2 \alpha} (v_1 \cos \alpha + v_2).$$

The constraint reaction momentums in the subsequent hinges can be determined by the formula (3.35). If the rods were not connected by hinges, the system would have $3n$ degrees of freedom. Introduce it generalized velocities as follows:

$$\begin{aligned}
\widehat{v}^\lambda &= v_{\lambda-1}, \quad \lambda = \overline{1, n+2}, \quad n+2 = l, \\
\widehat{v}^{l+2j-1} &= u_{j+1} - \frac{v_{j+1} \cos \alpha - v_j}{\sin \alpha}, \\
\widehat{v}^{l+2j} &= w_j - \frac{v_j \cos \alpha - v_{j+1}}{\sin \alpha}, \\
&\quad j = \overline{1, n-1}.
\end{aligned} \qquad (3.66)$$

In this case the equations of two constraints in the jth hinge connecting the jth rod with the $(j+1)$th one have the form

$$\begin{aligned}
\varphi^{2j-1} &= u_{j+1} - \frac{v_{j+1} \cos \alpha - v_j}{\sin \alpha} = 0, \\
\varphi^{2j} &= w_j - \frac{v_j \cos \alpha - v_{j+1}}{\sin \alpha} = 0, \\
&\quad j = \overline{1, n-1}.
\end{aligned}$$

Denote $\widehat{\mathbf{R}}_j$ the momentum applied to the end of the $(j+1)$th rod through the jth hinge. The orthogonal projection of the momentum $\widehat{\mathbf{R}}_j$ on the direction of the velocity u_{j+1} according to the formula (3.35) is equal to the generalized momentum $\widehat{\Lambda}_{2j-1}$, since $\partial \varphi^{2j-1} / \partial u_{j+1} = 1$. Similarly, we can show that $\widehat{\Lambda}_{2j}$ is the projection of the momentum $(-\widehat{\mathbf{R}}_j)$ on the direction w_j. The vector $\widehat{\mathbf{R}}_j$ can be found by a simple geometrical construction from the given projections $\widehat{\Lambda}_{2j-1}$ and $(-\widehat{\Lambda}_{2j})$ (Fig. 20).

The values $\widehat{\Lambda}_{2j-1}$ and $\widehat{\Lambda}_{2j}$, by definition equal to

$$\widehat{\Lambda}_{2j-1} = \frac{\partial T}{\partial \widehat{v}^{l+2j-1}}, \quad \widehat{\Lambda}_{2j} = \frac{\partial T}{\partial \widehat{v}^{l+2j}}, \quad j = \overline{1, n-1}, \qquad (3.67)$$

are given by the expression (3.20) in the detailed form. However, in our case it is easier to use directly the definitions (3.67).

If we consider the kinetic energy given under the form

Fig. 20 Rod number $(j + 1)$

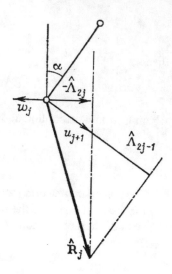

$$T = \sum_{i=1}^{n} \frac{M}{6} \left(u_i^2 - u_i w_i + w_i^2 + 3v_i^2 \right),$$

as a composite function of the new velocities $\widehat{v}^{\sigma}$ introduced by the formulas (3.66), we obtain

$$\widehat{\Lambda}_{2j-1} = \sum_{i=1}^{n} \left(\frac{\partial T}{\partial u_i} \frac{\partial u_i}{\partial \widehat{v}^{l+2j-1}} + \frac{\partial T}{\partial v_i} \frac{\partial v_i}{\partial \widehat{v}^{l+2j-1}} + \frac{\partial T}{\partial w_i} \frac{\partial w_i}{\partial \widehat{v}^{l+2j-1}} \right) =$$

$$= \frac{\partial T}{\partial u_{j+1}} = \frac{M}{6} (2u_{j+1} - w_{j+1}) = \frac{M(v_{j+1} \cos \alpha - 2v_j + v_{j+2})}{6 \sin \alpha}.$$

Similarly, it can be shown that

$$\widehat{\Lambda}_{2j} = \frac{\partial T}{\partial w_j} = \frac{M(v_j \cos \alpha - 2v_{j+1} + v_{j-1})}{6 \sin \alpha}.$$

Momentum constraints. Until now we considered the constraints existing both before and after the impact. Consider now the case of the constraints imposed on the system instantly. The simplest example is a constraint corresponding to the assumption of an absolutely inelastic collision of a material point against a plane. It can be defined by the equation $\dot{y} = 0$ assuming that the axis y is directed along the straight line along which the point falls on the plane. The momentum of the reaction of this constraint $\widehat{R}$ is evidently equal to the kinetic momentum mv of the point before the impact. The impact of the point on the plane can thus be considered as an instant imposing of the constraint given by the equation $\dot{y} = 0$.

Passing from this example to the general case, we suppose that the generalized velocities $\widehat{v}^{\sigma}$ of the mechanical system at $t = 0$ are independent and at $t = \tau$ satisfy the equations

$$\varphi^{\varkappa} = a_{\sigma}^{l+\varkappa}\widehat{v}^{\sigma} + a_{0}^{\lambda+\varkappa} = 0, \quad \varkappa = \overline{1,k}, \quad l = s - k.\tag{3.68}$$

If the displacements of the system in a time τ can be neglected, we will assume that the system is subject to *momentum constraints*.

The generalized velocities corresponding to the time moment $t = 0$ will be denoted as $\widehat{v}_{0}^{\sigma}$ to be distinguished from the velocities related by Eq. (3.68). Introduce the new velocities by the formulas

$$\widehat{v}_{*}^{\lambda} = a_{\sigma}^{\lambda}\widehat{v}^{\sigma} + a_{0}^{\lambda}, \quad \lambda = \overline{1,l},$$
$$\widehat{v}_{*}^{l+\varkappa} = a_{\sigma}^{l+\varkappa}\widehat{v}^{\sigma} + a_{0}^{l+\varkappa}, \quad \varkappa = \overline{1,k}.$$

By our assumption this is a biunique correspondence between $\widehat{v}^{\sigma}$ and $\widehat{v}_{*}^{\rho}$, hence we can also write

$$\widehat{v}^{\sigma} = b_{\rho}^{\sigma}\widehat{v}_{*}^{\rho} + b_{0}^{\sigma}, \quad \rho,\sigma = \overline{1,s}.$$

With the constraints (3.68) imposed the new velocities $v_{*}^{l+\varkappa}$ acquire the increments

$$\Delta\widehat{v}_{*}^{l+\varkappa} = -a_{\sigma}^{l+\varkappa}\widehat{v}_{0}^{\sigma} - a_{0}^{l+\varkappa}.$$

As has been shown, these specified increments of the generalized velocities $\widehat{v}_{*}^{l+\varkappa}$ are ensured by the additional generalized momentums $\widehat{\Lambda}_{\varkappa}$ applied to the system. The momentum constraints defined by Eq. (3.68) and equivalent to the generalized momentums $\widehat{\Lambda}_{\varkappa}$ applied will be called *ideal*. The most characteristic ideal momentum constraint is expressed in the fact that the distance between two points of the system becomes instantly permanent. The impact momentums ensuring such a constraint are directed along the straight line connecting these points.

Under ideal constraints the algebraic Lagrange equations for the new velocities take the form

$$\left.\frac{\partial T}{\partial\widehat{v}_{*}^{\lambda}}\right|_{0}^{\tau} = \widehat{V}_{\lambda}^{*}, \quad \lambda = \overline{1,l}, \quad l = s - k,$$
$$\left.\frac{\partial T}{\partial\widehat{v}_{*}^{l+\varkappa}}\right|_{0}^{\tau} = \widehat{V}_{l+\varkappa}^{*} + \widehat{\Lambda}_{\varkappa}, \quad \varkappa = \overline{1,k}.$$

Returning back to the initial velocities $\widehat{v}^{\sigma}$, i.e., from the algebraic Lagrange equations of second kind to the equations of first kind, we obtain

$$\left.\frac{\partial T}{\partial\widehat{v}^{\sigma}}\right|_{0}^{\tau} = \widehat{V}_{\sigma} + \widehat{\Lambda}_{\varkappa}\frac{\partial\varphi^{\varkappa}}{\partial\widehat{v}^{\sigma}}, \quad \sigma = \overline{1,s}.\tag{3.69}$$

In the case of an abrupt imposition of constraints the external impact momentums $\widehat{V}_{\sigma}$ are usually null, and Eq. (3.69) take the form

$$\left.\frac{\partial T}{\partial \widehat{v}^{\sigma}}\right|_{0}^{\tau} = \widehat{\Lambda}_{\varkappa}\frac{\partial \varphi^{\varkappa}}{\partial \widehat{v}^{\sigma}}, \quad \sigma = \overline{1, s}. \tag{3.70}$$

The unknown values are the generalized momentums $\widehat{\Lambda}_{\varkappa}$ of the constraint reactions and the velocities $\widehat{v}^{\sigma}$ at the time moment $t = \tau$. Completing the system (3.70) by the constraint Eq. (3.68), we obtain a closed system of equations for the unknown $\widehat{\Lambda}_{\varkappa}$ and $\widehat{v}^{\sigma}$.

Recall that in accordance with the definition (3.17) the generalized momentum $\widehat{\Lambda}_{\varkappa}$ is related to the momentums $\widehat{R}_{\mu}$ applied to the points of the system by the formulas

$$\widehat{\Lambda}_{\varkappa} = \widehat{R}_{\mu}\frac{\partial \dot{x}_{\mu}}{\partial \widehat{v}_{*}^{l+\varkappa}}. \tag{3.71}$$

The summation over μ means the summation over the points to which are really applied the impact momentums due to the constraint expressed by the equation $\varphi^{\varkappa} = 0$.

If the generalized velocity $\widehat{v}^{\rho}$ entering into the constraint equation $\varphi^{\varkappa} = 0$ is a projection to a certain direction of the velocity of the point covered by this constraint, then the projection to this direction of the specified constraint reaction momentum applied to this point, by analogy with (3.35), can be calculated by the formula (no summation over $\varkappa$!)

$$\widehat{R}_{\rho}^{v} = \widehat{\Lambda}_{\varkappa}\frac{\partial \varphi^{\varkappa}}{\partial \widehat{v}^{\rho}}. \tag{3.72}$$

Consider the two following problems as the examples of application of Eq. (3.70) and the formulas (3.71) and (3.72).

Problem 1 A tripod with uniform equal legs disposed at right angles to each other falls down with the velocity v_0 and strikes on a smooth inelastic floor with all the legs. Show that if the legs are connected at the upper point by a ball hinge, the velocity of the mass center of the tripod will be reduced twice at the instant of the impact.

The friction in the hinge which holds the legs at right angles to each other until the contact to the floor surely remains at the impact. However, in the approximate approach to this problem the momentum of the friction forces can be neglected, and the constraint in the ball hinge at the strike can be considered ideal. The motion of the three legs is assumed to be the same. Hence we can consider only one leg AB (Fig. 21a). Let it be a thin uniform rod, then the kinetic energy of the system is

$$T = 3\left(\frac{mv_c^2}{2} + \frac{J_c\omega^2}{2}\right), \quad J_c = \frac{ml^2}{3}.$$

Here m is the mass of the rod, $2l$ its length, v_c the velocity of the mass center of the rod AB, ω its angular velocity.

The velocity of the upper point B directed vertically down will be denoted as v. This velocity and the angular velocity ω will be considered as generalized velocities.

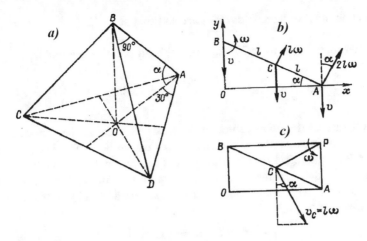

Fig. 21 Tripod fall

The square of the velocity of the mass center of the rod AB, as seen from Fig. 21b, is

$$v_c^2 = v^2 - 2l\,\omega v \cos\alpha + l^2\omega^2\,,$$

hence,

$$T = \frac{3m}{2}\left(v^2 - 2l\,\omega v \cos\alpha + \frac{4}{3}l^2\omega^2\right).$$

The projection of the velocity of the point A on the axis y is

$$v_* = 2l\,\omega\cos\alpha - v\,. \tag{3.73}$$

By assumption, the floor is inelastic. That means that the legs are not detached from the floor. Therefore, the equation of the constraint takes the form

$$\varphi = 2l\,\omega\cos\alpha - v = 0\,. \tag{3.74}$$

Applying Eq. (3.70) to this problem, we have

$$\left.\frac{\partial T}{\partial v}\right|_0^\tau = \Lambda\frac{\partial\varphi}{\partial v}\,,\quad \left.\frac{\partial T}{\partial\omega}\right|_0^\tau = \Lambda\frac{\partial\varphi}{\partial\omega}\,,$$

or in the explicit form

$$3m(v - l\,\omega\cos\alpha) - 3mv_0 = -\Lambda\,,$$

$$3m\left(\frac{4}{3}l^2\omega - lv\cos\alpha\right) + 3mlv_0\cos\alpha = 2l\,\Lambda\cos\alpha\,. \tag{3.75}$$

In these equations we have taken into account that at $t = 0$ the velocity is $v = v_0$, and $\omega = 0$.

Considering the system (3.75) together with Eq. (3.74), we found that

$$v = \frac{3}{2} v_0 \cos^2 \alpha, \quad \omega = \frac{3}{4} \frac{v_0 \cos \alpha}{l}, \quad \Lambda = 3mv_0 \left(1 - \frac{3}{4} \cos^2 \alpha \right).$$

The velocity of the mass center of the tripod is equal to the projection of the velocity of the mass center of the rod AB to the vertical. The instant center of velocities P of the rod AB is in the vertex of the rectangle with the diagonal AB (Fig. 21c). It follows that the velocity v_c of the mass center of the rod is ωl, and it forms the angle α with the vertical. Hence the required velocity of the mass center v' of the tripod is

$$v' = \omega l \cos \alpha = \frac{3}{4} v_0 \cos^2 \alpha.$$

One can see directly from Fig. 21a that

$$\cos \alpha = \frac{OA}{2l} = \frac{DA}{2 \cos 30° \cdot 2l} = \frac{2l\sqrt{2}}{2 \cos 30° \cdot 2l} = \sqrt{\frac{2}{3}}.$$

Hence,

$$v' = \frac{3}{4} v_0 \cos^2 \alpha = \frac{v_0}{2}.$$

Thus the velocity v' is indeed twice lower than the velocity v_0.

Consider now the physical meaning of the generalized momentum Λ. The constraint expressed by Eq. (3.74) leads to the appearance of three equal momentums $\widehat{R}_\mu = R$ applied from the floor to the ends of the legs. These momentums are directed vertically upwards. The velocities $\dot{x}_\mu$ of the points of their application are equal to the velocity v_* introduced by the formula (3.73). It follows that, in accordance with Eq. (3.71)

$$\Lambda = \sum_{\mu=1}^{3} \widehat{R}_\mu \frac{\partial \dot{x}_\mu}{\partial v_*} = 3R.$$

Hence,

$$R = \frac{1}{3} \Lambda = mv_0 \left(1 - \frac{3}{4} \cos^2 \alpha \right) = \frac{mv_0}{2}.$$

Problem 2 Four uniform thin rods with the mass M and the length l are connected in a frame. Two opposite vertices A and B of this frame are connected by a flexible weightless inextensible string of length $l\sqrt{2}$. At the time when the frame moving arbitrarily along a plane takes the form of a square, the thread is pulled instantly. Determine the momentum S of the tension of the string.

Fig. 22 Frame motion

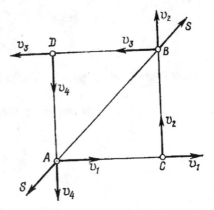

With the string unstrained the system has four degrees of freedom. The components of the velocity of vertices A and B can be taken as the generalized velocities. At the instant when the frame becomes a square, the velocity of two other vertices are easily expressed in terms of the velocities of the vertices A and B (Fig. 22).

Given the notations in Figs. 18 and 22 and the formula (3.50), we find that the kinetic energy of the system in this case is

$$T = \frac{M}{6}\left[5(v_1^2 + v_2^2 + v_3^2 + v_4^2) - 2(v_1 v_3 + v_2 v_4)\right].$$

The velocity of the point B with respect to the point A along AB is

$$v_* = \frac{\sqrt{2}}{2}\,(v_2 - v_3 - v_1 + v_4)\,. \tag{3.76}$$

With the string strained $v_* = 0$, hence the equation of the momentum constraint in this case takes the form

$$\varphi = v_2 - v_3 - v_1 + v_4 = 0\,. \tag{3.77}$$

Applying Eq. (3.70) to this problem we have

$$\left.\frac{\partial T}{\partial v_\sigma}\right|_0^\tau = \Lambda\frac{\partial \varphi}{\partial v_\sigma}\,, \quad \sigma = \overline{1,4}\,,$$

or in the explicit form

$$5v_1 - v_3 - (5v_{10} - v_{30}) = -3\Lambda/M\,,$$

$$5v_2 - v_4 - (5v_{20} - v_{40}) = 3\Lambda/M\,,$$

$$5v_3 - v_1 - (5v_{30} - v_{10}) = -3\Lambda/M \,,$$

$$5v_4 - v_2 - (5v_{40} - v_{20}) = 3\Lambda/M \,.$$

Denote the values of the velocities at $t = 0$ as $v_{\sigma 0}$. Then we obtain

$$\Lambda = M(v_{10} - v_{20} + v_{30} - v_{40})/3 = -M\sqrt{2}\,v_{*0}/3 \,,$$

$$v_1 = v_{10} + \frac{\sqrt{2}}{4}\,v_{*0} \,, \quad v_3 = v_{30} + \frac{\sqrt{2}}{4}\,v_{*0} \,,$$

$$v_2 = v_{20} - \frac{\sqrt{2}}{4}\,v_{*0} \,, \quad v_4 = v_{40} - \frac{\sqrt{2}}{4}\,v_{*0} \,.$$

Determine now the momentum S of the tension of the string. In the notation (3.76) the equation of the constraint (3.77) can be rewritten as

$$\varphi = \sqrt{2}\,v_* = 0 \,.$$

According to the formula (3.72), the momentum R applied to the point B in the direction of the velocity v_* is

$$R = \Lambda\frac{\partial\varphi}{\partial v_*} = \sqrt{2}\,\Lambda \,.$$

This momentum is applied to the point B from the side of the string. According to the third Newton law the required momentum is $S = -R$, i.e.,

$$S = -\sqrt{2}\,\Lambda = 2\,Mv_{*0}/3 \,.$$

Application of the impact theory to the controlled motion. The equations of the momentum constraints can be considered as an assigned program of motion. The problem is formulated in this case as follows: *In the time τ, during which the system is almost unmoved, the control momentums $\widehat{\mathbf{R}}_\mu$ are to be applied ensuring at the instant $t = \tau$ the velocities satisfying the relations* (3.68). The momentums of all the other forces X_μ in the time τ are neglected. As follows from the basic theorem of the impact theory, the sufficient control momentums $\widehat{R}_\mu$ satisfy the relations

$$\widehat{R}_\mu\frac{\partial\dot{x}_\mu}{\partial\widehat{v}_*^{\,l+\varkappa}} = \widehat{\Lambda}_\varkappa \,, \quad \varkappa = \overline{1,k} \,,$$

$$\widehat{R}_\mu\frac{\partial\dot{x}_\mu}{\partial\widehat{v}_*^{\,\lambda}} = 0 \,, \quad \lambda = \overline{1,l} \,.$$

$$(3.78)$$

Note that the forcing

$$G = \frac{1}{2}\sum_\mu m_\mu(\dot{x}_{\mu 1} - \dot{x}_{\mu 2})^2$$

in this case is minimal. If before the application of the momentums $\widehat{R}_\mu$ the system was in rest ($\dot{x}_{\mu 0} = 0$ at all μ), then the forcing G is equal to the kinetic energy of the system. Therefore, in this case, the control satisfying the conditions (3.78) is optimal in the sense that the kinetic energy is minimal, i.e., the energy costs are minimal.

Chapter 9
Statics and Dynamics of Thin Rod

A. K. Belyaev①, **N. F. Morozov**①, **P. E. Tovstik, and T. P. Tovstik**①

This chapter presents the classic results by Euler on nonlinear static deforma-
tion of the axially compressed rod, the results of the work by M.A. Lavrentiev and
A.Yu. Ishlinsky on rod buckling under dynamic axial compression and the recent
results mainly related to the study of interaction of longitudinal and transverse vibra-
tions. The chapter introduces readers to the basic research methods which are the
D'Alembert and Fourier methods, the methods of studying parametric resonances,
the asymptotic method of two-scale expansion.

1 Introduction

The chapter contains a brief overview of the static and dynamic axial compression
of a thin rod. Under static longitudinal axial compression the straight rod can buckle
and take adjacent equilibrium forms. This is Euler's classical problem, which he
began to engage in as early as 1744.[1]. The problem is much more difficult for the
case of axial dynamic loading. In the framework of a rigorous problem statement the
longitudinal elastic waves propagate in the rod, which in turn can generate intense
transverse vibrations. However, the propagation time of the longitudinal wave along
the rod length is significantly less than the smallest period of transverse vibration,
the approximate model was originally used based on the assumption that the longitudinal wave instantaneously propagates in the rod, i.e., the axial compressive
force is constant in all cross sections. With such a statement M.A. Lavrentiev and

---

[1] *L. Euler* The method of finding curves, possessing the properties of maximum either minimum, or
solution of isoperimetric problem taken in the broadest sense. M.-L.: Gostekhizdat. 1934. 600 p. [in
Russian].

© Springer Nature Switzerland AG 2021
N. N. Polyakhov et al., *Rational and Applied Mechanics*,
Foundations of Engineering Mechanics,
https://doi.org/10.1007/978-3-030-64118-4_9

A.Yu.Ishlinsky.[2] solved the problem for compressive force exceeding the critical static value. It was found out that the higher instability modes have rapid increase in deflection than the lower ones.

The papers by authors of this chapter[3] reported the problem solution in a more rigorous formulation, which takes into account the finite velocity of propagation of the longitudinal wave. We considered both a short-term longitudinal impact on the rod end and suddenly applied longitudinal load which then remains constant. The applied axial force is assumed to be constant and less than Euler's critical force. The problem is reduced to successive solving two linear problems. First, the longitudinal waves are considered and they generate a periodic system of axial stresses in the rod. These stresses can excite transverse oscillations associated with the appearance of parametric resonances. A shortcoming of the linear approach was mentioned, namely, the resistance forces cannot prevent unlimited growth of amplitudes of the transverse vibrations.

For this reason, a quasilinear approach was proposed that takes into account the influence of transverse vibration on longitudinal one. To solve the resulting nonlinear partial differential equations system, the Bubnov-Galerkin approach was first applied, leading to a system of ordinary differential equations that admits only numerical solution. Then the method of two-scale expansion was applied which resulted in a qualitative analysis. It turned out that the rod motion has the character of beats, in which the energy of longitudinal vibration goes into the transverse vibration energy and vice versa. The beating attenuates over time when the resistance is taking into account.

If a load exceeding the Euler critical one is suddenly applied and then remains constant, the parametric resonances also appear, however the analysis takes a back seat, since the growth of amplitude rate is less than that of the Euler buckling rate. Therefore, of particular interest is a more detailed consideration of instability in the cases in which the suddenly applied load is less than the Euler load. It is established that the instability region occupy a relatively small part in the parameter plane. A maximum amplitude estimate for bending vibration was obtained by means of the quasilinear approach.

The question of the final shape of the rod subjected to an axial force that exceed the Eular critical load was resolved by L.Euler himself. These shapes are referred to as *Euler elastica*. The evolution of the rod deflection over time is considered in what follows. It is interesting to note the time history of the rod deflection: as time progresses the rod consistently takes forms close to those predicted in the classic work by M.A.Lavrentiev and A.Yu.Ishlinsky (see footnote above), and then takes the form of an Euler elastica.

---

[2] *M.A.Lavrentiev, A.Yu.Ishlinsky* Dynamic forms of loss stability of elastic systems // Reports of the USSR Academy of Sciences. 1949. Vol.64. № 6. Pp.776–782 [in Russian].

[3] For a review of these works, see the article: *A.K.Belyaev, P.E.Tovstik, T.P.Tovstik.* Thin rod under longitudinal dynamic compression // Izv. RAS. Mechanics of Solids. 2017. № 4. Pp.19–34 [in Russian].

In what follows we consider the problem in a more realistic statement than the above one, namely, we consider an axial impact on the rod end by a solid body that flies up to the rod at a given speed and then bounces. In the simplest statement this problem was solved by Saint-Venant,[4] while a more accurate statement was suggested by Sears[5] who took into account the local deformations in the contact zone. A solution to this problems is given below along with discussion of excitation of transverse parametric resonances caused by the impact of a body.[6]

2 Axial Vibration of Rod. Linear Approximation

Let us consider a thin homogeneous elastic rod of length L and constant cross section S. The right end of the rod $x = L$ is fixed while the left end $x = 0$ is subjected to a time-dependent force $f(t)$. The trivial initial conditions are assumed. The equations of motion and the boundary conditions are given by

$$\frac{\partial^2 u}{\partial x^2} - c^2 \frac{\partial^2 u}{\partial t^2} = 0, \quad ES\frac{\partial u}{\partial x}\bigg|_{x=0} = -f(t), \quad u(L,t) = 0. \tag{2.1}$$

Here $u(x,t)$ denotes the axial displacement of the rod section x, $c = \sqrt{E/\rho}$ is the velocity of sound in the rod material, E is Young's modulus and ρ is the mass density.

Let us introduce the non-dimensional variables (with tilde sign ⌢), such that the rod length and the sound velocity are equal to unity

$$\widehat{x} = \frac{x}{L}, \quad \widehat{t} = \frac{tc}{L}, \quad \widehat{f(t)} = \frac{f(t)}{ES}.$$

In what follows, the sign ⌢ is omitted. The equation of motion, the boundary and initial conditions take the form

$$\frac{\partial^2 u}{\partial x^2} - \frac{\partial^2 u}{\partial t^2} = 0, \quad \frac{\partial u}{\partial x}\bigg|_{x=0} = -f(t),$$
$$u(1,t) = 0, \quad u(x,0) = u_t(x,0) = 0. \tag{2.2}$$

There exist two methods of obtaining solution of problem (2.2), which are the *D'Alembert method* and the *Fourier method*. We describe them since both methods are used in what follows.

---

[4] *A. Saint-Venant.* Sur le choc longitudinal de deux barres elastiques // J. de Math. (Liouville) Ser. 2. T. 12. 1867.

[5] *J.E. Sears.* On the longitudinal impact of metal rods with rounded ends // Proc. Cambridge Phil. Soc. 1908. Vol. 14.

[6] *A.K. Belyaev, C.-C. Ma, N.F. Morozov, P.E. Tovstik, T.P. Toвcmuk, A.O. Shurpatov.* Dynamics of rod under an axial body impact // Vestnik St. Petersburg University. Mathematics. Mechanics. Astronomy. 2017. Vol. 4(62). № 3. Pp. 506–515 [in Russian].

The *D'Alembert method* is related to the wave propagation. The general solution is superposition of a right-traveling wave and a left-traveling wave. Function $u(x, t) = g(x - t) + h(x + t)$ satisfies equation (2.1) for any functions g and h. It remains to find these functions so as to satisfy the boundary and initial conditions. For the problem in question, the displacement of the left end of the rod has the form of a convolution integral

$$u(0, t) = \int_0^t G(t - \tau) f(\tau) \, d\tau,$$

where $G(t)$ is called the unit-impulse response function, i.e., the response on $f(t) = \delta(t)$. Function $G(t)$ is a periodic piecewise constant function with period $4 (G(t + 4) = G(t))$. The explanation is as follows. When a wave is reflected from a fixed end $x = 1$, the sign of u does not change; when a wave is reflected from the free end $x = 0$, the sign changes to the opposite[7]

$$G(t) = \begin{cases} +1, & 0 < t < 2, \\ -1, & 0 < 2 < 4, \end{cases} \quad \text{ИЛИ} \quad G(t) = (-1)^{[t/2]}, \tag{2.3}$$

where $[z]$ denotes the integer part of a number z.

The *Fourier method* is a series expansion of the solution

$$u(x, t) = \sum_{k=1}^{\infty} u_k(x) \varphi_k(t) \tag{2.4}$$

in terms of the eigenfunction $u_k(x)$ of the boundary-value problem

$$\frac{d^2 u_k}{dx^2} + \nu^2 u_k = 0, \quad \frac{du_k}{dx}\bigg|_{x=0} = 0, \quad u_k(1) = 0.$$

The solutions are given by

$$u_k(x) = \cos \nu_k x, \quad \nu_k = (k - 0.5)\pi, \quad k = 1, 2, \dots.$$

Functions $\varphi_k(t)$ satisfy the equation

$$\frac{d^2 \varphi_k}{dt^2} + \nu_k^2 \varphi_k = 2 f(t), \quad \varphi_k(0) = 0, \quad \frac{d\varphi_k}{dt}\bigg|_{t=0} = 0$$

that yields

$$\varphi_k(t) = \frac{2}{\nu_k} \int_0^t f(t) \sin(\nu_k(t - \tau)) d\tau.$$

We are interested in the (dimensionless) axial force of the rod compression

---

[7] *V.I. Smirnov.* The course of higher mathematics. Vol. 2. M.: Nauka. 1974. 656 p. [in Russian].

$$\varepsilon(x, t) = \frac{P}{ES} = -\frac{\partial u}{\partial x} = \sum_{k=1}^{\infty} \nu_k \sin \nu_k x \, \varphi_k(t) \,. \tag{2.5}$$

We note an important fact for the further study that functions $\varphi_k(t)$ along with function $\varepsilon(x, t)$ are periodic with respect to t with period 4 after the impact is over (for $t > T$, where $f(t) \equiv 0$ for $t > T$). Indeed, in this case, for $t > T$

$$\varphi_k(t) = a_k \cos \nu_k t + b_k \sin \nu_k t \,,$$

with the integration constants a_k and b_k

$$a_k = -\frac{2}{\nu_k} \int_0^T f(t) \sin \nu_k t \, dt \,, \quad b_k = \frac{2}{\nu_k} \int_0^T f(t) \cos \nu_k t \, dt \,.$$

The periodicity follows from the equation $\nu_k = (k - 0.5)\pi$.

3 Bending and Transverse Vibration of Rod

Let us consider the rod of Sect. 2 and assume that the rod ends $x = 0$ and $x = L$ are simply supported. The small transverse vibrations of the rod compressed by axial force P are described by the equation

$$\rho S \frac{\partial^2 w}{\partial t^2} + \frac{\partial}{\partial x} \left(P \frac{\partial w}{\partial x} \right) + E J \frac{\partial^4 w}{\partial x^4} = 0 \,,$$
$$w = \frac{\partial^2 w}{\partial x^2} = 0 \,, \quad x = 0, L \,, \tag{3.1}$$

where $w(x, t)$ is the rod deflection and J is the moment of inertia of the cross section in the plane of least bending rigidity (see Fig. 1).

In terms of the dimensionless variables (2.1) problem (3.1) is rewritten in the form

Fig. 1 Rod deformation under axial compression

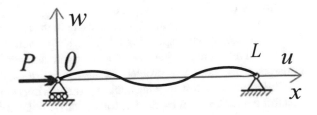

$$\frac{\partial^2 w}{\partial t^2} + \frac{\partial}{\partial x}\left(\varepsilon\frac{\partial w}{\partial x}\right) + \mu^2\frac{\partial^4 w}{\partial x^4} = 0,$$

$$w = \frac{\partial^2 w}{\partial x^2} = 0, \quad x = 0, 1,$$

(3.2)

where

$$\varepsilon = \frac{P}{ES}, \quad \mu = \frac{r}{L}, \quad J = r^2 S,$$

r is the radius of inertia of the cross section and μ is a small parameter of the rod thickness. Further we also use $\ell = 1/\mu$ which is a large parameter of the rod length.

4 Classical Solutions of Euler and Lavrentiev–Ishlinsky

The static problem of bifurcation of equilibrium of a rod with simply supported ends compressed by a constant force ($\varepsilon(x, t) = \varepsilon_0 = $ const) was solved by L. Euler. The adjacent stable equilibrium modes $w(x) = W \sin m\pi x$ have m half-waves and correspond to the axial compression strains $\varepsilon_m = \mu^2 m^2 \pi^2$. For $m = 1$ we have the classical Euler critical load

$$\varepsilon_{cr} = \mu^2 \pi^2.$$

For a sufficiently large value of the initial deformation ε_0, a simultaneous buckling is possible in several first modes, namely, with $m \leqslant m_0 = [\sqrt{\varepsilon_0/(\mu\pi)^2}]$, where $[z]$ is the integer part of z.

M.A. Lavrentiev and A.Yu. Ishlinsky drew attention to the fact that for $m_0 > 1$ the growth rate of the instability amplitude is maximum for one of the higher forms of buckling ($m \approx m_0\sqrt{2}$). Indeed, a particular solution of equation (3.2) for $m \leqslant m_0$, $\delta_w = 0$ is

$$w(x, t) = w_{0m} \sin m\pi x \, e^{\alpha_m t}, \quad \alpha_m = \sqrt{\varepsilon_0 m^2 \pi^2 - \mu^2 m^4 \pi^4},$$

where w_{0m} depends on the initial rod deflection $w_0(x)$, parameter α_m characterizes the rate of amplitude growth of mth buckling shape. Figure 2 displays the plots of function $\alpha_m(\varepsilon_0)$ for $m = 1, 2, ... , 8$.

As the compression strain ε_0 increases, both the maximum amplitude growth rate and the number of corresponding buckling shape increase.

We will return to the question of growth rate of buckling modes in Sect. 10 when considering the nonlinear problem of large post-critical deformations of the rod.

Choosing the fastest growing buckling shape is important for a number of dynamic instability problems, in particular, for various problems of stability of shells which are characterized by instability due to many buckling modes.

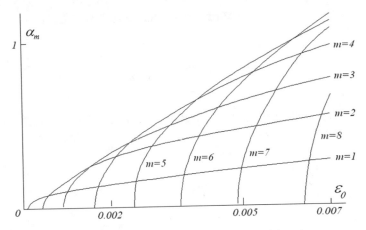

Fig. 2 Dependence of the growth rate of mth buckling mode on axial strain

5 Axial-Transverse Vibration. Linear Approximation

It was assumed in Sect. 4 that the longitudinal compression foce ε_0 is constant. Here we turn to a more rigorous formulation of the problem which takes into account the propagation of longitudinal waves. In the linear approximation we set $\varepsilon = \varepsilon(x, t) = -\partial u/\partial x$, eq. (2.5), in Eq. (3.2) where $u(x, t)$ is the solution to problem (2.2). As a result, the problem is reduced to a consistent solving two linear boundary-value problems.

The equation for bending vibration takes now the form

$$\frac{\partial^2 w}{\partial t^2} + \frac{\partial}{\partial x}\left(\varepsilon(x.t)\frac{\partial w}{\partial x}\right) + \mu^2 \frac{\partial^4 w}{\partial x^4} = 0,$$

$$\varepsilon(x, t) = \sum_{k=1}^{\infty} \nu_k \sin \nu_k x \, \varphi_k(t),$$

(5.1)

where coefficients $\varphi_k(t)$ depend on the time-dependent force $f(t)$.

In particular, for a rectangular pulse of duration T

$$f(t) = \begin{cases} \varepsilon_0, & t \leqslant T, \\ 0, & t > T, \end{cases} \quad \rightarrow$$

$$\rightarrow \quad \nu_k^2 \varphi_k(t) = \begin{cases} 1 - \cos(\nu_k t), & t \leqslant T, \\ \cos \nu_k(t - T) - \cos \nu_k t, & t > T, \end{cases}$$

and for a discontinuously applied long pulse

$$f(t) = \varepsilon_0, \quad t < T \quad \rightarrow \quad \nu_k^2 \varphi_k(t) = 1 - \cos(\nu_k t).$$

We turn to the consideration of transverse vibrations. We look for the solution of equation (5.1) as a series

$$w(x, t) = \sum_{n=1}^{\infty} T_n(t) \sin n\pi x .$$

(5.2)

Functions $T_m(t)$ satisfy the system of equations obtained by inserting (5.2) into (5.1), multiplying the result by $\sin m\pi x$ and integrating over x within $(0,1)$:

$$\frac{d^2 T_m}{dt^2} + \omega_m^2 T_m(t) - \sum_{n=1}^{\infty} a_{mn}(t) T_n(t) = 0 ,$$

$$m = 1, 2, \ldots , \qquad \omega_m = \mu m^2 \pi^2 ,$$

$$a_{mn}(t) = 2mn\pi^2 \int_0^1 \varepsilon(x, t) \cos m\pi x \, \cos n\pi x \, dx .$$

(5.3)

After the end of a short pulse, the periodic functions $a_{mn}(t)$ have the period $T = 2\pi/\nu_1 = 4$ and zero mean value.

For a long pulse, the mean value of strain $\varepsilon(x, t)$ is ε_0. Separating it, we write equation (5.3) in the form

$$\frac{d^2 T_m}{dt^2} + (\omega_m^2 - m^2 \pi^2 \varepsilon_0) T_m(t) - \sum_{n=1}^{\infty} \widehat{a}_{mn}(t) T_n(t) = 0 ,$$

$$m = 1, 2, \ldots ,$$

where coefficients $\widehat{a}_{mn}(t) = a_{mn}(t) - \varepsilon_0 \delta_{mn} m^2 \pi^2$ have zero mean value (δ_{mn} is Kronecker delta). At $a_{mn}(t) = 0$ the statical buckling by mth mode occurs if $\omega_m^2 - m^2 \pi^2 \varepsilon_0 < 0$, that is if $\varepsilon_0 > m^2 \varepsilon_{cr}$. The presence of periodic functions $a_{mn}(t)$ means that instability mth mode can also occur for small values of ε_0, in particular, for $\varepsilon_0 < \varepsilon_{cr}$, that is, under the load which is smaller than the Euler critical load (see Sect. 7).

6 Parametric Resonances

Consider stability of the trivial solution of system (5.3). This system contains two dimensionless parameters: the small amplitude of axial strain ε_0 and the small parameter μ which is the relative thickness of rod or the inverse of its large relative length $\ell = \mu^{-1}$.

At $\varepsilon_0 = 0$ the natural frequencies of system (5.3) are equal to $\omega_m = \mu m^2 \pi^2$, while the frequency of parametric excitation is $\nu_1 = \pi/2$. The methods studying parametric oscillations are well known.[8] For small ε_0 the instability regions are possible only in vicinity of the values

$$|\omega_m \pm \omega_n| = k\nu_1, \quad m, n, k = 1, 2, \dots.$$

Provided that the rod has zero damping, the regions of parametric instability touch the axis $\varepsilon_0 = 0$ in the parameter plane (ℓ, ε_0) for

$$\ell_{mnk} = \frac{2\pi(m^2 \pm n^2)}{k}, \quad m, n, k = 1, 2, \dots. \tag{6.1}$$

The set of critical lengths given by formula (6.1) is everywhere dense, and the resonances themselves are excited with different growth rates of the vibration amplitudes. When $m = n$, $k = 1$ we have principle resonances, when $m \neq n$ we have combination resonances and when $k > 1$ we have overtone resonances.

Consideration of the principle resonances for $\ell_m = 4\pi m^2$ reduces to Hill's equation

$$\frac{d^2 T_m}{dt^2} + \left(\omega_m^2 - \varepsilon_0 a_{mm}(t)\right) T_m(t) = 0, \quad m = 1, 2, \dots, \tag{6.2}$$

provided that the interaction of vibration modes is neglected. Equation (6.2) transforms into the Mathieu equation when a single term in the series (5.2) is preserved. It will be shown in Sect. 9 that this transformation gives the first approximation in ε_0 to the boundaries of principal resonances of the instability region in the (ℓ, ε_0) plane. As a result of numerical integration of Eq. (6.2), the characteristic exponents ρ_j of the monodromy matrix[8] are found. The growth rate of deflection α_m^* introduced in Sect. 4 is related to them by the formula $\alpha_m^* = (1/4) \ln(\max_j |\rho_j|)$.

Figure 3 shows rates of the amplitude growth $\alpha_m(\varepsilon_0)$ for a rod of dimensionless length $\ell = 100\pi$ at suddenly applied long loading $0 < \varepsilon_0 \leqslant 0.007$ for Euler's buckling modes (the same as in Fig. 2) and parametric resonances (with sign *). We can see that Euler's buckling modes grow faster than the parametric ones. The taken length is equal to the critical one for $m = 5$, that is why the parametric resonances are not excited at $m \leqslant 5$.

Let us carry out analysis of the instability regions in the plane (ℓ, ε_0). Typical instability regions are displayed in Fig. 4. These regions touch axis $\varepsilon_0 = 0$ for no damping and, if any, they are separated from this axis by dotted line. Short-term impulse generates regions symmetric with respect to the vertical straight line, see Fig. 4a and b), while these regions are tilted to the left for a long compression, see Fig. 4c). The main and some combination resonances are "wide", that is, the tangents to boundary of the stability region drawn from the point $\varepsilon_0 = 0$ form a nonzero angle α (Fig. 4a and c), while the tangents coincide for the "narrow" regions (Fig. 4b).

[8] See, e.g., V.A. Yakubovich, V.M. Starzhinskiy. Linear differential equations with periodic coefficients and their applications. M.: Nauka. 1972. 720 p. [in Russian].

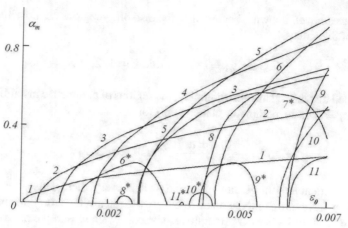

Fig. 3 Dependence of growth rate of the rod deflection on axial strain due to mth mode of Euler buckling and for parametric resonances (marked by *)

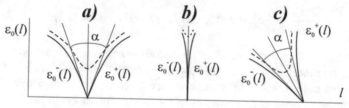

Fig. 4 Broad (a, c) and narrow (b) instability regions at parametric resonances

7 Instability at Load Less Than Euler Load

A jump loading by a force smaller than the Euler critical force causes axial vibration which, in turn, can lead to parametric resonances and instability. Calculations showed that instability is possible only in narrow intervals of lengths $\ell_m^* < \ell < \ell_m = 4m^2\pi$. In particular, this interval is $40.6 < \ell < 50.3$ for $m = 2$.

Consider a rod of length $\ell = 0.45$ under compression $\varepsilon_0 = 0.0035$, then $\varepsilon_{cr} = \pi^2/\ell^2 = 0.00487 > \varepsilon_0$. Parameter α_m of the amplitude growth rate for Euler buckling ($m = 1$) and parametric resonance ($m^* = 2$) are shown in Fig. 5.

Parametric resonances without damping lead to an unbounded increase in the amplitude of transverse vibration which contradicts the conservative nature of the problem. Calculations have shown that resistance forces can only reduce the instability regions. For this reason we proceed to a more accurate quasilinear formulation of the problem.

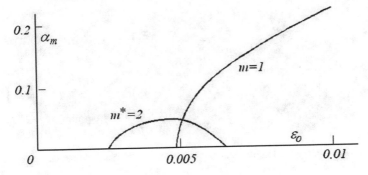

Fig. 5 Illustration of instability under load less than Euler critical one

8   Quasilinear Approximation

The reason for unbounded growth of the amplitude at parametric resonances in a linear approximation is that the energy is transmitted from the longitudinal vibration to the transverse vibration while the amplitude of longitudinal vibration remains unchanged. The quasilinear approach also takes into account the effect of transverse vibration on longitudinal one. In the dimensionless form, the equations of motion and the boundary conditions have the form:

$$\frac{\partial^2 u}{\partial t^2} + \frac{\partial \varepsilon}{\partial x} = 0, \quad \varepsilon = -\frac{\partial u}{\partial x} - \frac{1}{2}\left(\frac{\partial w}{\partial x}\right)^2,$$

$$\left.\frac{\partial u}{\partial x}\right|_{x=0} = -f(t), \quad u(1, t) = 0, \tag{8.1}$$

$$\frac{\partial^2 w}{\partial t^2} + \frac{\partial}{\partial x}\left(\varepsilon \frac{\partial w}{\partial x}\right) + \mu^2 \frac{\partial^4 w}{\partial x^4} = 0,$$

$$w = \frac{\partial^2 w}{\partial x^2} = 0, \quad x = 0, 1. \tag{8.2}$$

Equation (8.1) differs from eq. (2.2) in the accurate expression for axial strain $\varepsilon = -u_x - (1/2)w_x^2$. Equation (8.2) is identical to Eq. (3.2).

The system of nonlinear equations (8.1), (8.2) is not exactly integrated. Its approximate solution can be obtained using the Bubnov-Galerkin method. Representing the unknown functions as

$$u(x, t) = \varepsilon_0 \sum_{k=1}^{\infty} \varphi_k(t) \cos(\nu_k x),$$

$$w(x, t) = \sum_{m=1}^{\infty} T_m(t) \sin(m\pi x),$$

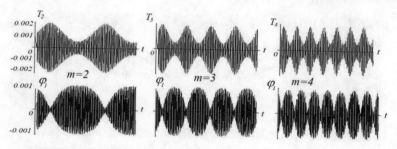

Fig. 6 Transfer of energy of longitudinal and transverse vibrations at parametric resonances generated by a short-term shock pulse

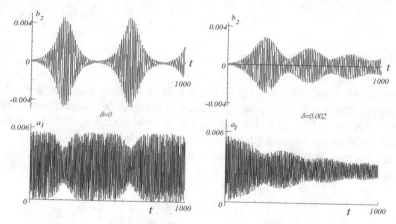

Fig. 7 Transfer of vibration energy during long-term compression without and with damping

we arrive at a system of ordinary differential equations for the unknown coefficients $\varphi_k(t)$ and $T_m(t)$. This system is integrated numerically. Typical results for a short-term shock pulse are shown in Fig. 6 for the case of zero damping. This figure displays the parametric resonances for $m = 2, 3, 4$.

Motion is a beating with the successive transfer of energy of longitudinal vibration in the transverse one and vice versa.

Parametric resonance turns into beats under a suddenly applied load, less than the Euler critical one. Figure 7 shows the evolution of vibration in the rod considered in Sect. 7 without (left) and with (right) small damping. Here $u(x, t) = a_1(t) \sin \nu_1 x$, $w(x, t) = b_2(t) \sin 2\pi x$.

In contrast to Fig. 6, the longitudinal vibration is superimposed on a constant compression which does not attenuate when the damping is present.

9 Asymptotic Integration of Quasilinear Equation of Rod Motion

The character of curves in Figs. 6 and 7 leads on thought of using the two-scale expansion[9] which are convenient for consideration of rapidly oscillating functions with slowly varying amplitudes. We present the method in the case of mth principal parametric resonance for a short-term shock pulse.

The solution is sought in the form

$$u(x, t) = \sum_{k=1}^{\infty} \varphi_k(t) \cos(\nu_k x),$$

$$w(x, t) = \sum_{m=1}^{\infty} T_m(t) \sin(m\pi x).$$
(9.1)

Multiplying Eqs. (3.1), (3.2) by functions $\cos(\nu_k x)$ and $\sin(m\pi x)$, respectively, and integrating over x within $(0,1)$, we have

$$\frac{d^2 \varphi_k}{dt^2} + \nu_k^2 \varphi_k + \sum_{m,n=1}^{\infty} D_{mnk} T_m T_n = 0, \quad k = 1, 2, \ldots,$$

$$\frac{d^2 T_m}{dt^2} + \omega_m^2 T_m - L_{mn}(T) - \sum_{k,n=1}^{\infty} C_{mnk} T_n \varphi_k = 0,$$
(9.2)

$$m = 1, 2, \ldots.$$

If $\varepsilon_0 \ll 1$ we look for the solution of system (9.2) with slowly varying amplitudes

$$\varphi_k = \varepsilon_0 \varphi_k^{(0)}(t, \theta) + \varepsilon_0^2 \varphi_k^{(1)}(t, \theta) + \cdots,$$
$$T_m = \varepsilon_0 T_m^{(0)}(t, \theta) + \varepsilon_0^2 T_m^{(1)}(t, \theta) + \cdots,$$

where the slow time is denoted by $\theta = \varepsilon_0 t$. Then in system (9.2)

$$\frac{d}{dt} = \frac{\partial}{\partial t} + \varepsilon_0 \frac{\partial}{\partial \theta}.$$

In addition, we set $\omega_m^2 = \nu_1^2/4 + \varepsilon_0 \delta$, where δ is a parameter of the frequency detuning at parametric resonance. The term $L_{mn}(T)$ in eq. (9.2) has the order ε_0^3 and is not included in the approximations considered.

In the zero approximation we have

---

[9] See monograph: *N.N. Bogolyubov, Yu.A. Mitropolsky*. Asymptotic methods in theory of nonlinear oscillations. M.: Nauka. 1974. 503 p. [in Russian].

$$\varphi_k^{(0)}(t, \theta) = a_{kc}(\theta) \cos(\nu_k t) + a_{ks}(\theta) \sin(\nu_k t),$$
$$T_m^{(0)}(t, \theta) = b_{mc}(\theta) \cos(\nu_1 t/2) + b_{ms}(\theta) \sin(\nu_1 t/2),$$

where the unknown functions $a_{kc}(\theta)$, $a_{ks}(\theta)$, $b_{mc}(\theta)$, $b_{ms}(\theta)$ are determined from the compatibility conditions for the first approximation equations

$$2\nu_1 \frac{da_{1c}}{d\theta} - D_{1m} b_{mc} b_{ms} = 0,$$

$$-2\nu_1 \frac{da_{1s}}{d\theta} - \frac{1}{2} D_{1m} \left(b_{mc}^2 - b_{ms}^2 \right) = 0,$$

$$2\frac{da_{kc}}{d\theta} = 0, \quad 2\frac{da_{ks}}{d\theta} = 0, \quad k = 2, 3, \dots, \tag{9.3}$$

$$\nu_1 \frac{db_{mc}}{d\theta} - \delta b_{ms} + \frac{1}{2}\widehat{C}_{1m} (b_{mc} a_{1s} - b_{ms} a_{1c}) = 0,$$

$$-\nu_1 \frac{db_{ms}}{d\theta} - \delta b_{mc} + \frac{1}{2}\widehat{C}_{1m} (b_{mc} a_{1c} + b_{ms} a_{1s}) = 0.$$

As follows from this system, functions $a_{kc}(\theta)$, $a_{ks}(\theta)$, $k \geqslant 2$, are not included in the last two equations in the zero approximation (to which we restrict ourselves). If there is no damping, these functions are constant, and if present, they slowly fade out. Thus, the problem is reduced to equations (9.3) with four unknown functions a_{1c}, a_{1s}, b_{mc}, b_{ms}.

Consider the initial conditions for system (9.3). The initial time instant is that time moment when action of the shock pulse is complete. Let $a_{1c}(0) = a$, $a_{1s}(0) = 0$ and the initial values $b_{mc}(0)$ and $b_{ms}(0)$ are negligibly small.

An approximate formula for the boundaries of instability region is obtained from system (9.3)

$$|\ell - \ell_m| < \ell_m k\varepsilon_0 a, \quad k \approx 4m^2, \quad \ell_m = 4\pi m^2,$$

indicating that the instability region expands with increasing m.

Let us consider the function

$$V(\theta) = b_m^2 + \alpha a_1^2,$$
$$b_m^2 = b_{mc}^2 + b_{ms}^2, \quad a_1^2 = a_{1c}^2 + a_{1s}^2, \quad \alpha \approx 4.$$

As follows from system (9.3)

$$\frac{dV}{d\theta} = 0. \tag{9.4}$$

Hence, $V(\theta) \leqslant V(0) = \alpha a^2$. From here we have an important conclusion that the amplitude of transverse vibration during beats does not exceed the double amplitude of the first mode of longitudinal vibration generating this transverse vibration (see also Fig. 8)

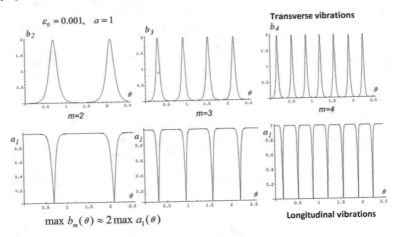

Fig. 8 Vibration amplitude at parametric resonances obtained by two-scale expansion

$$\max\{|w(x,t)|\} \leqslant 2a . \qquad (9.5)$$

Numerical integration of system (9.3) yields the time dependence of amplitudes of longitudinal and transverse vibrations with beats without taking into account the damping (Fig. 8). This dependence was previously obtained in Sect. 8 (see Fig. 6).

Consideration of a suddenly applied long-term load, less than the Euler critical one, results in a system similar to Eq. (9.3). Taking into account equality (9.4), we obtain from this system an estimate for the maximum deflection at parametric resonance $|w(x,t| < 16\varepsilon_0/\pi^2$. Since $\varepsilon_0 < \varepsilon_{kr} = \mu^2\pi^2$ we have an estimate $|w(x,t)| < 16\mu^2$ or in terms of mechanical parameters $|\widetilde{w}| < 16r^2/L$, where r and L denote the inertia radius of the cross section and length of the rod, respectively. Smoothness of the load application significantly reduces the maximum deflection.

The consideration of combinational resonances is carried out by analogy. Instead of the fourth-order system (9.3) we arrive at the sixth-order system of equations for amplitudes $a_{kc}, a_{ks}, b_{mc}, b_{ms}, b_{nc}, b_{ns}$. Here we confine ourselves to a single example, namely, we consider the principal resonance at $m = 4$ and three combinational resonances at $m = 4$, $n = 1, 2, 3$. We take the amplitude of excitation of longitudinal vibration $a = 1$ and $\varepsilon_0 = 0.001$. Functions $a_1(\theta)$, $T_m(\theta)$ and $T_n(\theta)$ are shown in Fig. 9. Damping is assumed to be absent.

In the four upper plots the rod length is equal to the critical one and there is no mismatch of length ($\delta = 0$). In this case we observe full transfer of energy of the longitudinal vibration to transverse one and vice versa. This follows from the fact of zero amplitude of longitudinal vibration at some time instants whereas the amplitude of transverse vibration vanishes at other instants of time.

In the four lower plots, a mismatch of δ is equal to 0.9 of the maximum value. As follows from Fig. 9, only a small part of the energy of longitudinal vibration participates in the energy exchange with transverse vibration excited substantially less than at $\delta = 0$.

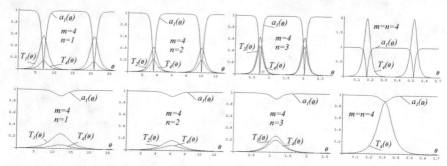

Fig. 9 Vibration amplitudes at combinational parametric resonances obtained by two-scale expansion

10 Post-Critical Deformation of Rod

When the axial force is greater than the Euler critical force the rod buckles out. The evolution of transverse deformation in the linear approximation was discussed in Sect. 4. Here we consider the nonlinear statement of the problem.

We consider deformation of a rod with simply supported ends A and B. End B is fixed, and end A can move along the straight line AB. The constant force P is a dead load, i.e., it is directed along the same straight line, see Fig. 10.

In terms of dimensionless variables, the equilibrium equations for inextensible rod have the form:

$$\frac{d\varphi}{ds} + py = 0, \quad \frac{dx}{ds} = \cos\varphi, \quad \frac{dy}{ds} = \sin\varphi, \quad 0 \leqslant y \leqslant \pi, \tag{10.1}$$

where the rod length is equal to π, while the compression force is normalized so that the critical compression is equal to $p = 1$.

L. Euler constructed possible forms of equilibrium of the elastic line under the action of a self-balanced system of forces and moments applied to its ends (Euler's elastica). Figure 11 displays the solutions of system (10.1) for $p_1 = 1.02$, $p_2 = 1.1$, $p_3 = 1.5$, $p_4 = 2.184$, $p_5 = 3$, $p_6 = 10$. With growth of p the rod end A

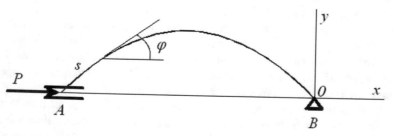

Fig. 10 Nonlinear buckling form of axially compressed rod

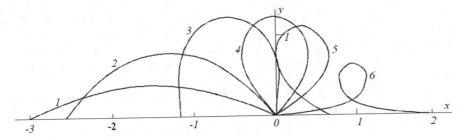

Fig. 11 Euler's elastica

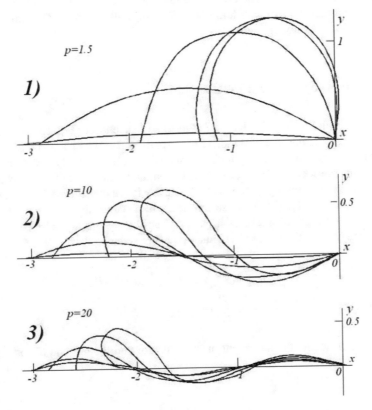

Fig. 12 Evolution of shape of the rod deflection under jump of axial compression

approaches the motionless end B, and at $p = p_4$ the rod ends coincide. A loop is formed if $p > p_4$.

For the one-dimensional dynamic problem shown in Fig. 10, the finite element method is used to construct a solution for three values of the compression force $p = 1.5$, $p = 10$ and $p = 20$. At the initial time instant, the rod is uniformly compressed and small transverse perturbation is applied. Figure 12 presents the form of elastic line at successive time instants. For $p = 1.5$ instability is possible only in the first form. The amplitude grows and coincides with the Euler elastic 3 in Fig. 11 in the limit. For $p = 10$, buckling is possible in the first three modes, the second mode having the greatest rate of amplitude growth $w = \sin 2x$. Growth of this mode is observed in Fig. 12(2) at the beginning of motion. Similarly, at $p = 20$, the third mode grows fastest. For $p = 10$ and $p = 20$, point A passes a fixed point B forming a loop as time progresses (not shown in Fig. 12).

Note, that the static self-balanced modes of a rod of the type $y = \sin mx$, $m \geqslant 2$ are unstable because the curvature of elastic line must retain a sign for stable modes. In Figs. 10, 11 and 12(1) curvature retains the sign, whereas in Fig. 12(2) and (3) the curvature changes the sign at intermediate time instants but in the final position the curvature sign does not change, see Fig. 11(6).

11 Axial Body Impact on the Rod

We consider[10] an elastic rod of length L with fixed right end, see Fig. 13). At the initial instant of time, an impactor of mass m flies up to the left end with a speed v_0 resulting in a contact interaction. The impactor is assumed to be a rigid body however we take into account the local deformation α in the contact zone. Plane wave propagates along the rod with velocity c (see Sect. 2). Gravity is not considered. The magnitude of contact force $P(t)$ and the collision time are to be determined.

This is the classic problem of impact on the rod by a body. The condition of coincidence of coordinates of the contact points is given by[11, 12]

$$u_0 - \alpha = u\,, \quad u_0 = v_0 t - \frac{1}{m} \int_0^t (t - \theta) P(\theta) d\theta\,,$$

$$u = \int_0^t P(\theta) Y_1(t - \theta) d\theta\,. \tag{11.1}$$

Here u_0 and u denote displacement of the impactor and the left end of rod, respectively,

---

[10] See article: *A. K. Belyaev, C. C. Ma, N. F. Morozov, P. E. Tovstik, T. P. Tovstik, A. O. Shurpatov.* Dynamics of rod at axial body impact // Vestink St.Petersburg State University. Mathematics, Mechanics, Astronomy. 2017. Vol. 4(62). № 3. Pp. 506–515 [in Russian].

[11] *J. E. Sears.* On the longitudinal impact of metal rods with rounded ends // Proc. Cambridge Phil. Soc. 1908. Vol. 14.

[12] *S. A. Zegzhda.* Collision of elastic bodies. Publishers of St.Petersburg State University. 1997. 316 p. [in Russian].

Fig. 13 Axial impact on rod

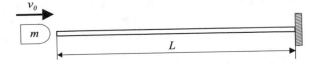

$$Y(t) = \frac{c}{ES} \begin{cases} 1, & 0 < t < 2L/c, \\ -1, & 2L/c < t < 4L/c, \end{cases} \qquad Y(t + 4L/c) = Y(t),$$

$$P = k\alpha^{3/2}, \quad k = \left(\frac{3(1-\nu^2)}{4E} + \frac{3(1-\nu_0^2)}{4E_0}\right)^{-1} \sqrt{\varkappa}.$$

Here $Y(t)$ is the rod reaction of the applied unit pulse, the contact force $P(t)$ is related to the local deformation α by Hertz's law,[13] E_0, E and ν_0, ν are Young modulus and Poisson ratio, respectively, S is cross-sectional area, $c = \sqrt{E/\rho}$ is velocity of sound in the rod, ρ is mass density of the rod material, $\varkappa$ is sum of curvatures of the colliding bodies (these bodies are assumed to be axisymmetric in the contact zone). Provided that the impactor hits the flat end of rod and the colliding bodies are made of the same material we have

$$P = k\alpha^{3/2}, \qquad k = \frac{2E\sqrt{R}}{3(1-\nu^2)},$$

where R stands for curvature radius of impactor in the contact zone.

As a result, Eq. (11.1) becomes the following nonlinear integral equation for force $P(t)$

$$\frac{1}{m} \int_0^t P(\theta)(t-\theta)d\theta + \int_0^t P(\theta)Y(t-\theta)d\theta+ \tag{11.2}$$
$$+(P(t)/k)^{2/3} = v_0 t.$$

The impact ends when force $P(t)$ goes to zero.

Similar to Sects. 2–9 we introduce the non-dimensional variables by relating force P to ES and assuming the unit rod length and unit sound velocity. Equation (11.2) takes the form

$$\xi \int_0^{\widehat{t}} \widehat{P}(\tau)(\widehat{t}-\tau)d\tau + \int_0^{\widehat{t}} \widehat{P}(\tau)G(\widehat{t}-\tau)d\tau+ \tag{11.3}$$
$$+r\widehat{P}(\widehat{t})^{2/3} = \widehat{v_0}\widehat{t}, \qquad G(t) = (-1)^{[t/2]},$$

where

$$\widehat{P} = \frac{P}{ES}, \quad \widehat{t} = \frac{ct}{L}, \quad \widehat{v_0} = \frac{v_0}{c},$$
$$\xi = \frac{M}{m}, \quad r = \frac{1}{L}\left(\frac{ES}{k}\right)^{2/3}. \tag{11.4}$$

[13] H. Hertz. Über die Berührung fester elastischer Körper // Z. f. Math. (Crelle). 1881. Bd 92.

Here, function $G(t)$ coincides with the previously introduced function (2.3), parameter ξ is equal to ratio of the rod mass $M = \rho SL$ to the impactor mass m, parameter r takes into account the local contact deformation. In what follows, the tilde sign $\widehat{}$ is omitted.

Equation (11.3) contains three dimensionless parameters ξ, v_0 and r. If $r = 0$, the local contact deformation is not taken into account, Eq. (11.3) becomes linear, the impact time depends only on ξ, and force $P(t)$ depends on ξ and linearly depends on v_0.

If $r > 0$, two parameters v_0 and r can be reduced to a single one by additional scaling. We set $P = r^3 \widehat{P}$, $v_0 = r^3 V$. In Eq. (11.3) the quantities P and v_0 are replaced by $\widehat{P}$ and $V = v_0/r^3$, respectively, and r is replaced by 1.

Let us return to Eq. (11.3). Differentiation with respect to t yields (see the first and second references in this chapter):

$$\frac{d\widehat{P}}{d\tau} = \frac{3}{2r} \widehat{P}(\tau)^{1/3} \left(\widehat{v}_0 - \xi \int_0^\tau \widehat{P}(\tau_1)d\tau_1 - \widehat{P}(\tau) + \right.$$
$$\left. +2\widehat{P}(\tau - 2) - 2\widehat{P}(\tau - 4) + \dots \right). \tag{11.5}$$

As a result, we arrived at the Sears differential equation with delayed arguments. The appearance of the delayed arguments is associated with propagation of longitudinal waves along the rod and with reflection from the ends.

Equation (11.5) is integrable under the condition $P(t) = 0$ for $t < 0$. To avoid the trivial solution $P(t) \equiv 0$ by numerical integration, one should set a small positive value of $P(0)$, for example, 10^{-10}. We integrate eq. (11.5) sequentially on the intervals $(0, 2)$, $(2, 4)$, $(4, 6)$ and so on using the value of function $P(t)$ determined in the previous interval. In this case, we take into account the continuity of function $P(t)$ for $t = 2, 4, \dots$. Note that in the Wolfram Mathematica 9 package, direct integration of Eq. (11.5) is possible without splitting time into intervals.

Above, function $G(t)$ and Eq. (11.5) were obtained based on the analysis of propagation of longitudinal waves along the rod. The same results can be obtained by Fourier expansion (2.4) in terms of the eigenfunctions $u_k(x) = \cos \nu_k x$ of the boundary-value problem described in Sect. 2. Given that $u_k(0) = 1$, displacement of the left end of rod can be represented as

$$u(0, t) = u(t) = \sum_{k=1}^\infty \varphi_k(t) = \sum_{k=1}^\infty \frac{2}{\nu_k} \int_0^t P(\tau) \sin(\nu_k(t - \tau))d\tau =$$
$$= \int_0^t P(\tau)G(t - \tau)d\tau \, ,$$

whence the representation of function $G(t)$ is taken in the form of a slowly converging series

$$G(t) = \sum_{k=1}^\infty \frac{2}{\nu_k} \sin \nu_k t \, . \tag{11.6}$$

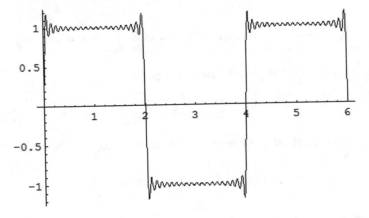

Fig. 14 Sum of a slowly converging series

Fig. 15 Impact force versus time

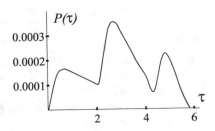

Figure 14 shows sum of the first 20 terms of series (11.6) for $0 \leqslant \tau \leqslant 6$. One can see a slowing the rate of convergence of series when approaching the discontinuity points $t = 0, 2, 4, \ldots$ of function $G(t)$ (*Gibbs phenomen*).

We give an example of the calculation. Consider a longitudinal impact by a steel body of the curvature radius R on the flat end of a rod of the same material. Let $L = 0.5\,\mathrm{m}$, $S = 5 \cdot 10^{-5}\,\mathrm{m}^2$, $R = 0.01\,\mathrm{m}$, $m = 0.5\,\mathrm{kg}$, $v_0 = 1\,\mathrm{m/s}$, $E = 2.1 \cdot 10^{11}\,\mathrm{N/m}^2$, $\nu = \nu_0 = 0.3$, $\rho = \rho_0 = 7.8 \cdot 10^3\,\mathrm{kg/m}^3$. We obtain $c = 5189\,\mathrm{m/s}$, $M = 0.195\,\mathrm{kg}$ and the non-dimensional parameters $v_0/c = 0.0001927$, $\xi = M/m = 0.39$, $r = 0.0155$, appearing in Eq. (11.5).

Function $P(\tau)$ of variables (11.4) is shown in Fig. 15. Curve $P(\tau)$ has three maxima. The impact duration is about 6 times longer than the travel time of the wave along the rod length.

After the body bounces at time instant T, the rod performs free periodic oscillations of period 4, considered in Sect. 2

$$u(x, t) = \sum_{k=1}^{\infty} (b_k \cos \nu_k t + c_k \sin \nu_k t) \cos \nu_k x =$$

$$= \sum_{k=1}^{\infty} a_k \sin(\nu_k t + \alpha_k) \cos \nu_k x \,,$$

where

$$b_k = -\frac{2}{\nu_k} \int_0^T P(t) \sin \nu_k t \, dt \,, \quad c_k = \frac{2}{\nu_k} \int_0^T P(t) \cos \nu_k t \, dt \,,$$

with the amplitude of kth vibration mode is equal to

$$a_k = \sqrt{b_k^2 + c_k^2} \,. \tag{11.7}$$

The dimensionless rebound velocity of the impactor v_T and the restitution coefficient K are

$$v_T = v_0 - \int_0^T P(t) dt \,, \quad K = \frac{|v_T|}{v_0} \,.$$

The dimensionless energy U_0 of body before the impact and its energy U_T after the impact are equal to

$$U_0 = \frac{v_0^2}{2\xi} \,, \quad U_T = \frac{v_T^2}{2\xi} = K^2 U_0 \,,$$

where energy U is related to value of ESL.

For $\tau > T$, energy U_s of axial vibration of rod is constant and equals $U_s = \widehat{U}_0 - \widehat{U}_T = U_0(1 - K^2)$.

Let us find the energy distribution U_s of the rod vibration after the end of impact over the vibration modes $\cos \nu_k x$. By virtue of orthogonality of the vibration modes we have

$$U_s = \frac{1}{2} \int_0^1 \left(\frac{\partial u}{\partial t} \right)^2 dx = \sum_{k=1}^{\infty} U_k \,, \quad U_k = \frac{1}{4} \nu_k^2 a_k^2 \,, \quad k = 1, 2, \dots \,.$$

The identity

$$U_0 = U_T + \sum_{k=1}^{\infty} U_k$$

serves to control the calculations.

Axial post-impact vibration can generate parametric transverse vibration. Using formula (9.5), it is easy to estimate the maximum possible amplitude of beats that occur during parametric resonance

$$\max\{|w(x, t)|\} \leqslant 2a_1 \,,$$

where the amplitude of axial vibration a_1 is calculated by Eq. (11.7) in which the values of w and a_1 are related to the rod length L.

The restitution coefficient K depends on two dimensionless parameters which are ratio of the body mass to the rod mass ξ and parameter V describing the impact

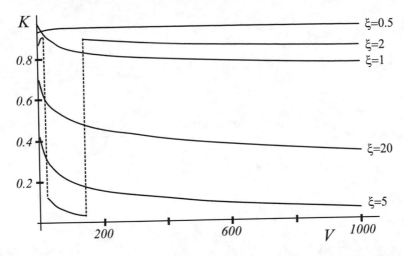

Fig. 16 Dependence of the restitution coefficient on the problem parameters at the axial impact on rod

velocity and local collapse. The results of calculations of function $K(\xi, V)$ are shown in Fig. 16.

The restitution coefficient K is seen to depend rather irregularly on parameters ξ and V. This is explained by the fact that the number of maxima of curve $P(t)$ changes abruptly with change in these parameters changing the impact time and the restitution coefficient.

Chapter 10
Flight Dynamics

N. N. Polyakhov and M. P. Yushkov ⓘ

In the present chapter we introduce basic coordinate systems used in flight dynamics, investigate motion equations of an aircraft in the body-axes system, discuss the forces acting on an aircraft, consider problems of motion of systems of variable mass, give formula for the jet engine thrust, study aircraft motion in the launch coordinate system, and apply methods of nonholonomic mechanics for aircraft targeting. It is also pointed out that flight dynamics is concerned with motion of airplane and launch vehicles in the Earth atmosphere and its methods can be applied to the study of motion of submarines and surface ships.

1 Basic Coordinate Systems Used in Flight Dynamics. Kinematic Equations of Motion

In flight dynamics, in solving its basic problems, the system, the origin of which is located at the Earth's center and the axes directed on visually stationary stars, is known as an inertial coordinate system Earth-centered noninertial coordinate systems are also widely useful. Among these systems frequent use is made of the so-called *launch coordinate system* with origin O_{Lcs} at the lift-off point, the y_{Lcs}-axis is the local vertical (straight up), the x_{Lcs}-axis points in the horizontal plane toward the target, and the z_{Lcs}-axis is chosen to complete the right-handed coordinate system.[1]

In several calculations along with the $O_{Lcs}x_{Lcs}y_{Lcs}z_{Lcs}$ coordinate system it is convenient to use the so-called *Earth-based coordinate system* $O_{Lcs}x_{Ecs}y_{Ecs}z_{Ecs}$ with origin at the lift-off point O_{Lcs}, whose y_{Ecs}-axis passes through the Earth's center and

---

[1] Throughout this chapter the directions of axes follow the convent adopted in the Soviet aerodynamics school. See, for example, the books by *Yu. G. Sirakhulidze*. Ballistics of aircrafts. Moscow: Nauka. Fizmatlit. 1982. 352 p. [in Russian] and *G. S. Byushgens, R. V. Studnev*. Aircraft dynamics. Spatial motion. Moscow: Mashinostroyeniye. 1983. 320p. [in Russian]. It is also worth noting in the study of the Stewart platform in Chap. 3 the directions of axes will be in line with those adopted in Europe and USA.

© Springer Nature Switzerland AG 2021
N. N. Polyakhov et al., *Rational and Applied Mechanics*,
Foundations of Engineering Mechanics,
https://doi.org/10.1007/978-3-030-64118-4_10

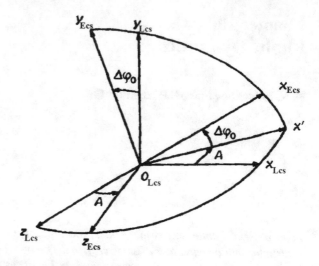

Fig. 1 The launch and Earth-based coordinate systems

is directed upwards, the x_{Ecs}-axis lies in the meridian plane and is directed north, and the z_{Ecs}-axis completes the right-handed coordinate system. The angle φ_0 between the y_{Ecs}-axis and the Earth equatorial plane measures the *geographic (geocentric) latitude* of the lift-off point (the point O_{Lcs}). The angle between the y_{Lcs}-axis and the equatorial plane is called the *geodetic latitude*. Clearly, the angle $\Delta\varphi_0$ between the y_{Ecs}- and y_{Lcs}-axes equals the difference between the geodetic and geocentric latitudes at the lift-off point (Fig. 1). The angle A between the x'-axis (which is the intersection of the planes $O_{Lcs}x_{Ecs}y_{Ecs}$ and $O_{Lcs}x_{Lcs}z_{Lcs}$) and the x_{Lcs}-axis is called the *target azimuth*.

The relation between the coordinates of a point x_{Lcs}, y_{Lcs}, z_{Lcs} and x_{Ecs}, y_{Ecs}, z_{Ecs} in the launch and Earth-based coordinate systems may be conveniently written in matrix form

$$\bar{x}_{Lcs} = N\bar{x}_{Ecs}, \qquad \bar{x}_{Ecs} = N^{-1}\bar{x}_{Lcs}, \tag{1.1}$$

where

$$\bar{x}_{Lcs} = \begin{pmatrix} x_{Lcs} \\ y_{Lcs} \\ z_{Lcs} \end{pmatrix}, \qquad \bar{x}_{Ecs} = \begin{pmatrix} x_{Ecs} \\ y_{Ecs} \\ z_{Ecs} \end{pmatrix}.$$

Recall that the inverse N^{-1} of an orthogonal matrix is equal to its transposed matrix N^T, and so it suffices to find the matrix N.

The coordinate system $O_{Lcs}x_{Lcs}y_{Lcs}z_{Lcs}$ may be superposed with the $O_{Lcs}x_{Ecs}y_{Ecs}z_{Ecs}$ coordinate system by two successive rotations. First the $O_{Lcs}x_{Lcs}y_{Lcs}z_{Lcs}$ coordinate system should be rotated along the y_{Lcs}-axis by the angle A to superimpose it with the $O_{Lcs}x'y_{Lcs}z_{Ecs}$ coordinate system, and then the $O_{Lcs}x'y_{Lcs}z_{Ecs}$ should be rotated by the angle $\Delta\varphi_0$ along z_{Ecs}-axis (see Fig. 1). In analogy with formula (1.10) of Chap. 2 of Vol. I. It follows that the required matrix N may be written as follows:

$$N = N_2(A)N_3(\Delta\varphi_0) \, .$$

Here,

$$N_2(A) = \begin{pmatrix} \cos A & 0 & \sin A \\ 0 & 1 & 0 \\ -\sin A & 0 & \cos A \end{pmatrix},$$

$$N_3(\Delta\varphi_0) = \begin{pmatrix} \cos\Delta\varphi_0 & -\sin\Delta\varphi_0 & 0 \\ \sin\Delta\varphi_0 & \cos\Delta\varphi_0 & 0 \\ 0 & 0 & 1 \end{pmatrix}.$$

Hence,

$$N = \begin{pmatrix} \cos A \cos\Delta\varphi_0 & -\cos A \sin\Delta\varphi_0 & \sin A \\ \sin\Delta\varphi_0 & \cos\Delta\varphi_0 & 0 \\ -\sin A \cos\Delta\varphi_0 & \sin A \sin\Delta\varphi_0 & \cos A \end{pmatrix}. \tag{1.2}$$

As in Chap. 8 of Vol. I we attach with a body the reference system x, y, z with origin at the center of mass of the body (the so-called *body-axes system*). The moving bodies encountered in flight dynamics have the following characteristic feature: they have a symmetry plane (ships, airplanes) or a symmetry axis (ballistic missiles). Usually, the x-axis points along the thrust axis of the body in the direction of motion, and the y-axis is chosen in the symmetry plane of the body to be orthogonal to the x-axis and to be pointed up in a natural motion of the body along the Earth's surface. The z-axis completes the x- and y-axes to right-handed coordinate system.

To describe the orientation of the x, y, z-axes relative to the ξ', η', ζ' axes (that are parallel to the axes of the inertial coordinate system $O_1\xi\eta\zeta$; see Fig. 1 of Chap. 8, Vol. I) it is sometimes more convenient to use, in lieu of the Euler angles, the *airplane angles* as defined in Fig. 2.

Assume that originally the x, y, z-axes coincide with the ξ', η', ζ'-axes. We rotate the body-axes system along the η'-axis by the *yaw angle* ψ to obtain the $Ox'y'z'$ system. The next rotation by the *yaw angle* θ will be effected along the z'-axis; this gives the $Ox''y''z''$ coordinate system. The last rotation will be carried out along the x''-axis by the *roll angle* γ to obtain the $Oxyz$ coordinate system in final position.

We note that $\theta = \pm\pi/2$ are special values of the pitch angle—at these values the roll deviation ceases to be different from the yaw deviation (it is useful to compare these cases with the values 0 or π of the nutation angle; see Chap. 2, of Vol. I).

The airplane angles introduced above are fairly transparent for the purpose of describing the body position in move: the yaw angle indicates the deviation of the body longitudinal axis from some selected direction in a horizontal plane, the pitch angle (in maritime it is known as the *trim angle*) describes the rise of the longitudinal axis along the horizontal plane, and the roll angle equals the angle of rotation of a body about the longitudinal axis.

Fig. 2 Airplane angles

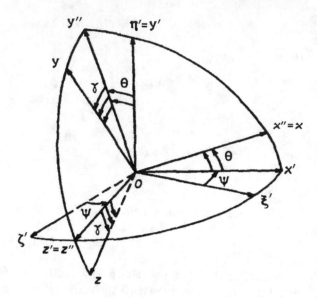

We now give the expressions for the direction cosines between the axes of the $Oxyz$ and $O\xi'\eta'\zeta'$ systems ($c_\psi = \cos\psi$, $s_\theta = \sin\theta$, etc.):

The axes	$O\xi'$	$O\eta'$	$O\zeta'$
Ox	$c_\psi c_\theta$	s_θ	$-s_\psi c_\theta$
Oy	$s_\psi s_\gamma - c_\psi s_\theta c_\gamma$	$c_\theta c_\gamma$	$c_\psi s_\gamma + s_\psi s_\theta c_\gamma$
Oz	$s_\psi c_\gamma + c_\psi s_\theta s_\gamma$	$-c_\theta s_\gamma$	$c_\psi c_\gamma - s_\psi s_\theta s_\gamma$

These formulas may be obtained similarly to the derivation of matrix (1.11) in Chap. 2, Vol. I. A different approach for derivation of these formulas uses the fact that the direction cosines are projections of the $\mathbf{i}, \mathbf{j}, \mathbf{k}$ unit vectors of the body-axes system to the $O\xi'$, $O\eta'$, $O\zeta'$-axes, which translates at constant speed relative to the launch coordinate system.

The angular velocity of the body in question may be written in analogy with formula (2.5) of Chap. 2, Vol. I, as follows:

$$\boldsymbol{\omega} = \dot{\psi}\mathbf{j}' + \dot{\theta}\mathbf{k}' + \dot{\gamma}\mathbf{i};\tag{1.3}$$

here $\mathbf{j}'$, $\mathbf{k}'$, $\mathbf{i}$ are, respectively, the unit vectors on the axes $y' = \eta'$, $z' = z''$, $x = x''$.

To find the projections of the angular velocity p, q, r to the axes of the body-axes system is it convenient to expand the vector $\dot{\psi}\mathbf{j}'$ in its components along the $x = x''$ and y'' axes with unit vectors $\mathbf{i}$ and $\mathbf{j}''$:

$$\dot{\psi}\mathbf{j}' = \dot{\psi}\sin\theta\,\mathbf{i} + \dot{\psi}\cos\theta\,\mathbf{j}''.$$

Hence, using (1.3) and the relations

Fig. 3 The wind
(aerodynamic) coordinate
system

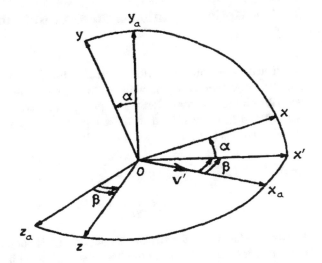

$$\mathbf{j}'' = \cos\gamma\,\mathbf{j} - \sin\gamma\,\mathbf{k}, \qquad \mathbf{k}' = \sin\gamma\,\mathbf{j} + \cos\gamma\,\mathbf{k},$$

it follows that

$$
\begin{aligned}
p &= \dot\psi\sin\theta + \dot\gamma, \\
q &= \dot\psi\cos\theta\cos\gamma + \dot\theta\sin\gamma, \\
r &= -\dot\psi\cos\theta\sin\gamma + \dot\theta\cos\gamma.
\end{aligned}
\tag{1.4}
$$

These equations, like Eq. (2.7) of Chap. 8, Vol. I, are called the *kinematic Euler equations.*

To conclude this section we describe the *wind (aerodynamic) coordinate system* $Ox_a y_a z_a$, which is mainly used to specify the forces acting on a moving body from the incoming water or airflow. The x_a-axis is directed along the air velocity $\mathbf{v}'$ (i.e., along the velocity vector of the body relative to the approach flow (Fig. 3). The y_a-axis, known as the *lift axis*, is directed in the body symmetry plane and points to its upper part. The z_a-axis (the *lateral axis*) completes the x_a- and y_a-axes to right-handed coordinate system. The body symmetry plane, which contains the x, y, y_a-axes, and the orthogonal plane through the velocity $\mathbf{v}'$ and the z-axis intersect along the x'-axis. The angle α between the x- and x'-axes is called the *incidence angle*. This angle is assumed to be positive if the projection of the air velocity to the y-axis is negative. The angle β between the x_a- and x'-axes is called the *gliding angle*. This angle is positive if the projection of the air velocity to the lateral axis z is positive.

2 The Motion Equation of an Aircraft in the Body-Axes System

Nonlinear equations of motion of an aircraft. Henceforth, by an aircraft we shall understand any solid body moving in a resisting medium. The motion of an aircraft is described by the vector differential equations (2.1) of Chap. 8, Vol. I. In accordance with (2.3)–(2.5), the second of these equations may be reduced to the dynamical Euler equations (2.6) of Chap. 8 of Vol. I:

$$A\frac{dp}{dt} + (C - B)qr = L_x, \quad B\frac{dq}{dt} + (A - C)rp = L_y,$$
$$C\frac{dr}{dt} + (B - A)pq = L_z. \tag{2.1}$$

However, when using the airplane angles, Eq. (2.1) should be augmented, instead of the kinematic equations (2.7) of Chap. 8 of Vol. I, with the kinematic equations of motion (1.4) considered in the present chapter. The x, y, z-axes are the axes of the body-axes system with origin at the aircraft center of mass, but now we assume that they coincide with the principal axes of inertia.

Similarly to formula (2.5) of Chap. 8 of Vol. I one may write

$$\frac{d\mathbf{v}_c}{dt} = \frac{d^*\mathbf{v}_c}{dt} + \boldsymbol{\omega} \times \mathbf{v}_c,$$

and hence the first formula in (2.1) of Chap. 8 of Vol. I, which is an expression of the law of motion of the aircraft center of mass, may be written in projections to the body-axes system as follows:

$$M\left(\frac{dv_{cx}}{dt} + qv_{cz} - rv_{cy}\right) = F_x,$$

$$M\left(\frac{dv_{cy}}{dt} + rv_{cx} - pv_{cz}\right) = F_y, \qquad \bullet \tag{2.2}$$

$$M\left(\frac{dv_{cz}}{dt} + pv_{cy} - qv_{cx}\right) = F_x.$$

Equations (2.1), (2.2), and (1.4) constitute the system of differential equations of motion of an aircraft in the body-axes system. This system of differential equations is nonlinear.

Let us now consider a given (*programmed*) motion of an aircraft in the vertical plane. Sometimes this motion is called *reference*. We shall assume that an aircraft moves in accordance with the rules

$$v_{cx}^{\mathrm{pr}} = v_{cx}^{\mathrm{pr}}(t), \quad v_{cy}^{\mathrm{pr}} = v_{cy}^{\mathrm{pr}}(t), \quad v_{cz}^{\mathrm{pr}} = 0,$$
$$\psi^{\mathrm{pr}} = 0, \quad \theta^{\mathrm{pr}} = \theta^{\mathrm{pr}}(t), \quad \gamma^{\mathrm{pr}} = 0. \tag{2.3}$$

By Eq. (1.4), the projections of the angular velocity of programmed motion should vary as follows:

$$p^{\mathrm{pr}} = 0, \quad q^{\mathrm{pr}} = 0, \quad r^{\mathrm{pr}} = \dot{\theta}^{\mathrm{pr}}(t). \qquad (2.4)$$

The programmed motion of the aircraft in the vertical plane satisfies the equations

$$M\left(\frac{dv_{cx}^{\mathrm{pr}}}{dt} - r^{\mathrm{pr}}v_{cy}^{\mathrm{pr}}\right) = F_x^{\mathrm{pr}}, \quad M\left(\frac{dv_{cy}^{\mathrm{pr}}}{dt} + r^{\mathrm{pr}}v_{cx}^{\mathrm{pr}}\right) = F_y^{\mathrm{pr}},$$

$$F_z^{\mathrm{pr}} = 0, \quad L_x^{\mathrm{pr}} = 0, \quad L_y^{\mathrm{pr}} = 0, \quad C\frac{dr^{\mathrm{pr}}}{dt} = L_z^{\mathrm{pr}}. \qquad (2.5)$$

Equations (2.5) solve the direct problem in mechanics: from the given (programmed, in our case) motion using these equations one may determine the forces and moments producing such a motion.

Linearization of equations of perturbed motion. As was pointed out, the system of differential motion equations of an aircraft is essentially nonlinear. In practical problems it is unfeasible to solve this system analytically, and so one needs to seek a numerical solution using computers. However, even this approach involves considerable difficulties due to essential nonlinearities of the differential equations. Mathematically this manifests itself in the appearance of singular points of the system of differential equations, which may result in several equilibrium states of an aircraft. Implementation of a concrete motion near singular points is determined by the aircraft history. Therefore it is of practical interests to be able to predict possible motions of an aircraft in order to be able to correctly perform numerical integration. For this purpose we simplify the original differential equations by neglecting the lower order terms. This enables one to qualitatively understand the properties of motion. Afterwards, the actual motion is numerically refined from the complete equations. This approach is of particular importance in the study of spatial maneuvers of airplanes.[2]

One fairly common method of studying the nonlinear systems (2.1), (2.2), (1.4) is the linearization of differential equations of perturbed motion (for more details, see Chap. I "Stability of motion"), which enables one to efficiently investigate problems of stability and controllability of aircrafts. Perturbations due to deviations of the initial data from the given ones and departures of power characteristics from the values assumed for a programmed motion result in a perturbed (rather than programmed) motion of an aircraft. In this connection, linearization of differential equation is commonly useful for the study of departures of the perturbed motion from the programmed motion on an aircraft.

As an example, let us consider the motion equations of an aircraft, assuming that the motion is perturbed, rather than programmed (see (2.3), (2.4)). Assume that the

---

[2] An example of successful application of methods of qualitative theory of differential equations to the study of singularities of spatial motion of an airplane is given in the book by *G. S. Byushgens, R. V. Studnev*. Aircraft dynamics. Spatial motion. Moscow: Mashinostroyeniye. 1983. 320 p. [in Russian], which was mentioned above.

perturbed motion is given by

$$v_{cx} = v_{cx}^{\text{pr}} + \Delta v_{cx}, \quad v_{cy} = v_{cy}^{\text{pr}} + \Delta v_{cy}, \quad v_{cz} = \Delta v_{cz},$$
$$p = \Delta p, \quad q = \Delta q, \quad r = r^{\text{pr}} + \Delta r \tag{2.6}$$

under the following values of the resultant external force and torque:

$$F_x = F_x^{\text{pr}} + \Delta F_x, \quad F_y = F_y^{\text{pr}} + \Delta F_y, \quad F_z = \Delta F_z,$$
$$L_x = \Delta L_x, \quad L_y = \Delta L_y, \quad L_z = L_z^{\text{pr}} + \Delta L_z. \tag{2.7}$$

Substituting (2.6) and (2.7) into (2.1) and (2.2), this gives

$$A\frac{d\Delta p}{dt} + (C - B)\Delta q\,(r^{\text{pr}} + \Delta r) = \Delta L_x,$$

$$B\frac{d\Delta q}{dt} + (A - C)\,(r^{\text{pr}} + \Delta r)\,\Delta p = \Delta L_y,$$

$$C\frac{d\,(r^{\text{pr}} + \Delta r)}{dt} + (B - A)\Delta p\,\Delta q = L_z^{\text{pr}} + \Delta L_z, \tag{2.8}$$

$$M\left[\frac{d\left(v_{cx}^{\text{pr}} + \Delta v_{cx}\right)}{dt} + \Delta q\,\Delta v_{cz} - (r^{\text{pr}} + \Delta r)\left(v_{cy}^{\text{pr}} + \Delta v_{cy}\right)\right] = F_x^{\text{pr}} + \Delta F_x,$$

$$M\left[\frac{d\left(v_{cy}^{\text{pr}} + \Delta v_{cy}\right)}{dt} + (r^{\text{pr}} + \Delta r)\left(v_{cx}^{\text{pr}} + \Delta v_{cx}\right) - \Delta p\,\Delta v_{cz}\right] = F_y^{\text{pr}} + \Delta F_y,$$

$$M\left[\frac{d\Delta v_{cz}}{dt} + \Delta p\left(v_{cy}^{\text{pr}} + \Delta v_{cy}\right) - \Delta q\left(v_{cx}^{\text{pr}} + \Delta v_{cx}\right)\right] = \Delta F_z.$$

For linearization of these nonlinear equations it is assumed that the perturbations $\Delta v_{cx}, \Delta v_{cy}, \Delta v_{cz}, \Delta p, \Delta q, \Delta r$ are much smaller than the corresponding characteristics of the programmed motion. Hence, one may neglect the terms of second and higher orders with respect to these values and their derivatives in time. In addition to this, thanks to specifics of the aircraft motion, the angle θ^{pr} varies slowly. Hence, by formula (2.4), the quantity r^{pr} is also small. Under these assumptions, using the equations of programmed motion (2.5) and system (2.8), we have the following linearized equations of perturbed motion:

$$A\frac{d\Delta p}{dt} = \Delta L_x, \quad B\frac{d\Delta q}{dt} = \Delta L_y, \tag{2.9}$$

$$C\frac{d\Delta r}{dt} = \Delta L_z, \tag{2.10}$$

$$M \left(\frac{d\Delta v_{cx}}{dt} - v_{cy}^{\mathrm{pr}} \Delta r \right) = \Delta F_x, \tag{2.11}$$

$$M \left(\frac{d\Delta v_{cy}}{dt} + v_{cx}^{\mathrm{pr}} \Delta r \right) = \Delta F_y, \tag{2.12}$$

$$M \left(\frac{d\Delta v_{cz}}{dt} + v_{cy}^{\mathrm{pr}} \Delta p - v_{cx}^{\mathrm{pr}} \Delta q \right) = \Delta F_z. \tag{2.13}$$

We now linearize the system of kinematic equations of motion (1.4). This system may be written as

$$\Delta p = \frac{d\Delta \psi}{dt} (\sin \theta^{\mathrm{pr}} \cos \Delta \theta + \cos \theta^{\mathrm{pr}} \sin \Delta \theta) + \frac{d\Delta \gamma}{dt},$$

$$\Delta q = \frac{d\Delta \psi}{dt} (\cos \theta^{\mathrm{pr}} \cos \Delta \theta - \sin \theta^{\mathrm{pr}} \sin \Delta \theta) + \left(\dot{\theta}^{\mathrm{pr}} + \frac{d\Delta \theta}{dt} \right) \sin \Delta \gamma,$$

$$r^{\mathrm{pr}} + \Delta r = -\frac{d\Delta \psi}{dt} (\cos \theta^{\mathrm{pr}} \cos \Delta \theta - \sin \theta^{\mathrm{pr}} \sin \Delta \theta) \sin \Delta \gamma +$$
$$+ \left(\dot{\theta}^{\mathrm{pr}} + \frac{d\Delta \theta}{dt} \right) \cos \Delta \gamma.$$

Expanding the sines and cosines in Maclaurin series and keeping only the terms which are linear relative to small quantities and their derivatives, it follows by the last formula of (2.4) that

$$\Delta p = \frac{d\Delta \psi}{dt} \sin \theta^{\mathrm{pr}} + \frac{d\Delta y}{dt}, \quad \Delta q = \frac{d\Delta \psi}{dt} \cos \theta^{\mathrm{pr}}, \tag{2.14}$$

$$\Delta r = \frac{d\Delta \theta}{dt}. \tag{2.15}$$

The programmed motion of an aircraft in the vertical plane is specified by the quantities $v_{cx}^{\mathrm{pr}}, v_{cy}^{\mathrm{pr}}, r^{\mathrm{pr}}, \theta^{\mathrm{pr}}$, which vary according to (2.3), (2.4). The departures $\Delta v_{cx}, \Delta v_{cy}, \Delta r, \Delta \theta$ from these quantities are responsible for the so-called *longitudinal perturbed motion of an aircraft*. In turn, the departures $\Delta v_{cz}, \Delta p, \Delta q, \Delta \psi, \Delta \gamma$ describe the *transverse (lateral) perturbed motion*. From the system of differential Eqs. (2.9)–(2.15) it follows that these linearized equations of motion split into two independent subsystems: Eqs. (2.9), (2.13), (2.14) describe the transverse perturbed motion, and Eqs. (2.10)–(2.12), (2.15), the longitudinal perturbed motion.

As is known, the solution of a linear system of differential equations is represented as the sum of the fundamental solution of the corresponding homogeneous system and a particular solution of the inhomogeneous system. For our system (2.9)–(2.13), the latter solution characterizes the perturbed motion due to departures of external forces and torques from values of the programmed motion. This may result from various perturbations of force terms—for example, when encountering wind gusts or application of dedicated control inputs. From the analysis of such forced motion one may determine the aircraft reaction to control inputs and thereby characterize its controllability. The controllability is commonly estimated by considering the aircraft (*prima facie*, airplane) reaction to discontinuous (stepwise) departure of control elements and harmonic variation of departures.

Unlike this, the solution of the corresponding homogeneous system (2.9)–(2.13) of differential equations describes the proper perturbed motion of an aircraft. Such a motion is consequent on the presence of nonzero initial perturbations of the linear and angular velocities. From the study of such motions one may estimate the stability of the longitudinal programmed motion (for more details, see Chap. 1).

It is worth pointing out that in the study of the steady horizontal flight of an actual airplane, the characteristic equation of the corresponding system of differential equations of perturbed motion usually has two complex conjugate roots, whose absolute values are scores of times greater than those of the other pair of complex conjugate roots. On this account, the proper perturbed motion is a composition of rapidly decaying oscillations with high frequency (short-period motion) and slowly decaying oscillations with small frequency (long-period motion). Since the short-period and long-period components appear as separated in time, they can be treated separately. This is a striking simplification in the study of the perturbed motion for horizontal flight. Departures from the programmed motion due to the long-period motion can be easily counteracted by the pilot. Unlike this, the pilot is usually unable to offset short-periodic motions by the properly engaging control elements. Moreover, the pilot's action may even inadvertently rock the airplane. Hence, the most severe requirement is imposed on the stability of an airplane for short-period motion.

3 Forces Acting on an Aircraft

The resultant external force $\mathbf{F}$ acting on an aircraft is basically formed from the Earth's gravitational force $\mathbf{G}$, the propulsive force $\mathbf{P}$, the aerodynamic force $\mathbf{R}^a$, the control forces $\mathbf{F}^{contr}$, and possibly, from some other forces (for example, from the electromagnetic ones). Let us briefly focus on the principal forces that have effect on the aircraft motion.

Depending on the precision required from the solution of a problem, one considers various models of attraction of a moving body to the Earth. The simplest model assumes a uniform plane-parallel gravitational field, when in the formula

$$\mathbf{G} = m\mathbf{g}$$

the constant gravitational acceleration g is independent of an altitude, is directed along the normal vector to the Earth's surface and is taken to be $9.81 \, \text{m/s}^2$. Such a rough model of the gravitational field proves applicable in problems of shooting at targets at distances of about several hundredth kilometers, motion of an aircraft relative to its center of mass, and in building iteration guidance algorithm, etc.

In a more refined theory of Earth's gravitation it is assumed that the shape of the Earth is reflected by the *geoid* (the equipotential surface at mean sea level), which corresponds to the free unperturbed surface of a hypothetical ocean at rest (virtually extended under or above the surface of all continents). It is worth noting that in the gravity force, which is the resultant of the Earth's gravitational force and the centrifugal force associated with the Earth's rotation, it proves impossible to experimentally separate these two forces. If one additionally takes into account that the gravity force also depends on the inhomogeneity of the Earth inner structure, it becomes clear that the form of the geoid is heavily complicated.

As a good approximation to the geoid one may consider the *general Earth ellipsoid* (or the *spheroid*), which is an ellipsoid of revolution obtained by rotating an ellipse about its semi-minor axis. The center of the general Earth ellipsoid coincides with the Earth's center, and the equatorial plane coincides with the Earth's equatorial plane. The sizes of the general Earth ellipsoid are taken to minimize the sum of squared distances from the center to the surfaces of the spheroid and the geoid. In 1964, the International Astronomical Union adopted for standard purposes the following parameters of the spheroid: the semi-major axis (the equatorial radius) $a = 6378137$ m, the flattening (the geometric ellipticity of the Earth) $\alpha = (a - c)/a = 1/298.25$, where c is the semi-minor axis. Subsequently, these parameters were refined. Starting from 1946, in the USSR use was made of the Krasovskii Earth ellipsoid with $a = 6378245$ m and $\alpha = 1/298.3$. One should note that d'Alembert (the early 19th century) was the first to introduce the concept of spheroid (with $a = 6375653$ m, $\alpha = 1/334.0$).

From the measurements used in developing the Krasovskii Earth ellipsoid one may approximate the Earth also by a triaxial ellipsoid, as was proposed by the Central scientific-research institute for geodesy, aerial photography, and cartography (Moscow). It has[3] the semi-axes $a = 6378351.50$ m, $b = 6378137.70$ m, $c = 6356863.02$ m.

One may show[4] that, with accuracy acceptable for ballistic analysis, the projections of the acceleration of gravity, g_r and g_φ from the spheroid (r is the geocentric radius, φ is the latitude) may be found from the formulas:

$$g_r = -g_{av} \left(R_{av}/r\right)^2 \left[1 + \left(R_{av}/r\right)^2 \left(\alpha - q/2\right)\left(1 - 3\sin^2\varphi\right)\right],$$

$$g_\varphi = -g_{av} \left(R_{av}/r\right)^4 \left(\alpha - q/2\right)\sin 2\varphi, \tag{3.1}$$

$$R_{av} = a\left(1 - \alpha/3\right), \quad g_{av} = \mu/R_{av}^2, \quad q = \omega_{Ecs}^2 a^3/\mu.$$

---

[3] See *D. V. Zagrebin*. Introduction to theoretical gravimetry. Moscow: Nauka. 1976 [in Russian].

[4] See *L. M. Lakhtin*. Free motion in the field of the Earth's spheroid. Moscow: Fizmatgiz. 1963 [in Russian].

Here, μ is the gravitational parameter of the attracting spheroid, ω_{Ecs} is the Earth's angular velocity.

The concept of the spheroid is usually used in the analysis of global ballistic problems, when the trajectory length is commeasurable with the Earth's size. An increased accuracy is required in the calculation of motion by significantly smaller distances calls for even more perfect local description of the gravitational field and the Earth's figure. To this aim one uses the concept of the *reference ellipsoid*, which is the ellipsoid whose selected part of the surface (for example, within one or several states) agrees in the best way with the selected surface on the geoid. Under this approach, the center of the ellipsoid may not coincide with the Earth's ellipsoid, but their axes of revolution must be parallel.

A fairly accurate model of the Earth's gravitational field is frequently given by a sufficiently large number of the geopotential series in spherical functions or as a sum of the potential of the general Earth ellipsoid of revolution with uniformly distributed mass and the potential of a system of point masses. It is also worth noting that a super accurate model of the geoid was constructed at the end of 2013.[5]

In the spherical model, it is usually assumed that the Earth's radius is 6 371 110 m, in this case the volume of the sphere equals the volume of the spheroid. To this model there corresponds the central (or Newtonian) gravitational field, which was discussed in detail in Sect. 9 of Chap. 4 "Dynamics of point" of Vol. I. Using this model one may fairly accurately calculate the trajectories of spacecrafts that move at distances over several thousand kilometers from the Earth and perform preliminary ballistic calculations of near-Earth objects.

For an aircraft the propulsive force **P** may be produced by engines of two classes. The engines of first class require the atmospheric oxygen and hence operate only in the atmosphere. In these engines the fuel energy is transformed either directly into the propulsive force via the jet reaction of combustion products or is used for propeller rotation. The second class includes the jet engines whose propulsive force is produced independently of the ambient environment. Such engines may be solid- or liquid-propelled. There are also ion and nuclear engines, capable of producing the highest specific thrust. However, ion engines may produce only minor accelerations, and hence may be applied only for spacecrafts launched from orbits.

Characteristics of the propulsive force **P**, which are usually determined during bench tests, are environment-specific. The question of analytical expression of the

---

[5] To this aim a Rockot launch vehicle carrying the GOCE (Gravity Field and Steady-State Ocean Circulation Explorer) satellite was launched from the Plesetsk cosmodrome on March 17, 2009 under the framework of ESA Earth Explorer Program. The spacecraft weighing 1 metric ton was injected into the low Earth orbit of altitude 260 km (this orbit is by ca. 500 lower than those used for conventional scientific satellites). Such a choice of the orbit proved instrumental in obtaining the mapping of the Earth's gravitational field with due account of nonuniform distribution of mass inside our planet. A several programmed maneuvers were carried out in order to refine the results obtained—an orbit of altitude 235 km was eventually reached. At these altitudes one ought to take into account the resistance of the Earth's upper atmosphere. The aforementioned super accurate model of the geoid was built from the measurements obtained from GOCE instruments. On 21 October 2013, the mission came to a natural end when it ran out of fuel. Three weeks later, on 11 November, the satellite disintegrated in the lower atmosphere over the western Pacific Ocean.

propulsive force of a jet engine will be examined in the next Sect. 4 (see Eq. (4.1) and formula (4.18)).

The aero(hydro)dynamic force excreted on a body from the incoming air or fluid is uniformly distributed over the hull, wings, and the tail unit. As a rule, this force is referenced to the center of mass and is replaced by the resultant force $\mathbf{R}^a$ and the resultant torque $\mathbf{L}^a$, which in-flight dynamics are called, respectively, the *total aerodynamic force* and the *total aerodynamic torque*. Traditionally, these vectors are represented as

$$R^a = c_R S \rho v^2 / 2, \qquad L^a = c_m l S \rho v^2 / 2. \tag{3.2}$$

Here, c_R, c_m are the dimensionless aerodynamic coefficients, l is the characteristic length (for example, the length of a launch vehicle or of a wing of an airplane), S is the characteristic area (for example, the *frontal area*[6] of a launch vehicle structure or the wing area of an airplane), ρ is the density of environment in which the motion occurs, and v is the free-stream velocity.

In studying the motion of an aircraft in the body-axes system, the vector $\mathbf{R}^a$ is specified in terms of its components,

$$R_1^a = c_x S \rho v^2 / 2, \quad R_2^a = c_y S \rho v^2 / 2, \quad R_3^a = c_z S \rho v^2 / 2$$

which correspond to the so-called *longitudinal, normal* and *transverse aerodynamic forces*. For positive incidence and gliding angles, the force $\mathbf{R}_2^a$ points in positive direction of the y-axis, while the forces $\mathbf{R}_1^a$ and $\mathbf{R}_3^a$ point in the negative direction, respectively, along the x- and z-axes. Similarly, the torque is expanded into the aerodynamic torques in the roll, yaw, and pitch axes, which may be found from the projections of the vector $\mathbf{L}^a$:

$$L_1^a = c_{mx} l S \rho v^2 / 2, \quad L_2^a = c_{my} l S \rho v^2 / 2, \quad L_3^a = c_{mz} l S \rho v^2 / 2.$$

It is also worth noting that the aerodynamic forces acting on the structure, wings, and the tail unit of an aircraft have a very complicated dependence on the incidence and gliding angles, on the Mach number (which is the ratio of the free-stream velocity to the velocity of sound), the environment viscosity, the flight altitude, the aircraft configuration, and a number of different additional factors. These dependencies are reflected in the corresponding functional form of the aerodynamic coefficients. In the majority of cases, one is able to find these coefficients only from experiments in wind tunnels or from flight testing. Due to conditions peculiar to such experiments, in these tests one measures the forces in the so-called *stability-axes system*, whose axes $x_{s.a.s}$, $y_{s.a.s}$, $z_{s.a.s}$ coincide with the axes x', y_a, z, which are shown in Fig. 3 (clearly, for $\beta = 0$ the stability-axes system agrees with the wind system). The corresponding forces (which are called the *drag force*, the *lift force*, and the *side force*) are given by

---

[6] This is the largest transverse section (from the Dutch word *middel*, which means in the marine sciences the middle (the most wide) transverse section of a ship).

$$R^a_{1,\text{s.a.s}} = c^{\text{s.a.s}}_x S\rho v^2/2, \quad R^a_{2,\text{s.a.s}} = c^{\text{s.a.s}}_y S\rho v^2/2, \quad R^a_{3,\text{s.a.s}} = c^{\text{s.a.s}}_z S\rho v^2/2.$$

In view of Fig. 3 there is the following relation between the aerodynamic coefficients:

$$c_x = c^{\text{s.a.s}}_x \cos\alpha - c^{\text{s.a.s}}_y \sin\alpha,$$

$$c_y = c^{\text{s.a.s}}_x \sin\alpha + c^{\text{s.a.s}}_y \cos\alpha, \quad c_z = c^{\text{s.a.s}}_z.$$

A programmed motion of aircraft is provided by the control forces $\mathbf{F}^{\text{contr}}$ and the control torques $\mathbf{L}^{\text{contr}}$ applied by the pilot or by an automatic control system. Consequent on this the magnitude and direction of the propulsive force and aerodynamic force may vary. Technically this is achieved by operations of main engine, swiveling engines and nozzles, special steering engines, or by additional gas or fluid injection into the exhaust plume, by changing the aircraft configuration (extending flaps, trimming stabilizer), by changing the position of air and gas vanes[7] and so on.

4 Motion of a Systems of Variable Mass. Thrust of a Jet Engine

Motion of a point of variable mass. By a point of variable mass we shall mean a material point of negligible extension and whose mass varies in time. In order to be able to apply theorems of classical mechanics, we consider the system consisting of a material point of mass m and an elementary mass dm joining the system during the time dt. Assume that prior to the joining, the velocities of masses m and dm are, respectively, $\mathbf{v}$ and $\mathbf{u}$, so after the joining the velocity of the total mass $(m + dm)$ becomes $(\mathbf{v} + d\mathbf{v})$. Hence, by the law of moment of momentum, we may write

$$(m + dm)(\mathbf{v} + d\mathbf{v}) - (m\mathbf{v} + dm\mathbf{u}) = \mathbf{F}dt,$$

where $\mathbf{F}$ is the resultant external force. Making $dt \to 0$, this gives

$$m\frac{d\mathbf{v}}{dt} = \mathbf{F} + \frac{dm}{dt}(\mathbf{u} - \mathbf{v}). \tag{4.1}$$

This equation is usually called the *Meshcherskii equation*.[8] This equation describes the motion of a point of variable mass $m(t)$. Note that for $dm > 0$ this equation

---

[7] Gas vans are similar to air vanes, but unlike those, they are installed to operate in the exhaust plume, and hence may not work in the air-free environment.

[8] This equation was first published by I. V. Meshcherskii (1859–1935) in his master thesis (1897). According to G. K. Michailov, this equation was first obtained in the 1812–1814 by the Czech scientist Georg von Buquoy. His papers on dynamics of bodies of variable mass, which left far ahead of all relevant problems of this time, faded into obscurity.

characterizes the motion of a point with increasing mass, and for $dm < 0$, the motion of a particle with decreasing mass.

The second term on the right of (4.1) is called the *thrust*. Clearly, $\mathbf{v}_r = \mathbf{u} - \mathbf{v}$ is the relative velocity of an elementary particle relative to the main mass. So, a moving point of variable mass is subject to the thrust $\mathbf{v}_r dm/dt$, in addition to the force $\mathbf{F}$.

Example 1 Consider the vertical ascent of a launch vehicle in a homogeneous field of gravity with constant outflow velocity $\mathbf{v}_r = $ const (the so-called *second Tsiolkovsky problem*).[9] By (4.1), we have

$$m\frac{dv}{dt} = -mg - v_r\frac{dm}{dt}$$

in the projection to the motion direction. So,

$$m\frac{d(v + gt)}{dt} = -\frac{v_r}{m}\frac{dm}{dt}.$$

Assuming that $\mathbf{v}(0) = 0$, $m(0) = m_o$, this gives

$$v = v_r\ln\frac{m_o}{m(t)} - gt.$$

For the finite value of velocity v_k at the end of the powered flight of a launch vehicle with $t = t_k$, we may write

$$v_k = v_r\,\ln\frac{m_o}{m_k} - gt_k, \quad m_k = m(t_k). \tag{4.2}$$

In the particular case of motion without external forces (*the first Tsiolkovsky problem*), we have

$$v_k = v_r\ln\frac{m_o}{m_k}. \tag{4.3}$$

From this formula it follows that v_k is directly proportional to v_r, is independent of the law of the variability of mass, and increases with increasing $Z = m_o/m_k$. This relation, known as the *Tsiolkovsky number*, is used in modern ballistic analysis.

Formula (4.2) refines these conclusions by showing that in order to produce the same value of v_k as in formula (4.3) (with constant gravity field) it requires to decrease the quantity m_k. Besides, (4.2) shows that increasing v_r and decreasing t_k have greater effect on the increase of the finial velocity than an increase of the Tsiolkovsky number. This is so because the log-function growths slower than a linear function.

---

[9] K. E. Tsiolkovsky (1857–1935) was a famous popularizer of the possibility of space flights. He solved some theoretical problems and proposed a number of original ideas—for example, the advisability of using multi-staged launch vehicles to achieve high velocities, construction of near-Earth space stations, etc.

Main theorems for a system of points of variable masses. The momentum $\widetilde{\mathbf{K}}$ of a system of points of variable masses is given by

$$\widetilde{\mathbf{K}} = \sum_k m_k(t)\,\mathbf{v}_k.$$

Hence,

$$\frac{d\widetilde{\mathbf{K}}}{dt} = \sum_k \frac{dm_k}{dt}\,\mathbf{v}_k + \sum_k m_k \frac{d\mathbf{v}_k}{dt}.$$

Replacing the terms $m_k d\mathbf{v}_k/dt$ in view of the Meshcherskii Eq. (4.1), we have the *law of moment of momentum for a system of points of variable mass*:

$$\frac{d\widetilde{\mathbf{K}}}{dt} = \mathbf{F} + \sum_k \frac{dm_k}{dt}\,\mathbf{u}_k.$$

Here, $\mathbf{F} = \sum_k \mathbf{F}_k$ is the resultant external force on the system, and the sum on the right characterizes the resultant external force due to variable mass.

Now we consider the position of the center of mass C of a system of points of variable composition. At this time instant, this position is given from the conventional formula

$$m(t)\mathbf{r}_c = \sum_k m_k(t)\,\mathbf{r}_k, \quad m(t) = \sum_k m_k(t).$$

Differentiating this formula with respect to time, this gives

$$\frac{d^2 m}{dt^2}\mathbf{r}_c + 2\frac{dm}{dt}\mathbf{v}_c + m\frac{d\mathbf{v}_c}{dt} =$$

$$= \sum_k \frac{d^2 m_k}{dt^2}\mathbf{r}_k + 2\sum_k \frac{dm_k}{dt}\mathbf{v}_k + \sum_k m_k \frac{d\mathbf{v}_k}{dt}.$$

Using the Meshcherskii formula (4.1) to replace the terms in the last sum, we arrive at the *law of moment of momentum of the center of mass of a system of material points of variable masses*:

$$m\frac{d\mathbf{v}_c}{dt} = \mathbf{F} + \sum_k \frac{dm_k}{dt}(\mathbf{u}_k - \mathbf{v}_k) +$$

$$+2\left(\sum_k \frac{dm_k}{dt}\mathbf{v}_k - \frac{dm}{dt}\mathbf{v}_c\right) + \sum_k \frac{d^2 m_k}{dt^2}\mathbf{r}_k - \frac{d^2 m}{dt^2}\mathbf{r}_c.$$

4 Motion of a Systems of Variable Mass. Thrust of a Jet Engine

453

Here, $\mathbf{F}$ is the resultant external force on the system, $\sum_k (dm_k/dt)(\mathbf{u}_k - \mathbf{v}_k)$ is the resultant thrust on points of variable masses, and the remaining terms characterize the additional forces due to variability of masses of the points in the system.

Let us now dwell on the absolute angular momentum $\tilde{\mathbf{l}}_o$ of a system of points of variable masses relative to the origin

$$\tilde{\mathbf{l}}_o = \sum_k \mathbf{r}_k \times m_k(t)\,\mathbf{v}_k \, .$$

The derivative is given by

$$\frac{d\tilde{\mathbf{l}}_o}{dt} = \sum_k \mathbf{r}_k \times \frac{dm_k}{dt}\,\mathbf{v}_k + \sum_k \mathbf{r}_k \times m_k \frac{d\mathbf{v}_k}{dt} \, . \tag{4.4}$$

By taking the vector product of the Meshcherskii Eq. (4.1) for the kth point on $\mathbf{r}_k$ and summing the so-obtained expressions over all points of the system, this gives

$$\sum_k \mathbf{r}_k \times m_k \frac{d\mathbf{v}_k}{dt} = \sum_k \mathbf{r}_k \times \mathbf{F}_k + \sum_k \mathbf{r}_k \times \frac{dm_k}{dt}(\mathbf{u}_k - \mathbf{v}_k) \, .$$

Using formula (4.4) to single out the expression $d\tilde{\mathbf{l}}_o/dt$, we obtain the *law of moment of absolute angular momentum for a system of points of variable masses relative to the origin at rest*:

$$\frac{d\tilde{\mathbf{l}}_o}{dt} = \mathbf{L} + \sum_k \mathbf{r}_k \times \frac{dm_k}{dt}\,\mathbf{u}_k \, .$$

It follows that a variation of the absolute angular momentum of a system of points of variable masses has effect not only on the resultant external torque $\mathbf{L} = \sum_k \mathbf{r}_k \times \mathbf{F}_k$, but also on the resultant torque $\sum_k \mathbf{r}_k \times (dm_k/dt)\,\mathbf{u}_k$ due to the variability of masses of points of the system.

Lastly, let us consider the kinetic energy theorem. Multiplying the Meshcherskii Eq. (4.1) by elementary length $d\mathbf{r} = \mathbf{v}\,dt$, this gives

$$m\mathbf{v} \cdot d\mathbf{v} + v^2\,dm = \mathbf{F} \cdot d\mathbf{r} + \mathbf{u} \cdot \mathbf{v}\,dm \, .$$

Since

$$d\left(\frac{mv^2}{2}\right) = m\mathbf{v} \cdot d\mathbf{v} + \frac{v^2}{2}\,dm \, ,$$

it follows that

$$d\left(\frac{mv^2}{2}\right) + \frac{v^2}{2}\,dm = \mathbf{F} \cdot d\mathbf{r} + \mathbf{u} \cdot \mathbf{v}\,dm \, . \tag{4.5}$$

This formula expresses the content of the *kinetic energy theorem for a point of variable mass*.

Note that the second term on the right of (4.5) may be expressed as

$$\mathbf{u} \cdot \mathbf{v} \, dm = \frac{dm}{dt} \mathbf{u} \cdot \mathbf{v} \, dt = \frac{dm}{dt} \mathbf{u} \cdot d\mathbf{r},$$

which characterizes the elementary work on the displacement $d\mathbf{r}$ by the force $(dm/dt)\,\mathbf{u}$, which is consequent on changes of the mass $m(t)$.

Expressing (4.5) for the kth point in the system and then summing in k, we obtain the *kinetic energy theorem for a system of points of variable masses*:

$$d\widetilde{T} + \sum_k dm_k \frac{v_k^2}{2} = \sum_k \delta A_k + \sum_k \mathbf{u}_k \cdot \mathbf{v}_k \, dm_k ; \qquad (4.6)$$

here

$$\widetilde{T} = \sum_k m_k(t) \frac{v_k^2}{2}, \quad \delta A_k = \mathbf{F}_k \cdot d\mathbf{r}_k .$$

Formula (4.6) states that the sum of the differential of the kinetic energy of a system of points of variable masses and the kinetic energy of the increments of their masses equals the sum of elementary works of forces acting on the points of the system and the sum of elementary works of forces depending on the variation of masses of points of the system.

The laws of moment of momentum and the theorem of the center of mass for a hard-shell body of variable mass. The principle of solidification. Many practical applications are concerned not with the motion of a finite system of points of variable masses, and even not the motion of a continuous medium, but rather with the motion of a hard-shell body with a flow of material particles flowing through it. In such a statement we are interested, first of all, in the study of motion of bodies like launch vehicles or jet-propelled aircrafts, when one is mostly interested in the motion of the hard-shell itself (the structure), rather with the motion of the exhaust plume and or with the motion of the mass center of the composite system consisting of a moving body and the stream of gaseous particles emitted. Hence, we shall be concerned with the motion of a variable mass body bounded by a closed surface S_o.[10] This surface coincides with the hard shell of the body under study (a launch vehicle, airplane, and so on) and contains in its interior the structure elements, payload, engines, pumps, etc., as well as high-speed inner flows of gas approaching the surface S_o and then exhausting through it. Note that together with the particles exhausting outside the surface S_o, there may be gas flows penetrating inside through it. For example, for airplanes with turbojet engines, there is a flow of air through the air intake supplying the oxygen necessary for the fuel to burn.

---

[10] See, for example, *K. A. Abgaryan, I. M. Rapoport*. Rocket dynamics. Moscow: Mashinostroyeniye. 1969. 379 p. [in Russian].

4 Motion of a Systems of Variable Mass. Thrust of a Jet Engine

455

Consider a system of variable composition $\widetilde{\Omega}$ confined inside a closed surface S_o and containing hard, liquid, and gaseous parts. The composition of the system varies with time: some particles escape or adhere to it. In addition to this, we shall be concerned with the motion of the masses comprising the system $\widetilde{\Omega}$ at a given time instant. We let Ω denote this mechanical system of constant composition.

In the case a mechanical system is composed of a finite number of material points, from the law of moment of momentum and the theorem of the center of mass, considered simultaneously, we have the following chain of equalities

$$\frac{d\mathbf{K}}{dt} = \frac{d}{dt} \sum_\nu m_\nu \mathbf{v}_\nu = \sum_\nu m_\nu \mathbf{w}_\nu = M\,\mathbf{w}_c = \mathbf{F}. \tag{4.7}$$

Note that the number of points of the system is assumed to be constant, and hence (4.7) may be extended only to a system Ω, but not to a system $\widetilde{\Omega}$.

Extending (4.7) to systems Ω, we consider, first of all, the equation

$$\frac{d\mathbf{K}}{dt} = \int_\Omega \mathbf{w}\,dm. \tag{4.8}$$

Since at a given time instant all masses of the system Ω are inside the hard shell, the integral appearing in (4.8) may be written as

$$\int_\Omega \mathbf{w}\,dm = \iiint_V \mathbf{w}(t, x, y, z)\,\rho(t, x, y, z)\,dx\,dy\,dz.$$

Here, ρ is the density, V is the constant volume confined inside the surface S_o, x, y, z are the coordinates of a point under study in the body-axes system $Cxyz$. This system is introduced as follows. Consider some fictitious body S, which would be obtained from the $\widetilde{\Omega}$ if it instantaneously solidifies at time t. We assume that the $Cxyz$ system will correspond to the given fictitious body S. The motion of particles of the system Ω relative to the $Cxyz$ system will be looked upon as a relative motion, while the motion of the $Cxyz$ system itself will be considered as a translational motion. Under this approach to the motion of particles of the system Ω, the vector $\mathbf{w}$ appearing in (4.8) will be written in accordance with Coriolis's theorem (see (1.10) of Chap. 3 of Vol. I) as the following sum:

$$\mathbf{w} = \mathbf{w}_a = \mathbf{w}_e + \mathbf{w}_r + \mathbf{w}_{\text{Cor}}, \qquad \mathbf{w}_{\text{Cor}} = 2\boldsymbol{\omega} \times \mathbf{v}_r; \tag{4.9}$$

here $\mathbf{w}_e$, $\mathbf{w}_r$, $\mathbf{w}_{\text{Cor}}$ are, respectively, the drag, relative, and Coriolis accelerations.

In view of (4.7) and (4.9), expression (4.8) may be written as follows:

$$\int_\Omega \mathbf{w}_e\,dm = \mathbf{F} + \mathbf{F}_{\text{Cor}} - \int_\Omega \mathbf{w}_r\,dm, \qquad \mathbf{F}_{\text{Cor}} = -\int_\Omega \mathbf{w}_{\text{Cor}}\,dm. \tag{4.10}$$

Here, $\mathbf{F}$, $\mathbf{F}_{Cor}$ are, respectively, the resultant external forces and the Coriolis inertial forces, acting on the system Ω. We note that the forces $\mathbf{F}$ and $\mathbf{F}_{Cor}$ may in fact be looked upon as forces applied to the fictitious body S, because at time t, for which Eq. (4.10) is written, the body S and the system Ω coincide.

All points of the fictitious body have only the drag acceleration $\mathbf{w}_e$, and hence, in view of (4.7), the acceleration of its center of mass may be written as

$$M^s \frac{d\mathbf{v}_c^s}{dt} = \int_\Omega \mathbf{w}_e \, dm . \tag{4.11}$$

Here, $\mathbf{v}_c^s$ is the velocity of the center of mass of the fictitious body, M^s is its mass, which equals the mass of the system Ω at a given time instant t.

From (4.11) it follows that Eq. (4.10) may be regarded as the equation of motion of the center of mass of a fictitious solid body. To be able to apply (4.10), it remains to find out what is the third term on the right of (4.10).

An observer, which is located in the $Cxyz$ system (that is, in the structure of a launch vehicle or an airplane), may notice only relative motion of particles of the system Ω. Hence, Eq. (4.8), as applied to the $Cxyz$ system, has the form

$$\frac{d^*\mathbf{K}_r}{dt} = \int_\Omega \mathbf{w}_r \, dm . \tag{4.12}$$

Recall that the star near the derivative sign means that the derivatives are taken in the $Cxyz$ system; that is, the derivative is local.

Consider the integral

$$\widetilde{\mathbf{K}}_r = \iiint_V \mathbf{v}_r \, \rho \, dx \, dy \, dz$$

and consider its local derivative with respect to time

$$\frac{d^*\widetilde{\mathbf{K}}_r}{dt} = \frac{d^*}{dt} \iiint_V \mathbf{v}_r \, \rho \, dx \, dy \, dz . \tag{4.13}$$

Next, we write the difference

$$\frac{d^*\mathbf{K}_r}{dt} - \frac{d^*\widetilde{\mathbf{K}}_r}{dt} = \frac{d^*\Delta\widetilde{\mathbf{K}}_r}{dt} , \tag{4.14}$$

where

$$\Delta\widetilde{\mathbf{K}}_r(t + \Delta t) = \mathbf{K}_r(t + \Delta t) - \widetilde{\mathbf{K}}_r(t + \Delta t) .$$

Difference (4.14) appears because the expression $\mathbf{K}_r(t + \Delta t)$ should take into account the momentum of those particles which in time Δt either enter the volume V or escape from it.

We now consider an $d\sigma$ on the surface S_o. An elementary mass Δm, which in time Δt leaves the volume V through the surface $d\sigma$, is $\rho\, v_{rn}\, d\sigma\, \Delta t$. Here, ρ is the density of matter flowing through the surface S_o and v_{rn} is the projection of the relative velocity $\mathbf{v}_r$ to the external normal vector $\mathbf{n}$ to the surface S_o. The variation $\Delta \widetilde{\mathbf{K}}_r$ of the momentum in time Δt due to elementary masses Δm escaping through the entire surface S_o is as follows:

$$\Delta \widetilde{\mathbf{K}}_r = \left(\iint\limits_{S_o} \mathbf{v}_r\, \rho\, v_{rn}\, d\sigma \right) \Delta t .$$

Hence,

$$\frac{d^* \Delta \widetilde{\mathbf{K}}_r}{dt} = \lim_{\Delta t \to 0} \frac{\Delta \widetilde{\mathbf{K}}_r}{\Delta t} = \iint\limits_{S_o} \mathbf{v}_r\, \rho\, v_{rn}\, d\sigma .$$

Applying the mean-value theorem, this gives

$$\frac{d^* \Delta \widetilde{\mathbf{K}}_r}{dt} = m_{\sec} \mathbf{v}_r^{av} . \qquad (4.15)$$

Here, $m_{\sec} = \int_{S_o} \rho\, v_{rn}\, d\sigma$ is the mass discharge per second through the surface S_o and $\mathbf{v}_r^{av}$ is the averaged relative velocity of continuous medium flowing through the surface.

Using expressions (4.11), (4.10), (4.12), and (4.14) it is found that the motion equation of the center of mass of a fictitious body S may be written as

$$M^s \frac{d\mathbf{v}_c^s}{dt} = \mathbf{F} + \mathbf{F}_{\mathrm{Cor}} + \left(-\frac{d^* \Delta \widetilde{\mathbf{K}}_r}{dt} \right) + \left(-\frac{d^* \widetilde{\mathbf{K}}_r}{dt} \right) . \qquad (4.16)$$

In accordance with (4.15), the third term on the right of Eq. (4.16) is the resultant thrust acting on the body S. As is seen from (4.13), the last term on the right of (4.16) will appear in the case when the relative velocity of the particles $\mathbf{v}_r$ and density ρ vary with time. This is why this term is called the *resultant variational force* acting on the body S.

In case a launch vehicle may be looked upon as a point of variable mass, the second and fourth terms on the right of (4.16) will disappear, and (4.16) will become the Meshcherskii Eq. (4.1), because in this case in (4.15) we have $m_{\sec} = -dm/dt$, $\mathbf{v}_r^{av} = \mathbf{u} - \mathbf{v}$.

From this derivation of Eq. (4.16) we may formulate the following *principle of solidification* for a hard-shell body of variable composition: the translation motion equation of a body of variable composition $\widetilde{\Omega}$ with hard shell S_o at a given time may be written as an equation of motion of the center of mass of some solid body of constant composition, provided that the system of variable composition $\widetilde{\Omega}$ at this time instant is solidified and the so-obtained fictitious solid body S is subject to external forces acting on the system $\widetilde{\Omega}$, the Coriolis inertial forces, as well as the thrusts and variational forces.

Thrust of a jet engine. Let us now focus on bench tests of a jet engine. As usual, we attach the $Cxyz$ coordinate system with the shell, which is now assumed to be at rest. Since the shell is stationary, we have $\mathbf{v}_c^s \equiv 0$, $\mathbf{F}_{\text{Cor}} \equiv 0$, and hence, by (4.16),

$$\mathbf{F} + \left(-\frac{d^* \Delta \widetilde{\mathbf{K}}_r}{dt} \right) + \left(-\frac{d^* \widetilde{\mathbf{K}}_r}{dt} \right) = 0. \tag{4.17}$$

Here, the resultant external force $\mathbf{F}$ involves the gravity force of the engine, the reaction at the supports of the test bench, the atmospheric pressure, and the gas pressure in the nozzle. For a horizontally positioned engine in tests, the gravity force acting on it may be excluded from consideration, because it is balanced by the vertical component of the reaction at the supports. The bench instruments measure the horizontal component $\mathbf{P}$ of the reaction at the supports, which equals the propulsive force of the engine. By formula (4.17), the thrust $\mathbf{P}$ of the engine is

$$\mathbf{P} = \mathbf{F}^{\text{press}} + \left(-\frac{d^* \Delta \widetilde{\mathbf{K}}_r}{dt} \right) + \left(-\frac{d^* \widetilde{\mathbf{K}}_r}{dt} \right), \tag{4.18}$$

where

$$F^{\text{press}} = S_{\text{n.e.}}(p_{\text{n.e.}} - p_{\text{a.p.}}).$$

is the resultant force of the atmospheric pressure and the gas pressure. Here, $S_{\text{n.e.}}$ is the area of the nozzle exit, $p_{\text{n.e.}}$ is the pressure at the nozzle exit, and $p_{\text{a.p.}}$ is the atmospheric pressure.

With the use of the above concept of the propulsive force of an engine, Eq. (4.16) of motion of the center of mass of a fictitious body S may be written as

$$M^s \frac{d\mathbf{v}_c^s}{dt} = \mathbf{F}^* + \mathbf{P} + \mathbf{F}_{\text{Cor}}, \tag{4.19}$$

where now $\mathbf{F}^*$ denotes the resultant external force, from which the atmospheric pressure and the pressure at the nozzle exit are excluded.

The law of resultant moment of momentum for a hard-shell variable mass body relative to the center of mass. For simplicity, let us start with this law when the above system Ω is composed of a finite number of material points.

By formula (3.9) of Chap. 5 of Vol. I,

$$\frac{d\mathbf{l}'}{dt} = \mathbf{L}', \tag{4.20}$$

where

$$\mathbf{l}' = \sum_{\nu} \mathbf{r}'_{\nu} \times m_{\nu} \mathbf{v}'_{\nu}, \quad \mathbf{L}' = \sum_{\nu} \mathbf{r}'_{\nu} \times \mathbf{F}^{(e)}_{\nu}.$$

We recall that here $\mathbf{r}'_{\nu}$ is the radius vector of a point M_{ν} relative to the center of mass, $\mathbf{r}'_{\nu} = \overrightarrow{CM}_{\nu}$, and $\mathbf{v}'_{\nu}$ is the velocity of a point M_{ν} relative to the moving frame $Cx'_1 x'_2 x'_3$ (see Fig. 2 of Chap. 5 of Vol. I), whose axes, throughout the motion, are parallel to the axes of that system of coordinate that is assumed to be inertial in this problem.

Differentiating with respect to time the vector $\mathbf{l}'$, this gives

$$\frac{d\mathbf{l}'}{dt} = \sum_{\nu} \frac{d\mathbf{r}'_{\nu}}{dt} \times m_{\nu} \mathbf{v}'_{\nu} + \sum_{\nu} \mathbf{r}'_{\nu} \times m_{\nu} \frac{d\mathbf{v}'_{\nu}}{dt}.$$

Setting $\mathbf{w}'_{\nu} = d\mathbf{v}'_{\nu}/dt$ and taking into account that $d\mathbf{r}'_{\nu}/dt = \mathbf{v}'_{\nu}$, this establishes

$$\frac{d\mathbf{l}'}{dt} = \sum_{\nu} \mathbf{r}'_{\nu} \times m_{\nu} \mathbf{w}'_{\nu}. \tag{4.21}$$

Extending expressions (4.20), (4.21) to the case when a system Ω is a continuous medium, we have

$$\int_{\Omega} \mathbf{r}' \times \mathbf{w}' dm = \mathbf{L}'. \tag{4.22}$$

The motion of the body-axes system $Cxyz$ relative to the $Cx'_1 x'_2 x'_3$ system (see Fig. 2 of Chap. 5 of Vol. I) will be assumed to be translational, and the motion of particles with respect to the $Cxyz$ system will be assumed to be relative. Under this approach, the law of momentum (4.22) is written as

$$\int_{\Omega} \mathbf{r}' \times \mathbf{w}'_e dm = \mathbf{L}' + \mathbf{L}_{\text{Cor}} - \int_{\Omega} \mathbf{r}' \times \mathbf{w}_r dm, \tag{4.23}$$

where $\mathbf{L}_{\text{Cor}} = -\int_{\Omega} \mathbf{r}' \times \mathbf{w}_{\text{Cor}} dm$ is resultant torque of the Coriolis inertial forces relative to the center of mass.

In analogy with the expressions (4.12)–(4.14), the third term on the right of (4.23) may be written

$$-\int_{\Omega} \mathbf{r}' \times \mathbf{w}_r \, dm = -\frac{d^* \Delta \widetilde{\mathbf{l}}_r}{dt} - \frac{d^* \widetilde{\mathbf{l}}_r}{dt}, \tag{4.24}$$

where

$$\frac{d^* \mathbf{l}_r}{dt} = \frac{d^*}{dt} \iiint_V \mathbf{r}' \times \mathbf{v}_r \, \rho \, dx \, dy \, dz.$$

The first and second terms on the right of (4.24) are, respectively, the resultant thrusts and variational forces relative to the center of mass.

Now let us consider the left-hand side of Eq. (4.23). First of all, we note that it may be written as

$$\int_{\Omega} \mathbf{r}' \times \mathbf{w}'_e \, dm = \iiint_V \mathbf{r}' \times \mathbf{w}'_e \, \rho \, dx \, dy \, dz, \tag{4.25}$$

because at time t, for which Eq. (4.23) is written, the system Ω coincides with with fictitious solid body. The acceleration $\mathbf{w}'_e$ appears because this body rotates about the center of mass with angular velocity ω. Hence,

$$\mathbf{w}'_e = \frac{d}{dt}(\omega \times \mathbf{r}') = \frac{d\omega}{dt} \times \mathbf{r}' + \omega \times (\omega \times \mathbf{r}'). \tag{4.26}$$

On the other hand, for the same fictitious solid body we have, in accordance with (4.21) and (4.26),

$$\frac{d\mathbf{l}'}{dt} = \iiint_V \mathbf{r}' \times \left[\frac{d\omega}{dt} \times \mathbf{r}' + \omega \times (\omega \times \mathbf{r}') \right] \rho \, dx \, dy \, dz, \tag{4.27}$$

where

$$\mathbf{l}' = \iiint_V \mathbf{r}' \times (\omega \times \mathbf{r}') \, \rho \, dx \, dy \, dz.$$

The axes of the body-axes system $Cxyz$ are the principal axes of inertia, and hence it follows by formulas (2.4) and (2.5) of Chap. 8 of Vol. I, that

$$\mathbf{l}' = A \, p \, \mathbf{i} + B \, q \, \mathbf{j} + C \, r \, \mathbf{k},$$
$$\frac{d\mathbf{l}'}{dt} = A \, \dot{p} \, \mathbf{i} + B \, \dot{q} \, \mathbf{j} + C \, \dot{r} \, \mathbf{k} + \omega \times \mathbf{l}'. \tag{4.28}$$

Notice that the last formula may be verified by direct calculation of the right-hand side of (4.27).

Using (4.25)–(4.28) it follows that Eq. (4.23) may be written as the dynamical Euler equations:

$$A \, \dot{p} + (C - B) \, q \, r = L_x, \quad B \, \dot{q} + (A - C) \, r \, p = L_y, \quad C \, \dot{r} + (B - A) \, p \, q = L_z. \tag{4.29}$$

Here,

$$L_x = \mathbf{L} \cdot \mathbf{i}, \quad L_y = \mathbf{L} \cdot \mathbf{j}, \quad L_z = \mathbf{L} \cdot \mathbf{k},$$

$$L = L' + L_{Cor} + \left(-\frac{d^* \Delta \widetilde{I}_r}{dt}\right) + \left(-\frac{d^* \widetilde{I}_r}{dt}\right).$$

Similarly to formula (4.19), augmenting the thrust torque L_P with the torques $(-d^* \Delta \widetilde{I}_r/dt)$, $(-d^* \widetilde{I}_r/dt)$ and the torques of atmospheric pressure and the pressure at the nozzle exit relative to the center of mass, we see that

$$L = L^* + L_P + L_{Cor},$$

where L^* is the external torque acting on the body of variable composition, except for the atmospheric pressure and the pressure at the nozzle exit.

Equations (4.29) should be augmented with the kinematic Euler Eq. (1.4), which, after being solved relative to the time derivatives of the airplane angles, may be written as

$$\begin{aligned}
\dot{\psi} &= (q \cos \gamma - r \sin \gamma)/\cos \theta, \\
\dot{\theta} &= q \sin \gamma + r \cos \gamma, \\
\dot{\gamma} &= p + (r \sin \gamma - q \cos y) \tan \theta.
\end{aligned} \tag{4.30}$$

The nonlinear system of differential Eqs. (4.29), (4.30) can be integrated numerically.

Here the following fact should be pointed out. The thing is that Eq. (4.29), as written in accordance with the principle of solidification, are valid at a given time instant. However, in the process of motion of an aircraft, its center of inertia and principal axes of inertia change with the variation of its mass. Nevertheless, these translations are negligible in comparison with the trajectory size. Hence, they are usually neglected for practical purposes, assuming that the center of mass and the principal axes of inertia are at rest relative to the structure of a moving body.

5 Motion of an Aircraft in the Launch Coordinate System

Motion of an aircraft center of mass. Consider the inertial system of coordinate whose axes agree with the original axes of the launch coordinate system. This system will be called the *launch inertial coordinate system*. To maintain the directions of these axes an aircraft is equipped with the *gyro platform* (*the gyro unit*) involving a three-axis gyroscopic stabilizer.[11] For any motion of an aircraft this platform moves in translation, and hence it suffices to direct ab initio its axes to be parallel with the axes of the launch coordinate system. Motion of aircraft in the inertial coordinate system is called *inertial navigation*.

By (4.19) the motion equation of the center of mass of aircraft may be written as

---

[11] See A. Yu. Ishlinskii. Inertial guidance of ballistic missiles. Some theoretical aspects. Moscow: Nauka. 1968. 142 p. [in Russian], A. Yu. Ishlinskii. Orientation, gyroscopes and inertial navigation. Moscow: Nauka. 1976. 670 p. [in Russian]; D. M. Klimov. Inertial navigation on a sea. Moscow: Nauka. 1984. 111 p. [in Russian]; Yu. G. Sikharulidze. Ballistics of aircrafts. Moscow: Nauka. Fizmatlit. 1982. 352 p. [in Russian].

$$M^s \mathbf{w}^s_c = \mathbf{G} + \mathbf{P} + \mathbf{R}^a + \mathbf{F}^{contr}. \tag{5.1}$$

Here, for brevity, we omit the superscript "s" and take into account that $\mathbf{F}_{Cor} = 0$ for a motion in an inertial system.

In engineering practice a frequent use is made of the instrument called *accelerometer*, which consists of a mass attached to a spring. This mass may move along the spring mass, which results in its deformation. From the string deformation one measures the force acting by the mass and then, according to Newton's second law, the acceleration corresponding to this mass. At present accelerometers have been brought to great perfection and are capable of making fairly precise measurements.

Aircraft are equipped with three similar accelerometers located along the instrument axes of the gyro unit. If an aircraft is subject to forces $\mathbf{P}$, $\mathbf{R}^a$, $\mathbf{F}^{contr}$, this results in corresponding deformation of springs of accelerometers—from which one may measure the three components in the initial lift-off inertial coordinate system of the so-called *apparent acceleration* $\mathbf{w}^*$ adopted in flight dynamics:

$$w^*_{x\,\mathrm{Lcs}} = (P_{x\,\mathrm{Lcs}} + R^a_{x\,\mathrm{Lcs}} + F^{contr}_{x\,\mathrm{Lcs}})/M\,,$$
$$w^*_{y\,\mathrm{Lcs}} = (P_{y\,\mathrm{Lcs}} + R^a_{y\,\mathrm{Lcs}} + F^{contr}_{y\,\mathrm{Lcs}})/M\,,$$
$$w^*_{z\,\mathrm{Lcs}} = (P_{z\,\mathrm{Lcs}} + R^a_{z\,\mathrm{Lcs}} + F^{contr}_{z\,\mathrm{Lcs}})/M\,.$$

But if we rewrite Eq. (5.1) in the form

$$M\,(\mathbf{w}_c - \mathbf{g}) = \mathbf{P} + \mathbf{R}^a + \mathbf{F}^{contr},$$

it becomes clear that from the apparent acceleration $\mathbf{w}^*$ one may determine the sought-for acceleration of the center of mass

$$\mathbf{w}_c = \mathbf{w}^* + \mathbf{g}\,,$$

provided one knows the acceleration of gravity $\mathbf{g}$. So, for the acceleration $\mathbf{w}_c = (w_{x\,\mathrm{Lcs}}, w_{y\,\mathrm{Lcs}}, w_{z\,\mathrm{Lcs}})$, we may write

$$w_{x\,\mathrm{Lcs}} = w^*_{x\,\mathrm{Lcs}} + g_{x\,\mathrm{Lcs}}\,, \quad w_{y\,\mathrm{Lcs}} = w^*_{y\,\mathrm{Lcs}} + g_{y\,\mathrm{Lcs}}\,,$$
$$w_{z\,\mathrm{Lcs}} = w^*_{z\,\mathrm{Lcs}} + g_{z\,\mathrm{Lcs}}\,. \tag{5.2}$$

Here, the values $w^*_{x\,\mathrm{Lcs}}$, $w^*_{y\,\mathrm{Lcs}}$, $w^*_{z\,\mathrm{Lcs}}$ are measured by accelerometers, and the values $g_{x\,\mathrm{Lcs}}$, $g_{y\,\mathrm{Lcs}}$, $g_{z\,\mathrm{Lcs}}$ may be calculated in the adopted model of the Earth gravitation of an aircraft (see Sect. 3). So, uploading all these available values into an onboard computer, taking into account the initial data, and integrating Eq. (5.2), it proves possible to solve the inertial navigation problem; i.e., to determine the coordinates and the velocity of an aircraft in the launch coordinate system. From the so-obtained values one may find various motion characteristics, for example, the motion trajectory and the velocity components of an aircraft with respect to the Earth. Changing if required the values $w_{x\,\mathrm{Lcs}}$, $w_{y\,\mathrm{Lcs}}$, $w_{z\,\mathrm{Lcs}}$, which are obtained from Eq. (5.2), one may use Eq. (5.1) to find the required control $\mathbf{F}^{contr}$. The components of forces may be transformed from one coordinate system into a different coordinate system using the transformation formulas (1.1) and (1.2).

Inertial coordinate systems operate for several dozens of minutes on launch vehicles, several hours on airplanes, and several days or even months on ships. Hence, the most exacting requirements are placed on the operational accuracy and stability of inertial coordinate systems.

Let us dwell in more detail on the calculation of the quantities $g_{x\,Lcs}$, $g_{y\,Lcs}$, $g_{z\,Lcs}$, which enter into the right-hand sides of Eq. (5.2). We shall assume that the Earth is represented by the spheroid (the general Earth ellipsoid). In this case, the projections of the acceleration of gravity g_r and g_φ onto the directions of the y_3- and x_3-axes (the projection onto the z_3-axis is zero) may be calculated by formulas (3.1). Hence, in order to use them to update the projections of the gravitational acceleration with respect to the launch inertial coordinate system one should use formulas (1.1):

$$g_{x\,Lcs} = g_\varphi \cos A \cos \Delta\varphi_0 - g_r \cos A \sin \Delta\varphi_0 \,,$$
$$g_{y\,Lcs} = g_\varphi \sin \Delta\varphi_0 + g_r \cos \Delta\varphi_0 \,,$$
$$g_{z\,Lcs} = -g_\varphi \sin A \cos \Delta\varphi_0 + g_r \sin A \sin \Delta\varphi_0 \,.$$

But to be able to calculate using (3.1) it is required to know, at each time instant, first the distance r of the aircraft center of mass to the spheroid center, and second, its latitude φ. These quantities may be calculated as follows:

$$r = \sqrt{x_{Ecs}^2 + (R_0 + y_{Ecs})^2 + z_{Ecs}^2}\,, \quad \varphi = \arcsin(\omega_{Ecs}^0 \cdot \mathbf{r}^0); \qquad (5.3)$$

here R_0 is the distance from the lift-off point to the spheroid center, ω_{Ecs}^0 is the unit vector of Earth's angular velocity, $\mathbf{r}^0$ is the unit vector in the direction of the aircraft center of mass with respect to the Earth ellipsoid. The unit vector ω_{Ecs}^0 may be represented from its projections in terms of the latitude φ_0 of the lift-off point:

$$\omega_{Ecs}^0 = (\cos\varphi_0,\ \sin\varphi_0,\ 0)\,.$$

The coordinates of the aircraft center of mass may be calculated in terms of the known quantities x_{Lcs}, y_{Lcs}, z_{Lcs} via (1.1), (1.2), taking into account that $N^{-1} = N^*$:

$$x_{Ecs} = x_{Lcs} \cos \Delta\varphi_0 \cos A + y_{Lcs} \sin \Delta\varphi_0 - z_{Lcs} \cos \Delta\varphi_0 \sin A \,,$$
$$y_{Ecs} = -x_{Lcs} \sin \Delta\varphi_0 \cos A + y_{Lcs} \cos \Delta\varphi_0 + z_{Lcs} \sin \Delta\varphi_0 \sin A \,,$$
$$x_{Ecs} = x_{Lcs} \sin A + z_{Lcs} \cos A \,.$$

We note that the distances R of points of the spheroid to its center may be found from the formula

$$R^2 = \frac{a^2 \left(1 - e^2\right)}{1 - e^2 \cos^2 \varphi}\,, \qquad (5.4)$$

where a is the semi-major axis, e is the eccentricity of the ellipse, which, after rotating about its minor axis, gives the ellipsoid of revolution taken for the spheroid. The value of R_0 is obtained from (5.4) with $\varphi = \varphi_0$. Now, using formulas (5.3) one may find the values of r and φ, because the scalar product $\omega_{\text{Ecs}}^0 \cdot \mathbf{r}^0$ may be written as

$$\omega_{\text{Ecs}}^0 \cdot \mathbf{r}^0 = [x \cos \varphi_0 + (y + R_0) \sin \varphi_0] / r \, .$$

In particular, after calculating r and R at a given time instant one may find the altitude h of an aircraft over the surface of the spheroid:

$$h = r - R \, .$$

Aircraft motion relative to the center of mass. The study of an aircraft motion relative to the center of mass proved instrumental in determining its attitude (orientation in space). This together with information about the velocity of the center of mass enables one, in particular, to ascertain several characteristics of forces acting on a moving body. When the mass is taken to be variable, we use Stevin's principle of solidification. Hence, as follows from the previous section, one should investigate the integration of dynamic and kinematic Euler Eqs. (4.29), (4.30). As was pointed out above, in real-world problems of flight dynamics this system of differential equations may be solved only numerically.

6 Application of Methods of Nonholonomic Mechanics for Aircraft Targeting

Flight dynamics frequently encounters with the problem of aircraft targeting either for purposes of destruction or for delivering payload to the target (in-flight refueling, rendezvous of a ship and drift ice, etc.).

At present there many different methods of targeting. For definiteness, we consider one classical method called the pure pursuit method. Under this approach the velocity of a moving aircraft is always directed toward the target (similar to a dog chasing a rabbit).

Assume that a point $M(x, y)$ describes the motion of an aircraft in the horizontal plane Oxy, and a point $M_1(\xi, \eta)$ describes the motion of the target in this plane. It is assumed that

$$\xi = \xi(t), \quad \eta = \eta(t) \tag{6.1}$$

are given time functions.

Clearly, the condition

$$\frac{\dot{x}}{x - \xi} = \frac{\dot{y}}{y - \eta}$$

should be satisfied for this targeting method. Equivalently, we have

$$(y - \eta)\dot{x} - (x - \xi)\dot{y} = 0. \tag{6.2}$$

Thus, the problem of aircraft targeting may be looked upon as a problem in nonholonomic mechanics. In this case a point motion is subject to the nonholonomic constraint (6.2), the nonstationarity is determined by the target motion (6.1). To attack this problem we employ Maggi's equations (see Sect. 4, Chap. 6 of Vol. I), which are the most general equations of nonholonomic mechanics.[12]

As the generalized coordinates of a point M we take its Cartesian coordinates $(s = 2)$:

$$q^1 = x, \quad q^2 = y. \tag{6.3}$$

The motion of a point is subject to one $(k = 1)$ equation of nonholonomic constraint (6.2). Consider the new variables v_*^1, v_*^2 related with the generalized velocities $\dot{q}^1 = \dot{x}$, $q^2 = \dot{y}$ by the relations $(l = s - k = 1)$:

$$v_*^1 = \dot{x}, \quad v_*^2 = (y - \eta)\dot{x} - (x - \xi)\dot{y}.$$

The inverse transformation may now be easily calculated

$$\dot{x} = v_*^1, \quad \dot{y} = (y - \eta)v_*^1/(x - \xi) - v_*^2/(x - \xi). \tag{6.4}$$

For our purpose, Maggi's equations will assume the form (m is the aircraft mass):

$$(mw_1 - Q_1)\frac{\partial \dot{x}}{\partial v_*^1} + (mw_2 - Q_2)\frac{\partial \dot{y}}{\partial v_*^1} = 0, \tag{6.5}$$

$$(mw_1 - Q_1)\frac{\partial \dot{x}}{\partial v_*^2} + (mw_2 - Q_2)\frac{\partial \dot{y}}{\partial v_*^2} = \Lambda. \tag{6.6}$$

The pursuit law (6.2) will be looked upon as an ideal nonholonomic constraint. We recall that in accordance with the theory outlined in Sect. 4, Chap. 6 of Vol. I, Maggi's equations are convenient in that they properly contain the motion Eq. (6.5), which should be integrated jointly with the constraint Eq. (6.2), and separately from this system, the second equation in (6.6), from which, after the motion of a point is found, one may determine the generalized reaction Λ. From the targeting point of view, this reaction is a control force, which should be added to the given forces in order to have the pursuit law (6.2).

The kinetic energy of an aircraft, which is assumed to be a point-mass, is given by

[12] V. I. Kirgetov was one of the first to apply methods of nonholonomic mechanics to pursuit problems (see, for example, his paper "The motion of controlled mechanical systems with prescribed constraints (servoconstraints)," J. Appl. Math. Mech. USSR, **31**, 431–466 (1967) [in Russian]). In the problem of pursuit navigation in case of planar motion N. A. Kil'chevskii wrote down the generalized Chaplygin's equations and Schouten's equations (see his book A *Course Theoretical Mechanics* (Nauka, Moscow, 1977) [in Russian]).

$$T = m \frac{\dot{x}^2 + \dot{y}^2}{2}.$$

Hence,

$$mw_1 = m\ddot{x}, \quad mw_2 = m\ddot{y}.$$

Taking into account transformations (6.4), we rewrite Eqs. (6.5) and (6.6) as

$$m\ddot{x} - Q_1 + (m\ddot{y} - Q_2)(y - \eta)/(x - \xi) = 0, \tag{6.7}$$

$$(m\ddot{y} - Q_2)/(\xi - x) = \Lambda. \tag{6.8}$$

An aircraft moving horizontally will be assumed to be subjected to the propulsive force **P**, codirectional with its velocity **v** and the aerodynamic drag force **R**^{a}, of opposite direction. With our choice of the generalized coordinates (6.1) we have $e_1 = \mathbf{i}$, $e_2 = \mathbf{j}$, and hence,

$$\begin{aligned}
Q_1 &= (\mathbf{P} + \mathbf{R}^a) \cdot \mathbf{i} = P_x + R_x^a = (P - R^a) \dot{x}/\sqrt{\dot{x}^2 + \dot{y}^2}, \\
Q_2 &= (\mathbf{P} + \mathbf{R}^a) \cdot \mathbf{j} = P_y + R_y^a = (P - R^a) \dot{y}/\sqrt{\dot{x}^2 + \dot{y}^2}.
\end{aligned} \tag{6.9}$$

Recall that the aerodynamic drag force R^a may be found from the first formula in (3.2).

As was pointed out above, the motion of an aircraft may be determined by integrating Maggi's Eq. (6.7) jointly with the constraint Eq. (6.2). For numerical integration it is convenient to represent this systems in the normal form. To this aim we differentiate the constraint Eq. (6.2) to obtain

$$\ddot{x}(y - \eta) + \ddot{y}(\xi - x) = \dot{\eta}\dot{x} - \dot{\xi}\dot{y}. \tag{6.10}$$

In view (6.9) Eq. (6.7) may be written as

$$\begin{aligned}
\ddot{x} + \ddot{y}(y - \eta)/(x - \xi) &= (P - R^a)\dot{x}/(m\sqrt{\dot{x}^2 + \dot{y}^2}) + \\
&\quad + (P - R^a)\dot{y}(y - \eta)/[m\sqrt{\dot{x}^2 + \dot{y}^2}(x - \xi)].
\end{aligned} \tag{6.11}$$

Solving Eqs. (6.10) and (6.11) as a system of linear algebraic inhomogeneous equations in $\ddot{x}$ and $\ddot{y}$, we obtain

$$\begin{aligned}
\ddot{x} &= \frac{(\dot{\eta}\dot{x} - \dot{\xi}\dot{y})(y - \eta)}{(x - \xi)^2 + (y - \eta)^2} + \frac{(P - R^a)[\dot{x}(x - \xi) + \dot{y}(y - \eta)](x - \xi)}{m\sqrt{\dot{x}^2 + \dot{y}^2}[(x - \xi)^2 + (y - \eta)^2]}, \\
\ddot{y} &= \frac{(\dot{\xi}\dot{y} - \dot{\eta}\dot{x})(x - \xi)}{(x - \xi)^2 + (y - \eta)^2} + \frac{(P - R^a)[\dot{x}(x - \xi) + \dot{y}(y - \eta)](y - \eta)}{m\sqrt{\dot{x}^2 + \dot{y}^2}[(x - \xi)^2 + (y - \eta)^2]}.
\end{aligned} \tag{6.12}$$

System (6.12) may be written in the normal form as a system of four first-order differential equations and then integrated numerically.

After integrating (6.12), Eq. (6.8) may be used to determine the generalized reaction Λ, and from Λ one may find the components of the reaction of nonholonomic constraint or, in other words, the control forces that provide targeting according to (6.2). Indeed, in the general case, the reaction of nonholonomic constraint $\varphi(t, q, \dot{q}) = 0$ reads as

$$\mathbf{R} = \Lambda \nabla' \varphi + \mathbf{T}_0 \, .$$

We recall, that constraint (6.2) may be assumed to be ideal; that is, $\mathbf{T}_0 = 0$. Hence, the reaction in which we are interested in may be written as follows:

$$\mathbf{R} = \Lambda \left(\frac{\partial \varphi}{\partial \dot{x}} \mathbf{i} + \frac{\partial \varphi}{\partial \dot{y}} \mathbf{j} \right) .$$

As a result, we get
$$R_x = \Lambda(y - \eta), \quad R_y = -\Lambda(x - \xi) . \tag{6.13}$$

Attention should be drawn to a certain convenience in the use of Maggi's equations rather than other equations of nonholonomic mechanics (see Sect. 10, Chap. 6 of Vol. I) in solving pursuit problems in flight dynamics. Indeed, Maggi' equations may be written down following a fairly simple and consistent method of obtaining motion equations. Moreover, using Maggi's equations one may easily obtain the constraint force (the control force) without no additional requirements on the form of a nonholonomic constraint (the pursuit law). In particular, a constraint may be also nonlinear in the generalized velocities. In the latter case it is recommended to differentiate the constraint equation with respect to time to obtain an expression of form (6.10), which is linear in the generalized accelerations. This enables one to easily represent in the normal form the systems of differential equations for the motion of an aircraft.[13]

---

[13] For more details concerning the application of methods of analytical mechanics in problems of aircraft guidance on a target see the monograph: *Sh. Kh. Soltakhanov, M. P. Yushkov, S. A. Zegzhda. Mechanics of nonholonomic systems. A New Class of control systems.* Berlin Heidelberg: Springer-Verlag. 2009. 329 p.

Index

© Springer Nature Switzerland AG 2021
N. N. Polyakhov et al., *Rational and Applied Mechanics*,
Foundations of Engineering Mechanics,
https://doi.org/10.1007/978-3-030-64118-4